W9-BAD-947

ENVIRONMENTAL DESIGN OF URBAN BUILDINGS

ENVIRONMENTAL DESIGN OF URBAN BUILDINGS

An Integrated Approach

Edited by

Mat Santamouris

London • Sterling, VA

First published by Earthscan in the UK and USA in 2006

Environmental Design of Urban Buildings was supported by the European Commission's SAVE 13 Programme

ISBN-10: 1-902916-42-5 (hardback)
ISBN-13: 978-1902916-42-2 (hardback)

Typesetting by Mapset Ltd, Gateshead, UK
Printed and bound in the UK by Bath Press, Bath
Cover design by Paul Cooper

For a full list of publications please contact:

Earthscan
8–12 Camden High Street
London, NW1 0JH, UK
Tel: +44 (0)20 7387 8558
Fax: +44 (0)20 7387 8998
Email: earthinfo@earthscan.co.uk
Web: **www.earthscan.co.uk**

22883 Quicksilver Drive, Sterling, VA 20166-2012, USA

Earthscan is an imprint of James and James (Science Publishers) Ltd and publishes in association with the International Institute for Environment and Development

A catalogue record for this book is available from the British Library

Library of Congress Cataloging-in-Publication Data has been applied for

The paper used for the text of this book is FSC certified.
FSC (The Forest Stewardship Council) is an international network to promote responsible management of the world's forests.

Printed on totally chlorine-free paper

Contents

List of Tables

List of Figures

List of Boxes

List of Contributors

Spyros Amourgis is Vice President of the Hellenic Open University (EAP), Professor of Bioclimatic Architecture at EAP, Professor Emeritus of the California State Polytechnic University, Pomona (CSPU), a former Dean at the College of Environmental Design (CSPU) and former Secretary of the Board of the Collegiate Schools of Architecture (ACSA). He has taught design at the Architectural Association (AA) School of Architecture, London, was Professeur Invité at EPF Lausanne, and a Senior Fellow at the Center of Metropolitan Planning and Research, The Johns Hopkins University, Baltimore.

Ciril Arkar is a teaching assistant at the Faculties of Mechanical Engineering and Architecture, and at the University College of Health Care of the University of Ljubljana, Slovenia. His research areas are renewable energy sources and heat and mass transfer in buildings. He has cooperated on international projects CEC JOULE II, OPET and SAVE, and also on national projects.

Marc Blake is a licensed architect in California and Greece. He is currently a partner at AMK Architects and Designers in Athens, designing for the cruise and hotel industry. His undergraduate work at the California State Polytechnic University at Pomona included one year of studies in Florence, Italy and Athens, Greece. He won second place in the annual Paris Prize Competition and then obtained a masters degree in Architecture at UCLA while working with the Urban Innovations Group and the Architect Panos Koulermos. After graduation he moved to Athens in Greece and began working for OTOME and then AMK. He is also teaching in the CSPUP Summer Programme in Athens.

Evangelos Evangelinos is a professor at the National Technical University of Athens (NTUA). He holds a Diploma in Architectural Engineering from NTUA, and a Graduate Diploma in Energy Studies from the AA, London. He currently teaches courses on architectural technology and design, building construction, bioclimatic and sustainable architecture. He has participated as a researcher and head researcher in a number of projects on sustainable and bioclimatic architecture. He is an author on the topics of sustainable building design, as well as bioclimatic architecture for the Greek Open University.

Vassilios Geros is a research associate at the National and Kapodistrian University of Athens (NKUA). He holds a physics degree (from NKUA, Greece), a DEA in building design methods (ENTPE-INSA de Lyon, Université de Chambéry, France) and a PhD on the thermal performance of night ventilation techniques (INSA de Lyon, France). He has participated in several research and other projects concerning building energy design, building automation and building regulations. He has also participated in developing educational material and software tools for energy and environmental design and certification of buildings.

Samuel Hassid has been associate professor in the Civil and Environmental Engineering Department of Technion – Israel Institute of Technology, in Haifa, Israel since 1990. He has a BSc and an MSc in nuclear engineering from Queen Mary College of the University of London, and a DSc from the Technion. He is currently teaching climatology of buildings, heat transfer and computational fluid dynamics. He has participated in several research projects in passive solar buildings and energy in buildings in Israel, as well as the Energy Towers Project.

Stavroula Karatasou is a physicist. She graduated from the Physics Department of the University of Athens and holds an MSc in physics of the environment. She is currently working as a research associate in the Group of Building Environmental Studies (University of Athens) and has participated in a variety of national and European research, and in applied projects in energy conservation, integration of renewable energies in buildings, indoor air quality, thermal comfort and passive cooling.

Sašo Medved is an associate professor at the University of Ljubljana. His main research areas are heat and mass transfer in buildings, renewable energy sources and computer simulations of buildings and buildings service systems. He is Head of the Department of Renewable Energy Sources in the Laboratory for Heating, Sanitary and Solar Technology in the Faculty of Mechanical Engineering. He is author of five books, more then 30 scientific and research articles, and several software and multimedia tools.

Dana Raydan is a practising project architect within the award-winning multi-disciplinary consultancy RMJM Ltd, leading a UK£10 million new build School of Nursing and Midwifery at the University of East Anglia in Norwich (completion December 2005). She is a registered architect in the UK as well as in Lebanon where she practised before moving to the UK in 1998. She received her MPhil in Environmental Design at the Martin Centre for Architectural and Urban Studies in 1994, where she later returned as a research associate in 1998 to work on an EU-funded project coordinated by the Martin Centre (Koen Steemers), looking into the potential of renewable energies in the urban environment. She is the Passive and Low Energy Architecture (PLEA) Board representative for the 22nd international conference series taking place in Lebanon in 2005, working with the hosting university on the conference organization. Dana has published numerous papers on environmental architecture and design in international forum proceedings (REBUILD 99, PLEA 2000 and 2005) and in journals such as *Energy and Buildings* (vol 31, no 1, January 2003), as well a book chapter in *Courtyard Housing: Past, Present and Future* (2006). She is the editor of the *PLEA 2005 Conference Proceedings* (2005).

Mat Santamouris is an associate professor of energy physics at the University of Athens. He is associate editor of the *Solar Energy Journal* and a member of the editorial board of the *International Journal of Solar Energy*, the *Journal of Energy and Buildings*, and the *Journal of Ventilation*. He is editor of the series of books on Buildings, Energy and Solar Technologies published by James and James (Science Publishers)/Earthscan in London. He has published nine international books on topics related to solar energy and energy conservation in buildings. He has been guest editor of six special issues of various scientific journals. He has coordinated many international research programmes and he is author of almost 120 scientific papers published in international scientific journals. He is visiting professor at the Metropolitan University of London.

Koen Steemers was appointed the Director of the Martin Centre for Architectural and Urban Studies in 2002 (the Centre was founded by Sir Leslie Martin and Lionel March in 1967 and is the funded research wing of the University of Cambridge Department of Architecture). He is leading a team undertaking environmental building research projects in Europe, Australia, China and the US. He has produced over 100 publications, including books such as *Environmental Diversity in Architecture* (2004), *The Selective Environment* (2002), *Daylight Design of Buildings* (2002) and *Energy and Environment in Architecture* (2000). He coordinates a team of research staff and PhDs, and directs the MPhil course in Environmental Design in Architecture.
He is a registered architect (has practised in the UK, Germany and Holland); an environmental design consultant (as Director of CAR Ltd); President of PLEA (an international forum of over 2000 members); Fellow of Wolfson College, Cambridge; and guest professor at Chongqing University, China and Aalborg University, Denmark.

Elias Zacharopoulos is an architect and assistant professor at the NTUA. He studied at the NTUA (Diploma of Architect Engineer) and the University of Bristol (MSc in Advanced Functional Design Techniques for Buildings). He teaches courses on architectural technology and design, building construction and bioclimatic architecture.

List of Acronyms and Abbreviations

ACH	air changes per hour
AI	artificial intelligence
AIC	acceptable indoor concentration
ASHRAE	American Society of Heating, Refrigerating and Air-conditioning Engineers
AST	apparent solar time
BACnet	Building Automation and Control Network
BEMS	building energy management systems
BMS	building management systems
bps	bits per second
BRE	Building Research Establishment, Watford
BRI	building-related illness
cd	candelas
CEN TC247	European Committee for Standardization, Technical Committee 247
CFC	chlorofluorocarbon
CFD	computational fluid dynamics
CHP	combined heat and power
CLN	control and automation-level network
cm	centimetres
CO	carbon monoxide
CO_2	carbon dioxide
COHb	carboxihaemoglobin
COP	coefficient of performance
CPU	central processing unit
dB(A)	decibel
DCHP	district heating power plant
DDC	direct digital control
DF	daylight factor
DGI	daylight glare index
DSG	double strength glass
DSM	demand-side management
dT	temperature difference
EC	European Commission
EHS	European Home System
EIB	European Installation Bus
EIBnet	European Installation Bus Network
ELA	effective leakage area
EMAS	Eco-management and Audit Scheme
ETS	EIB tool software
ETS	environmental tobacco smoke
EU	European Union
FLN	field-level network
FND	Firm Neutral Datatransmission
GDP	gross domestic product
GI	glare index
GJ	gigajoules
ha	hectares
H/W	height-to-width ratio
HCHO	formaldehyde
HVAC	heating, ventilation and air conditioning system
HZ	hertz
IAQ	indoor air quality
IBDS	integrated building design system
IES	Illuminating Engineering Society
I/O	input/output
IPCC	Intergovernmental Panel on Climate Change
IRR	internal rate of return
kg	kilograms
KJ	kilojoules
km	kilometres
kWh	kilowatt hours
kWh/kg	kilowatt hours per kilogram
LAN	local area network
LCA	life-cycle assessment
LCC	life-cycle cost
LG	liquefied gas
L/H	length-to-height ratio
lm	lumens
LON	local operating network
LSM	local standard meridian
LST	local standard time
lx	lux
L/W	length-to-width ratio
m	metres
MAC	maximum allowable concentration
Mbps	megabits per second
ME	maximum environmental value

MJ	megajoules
MLN	management-level network
mm	millimeters
MRT	mean radiant temperature
MW	megawatts
NAAQS	National Ambient Air Quality Standards
NB	net benefits
NO_2	nitrogen dioxide
NO_x	nitrogen oxide
NMP	Dutch national environmental policy plan
NS	net savings
OECD	Organisation for Economic Co-operation and Development
P	proportional control action
Pb	lead
PB	payback period
PE	population equivalent
PI	proportional plus integral control action
PID	proportional plus integral plus differential control action
PLEA	passive and low-energy architecture
ppm	parts per million
PROFIBUS	Process Field Bus
PV	photovoltaics
RME	rapeseed oil methyl ester
SA	sustainability appraisal
SBS	sick building syndrome
SC	shading coefficient
SCADA	Supervisory Control and Data Acquisition
SHG	solar heat gain
SHGF	solar heat gain factor
SO_2	sulphur dioxide
SO_x	sulphur oxide
sr	steradian
TVOC	total volatile organic compound
UBL	urban boundary layer
UCL	urban canopy layer
UK	United Kingdom
URR	unadjusted rate of return
US	United States
USEPA	US Environmental Protection Agency
UTF	Urban Task Force
VAV	variable air volume
VOC	volatile organic compound
WHO	World Health Organization

1

Environmental Urban Design

Dana Raydan and Koen Steemers

Scope of the chapter

This chapter describes the wider context of urban environmental design issues, and outlines the existing related knowledge, research and experience. The emphasis is on the larger scale urban context, with particular reference to vernacular or traditional examples; experiences and issues raised by recent urban design practice; and a review of the current technical and social aspects that are related to urban environmental issues.

Learning objectives

When you complete the study of the chapter, you will:

- understand the historical, technical and social context of energy-efficient urban building design;
- become aware of the wider sphere related to this work.

Key words

Key words include:

- vernacular urban planning;
- urban microclimate;
- urban energy;
- urban design practice.

Introduction: Urban environmental facts today

Many studies examine the environmental problems that cities suffer from and how these contribute to the degradation of the global environment. The purpose of this chapter is not to discuss such studies. Instead, it is to establish a general background and introduction for this handbook, to set it in context and to provide a review of environmental and energy research with respect to urban building projects.

The structure of this chapter is broadly that it starts, after this introduction, with an overview of vernacular urban planning, followed by a series of sections that focus on current research into urban climatology, energy use, renewable energy and environmental potential related to building form. The final section raises issues related to urban planning, such as amenity, equity and aesthetics, in the context of energy-efficient urban design.

By 1980, it had been estimated that the total area of the Earth that had been converted to urban land use was approximately 1×10^6 square kilometres (0.2 per cent of the Earth's total area), with an estimated rate of change of 2×10^4 square kilometres per year (4×10^{-3} per cent per year) (Oke, 1988). According to 1991 statistics, 45 per cent of the world's population is living in cities, a proportion that is rising at the rate of 3 per cent per year (Sadik, 1991). The Brundtland Report predicted that by the year 2000, approximately half of the world's population would live in urban settlements (WCED, 1987), compared with 10 per cent of the world's population living in cities and towns at the start of the 20th century (UNCHS, 1996).

Consequently, a constant rise in urban population and land consumption has led to high demands for energy for lighting, heating, cooling and transport being concentrated in cities. Recent statistics show that 75 per cent of pollution is caused by urban environments – roughly 45 per cent from buildings and 30 per cent from transport (Rogers, 1995). Transport specifically is the cause of 20 per cent of carbon dioxide (CO_2) emissions, the latter constituting half of the total effect of global warming (see Figure 1.1) (H. Barton in Breheny, 1992).

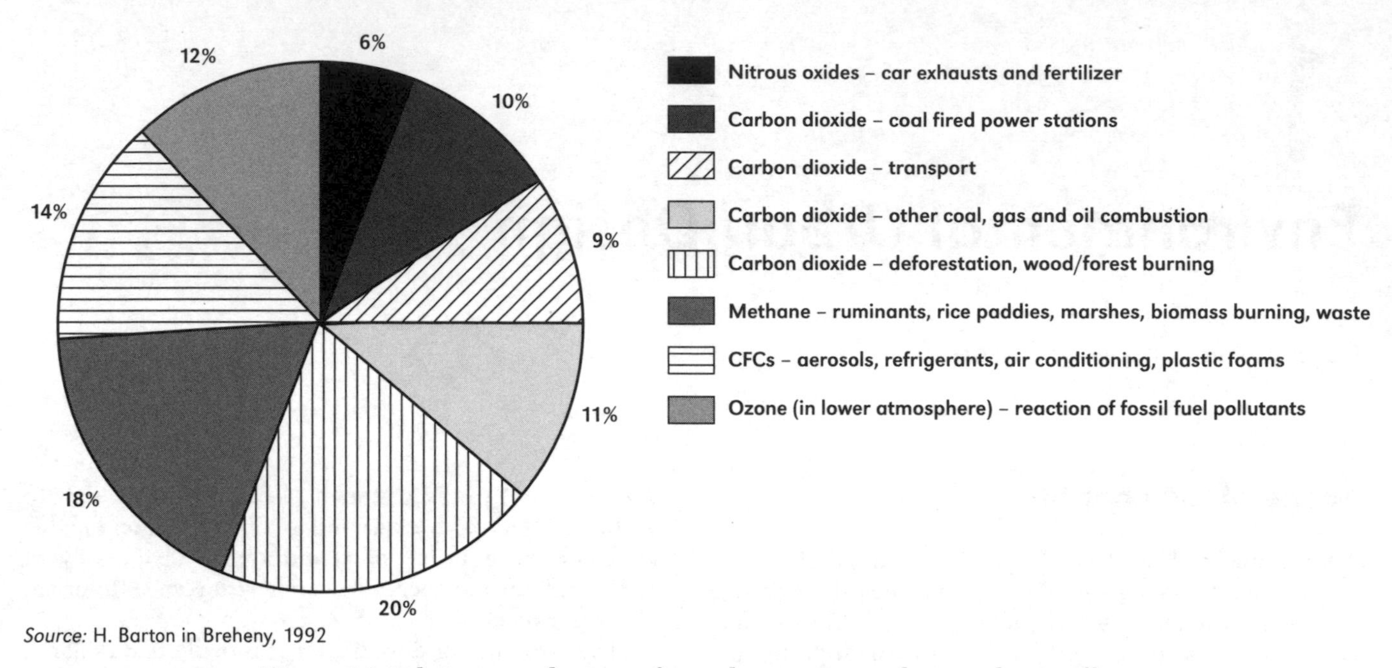

Source: H. Barton in Breheny, 1992

Figure 1.1 *Relative contributions of greenhouse gases to the greenhouse effect*

The urbanization process has been rapidly progressing since the Industrial Revolution, and health, as a consequence of the individual's environment and living conditions, became the focus of medical attention in a scientific and modern way as early as the 19th century (Davies and Kelly, 1993).

John T. Lyle (Lyle, 1993) sums up rather effectively the degradation of the environmental and health conditions of our urban milieu:

> *Cities of the industrial era have consciously excluded natural processes, substituting mechanical devices made possible by intensive use of fossil fuels. Rather than using the solar energy falling on their streets and buildings, they dissipate it as excess heat. At the same time they import immense quantities of concentrated energy in various forms, most of it derived from the petroleum coaxed from the ground in distant landscapes... Thus, we might see our overwhelming problems of depletion and pollution as largely outgrowths of our ways of shaping the urban environment.*

Arising urban sustainability needs

Urban sustainability has become a pressing task on worldwide environmental agendas due to the rise in fossil fuel costs; environmental damage and related health problems; and the depletion of non-renewables and, therefore, the need for alternative non-polluting and renewable sources.

How realistic is the target of urban sustainability? Owens claims that sustainable urban development is arguably a contradiction in terms, as urban areas require the resources of a wider environment for their survival (S. Owens in Breheny, 1992). Others maintain that the city itself as a spatial entity is an embodiment of sustainability (as quoted by A. Gillespie in Breheny, 1992):

> *It was the demand for ease of communications that first brought men into cities. The time-eliminating properties of long-distance communication and the time-spanning capacities of the new communication technologies are combining to concoct a solvent that has dissolved the core-oriented city in both time and space, creating what some refer to as an 'urban civilization without cities'.*

Vernacular urban planning: A lesson from the past?

> *Today, sensing the urgent need for more articulate structures, more harmonious landscapes and more agreeable communities,*

we are formulating an approach by which awareness of environmental factors is once again inherent in the planning/design process. The approach – instinctive for the earliest builders – has been repeatedly forgotten and repeatedly rediscovered (Simmonds, 1994).

The climate-responsive aspect of vernacular settlements has been highlighted by many scholars. For example, Morris (1994) compares the prevalence of courtyard planning in hot-arid climates and the resistance it received in northerly cool climates when Roman conquerors tried to introduce it in urban planning. It was believed that the absence of climatic stresses in moderate northern latitudes did not necessitate the clustering of shelters and introversion around a courtyard to filter the climate. The housing type that prevailed was, rather, extroverted and often free standing (Morris, 1994).

Cities today unconsciously strive to return to what could be perceived as 'vernacular urban living'. Solutions often consisted of selecting 'obvious', least resource-intensive scenarios (such as the location of settlements along easy transport routes – for example, canals and rivers – and climate-responsive urban planning principles). Examples of such vernacular settlements are the Pueblo Navajos and Anasazis Native American settlements (Golany, 1983), and Middle-Eastern settlements where compact urban planning defied the harshness of hot-arid climates (Rudofsky, 1964), such as in Ur, Olynthus and Cordoba (Morris, 1994).

Cities today as an inheritance from the past

Most of the cities we live in today have developed by incremental growth from a nucleus and an overlay of centuries of civilization. It would thus seem appropriate to review the origin of our cities in order to understand the problems of city design.

Cities of the great early civilizations adopted common physical features consisting of grids, straight axial streets, orientation of main buildings to the path of the sun and encircling fortifications. Hierarchy, geomancy and cosmology were also planning concepts that were seen in, respectively, Egyptian cities, Chinese cities and cities in Niger. The Egyptian social hierarchy is expressed by placing the Pharoah's pyramid at the centre of the settlement, surrounded by tombs of high officials, with less important tombs at the outskirts (see Figure 1.2) (Moughtin, 1996).

Chinese cities followed intricate geomancy (Feng Shui) for an environmental layout (Moughtin, 1996). The layout of ancient cities in Niger during pre-Islamic periods was ruled by cosmology. The expression of power, seen in Renaissance cities, is symbolized by mathematical order and unity. Baroque city planning exhibits the power of the Church, with its use of interconnected axes with vistas opening to religious buildings. This principle is also adopted in other cities as a device to symbolize power (e.g. L'Enfant in Washington and Hausmann in Paris). The modern city has the tall building – typically a financial or business institution – as the city landmark in order to dominate the city by its size. Since the earliest of times, political, religious and other vested interests have been glorified in cities and, often, physically raised (see Figure 1.3) (Morris, 1994).

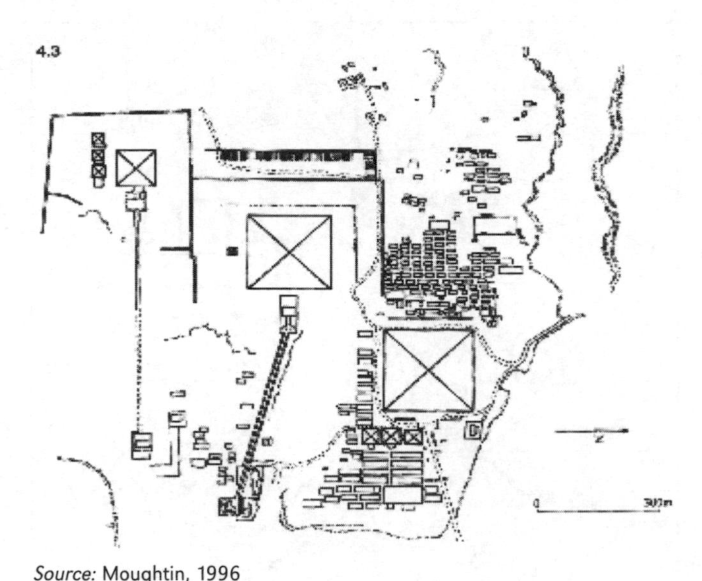

Source: Moughtin, 1996

Figure 1.2 *City of the Dead in Gizeh, Egypt*

Sustainable cities today: Lessons from the past

In an era where sustainability is becoming an urgent priority, Moughtin (1996) believes that a new symbolism is needed for the role of the sustainable city, which nurtures both man and the environment.

Various schools of thought tried to explain city form throughout history. Lynch (1960) identified three main metaphors that attempt this task:

- The first comprises a magical metaphor, where there is an attempt to link the city to the cosmos and the environment.
- The second is the analogy of the city to a machine. This notion existed in earlier civilizations. Modern examples are Le Corbusier's 'Cité Radieuse' (see

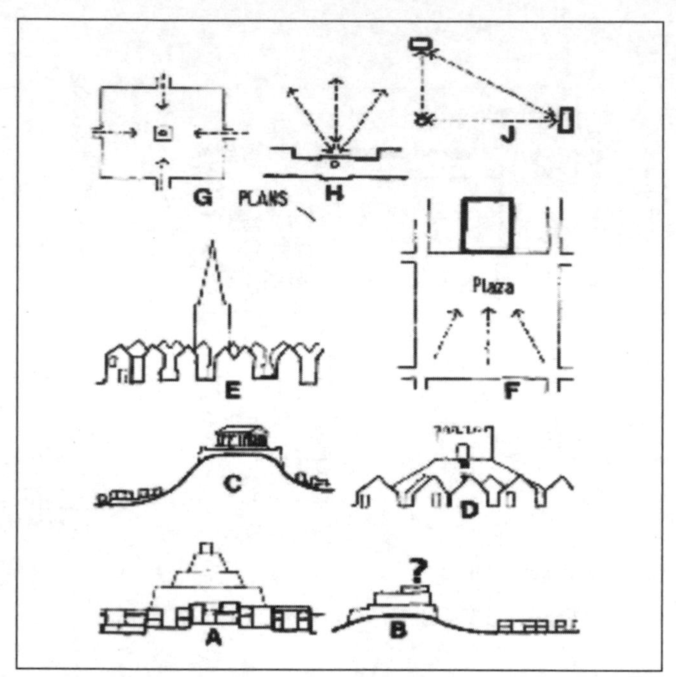

Note: diagrammatic sections through old towns are presented as follows:
A Sumerian city with ziggurat
B Harappan city with Western citadel
C Ancient Greek city with temple on its acropolis
D Norman castle in 11th-century England, dominant over conquered Saxon town
E Church in a medieval European village
F Church in a Latin American city
G Royal square with statue
H Royal aggrandizement at Versailles, France
J Democratic aggrandizement in Washington, DC

Source: Morris, 1994

Figure 1.3 *Sections through old towns*

Figure 1.4) and Garnier's 'Cité Industrielle' (see Figure 1.5). In older times, the city as a machine was encountered in Pharaonic Egypt, for workers' villages (see Figure 1.6), Greek cities (see Figure 1.8) and Roman camps (see Figure 1.7) (Moughtin, 1996). The model of the machine emphasizes the components of urban form rather than the city as an entity. Hence, this metaphor for the city is not ideal for the sustainable city, which must be holistic.

- The third comprises the analogy of a city to an organism composed of cells, which is thought to be 'most in tune with the ethos of sustainable development' (Moughtin, 1996). Early settlements strove for such an approach where, although 'designed and planned, they were constructed to respect rather than override the environment'. The principle of organic planning is 'structuring the city into communities, each of which is a self-contained unit, where cooperation is emphasized rather than competition' (see Figure 1.9) (Moughtin, 1996).

As explained by Alexander (Alexander et al, 1987), the most important goal of 'organic theory is its holistic view of the city as part of nature', where process and form are one, where:

> *... the pattern is in the seed, at the point of origin... With the sustainable city, pattern evolves from the principles used for the design and linkage of the parts... The urban structure of the organic city is non-geometrical: roads follow a curving path... The limitation, however, of organic cities is that they are not as their organic natural counter parts, self-reproducing and self-healing; the main element for their change is man* (Alexander et al, 1987).

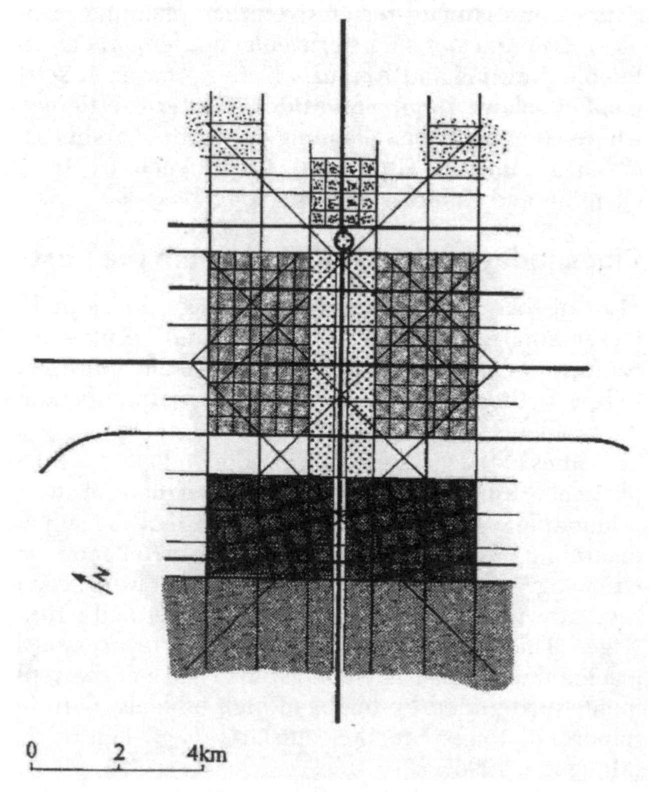

Source: Moughtin, 1996

Figure 1.4 *'Cité Radieuse' by Le Corbusier*

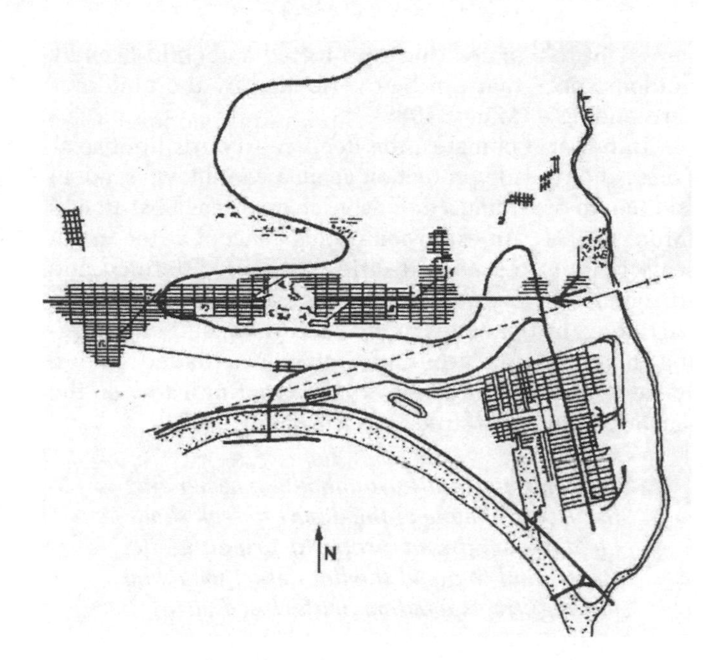

Source: Moughtin, 1996

Figure 1.5 *'Cité Industrielle' by Garnier*

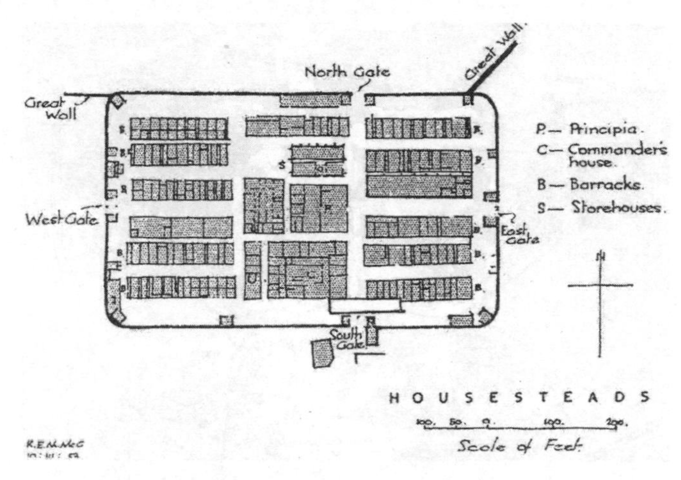

Source: Moughtin, 1996

Figure 1.7 *Plan of a Roman fort*

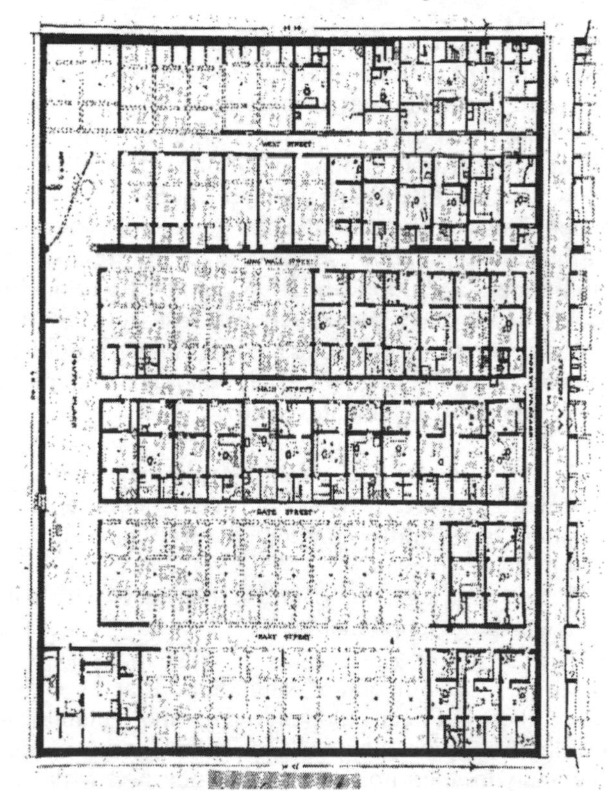

Source: Moughtin, 1996

Figure 1.6 *Pharaonic Egyptian workers' village as regular grid*

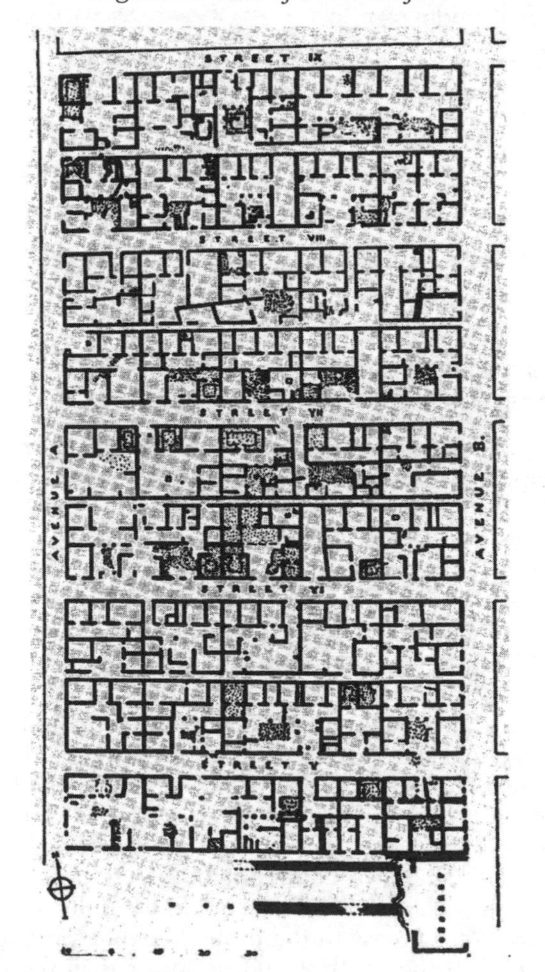

Source: Moughtin, 1996

Figure 1.8 *Plan of a Greek city*

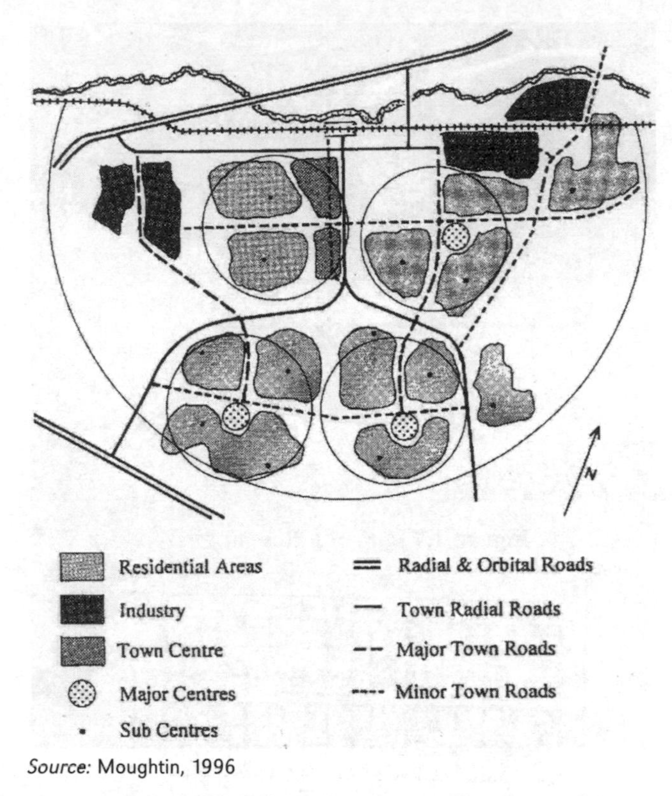

Source: Moughtin, 1996

Figure 1.9 *Harlow New Town, illustrating the principle of organic planning*

Conscious contemporary adoption of vernacular urban-planning principles

So far, there has been a conscious adoption of vernacular principles in urban design and planning in contemporary cities. In Finland, for example, following the age of functionalism where buildings were free standing within arteries of infrastructure, the courtyard house re-emerged. The revival of the vernacular south-oriented courtyard type of planning took place for several reasons. In addition to the creation of a pleasant and 'energy concentrating' sun pocket, formed at the corner of two buildings, the same configuration acts as a buffer against the wind and street pollution (e.g. traffic, noise, dirt and particles), as well as providing efficient land use. Instances of courtyard planning can be seen in vernacular villages in Norway, Sweden and Switzerland (Mänty, 1988). The shape of courtyard buildings in an urban context in such extreme climates was carefully designed to allow solar access to the protected courtyard space; buildings on the south would be lower than the other surrounding buildings. Specific proportions of the courtyard apply in order to preserve its sheltering effect (Mänty, 1988). Furthermore, the courtyard plays an important role in creating a protected and child-friendly outdoor space that can be overlooked by the buildings surrounding it (Mänty, 1988).

In hot-arid climates, the deep courtyards limit solar excess and provide protection against windblown sand, in addition to providing a defensive character against attacks (Morris, 1994). An extension of this concept at the urban scale is found in ancient cities that were defined and surrounded by high walls, which satisfied two purposes: a barrier against enemies, winds and sand, and encouraging high density (thus the narrow, shaded alleys; Rahamimoff and Bornstein, 1982). Quoting Fathy on the narrow and organic street patterns (Fathy, 1973):

> *It is only natural for anybody experiencing the severe climate of the desert to seek shade by narrowing and properly orienting the streets and to avoid the hot desert winds by making streets winding, with closed vistas.*

Practical research into urban climatology related to built form

Substantial work initiated and carried out by Oke provides insight into the interrelationship between urban form and environmental performance, specifically concerning street dimensions and building density (Oke, 1988). Such research, similar to but more theoretical and narrow than the PRECis project (an EU-funded research project assessing the potential for renewable energy in cities') (Steemers, 2000), aims at shedding light on the consequences and implications of various scenarios. Just as Oke demonstrated through his research, general, isolated recommendations to resolve isolated problems often end up conflicting with each other. For example, in mid- to high-latitude climates (cold):

- Density and compact urban morphology, which can provide protection from high winds, conflicts with openness, separation and low density, which allow for pollution dispersal.
- Warmth through compactness contrasts with solar access through openness.

As a result, a combination of configurations and scenarios is more likely in the city, depending upon orientation and prevailing winds (i.e. climate). For example, a north–south urban canyon need not be overly concerned with solar access to façades (such as east–west streets) as the sun is already low in the sky at sun rise and sun set; for pollution dispersal, prevailing winds should be taken into

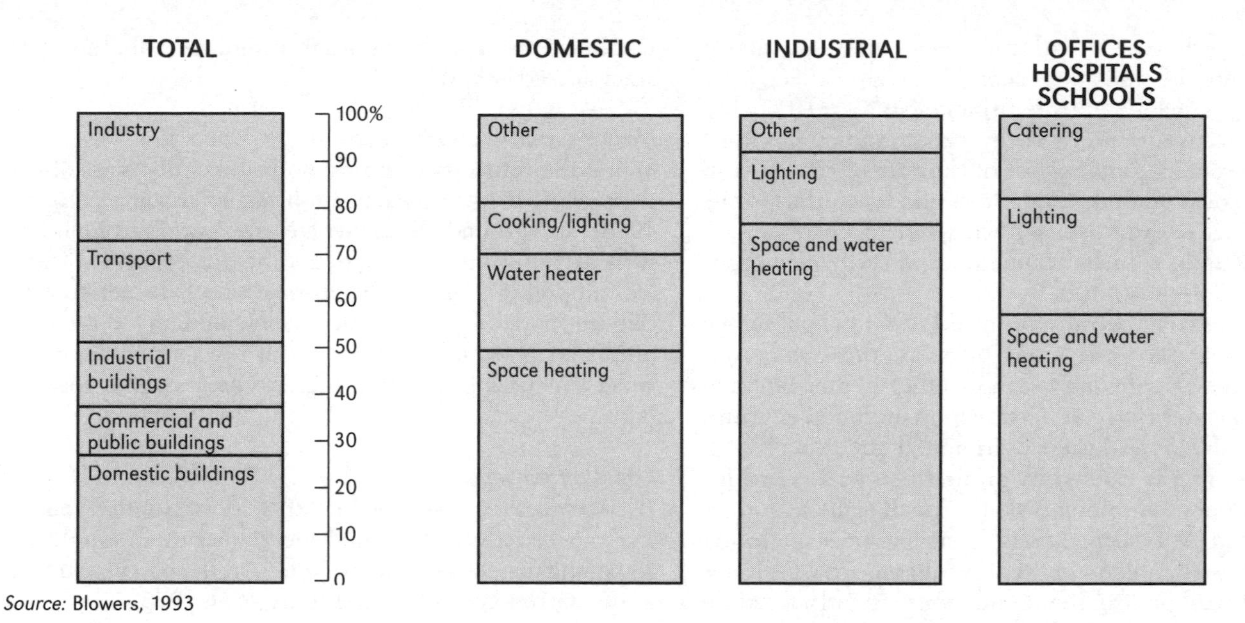

Figure 1.10 *Energy use by sector in the UK*

account. The choices made in designing cities should aim to fulfil a mix of objectives; but since not all can be accommodated, and could even be conflicting, there is a need to assess priorities. A '*zone of compatibility*' relates to identifying the range of compromises of the various climatic parameters discussed previously to arrive at a satisfactory design solution (i.e. achieving a solution that is 'good enough' rather than attempting to reach an impossible 'optimum' solution (Oke, 1988)).

These observations are among the preliminary, yet most useful, outlooks into the environmental performance of various urban-form configurations. More recent and scientifically elaborate work has been carried out within the framework of worldwide conferences on energy in the urban context (the Passive and Low-energy Architecture, or PLEA, conferences in 1998, 1999 and 2000, see www.plea-arch.org).

Energy consumption and urban spatial structure

Urban environments consume high levels of fossil fuel. In the developed world and, specifically, the UK, buildings account for 50 per cent of the total fossil fuel consumption (see Figure 1.10), with transport consuming as much as 21.2 per cent of the delivered energy (see Table 1.1).

This section deals with energy use in relation to the spatial structure of the city, particularly urban geometry and transport-related urban design configuration. A concise listing of recommendations for an optimal energy-efficient urban morphology reached by various research projects is provided at the end of the section.

Transport-related energy use and urban spatial structure

Transport is one of the major causes of the environmental problems in cities; in addition to being the source of intensive energy consumption, it causes air pollution, noise and traffic congestion. Therefore, attempting to solve the invasive transport-related problems would be a step towards reducing the environmental strain in cities. Warren argues that even with developed technologies in vehicle design (e.g. lighter, ultra-strong, non-fuel dependent or less fuel consuming and small) it is foreseen that the problems of traffic congestion, parking space requirements and others will remain, although there will be a reduction in air pollution (Warren, 1998). Several authors

Table 1.1 *Percentage delivered energy by end use in the UK*

End use	Delivered energy %
Low-temp heat (<80°C)	34.8
High-temp heat (>80°C)	25.0
Electricity	4.1
Transport	21.2
Non-energy uses	11.0

Source: Owens, 1986

have dealt extensively with the energy-intensive urban phenomenon of transport, such as Owens (1986), who looked specifically at the urban spatial relationship between transport and energy consumption in cities. Banister examines the extent of the role of transport in resource consumption, trying to assess how urban forms might affect energy use, with respect to, for example, journey length, vehicle occupancy and settlement types (Banister in Breheny, 1992).

Energy saving in transport could result in both spatial (representing city form and infrastructure) and non-spatial options (switching to a more efficient car). Building energy consumption due to transport-inflicted environmental problems and their detrimental effects on health and well-being is caused by an increase in demand for sealed indoor environments that are artificially lit and air conditioned. Mitigating atmospheric and noise pollution is therefore a double target that would ensure a healthier outdoor urban environment and, consequently, a reduction in relying upon artificial – and energy-intensive – means of indoor comfort provision. Owens acknowledges that the extent to which the transport energy market would influence or cause a change in the spatial structure in response to energy constraints remains unclear (Owens, 1986).

Energy use as a consequence of transport-inflicted environmental problems[1]

Traffic noise attenuation, in particular, involves an interdisciplinary approach based on the relationships between road traffic noise, human response, architectural characteristics of the buildings and urban morphology (Kihlman and Kropp, 1998). Traffic noise attenuation through urban form is an area of research where little work has been carried out. The need for such a study is summed up by Kihlman and Kropp (1998) in their paper presented at the 16th International Congress on Acoustics (1998), in which they recommend that buildings act as sound barriers by screening noise to create quiet areas. They propose that it is realistic to reach a two-figure target in noise reduction, allowing dwellings to have a noisy side as long as they also benefit from a quiet side.

It is important to understand how the propagation and transformation of traffic noise is influenced by a number of environmental and physical factors. Sound is naturally attenuated due to distance; the interaction of the propagating wave with the ground surface; screening provided by near-ground obstacles, such as noise filters or barriers (i.e. vegetation and buildings); and, for long-distance propagation, varying weather conditions. An initial step to mitigate traffic noise in the urban context would be to exploit potential attenuating features, as summarized below.

Distance and ground effects

While the intensity of traffic noise level decreases by 3 decibels (dB(A)) for each doubling of distance (Egan, 1988),[2] the ground effect involves absorption of sound as well as the acoustic impedance of the ground surface (Attenborough, 1998). The absorption characteristics of different ground surfaces vary widely and are considered either acoustically hard (i.e. with low porosity: sound reflective) or soft (i.e. with high porosity: sound absorbing).

Effect of barriers

Barriers that intercept the line of sight from sound source to receiver reduce the sound level, bearing in mind that screening depends on the frequency of the sound as much as the barrier geometry (Alexandre et al, 1975).

Effect of vegetation

It has been found that a tree-filled park should be at least 30 metres (m) wide to provide 7 to 11dB of sound attenuation for frequencies of between 125 to 8000 hertz (Hz), providing no attenuation to the low frequency content of traffic noise (Egan, 1988). Height and length of these plantings will further determine the attenuation provided.

Canyon effect: Reflection and scattering

The canyon effect on traffic noise propagation is caused by building façades, which cause multiple sound reflections as well as sound scattering within the urban canyon between the canyon edges (Wu et al, 1995). In general, the sound energy absorbed in the canyon ranges between 0 and 35 per cent of the incident energy, depending upon the frequency of the incident sound wave and its angle of incidence (Wu et al, 1995).

Effect of the atmospheric conditions

Atmospheric conditions relate to vertical temperature and wind gradients. The propagation and speed of sound vary with height above the ground due to these gradients which cause sound waves to be refracted (i.e. bent upward or downward) (Beranek, et al, 1982). Usually, it has been noted that a shadow zone is most commonly encountered upwind from a source, while downwind sound is bent downward with the absence of a shadow zone (Egan, 1988). As for the effect of temperature gradients on noise propagation, warmer air near the ground causes sound to bend upward; conversely, at times of air temperature inversions, sound will tend to bend downwards (Egan, 1988).

Urban spatial structure and travel-related energy requirements

In the short to medium term, the response of people to energy constraints would consist of a conscious reduction in energy consumption, with minimal effect on spatial structure. Concerning long-term adjustments, although no factual evidence exists, drastic relocation has been considered by people in order to minimize travel requirements, which, in turn, would affect the spatial structure of the city (Owens, 1986).

A simple consequence of adapting to energy constraints is re-centralization (as predicted by models that often omit social aspects and tend to oversimplify problems). However, it has been noted that energy constraints in cities did not result in centralization or suburbanization ('autonomy in suburban centres, yet links maintained with metropolitan centre'), but, rather, in counter-urbanization ('loss of people from the whole of the metropolitan area and growth of autonomous smaller towns = clean break'). The latter alternative represents a 'return to smaller place-bounded communities' and is a more energy-efficient pattern than suburbanization (see Figure 1.11) (Owens, 1986).

Various studies were carried out investigating the relationship between spatial structure and travel/transport energy requirements, using hypothetical urban forms (Owens, 1986). On the relationship between travel needs and urban size, preliminary observations in the UK showed that travel needs increase with a decrease in urban size (see Figure 1.12), with the exception of London (the analysis is based on journey-to-work data). Yet, contradictory results were found in studies of US states (Owens, 1986). Newman and Kenworthy (1989) also reached the conclusion that little correlation exists between city size and gasoline use.

What seemed probable to Owens was that 'travel needs and transport energy use are less dependent on the overall 'shape' of a settlement, defined in terms of its transport network, than on the internal arrangement and physical separation of activities, which are determined by density of cities and interspersion of activities' (Owens, 1986). Banister also noted that travel energy use is a function of density and intensity of land use (Banister in Breheny, 1992).

On density, several studies, factual and theoretical, suggest that city density is inversely proportional to transport energy consumption, as is illustrated in one of the most prominently quoted figures by Newman and Kenworthy (1989) (see Figure 1.13). Banister also found that petroleum use increases when population density reaches below 29 persons per hectare (Banister in Breheny, 1992).

On interspersion of activities or 'clustering' of functions and activities (residential, employment and services), some aspect of decentralization of activities is desirable, 'achieving a more effective integration at a smaller geographic scale' (see Figure 1.14) (Owens, 1986).

It is very difficult for land-use patterns to guide energy-efficient planning because of the inevitability of non-spatial variables: necessity of travel, choice of jobs

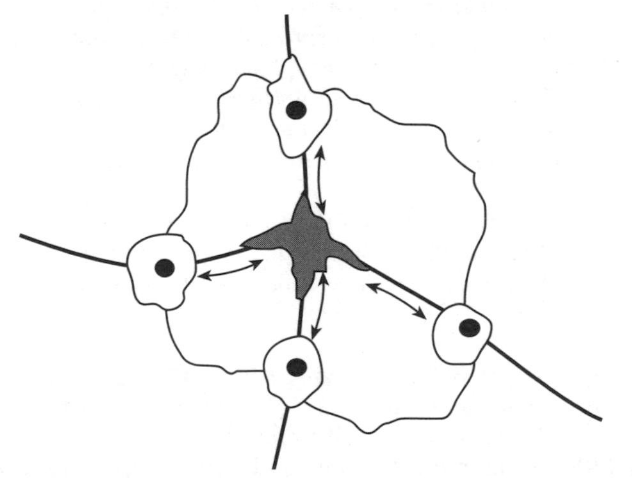

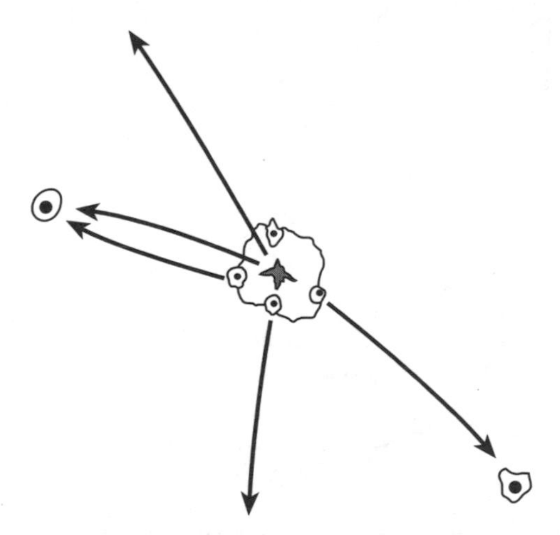

Note: Suburbanization is where suburban centres retain some autonomy, maintaining links with the metropolitan centre – believed to be energy intensive. Counter-urbanization is where there is loss of people and jobs from the metropolitan area and growth of autonomous towns – believed to be energy efficient.

Source: Owens, 1986

Figure 1.11 *Suburbanization* (left) *and Counter-urbanization* (right)

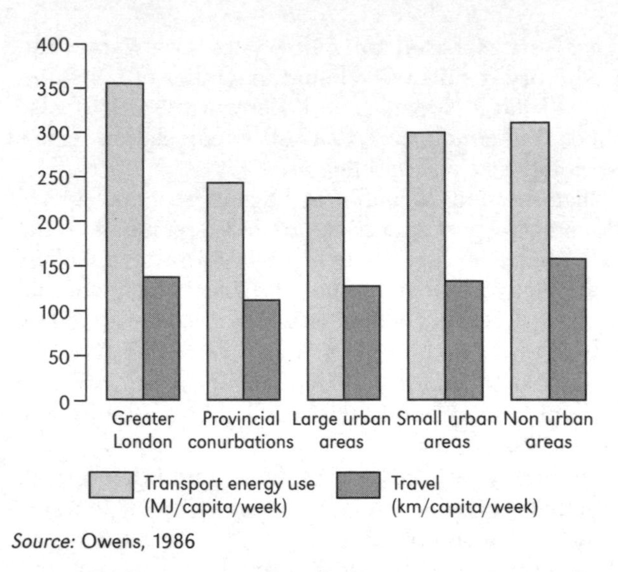

Source: Owens, 1986

Figure 1.12 *Travel and transport energy use for different area types in the UK*

and services, and the extent to which cars are used. Furthermore, ensuring 'efficient' spatial structure in a city will not guarantee transport energy savings (Owens, 1986). As noted by Barton, the reality is that 'current land-use trends are progressively undermining the function of public transport'. The reason is that the falling prices of energy and, therefore, the rise in car ownership during the post-war area have transformed cities (e.g. physical separation of activities, spread of urban hinterland, lower densities and dispersed employment) (Barton in Breheny, 1992). However, reducing the physical separation of activities is not a sufficient condition for reducing transport energy requirements. It should be accompanied by a decrease in *inclination* to travel long distances, in addition to an increase in people's interest in non-motorized travel (Owens, 1986).

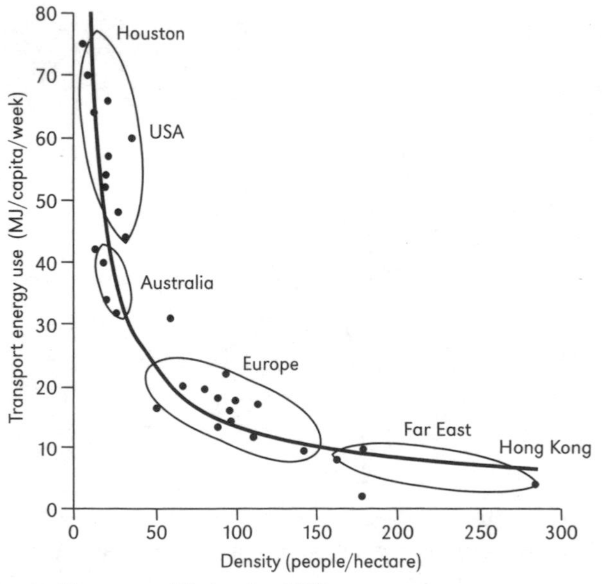

Source: Newman and Kenworthy, 1989

Figure 1.13 *Urban density versus annual gasoline use per capita*

Despite all these uncertainties and the dynamic nature of the variables involved, it is still possible to identify robust urban form options that perform relatively well within a range of possible future conditions. For example, a study in Norfolk showing 'a growth pattern involving the concentration of a new population and some degree of dispersal of new employment' was thought to be most robust. Furthermore, 'clustering' of certain functions and even 'dispersed clusters' (i.e. relative interspersion of land uses) were revealed to be effective in reducing travel needs by making trips multi-purpose (Owens, 1986).

Environmental response and urban physical properties

Research has varied between qualitative and quantitative approaches, and between generalized observations and detailed monitoring of the urban neighbourhood; but there has been increasing realization of the importance of climate specifics. In Erskine's comment on the built environment in cold climates, there is a conviction that:

> ... *houses and towns should open like flowers to the sun of spring and summer but, also like flowers, turn their backs on the shadows and the cold northern winds, offering sun warmth and wind protection to their terraces, gardens and streets. They should be most unlike the colonnaded buildings, the arcaded towns and matte-shadowed streets of the south Europeans and Arabs, but most similar in the basic function* (Erskine in Mänty, 1988).

In general, it is largely agreed that urban design with climatic consideration 'deals with the holistic morphology of the city, as well as with urban details, such as street width, form, configuration and orientation, building heights, city compactness, or dispersion, urban open space, integration or segregation of land uses, and other related physical issues' (Golany, 1995). In his 1998 publication, Givoni acknowledges the effect of urban morphology on the urban microclimate and, therefore, on energy consumption. The physical parameters that he was

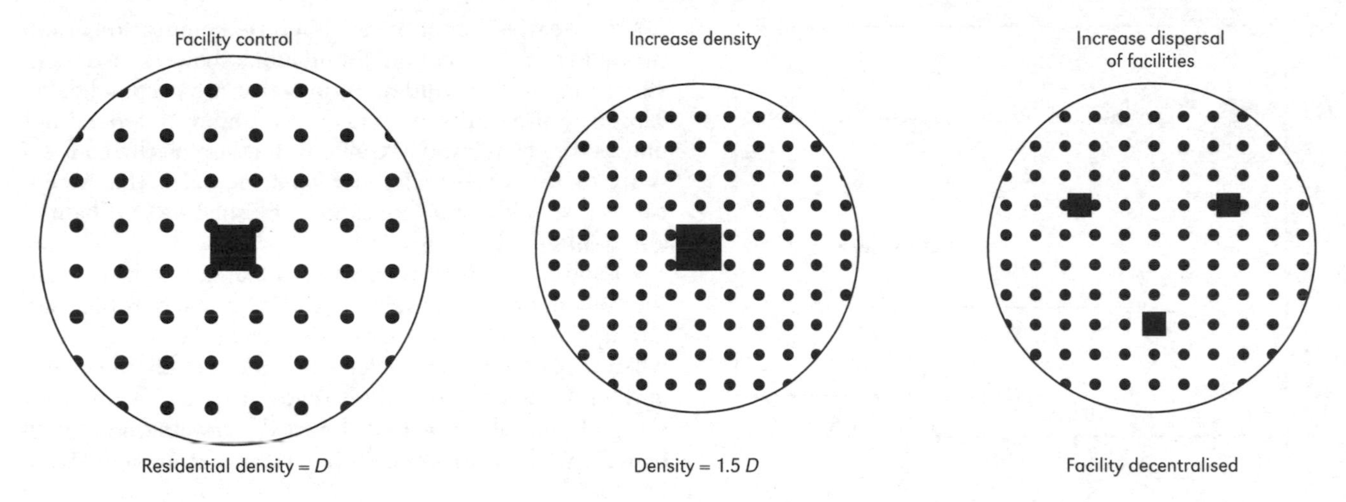

Source: Owens, 1986

Figure 1.14 *From left to right, gradual decrease in physical separation of activities*

able to extract were size of city, density of built-up area, land coverage, building height, orientation and width of streets, and building-specific design details affecting the outdoor conditions (Givoni, 1998). He proposes urban design recommendations in different climates. Below is a review of work conducted on the environmental response of various urban physical characteristics.

Streets or urban canyons

In his book *Great Streets*, Jacobs (1993) refers the feel of an urban street to its capacity in providing comfort and liveability:

> *Is there some point, some proportions or absolute height, at which the buildings are so high in relation to street that the building wall becomes oppressive?... It may be that the upper limit (of the building) is more appropriately determined by the impact of height on comfort and liveability of the street, as measured by sunlight, temperature and wind, than by absolute or proportional height* (Jacobs, 1993).

Jacobs further stresses the importance of the street in the urban context: 'Think of a city and what comes to mind? Its streets. If the city's streets look interesting, the city looks interesting; if they look dull, the city looks dull' (Jacobs, 1993).

Strong correlations have been established between urban street configuration and wind flow and, therefore, pollution dispersion.

Wind is believed to be among the most notorious alterations caused by urbanization. Although there is a good case to be made for designing cities that facilitate dispersion, Oke (1988) warns of the difficulties in designing streets purely for the general comfort of their citizens. There is a subtle trade-off in street design which aims to maximize ventilation, dispersion of pollutants and solar access, while not compromising shelter and urban warmth. Oke sets out guidelines based on relationships between these factors and urban geometry towards finding a 'zone of compatibility'. Urban canyons can channel wind and create an acceleration of wind speed, which poses a hazard to pedestrians. These phenomena can be remedied by adopting appropriate street width and building design (Landsberg, 1981).

In his 1988 paper, Oke stresses that the main concern is comfort and safety of pedestrians and heat loss from building envelope, and that both are concentrated on the sides of the urban canyons – not in the centre. He was able to identify wind flow depending upon the height-to-width ratio (H/W) of a street canyon (see Figure 1.15). Oke was able to define wind and turbulence diminution factors as a function of H/W. He discovered that a H/W ratio of about 0.65 ensures considerable protection (Oke, 1988).

On pollution dispersion, a small H/W was shown to be good for an exchange between ground-level air and cleaner air above. This stops being the case for H/W beyond the threshold for 'skimming flow' (0.65). In his conclusion to a detailed analysis of wind flow pattern and associated pollution dispersion, Oke notes that a H/W $\approx$ 0.65 and a building density of $\approx$ 0.25 may provide an

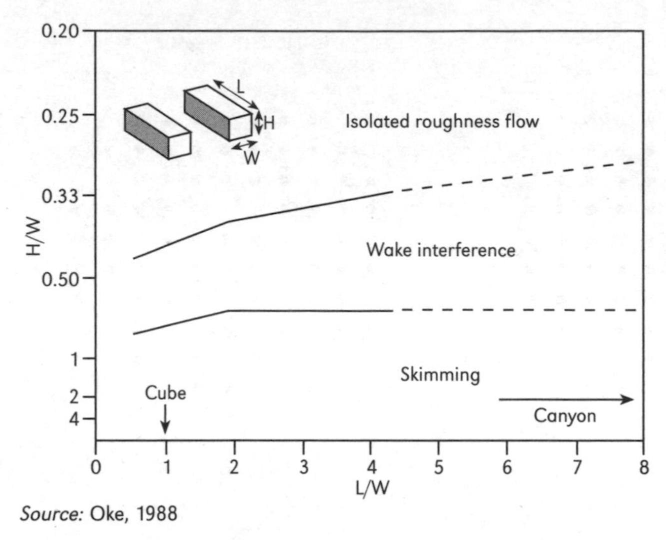

Source: Oke, 1988

Figure 1.15 *Height-to-width ratio (H/W) and length-to-width ratio (L/W) are critical to wind flow in the urban canyon where L indicates length of building perpendicular to wind direction*

upper limit to satisfactory dispersion from street canyons (Okc, 1988).

An interesting case study of London (Croxford et al, 1995; Croxford and Penn, 1995) was carried out with the main objective of visualizing flows in a street network where there has been consistent relationship between relative carbon monoxide (CO) concentrations and prevailing wind direction. Simulations showed vortices in street canyons for wind directions that are perpendicular. The structure of these vortices was found to depend not only on the H/W ratio of canyons, but also on the connectivity of the street canyons. This indicates that street canyons should not be considered in isolation (Ni Riain et al in Jenks et al, 1996).

Figure 1.16 illustrates where clean air is drawn into the canyon on the leeward side and back across the road and up the windward side of the street. For this reason, when the sensor is on the leeward side it reads 'clean air', and on the windward side it reads 'dirty air', caused by the pollution blown across the road. Simulation helps to explain why clean air is being recorded at pedestrian level.

The Urban Task Force has praised 'permeability' or interconnectivity of streets within the urban context, as this provides accessibility and increasing mobility, in contrast to cul-de-sac street configurations (tree-like structures), which are seen as inefficient (see Figure 1.17) (Urban Task Force, 1999).

On solar access and urban geometry, Oke observes tentatively that 'a H/W of approximately 0.6 seems to be a suitable upper limit to maintain solar access in a city at a latitude of 45 degrees' (Oke, 1988). Oke acknowledges that 'this conclusion still needs to be refined a lot on the basis of complete analysis of passive solar energy gain and

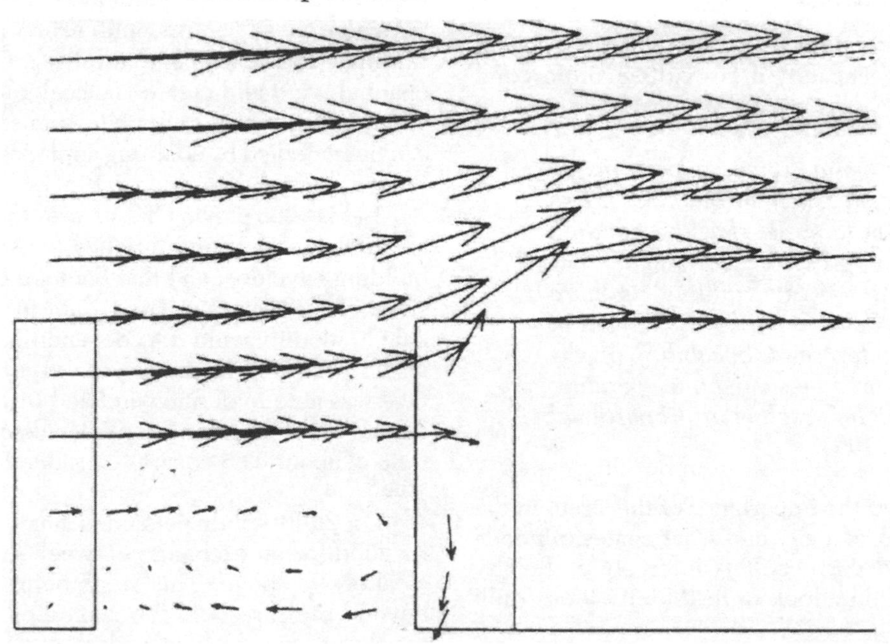

Source: Ni Riain et al in Jenks et al, 1996

Figure 1.16 *Model of the investigated area, showing the relationship between vortices and pollution dispersal*

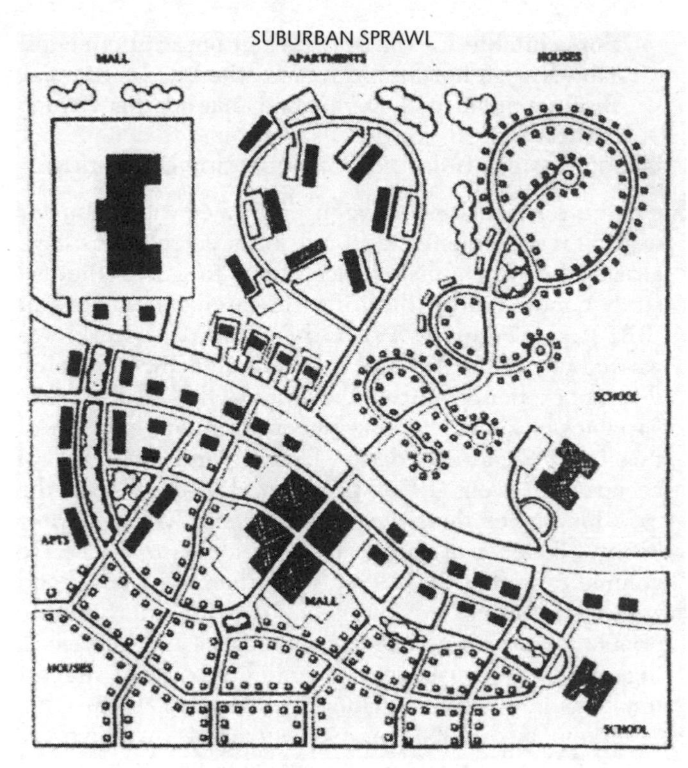

Source: Urban Task Force, 1999

Figure 1.17 *Permeable* (bottom) *versus impermeable* (top) *urban layouts*

illumination of the total urban system' (Oke, 1988). More recent and elaborate quantitative work on solar access has been taking place, taking into account previously unaccounted for multiple reflections within the urban canyon (Steemers, 2000).

On urban warmth and the urban heat island with respect to geometry, Oke observes that a city with an elevated occurrence of high H/W and, therefore, tending towards compactness promotes the trapping of solar radiation and urban warmth, especially at night. This is beneficial in mid-latitude countries as this reduces the space heating load required during cold seasons. The same observation was previously made by Ludwig (1970), whose figure illustrates solar radiation for different H/W ratios (see Figure 1.18). On the flat plane, most of the absorbed solar radiation is re-radiated to the sky as long wave radiation. In a medium density of H/W = 1, most of the reflected solar radiation hits other buildings, as well as the ground, until it is absorbed near or at the ground. For higher densities of H/W = 4, most of the absorption takes place at a high level of the canyon, reducing the amount of radiation that reaches the ground.

Empirical studies have shown that, due to the heat island effect, it is possible to save 5 to 7.5 per cent of space heating costs per 1 degree Celsius increase in mean daily temperature (from Oke, 1988). Benefits can also be claimed for outdoor comfort, vegetation growth and pollution dispersion through thermal turbulence and breezes. However, such advantages during the cold season can be outweighed by heat stress and pollution accumulation in the hot season. Oke defines the following simplified thresholds 'under ideal heat island conditions':

- one third of the maximum possible intensity is gained with H/W = 0.4;
- one half of the maximum possible intensity is gained with H/W = 0.7;
- two-thirds of the maximum possible intensity are gained with H/W = 1 (Oke, 1988).

The thermal comfort conditions of street canyons in hot-arid climates have been studied in order to test and

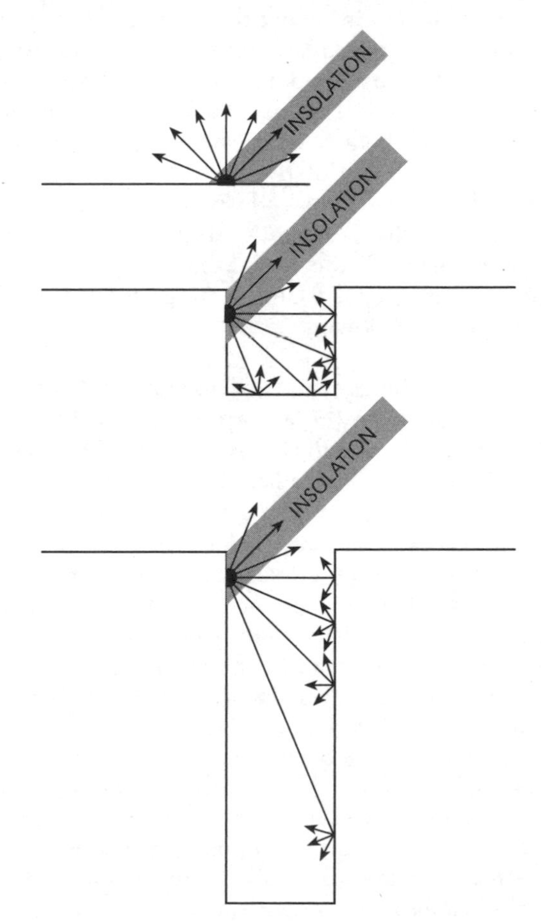

Source: Ludwig (1970)

Figure 1.18 *Solar radiation for different H/W ratios*

validate the appropriateness of dense and compact urban configuration (Pearlmutter, 1998). Several authors have drawn attention to the advantages, as well as the disadvantages, of compact city configuration (Jenks et al, 1996). It was, indeed, highlighted by Jenks et al (1996) that 'a dense urban fabric may provide solar shading of pedestrians within deep street canyons. On the other hand, such canyons may become relative 'heat traps' due to multiple solar reflection and reduced albedo, diminished night sky radiation and substantially restricted ventilation'. The innovation in this work lies in taking account of the overall heat balance within the canyon for a body (e.g. a pedestrian) located at a certain point within an arid climate, rather than for the urban canyon as a whole. This gives a more truthful indication of human/user comfort outdoors. The total rate of thermal energy flux between a pedestrian and the urban environment was calculated (Pearlmutter, 1998). On-site measurements showed that, unlike what theory might suggest, during daytime summer hours, pedestrians in the street canyon absorb less thermal energy from the environment than when in the open air. This refutes the original assumption that reduced wind speeds and elevated temperatures in the street canyon might result in an overheated environment. In winter, heat loss from the body to the surrounding area is reduced due to the protection of the body from wind. During summer, the overall energy loss during night hours is reduced in canyons (Pearlmutter, 1998).

In conclusion, it was noted that the compact street canyon acts as a 'cool island' during summer daylight hours due to the following:

- Shading of the canyon by flanking walls, protecting pedestrians from the absorption of short-wave and reflected radiation during summer daylight periods. The limited extent of radiant heat gain by a pedestrian in the canyon during daytime hours is due to the restricted exposure of a pedestrian to the horizontal ground surface. Of all surfaces, this is the greatest source of radiant heat because it intercepts the highest flux density of solar radiation, and it is often a dark colour – thus maintaining by far the highest surface temperature during the daytime. At night this relationship changes so that the magnitude of restricted radiant heat loss to the sky becomes dominant.
- Thermal inertia also plays a role as it tends to be high for most urban surfaces. Comfort in a desert environment depends largely upon the stabilization of thermal extremes, diurnally (between day and night) and seasonally (between winter and summer). With proper urban design, the 'compact' urban street canyon holds great potential in such thermal moderation, suitable for the specifics of hot-arid climates. However, in hot-humid regions thermal inertia may result in night-time overheating (Pearlmutter, 1998).

Urban design, building configuration and siting

On the relationship between energy consumption for heating requirements and built form, correlations have already been established according to a hypothetical study conducted by Building Research Establishment (BRE) (see Figure 1.19). Other empirical work was carried out in the US and UK, although such research does not explicitly indicate the scope of the effect of other variables on domestic energy consumption (e.g. size of dwelling, socio-economic factors and income of occupants) (Owens, 1986). However, at the local scale, the case for higher densities is reinforced by the greater energy efficiency of built forms with a low surface area to volume ratio (Owens, 1986). For urban neighbourhood zones that are disadvantaged (i.e. with unfavourable orientation for renewable energy reliance and, therefore, higher energy consumption) guidelines should suggest implementing super insulation (Bentley et al, 1985).

Simmonds (1994) praises attached dwellings as opposed to separate dwellings, which reduce the exposed external envelope to the outside environment and, thus, reduce heat loss/heat gain, thereby lowering heating and cooling loads. Other advantages include efficient use of space in the urban context and savings in construction cost and maintenance due to shared walls (see Figure 1.20).

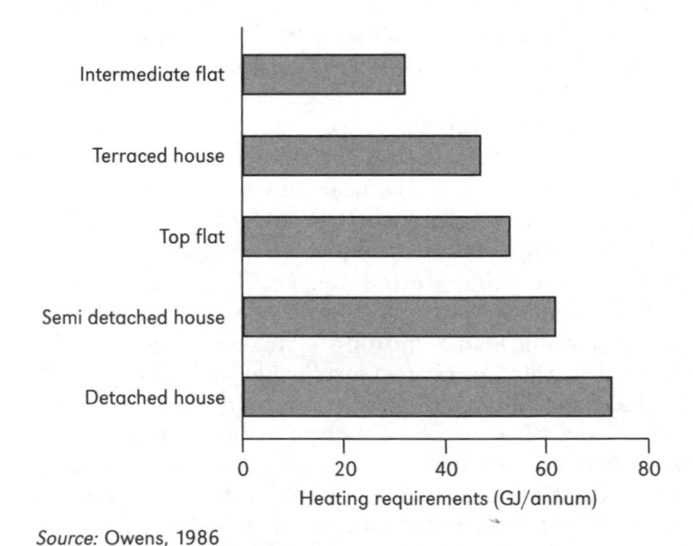

Source: Owens, 1986

Figure 1.19 *Influence of built form on heating requirement in the UK according to Building Research Establishment (BRE) data*

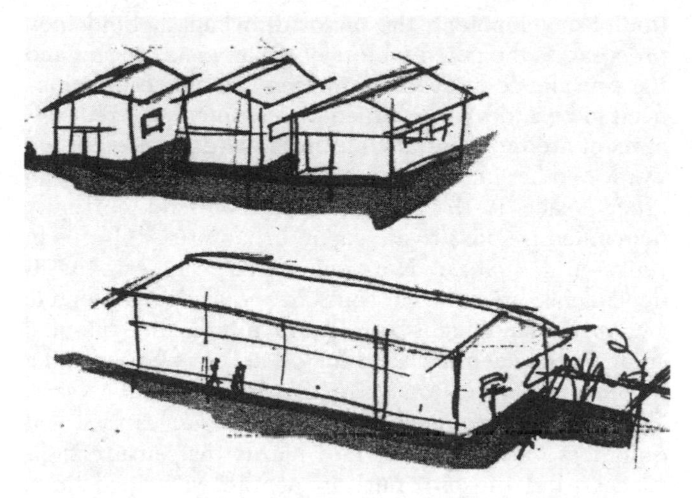

Source: Simmonds, 1994

Figure 1.20 *Attached versus detached dwellings; single-family detached houses seem to cause more loss of valuable space and privacy than attached houses, where side yards are used more efficiently*

Owens suggests that the use of renewable resources in buildings in cities would require relatively low density, maximizing exposure of structures to solar radiation (see Table 1.2). Yet, this aspect clashes with one of the conclusions for energy-efficient transport, as discussed in the previous sections; such trade-offs would require careful economic feasibility analysis (Owens, 1986).

In general, it is believed that high-rise buildings are non-energy efficient due to their high weather exposure and their necessity for heating/cooling. They also involve an energy-intensive construction industry, and have high maintenance and operating costs (Owens, 1986). Tall buildings were also noted to produce undesirable wind eddies. Some evidence points to the fact that uniform building height and uniform distance between buildings create less flow disturbance (Landsberg, 1981). It is generally agreed that airflow within the urban context around buildings depends upon the shapes, sizes and orientations of buildings. Although an average decrease in street-level winds is noticed when compared with wind in the countryside, 'increases may occur locally due to "jetting" down canyons aligned with the wind and in the vicinity of tall buildings, which deflect momentum from above' (Oke, 1980).

Building materials and colours have a great influence on the transformation of solar radiation through their absorptance, reflectance and transmittance, which, in turn, affects surface temperature and, therefore, outdoor comfort conditions. As demonstrated by Fezer (1982), the surface temperature of different types of pavements can vary significantly. Accordingly, recommendations for different climates could include light-coloured ground finishes, which are more suitable for hot climates, limiting the absorption of solar radiation; dark-coloured ground finishes are suitable in cold climates, maximizing solar radiation absorption (see Figure 1.21) (Fezer, 1982).

On mitigating glare caused by solar reflection on building façades, which causes visual discomfort for

Table 1.2 *Site orientation chart*

Adaptations	**Objectives (climate types)**			
	cool	**temperate**	**hot-humid**	**hot-arid**
Position on slope	Low for wind shelter	Middle-upper for solar radiation	High for wind	Low for cool air flow
Orientation on slope	South to south-east	South to south-east	South	East to south-east for afternoon shade
Relation to slope	Near large body of water	Close to water, but avoid coastal fog	Near any water	On lee side of water
Preferred winds	Sheltered from north and west	Avoid continental cold winds	Sheltered from north	Exposed to prevailing winds
Clustering	Around sun pockets	Around a common sunny terrace	Open to wind	Along east-west axis for shade and wind
Building orientation	Southeast	South to south-east	South towards prevailing wind	South
Tree forms	Deciduous trees near building, evergreens for windbreaks	Deciduous trees nearby west, no evergreen near south	High canopy trees, deciduous trees near building	Trees overhanging roof if possible
Road orientation	Crosswise to winter wind	Crosswise to winter wind	Broad channel, east–west axis	Narrow, east–west axis
Materials colouration	Medium to dark	Medium	Light, especially for roof	Light on exposed surfaces, dark to avoid reflection

Source: Owens, 1986

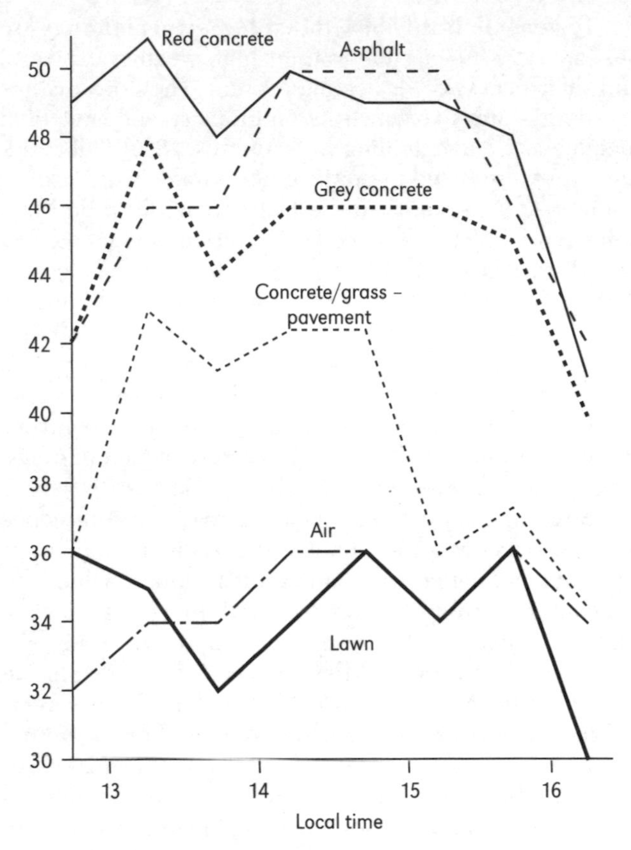

Source: Fezer, 1982

Figure 1.21 *Daily variation in surface temperature of various types of pavement finishes, measured in Stuttgart in July 1978*

pedestrians, Givoni (1998) gives various urban and building design recommendations, such as façade colour (light colours), geometric treatment (overhanging projections that intercept reflections) and vegetal wall cover (reducing wall reflection).

On pollution dispersion and urban geometry, Oke highlights that the urban canopy layer (UCL) and the urban boundary layer (UBL) are interrelated – having a low concentration of pollutants in the latter would be beneficial for the former (Oke, 1980, 1988). When air is clean, UBL provides a clean air source for ventilating urban canyons and buildings to the UCL. 'The lower the upper-level concentration, the greater is the vertical concentration gradient for upward turbulent diffusion' (Oke, 1988).

Urban green space is known to have beneficial effects with respect to traffic noise, pollution and heat island, as documented by various research studies, to date. Owens (1986) acknowledges the fact that although existing urban centres may be revitalized and boosted by a concentration of development, the major drawback behind such proposals is the potential loss of urban green spaces and the prevalence of 'town cramming'. These reservations need to be addressed if urban development and redevelopment are to be sustainable in the widest sense of the word. Another important factor affecting the loss of urban green space is the large amount of land currently demanded by the private car in urban areas (Owens in Jenks et al, 1996). A Norwegian study suggests that if dependence upon the car could be reduced, considerable amounts of land might be released, minimizing potential conflicts between the need for urban green space and a more compact pattern of urban development (Naess, 1991). Whyte suggests that it is not the size, leafiness and quietness of urban parks and plazas that ensure their success, but the proximity to people who need them (Whyte, 1990). Landsberg comments more generally on the existing natural flow patterns, such as proximity to lakes, rivers or mountain valleys, and states that these should be taken advantage of, where appropriate (Landsberg, 1981).

General recommendations for energy-efficient urban design

Based on the brief exposé of work carried out to date in the field of energy and environmental response to urban form and urban physical properties, recommendations were formulated by the respective authors. The compilation of these recommendations in one section offers the opportunity to compare the varying conclusions reached.

On transport

According to Owens, some of the key characteristics of an environment that is 'inherently' energy efficient in terms of transport can be summed up as follows. The environment:

- Is compact and mixes land uses, providing a choice of jobs and services, with a clustering of trip destinations. Mixed-use has also been promoted by the Urban Task Force (Urban Task Force, 1999) as forming one of the major attractions of urban living, providing proximity to work, shops and social, educational and leisure uses, and offering horizontal as well as vertical mixing (along street and in building). Careful design should govern this process in order to exclude noisy industries or industries that require heavy traffic (Owens, 1986).
- Avoids dispersed development and facilities that are not well served by public transport facilities.
- Facilitates public transport with the provision of facilities in such a way as to encourage walking and cycling, and discourages use of cars.

- Features higher densities in certain appropriate locations, such as along public transport routes (Breheny, 1992).

Yet, political barriers in terms of transport policies are numerous, and they should be overcome before a more beneficial spatial structure is implemented and for it to be effective (Owens, 1986).

Controlling the impact of traffic noise and air pollution in an urban context, and improving the quality of the outdoor environment, consists of two approaches. While the first involves dealing with the problem at its source, designing cleaner and quieter transport technologies (Urban Task Force, 1999) and controlling traffic flow and speed, the second entails limiting the spread of noise and pollution once generated, and is more remedial than preventive in nature. This second approach features varying scenarios. One scenario, for example, might involve making neighbourhoods more pedestrian friendly, even when traffic continues in nearby streets (Warren, 1998). Road and building design, as well as urban design, can contribute to attenuating traffic noise and promoting pollution dispersal. Examples are seen in Oke's research (Oke, 1980, 1988) where he recommends a certain canyon H/W ratio for a quicker purging of pollution. Road surfaces contribute to absorbing traffic noise, as well as noise barriers and façade design.[3]

On renewable energy potential and energy saving

Design ingenuity is required to harness passive solar energy in higher densities (more than 35 dwellings per hectare. Results, in theory, suggest a reduction in energy requirement for travel and heating (Owens, 1986). Energy-efficient urban design is feasible, which could result in substantial long-term energy savings (see Table 1.3) (Owens, 1992).

There have been several government planning policies discussed since the 1960s, particularly for extreme climatic contexts, such as in arctic settlements, where heat conservation was encouraged through compact building groups to take advantage of winter sunlight and to afford the greatest protection in storms (Mänty, 1988). Further recommendations for energy savings followed, specifically after the energy crisis during the early 1970s. Some proposals relate to restructuring entire urban environments in the context of winter months – avoiding north-facing windows, encouraging south-oriented greenhouses and wind shielding from buildings, structures and plants (Mänty, 1988).

Table 1.3 *Gasoline use versus urban density according to area types*

Area	Gasoline use (gallons per capita)	Urban density (persons per acre)
Outer area	454	5.3
Whole urban area	335	8.1
Inner area	153	48.3
Central city	90	101.6

Source: Owens, 1992

On optimal general environmental performance

Oke notes that for all four goals to be satisfied (pollution dispersal, solar access, shelter and heat island), a H/W of 0.4 to 0.6 offers a compatible range (upper limit for shelter and urban warmth and lower limit for pollution dispersal and solar access) (Oke, 1988).

Owens's (1986, 1992) observations on linking energy use and urban form are summarized as follows:

- Interspersion of activities can change trip requirements, especially length, bringing energy demand variations of up to 130 per cent.
- The shape of the urban area can lead to variations in energy demand of 20 per cent.
- Density or clustering of trip destinations can bring about energy savings of 20 per cent, mainly by facilitating public transport.
- Dense or mixed-use zones, facilitating combined heat and power systems, can increase the efficiency or primary energy use by 100 per cent.
- Layout and orientation of buildings can lead to energy savings of 12 per cent through passive solar gain.
- Siting, landscaping, layout and materials can produce energy savings of up to 5 per cent through modifying microclimates.

General recommendations, policy-making and implementation

As a result of Local Agenda 21 activities, many municipalities frequently assess the environmental functioning of their own offices, agencies and utilities. Such environmental audits represent attempts to comprehensively study and gauge the ways in which a local government's actions and policies affect the environment. They often lead to the preparation of a local state-of-the-environment report and an environmental action plan (Beatley, 2000). Within the European Union (EU), municipal governments now have the ability to participate in the Eco-management and Audit Scheme (EMAS) programme, a form of environmental auditing and environmental management system formerly available only to private companies. A number of localities are currently in the early stages of EMAS; but only a handful have completed the certification process. An example, a sustainability appraisal (SA) was prepared by Leicester city, compiling evaluative factors and organizing them into a matrix (see Table 1.4). This appraisal methodology is applied to proposed development sites within a city (Beatley, 2000).

In the UK, the Urban Task Force (UTF) has been fully involved in establishing 'a framework to deliver a new future to urban England, to use a projected 3.8 million households over a 25-year period as an opportunity to revitalize our towns and cities' (Urban Task Force, 1999). The concept of 'urban renaissance' emerged, to be founded on 'principles of design excellence, economic strength, environmental responsibility, good governance and social well-being'. Recommendations for this urban renaissance are aimed to encompass streets, as well as the larger scale of town. It is acknowledged that no one single solution exists. Instead, a framework for change is offered, where different places are given the opportunity to define and interpret their own priorities. On the argument about having to redress the quality of life in the urban environment, the UTF points out that the quality of urban design and architecture must be re-established as part of our everyday urban culture. In order to achieve this, we must realize that it is not a question of regulation or manuals – which often failed to provide quality products – but a question of investing in good designers, bearing in mind that good design is an investment in itself towards the long-term sustainability of the city (Urban Task Force, 1999). Therefore, it is recommended that concern for energy efficiency should be felt by planners, urban designers and architects at an urban neighbourhood scale, where intervention is made possible rather than enforced. However:

> *... the well-governed city must establish a clear vision, where all policies and programmes contribute to high-quality urban development. In partnership with its citizens and its business leaders, the city authorities have a flexible city-wide strategy which brings together core economic, social and environmental objectives. It is, therefore, a city characterized by strong political leadership, a proactive approach to spatial planning, effective management, and commitment to improve its skills base'* (Urban Task Force, 1999).

Table 1.4 *Impact and commentary for a proposed housing development in Hamilton, Leicester*

Sustainability impact criteria	Impact	Commentary
Quality of Life and Local Environment		
1 Open space	+	Opportunities to provide new public open space within development
2 Health	–	Emissions from new traffic
3 Safety and security		Covered by other policies
4 Housing	++	Meeting identified housing needs of city
5 Equity	+	Range/mix of housing together with ancillary community facilities
6 Accessibility	+	Urban fringe location currently not well served by public transport. Still transport choice location
7 Local economy		
8 Vitality of centres	+	Additional housing will support Hamilton District Council facilities
9 Built environment	+	High-quality design could contribute to appearance of development
10 Cultural heritage		
Natural Resources		
11 Landscape	–	Loss of open countryside, but structural planting often an important pre-development feature
12 Minerals		
13 Waste		
14 Water	–	Possible disruption to existing ground water/drainage, etc.
15 Land and soil	–	Loss of agricultural land
Global Sustainability		
16 Biodiversity	–	Loss of natural habitats (greenfield site), but new development will create parkland and water settings
17 Movement	–	Increased use of private car due to peripheral location
18 Transport mode	–	As above
19 Energy	?	Depends on detailed layout
20 Air quality	–	

Source: Beatley, 2000

According to the UTF, the key principles of urban design are:

- site and setting;
- context, scale and character;
- public realm;
- access and permeability;
- optimal land use and density;
- mixing activities;
- mixing tenures;
- building to last;
- sustainable buildings;
- environmental responsibility.

Along with these principles, design recommendations should advocate flexibility for the changing use of buildings and for reducing construction cost in new buildings – in other words, 'long-life, loose-fit, low-energy buildings' (Urban Task Force, 1999).

In conclusion and as Owens pointed out:

> *It should be stressed again that energy-conscious planning is not an exact science, especially at the urban and regional scales. Because of many intrinsic uncertainties in urban and energy systems, we are unlikely to be able to identify development patterns that would be the most energy efficient under all possible future circumstances. However, research suggests that it is possible to identify land-use patterns and built forms which are robust and flexible* (S. Owens, 1992).

Energy efficiency and renewable energy potential versus city texture and configuration

Urban morphology became a major topic of discussion at the end of the 1960s, when various comparative studies demonstrated how, at the same densities, various urban forms were possible, each providing different qualities (Martin and March, 1972). Later, Jacobs (1993) compiled a wide variety of urban textures from different parts of the world, showing an amazing diversity through time and space. However, his approach was more qualitative than quantitative. Urban texture was analysed in the Swiss context at the University of Geneva, where maps of various Swiss cities were gathered and analysed. A typological classification was carried out based on a purely morphological dimension (CETAT, 1986). Comparisons showed a great diversity within a relatively confined geographical context. The reason for this phenomenon was attributed to various impacts of norms and building regulations on urban forms. Interesting notions, such as 'building thickness', began to emerge from this research, constituting useful urban-form indicators. The textural characteristics of urban neighbourhoods are a result of the accumulation and overlaying of micro-scale building form characteristics and are therefore believed to have strong implications for the energy consumption and environmental performance of an urban neighbourhood. However, little attention was given to the relationship between grain size and city texture with general building use and mean energy use per square metre of the city (CETAT, 1986).

According to Golany (1995), there are basic principles that can guide the urban designer with climatic considerations in mind; the two levels to be considered are the city site selection and urban configuration. Golany was able to propose a 'preferred' urban design morphology (compact form, dispersed form, clustered form and combined form) in response to his categorization of climate in six major climatic types: hot humid, cold humid, hot dry, cold dry, seashore strips and mountain slopes. Yet, he stresses that design solutions may be suitable to more than one type of climatic area, in spite of distinct climatic differences (see Table 1.5).

For Owens, the biggest contribution that planning can make in reducing energy consumption and pollution is to design urban forms that minimize the need to travel, seeing travel as the major energy-consuming and pollution-generating aspects of urbanization (Owens, 1986). Cities can be designed in such a way as to encourage public transport at the expense of privately owned cars; therefore, appropriate land-use planning policies can contribute effectively to energy conservation. In addition to spatial modifications, non-spatial policies could be promulgated to encourage and promote public transport by local authorities in conjunction with central governments (e.g. pedestrianization of areas, cycling paths, car-pooling and parking control) (Owens, 1986). An overview of the research on transport-efficient urban structures as carried out by various scholars highlights various urban forms:

- Non-extreme 'compact city' (Dantzig and Saaty, 1973): these cities feature high densities, where employment and services are centralized, surrounded by high-density residential development. The limitation is in the use of renewable energy sources (see Figure 1.22).

Table 1.5 *Main climatic types, major resulting problems and corresponding urban design responses*

Main climate type	Major problems/issues	Basic urban design response
Hot-humid	Excessive heat High humidity	Ventilation: open ends & dispersed form. Widely open streets to support wind movement. Extensive shadow. Dispersion of high rise buildings to support ventilation. Combined variation of building heights. Wide, yet shadowy open spaces. Shadowing, planned tree zones
Cold-humid (temperate)	Low temperature Winter & summer high precipitation Windy	Heating (passive & active): Mixture of open & enclosure forms. Protected edges at windward side (with structure or trees). Uniformed building heights. Medium dispersed open space. Circumferential & intersecting tree strips
Hot-dry	Excessive dryness combined with high day temperatures Dusty and stormy	Compact forms: Shadowing. Evaporative cooling. Protected urban edges from hot winds. Windward location near a body of water. Narrow winding neighbourhood roads & alleys. Mix of building height to shadow the city. Small, dispersed & protected public open spaces. Circumferential & intersecting tree zones. Use of geospace city concept
Cold-dry	Excessive low temperature associated with dryness Stressful wind	Compact & aggregate forms, clustered forms. Protected urban edges. Narrow winding neighbourhood roads & alleys. Uniform city height. Small, dispersed & protected public open spaces. Circumferential & intersecting tree zones. Use of geospace city concept
Seashore strips	High humidity Windy	*In humid region:* Moderately dispersed form. Open urban edges. Wide streets perpendicular to the shore to receive breeze. Dispersed high rise buildings to receive ventilation. Variety of building heights. Wide public open space. Shadowing planed tree zones *In dry region:* Open toward the sea, compact and protected toward the inland. High rise buildings mixed with low height. Small protected dispersed public open spaces. Shadowing planned tree zones
Mountain slopes	Windy	Semi-compact form: mix of compact & dispersed. Horizontal streets & alleys to enhance the view. Low height buildings. Small, dispersed public open spaces. Non-obstructed protected tree zones. Use of geospace city concept

Source: Golany, 1995

- 'Archipelago' pattern (Magnan and Mathieu, 1975): this entails 'compact, nucleated urban sub-units' within close cycling or walking distance from each other. However, energy efficiency relies on the energy efficiency of mobility, where people would take advantage of proximity and not use vehicles. It has been found that decentralization of employment and services was more energy efficient with regard to travel requirements than concentration in one centre. The notion of 'neighbourhood', which was already advocated during the 1960s by urban planners, has been revived. The potential disadvantage would be a reduction in large open areas for recreation and aesthetics (see Figure 1.23).
- The linear grid structure (Martin and March, 1972), with high development densities and integration of diverse activities, as well as accessibility to open land. This configuration is characterized by high linear densities; destinations and origins of trips are concentrated along a defined number of routes, which is beneficial for public transport and potential renewable energy due to the 'holes' (see Figure 1.24). Yet, 'ribbon development' could abolish the notion of 'place' in terms of neighbourhood and liveable environments. Care should be taken not to confuse linear grid and unplanned linear development (Owens, 1986).

Other schools of thought proposed theoretical forms for the sustainable city, all based on the notion of reducing the need for movement by private car and reducing goods transportation by road. These forms comprise:

- a compact high-density city;
- low-density decentralized urban areas;
- an urban form based on policies for 'decentralized concentration';
- the concept of a sustainable city region (e.g. Howard's Garden City) (Moughtin, 1996).

However, there is little evidence to support the relationship between energy efficiency and city form. On the shape and size of our existing cities, Owens argues that they are more a reflection of 'the nature and availability of energy resources that influenced the distribution of human activities and urban form'. The clearest illustration of this observation is during the 20th century, when the lack of energy constraints and reduced energy prices allowed the physical separation of activities and the outward spread of urban areas at lower densities (Owens, 1992). Mumford (1961) has pointed out that the scale of our cities today depends upon mass communication. Even historically, the scale of the city was conditioned by what Mumford termed 'the range of collective communications systems':

> *Early cities did not grow beyond walking distance or hearing distance. In the Middle Ages, to be within the sound of Bow Bells defined the limits of the City of London; and until other systems of mass communication were invented in the nineteenth century, they were among the effective limits to urban growth. For the city, as it develops, becomes the centre of a network of communications ... the permissive size of the city partly varies with the velocity and the effective range of communication'* (Gillespie in Breheny, 1992).

For Goethert, energetically 'optimal' urban layouts are assessed in terms of incurred cost on land subdivision and the related basic utilities (water supply, sewage disposal, circulation/storm drainage, electricity/street lighting) (Goethert, 1978). He compares two models illustrating extreme conditions of land use, with a site area 400m × 400m and a lot area of 100 square metres. The difference involves building layouts, lot proportions and land-use patterns. Efficiency in utility use is compared between both models (water supply, sewage disposal, circulation/storm drainage, electricity/street lighting).

The same type of comparative study was carried out on existing settlements, highlighting deficiency in land and infrastructure utilization. Evaluation was based on a comparison of three layouts: (a) existing layout; (b) optimal

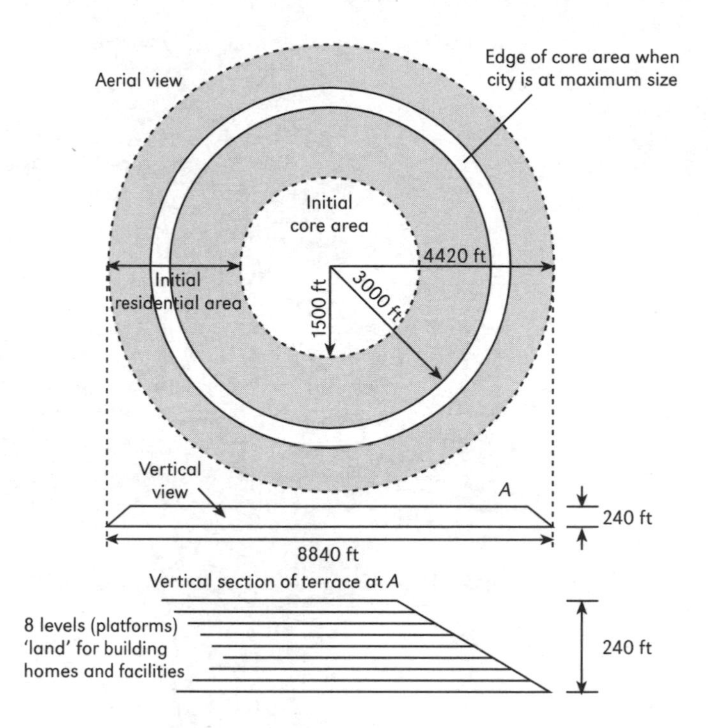

Source: Dantzig and Saaty, 1973

Figure 1.22 *'Compact city'*

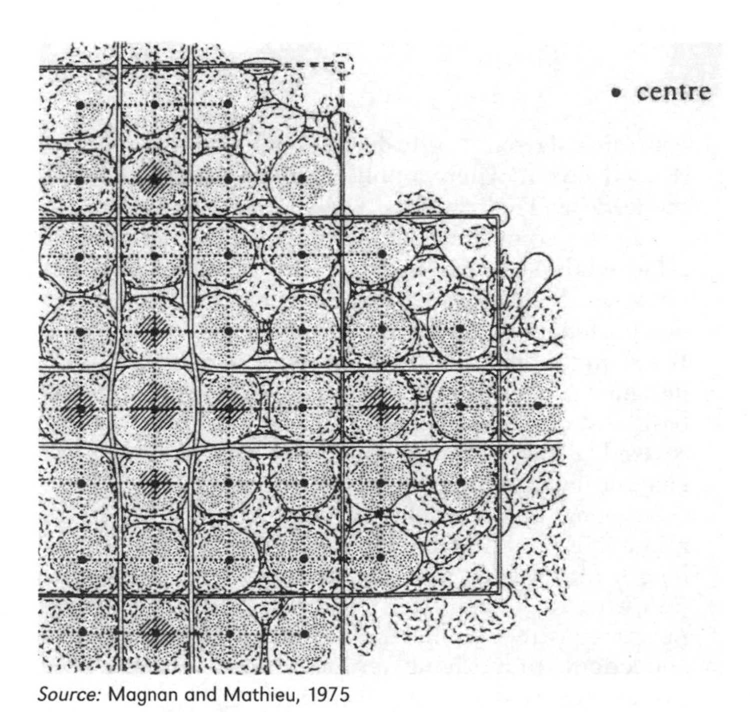

Source: Magnan and Mathieu, 1975

Figure 1.23 *'Archipelago' layout or nucleated urban sub-units*

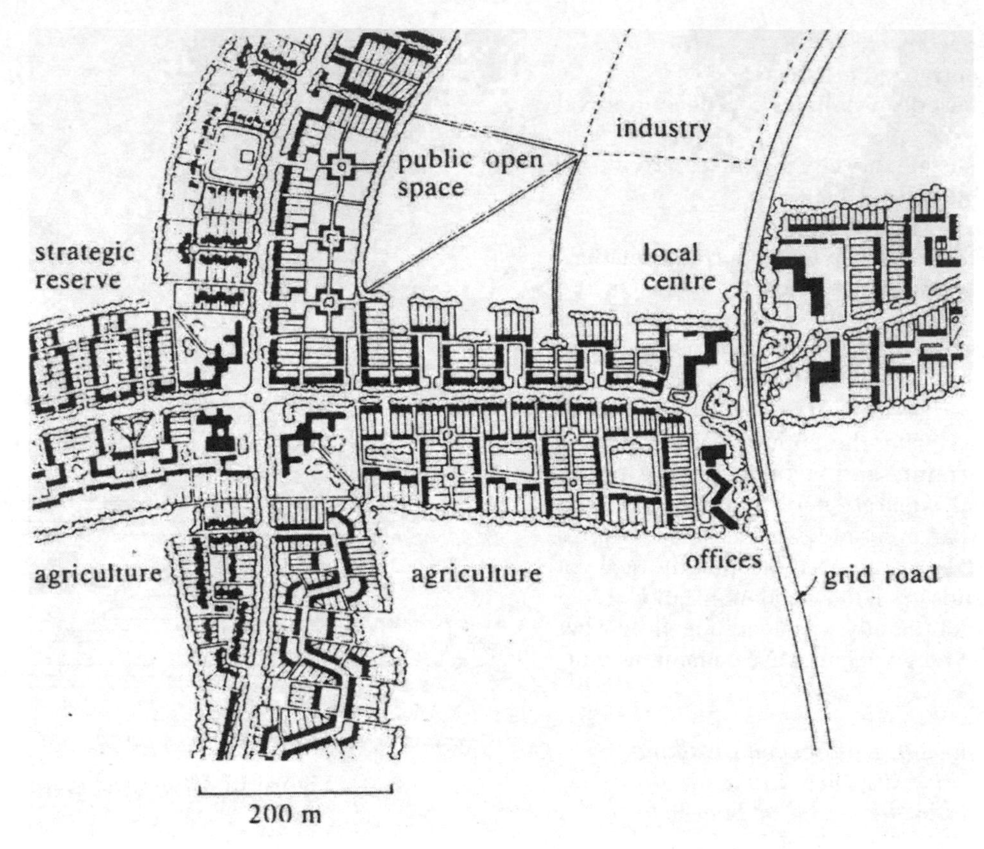

Source: Rickaby, 1979

Figure 1.24 *'Linear cruciform' layout*

equivalent layout in grid-iron format; and (c) proposed revised layout where public land for circulation was reduced (see Figure 1.25).

In the equivalent grid-iron layout, optimization is achieved through minimizing public circulation by eliminating small blocks and making larger blocks; gained area was pooled into expanding semi-public and private land. Based on the equivalent grid-iron plan, a revised plan was designed on the existing site as a 'remedial' proposal. The basic cost of utilities between the existing layout and the revised layout (based on equivalent grid iron) is almost half (see Table 1.6).

Among the case studies that follow, a general trend of minimizing public land for circulation and, therefore, length of infrastructure per area served was adopted (infrastructure comprised electricity; water; sewerage networks; street lights; police protection; and waste collection), thus saving on government construction, maintenance and operation (Goethert, 1978).

As a general trend, Goethert observes that '*compact shapes* are generally more apt for efficient development. *Irregular, dispersed shapes* may result in unusable areas and/or uneconomical/inefficient layouts'. The results of the analysis give an indication of utility costs per lot for each layout, with the obvious result that 'costs per lot are lower when there are more lots per unit area. However, it should be borne in mind that more lots result in smaller or narrower parcels of land' (Goethert, 1978).

In Figure 1.26 the notion of settlement edges is investigated – specifically, the deterministic role of these edges on architecture and planning in *extreme climatic conditions*. Edges should therefore follow careful design, responding to topography, wind pattern, humidity, temperature, rain and sun. Settlement edges could act as environmental filters, reducing the harshness of external environments. A chart of edge morphology is proposed.

Oke (1988) was able to give recommendations on an optimal combination of urban density and average H/W urban canyon aspect ratio, satisfying all four goals, which are pollution dispersal, solar access, shelter and heat island warmth. Concerning density, a range between 0.2 and 0.4 was suggested for an optimal roughness and absorptance. Although typical central areas in cities in both Europe and North America do not conform to the

BASIC LAYOUTS

Layout	Lot	Public		Semi-public		Private		Total	Street lengths (m)		
	no.	No	%	No	%	No	%	No	I–III	III–IV	total
Existing	573	5.97	38	2.32	15	7.47	47	15.76	2506	892	3398
Equivalent	732	4.69	29	2.60	16	8.71	55	16.00	3090	800	3890
Revised	687	5.19	20	2.35	15	10.15	65	15.76	1146	892	2038

Figure 1: TABLE OF LAYOUTS, COMPONENTS AND QUANTITIES
The table shows that the Existing layout devotes more public land to circulation and consequently offers less private land than the two others. The Revised layout reduces public circulation still firther by providing cluster lots or condominiums with semi-private courts; in addition, it offers the option of larger lots in the periphery of the blocks.

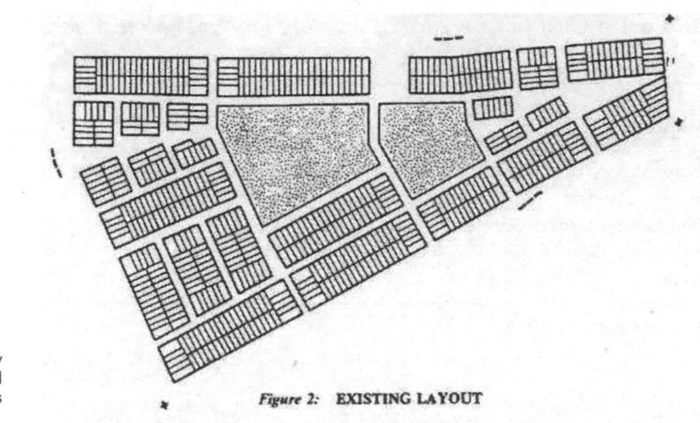

Figure 2: EXISTING LAYOUT

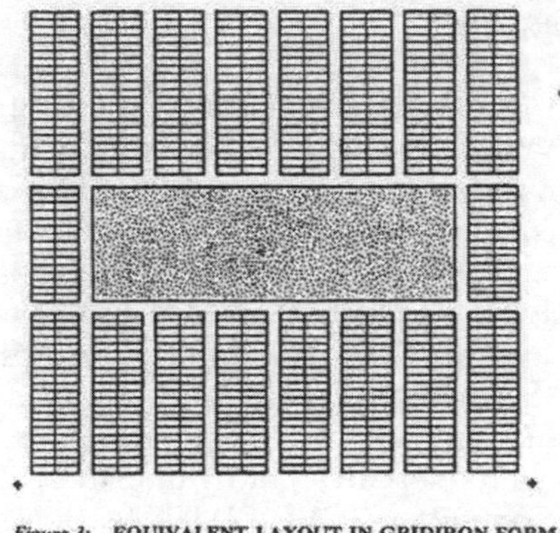

Figure 3: EQUIVALENT LAYOUT IN GRIDIRON FORMAT

Figure 4: REVISED LAYOUT

Note: Total site area = 15.76ha; lot area = 120 square metres (6m × 20m)

Source: Goethert, 1978

Figure 1.25 *Three urban layouts, where (a) illustrates existing situation, (b) optimal equivalent in gridiron format and (c) proposed revised layout*

above values (H/W in Europe = 0.75–1.7; in North America, H/W = 1.15–3.3), European cities are still closer to these figures than North American cities. This leads to the observation that 'European compact form of residential/suburban areas are more likely to conform to the suggested compatible ranges than the North American dense cores and scattered suburbs' (Oke, 1988). Comparing land consumption between US and European cities due to urban growth, Leinsberger (1996) supported this observation. He noted that US cities 'consume land and growth spatially at a much faster rate than population growth'. An example of a country with a relatively large population and one of the highest population densities in the world is The Netherlands, where percentage of land occupied by cities and developed areas is 13 per cent (Van der Brink, 1997). European cities have maintained compactness and density despite some degree of urban sprawl (Beatley, 2000). Several factors explain this difference in density between US and European cities, such as historic urban traditions, where several European cities evolved from an old centre originally built within historical defensive fortifications. Scarcity of land combined with population growth explains cautious land use. Conscious public policies and planning traditions also explain this trend in controlled urban sprawl in European cities. So, while the reason behind compact cities today could have originated historically from demographic and political reasons, its consequences are believed to be environmentally positive (Beatley, 2000).

Urban density has been identified as an urban-form indicator with environmental repercussions during the last few decades.

Table 1.6 *Cost per hectare of existing layout is almost double the cost of revised layout, with improvements consisting of major savings in circulation and storm drainage*

	Basic network: Cost per hectare (US dollars)		
Utilities	**Existing layout**	**Equivalent layout in gridiron format**	**Revised layout**
Water supply	3809	3680	1995
Sewage disposal	3489	2851	1612
Circulation/storm drainage	42,203	31,132	19,627
Electricity/street lighting	11,222	11,282	8358
Totals	60,723	48,945	31,592

Source: Goethert, 1978

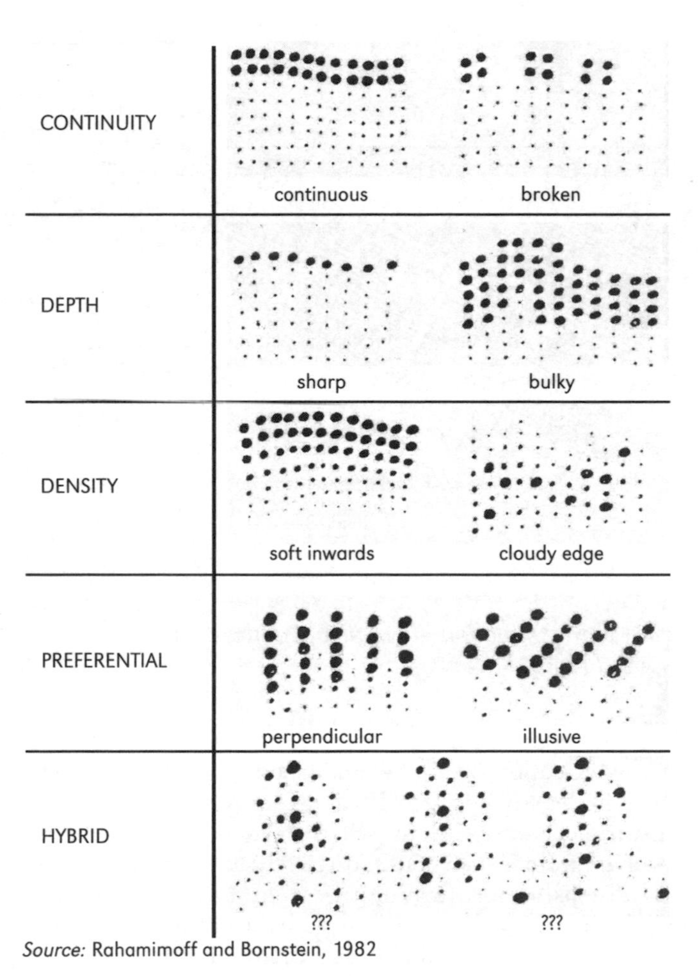

Source: Rahamimoff and Bornstein, 1982

Figure 1.26 *Morphology of edges*

Research into practice for environmental urban planning and design

Although principles of energy-conscious urban design seem to be well established as an issue for debate, implementation and dissemination would require, among other aspects, political will and policy coordination. Focus will then have to be given to action on building or neighbourhood scale, such as siting and the use of passive solar energy. However, there are limiting measures to applying these in the current planning and building regulations. An instance in the UK illustrates this resistance. Following a request sent out by the House of Commons Energy Committee (CEC, 1990) to the government, requesting that all planning applications should be assessed in terms of the consideration given to energy-efficiency aspects in layout and design, the government reply was that 'the energy implications of microclimate or orientation will seldom be sufficiently weighty ... to justify refusing planning permission'. In order to counteract such bureaucratic obstacles, Owens proposes that energy efficiency becomes an explicit consideration in urban planning policies (Owens, 1992).

Environmental policy-making

A lot has been said about the relationship between research and policy, and the translation of knowledge acquired through research into action. As the past has proven, the path of translating research knowledge into action is still clumsy and unsystematic. It is generally agreed that there is a wide gap between the expanding scope of environmental policy and its effectiveness. The process of implementing environmental regulations is therefore unsystematic, and large differences have often arisen between what is expected and the effect, in reality. Heavy criticism is particularly targeted at the irrational manner in which social policy progresses, at a time when it is commonly assumed that research can direct policy. It is a misconception to presume that policy-makers draw upon the most convincing and up-to-date findings and evidence on any topic. They tend to remedy current rather than future problems, thus working towards short-term instead of long-term goals, as some have observed (see Hunt in Davies and Kelly, 1993). The success of policy implementation depends upon the close cooperation between various groups of society, and between policy-makers and stakeholders; thus, research–community policy interfaces should be highlighted and encouraged (Landsberg, 1981).

It has been observed by Nijkamp and Perrels (1994) that the most adequate institutional level of environmental policy with regard to the use of renewable resources should remain at a centralized level. The reason is that 'there may be a danger that the closer the relationship between the decision-maker and the polluter, the more sensitive the decision-maker may be to social and economic arguments, and the less inclined he or she will be to take measures which benefit the environment' (Nijkamp et al in Breheny, 1992). However, most importantly, all institutional changes must be preceded by changes in attitude, behaviour and values at the level of the consumer (Blowers in Breheny, 1992). Community involvement and participation – when tackling issues relating to the urban context – is a crucial requirement. It has been proven that decisions made by those concerned are far better than decisions made extraneously by others. This was noticed in the Healthy Cities project,[4] where inhabitants involved in decision- and policy-making grew in confidence (Smithies et al in Davies and Kelly, 1993). This is further supported by the UTF, who found that different models of neighbourhood management bodies would be useful, involving local people in the decision-making process, 'relaxing regulations and guidelines to make it easier to establish devolved arrangements' (Urban Task Force, 1999).

Research into practice

Despite the past failures of environmental policy implementation, there are instances that prove the opposite concerning certain aspects of policy implementation, such as supporting sustainable land-use patterns in the European context, which demonstrates the political and governmental action-taking ability of national governments in Europe. For example, in Norway, in 1999, the government banned by royal decree new shopping malls located outside of city centres for a five-year period, a move seen as necessary to reduce traffic and the economic undermining of downtown areas (Associated Press, 1999; World Media Foundation, 1999). In the UK, steps to promote further compact urban growth pattern were taken by the national government. Its national strategy of sustainable development explicitly calls for efforts to 'further promote urban compaction' and numerous guidance documents were issued to encourage further compact growth. 'The UK national government has also issued recent policy papers proposing a target that 60 per cent of future residential growth [will] occur on reused urban land' (Breheny, 1997). National government statistics for 1993 show that 49 per cent of residential development occurred on redeveloped or brownfield sites, and another 12 per cent on vacant urban land. Thirty-nine per cent of residential development in the UK occurred on 'rural' land (Breheny, 1997; Beatley, 2000).

In the case of health, it is social processes that are of most critical importance in determining the health status of individuals and communities. This is clearly demonstrated through the Healthy Cities project that grew as a practitioner-led activity, consisting of a series of local community experiments rather than as a research project. The Healthy Cities philosophy is based on the conviction that cities should provide a clean and safe physical environment of a high quality, based upon sustainable ecosystems. In the context of this project, the promotion of health includes the adaptation and transformation of social structures that create ill health. The improvement of the health conditions of an environment would rely upon individuals' initiative, where people themselves must be empowered individually and through their local communities to take control of their health (Davies and Kelly, 1993). Preparing for the project has revealed surprising and unexpected realities, such as the lack of conviction that politicians and business leaders have about the impact of urban planning and housing policy, as well as traffic and pollution control, and general health in the urban context (Davies and Kelly, 1993).

Several European cities are cited in environmental literature as having been engaged in a variety of innovative sustainability initiatives, having adopted and implemented sustainability policies in a wide range of sectoral areas through holistic strategies (see Table 1.7) (Beatley, 2000).

Among the obvious energy features of green urban European cities are combined heat and power (CHP), sea water use for natural cooling and summer air conditioning, and solar-assisted district heating networks (where solar panels generate much of the heating of water circulated in district networks). In the case of Denmark, solar heating provides 12.5 per cent of the annual heating needs of the district heating of a town of 5000 people (Beatley, 2000). Other means of contributing to energy efficiency include better management of the buildings by

Table 1.7 *European cities where interviews and field visits were carried out*

Austria	Graz*, Linz, Vienna
Denmark	Albertslund*, Copenhagen, Herning, Kalundborg, Kolding, Odense
Finland	Helsinki, Lahti
France	Dunkerque*
Germany	Berlin, Freiburg, Heidelburg*, Muenster, Saarbruecken
Ireland	Dublin
Italy	Bologna
Switzerland	Zuerich
Sweden	Stockholm*
The Netherlands	Almere, Amersfoort, Amsterdam, Den Haag*, Groningen, Leiden, Utrecht, Zwolle, others.
United Kingdom	Leicester*, London (including boroughs)

Note: * Indicates recipients of the European Sustainable City Award.

Source: Beatley, 2000

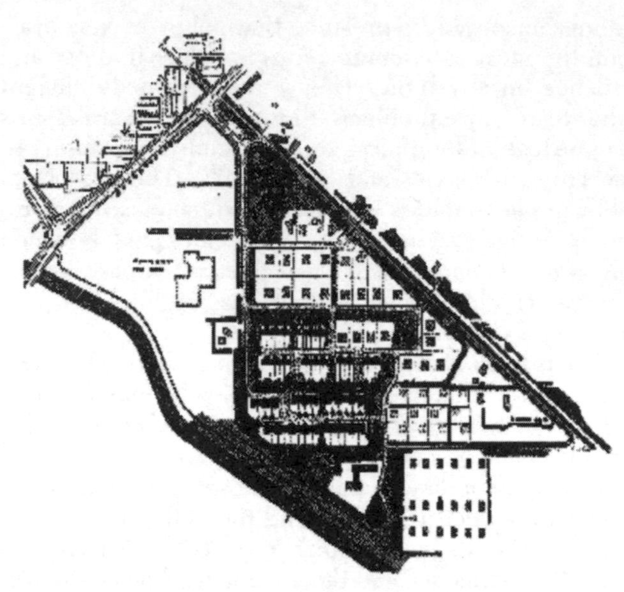

Source: Beatley, 2000

Figure 1.27 *Diagrammatic plan of Morra Park showing a closed-loop circulation*

trained individuals (building managers and housekeepers who maintain certain levels of control – for example, switching on and off heating/cooling/artificial lighting), and the upgrading of building envelopes (e.g. insulation of walls and better performing windows and glazing). Therefore, energy efficiency and reduction in energy consumption were targeted at the level of each individual building by applying the above measures. One example is the city of Saarbrücken, in Germany, where dramatic results were noted. Between 1981 and 1996, consumption of heating in the city's properties was reduced by 53 per cent. Corresponding reductions in CO_2 emissions were considerable: from 65,000 tonnes in 1981 to 35,000 tonnes in 1997. Investment in upgrading and implementation of the measures discussed above proved to be highly cost effective, where a yearly investment of 1 million Deutsch marks gave a return of 10 million Deutsch marks in energy savings (Beatley, 2000).

Much can be learned from ecological pilot building and neighbourhood projects on the degree of success, as well as failure, of environmental strategies. Examples from The Netherlands, Denmark and Germany show successful environmental performance and high energy conservation standards, as well as use of renewable energies. Important experimental projects cited for reaching these targets are found in The Netherlands (Ecolonia-Alphen a/d Rijn. Morra Park-Drachton, Eco-Dus-Delft). For example, Ecolonia combines a cluster of 101 residential units, designed by several architects, and the whole project is sponsored by the Dutch national government. The design brief clearly advocated ecological design (energy standard of 200 megajoules (Mj)/m^2). Ecolonia reached the target in the national environmental policy plan (NMP) of reducing household energy use by 25 per cent. However, according to the evaluation studies (carried out by Edwards, 1996), some problems were identified, such as the necessity for proper ventilation due to moisture build-up and rotting of insulation, and the expensive and unreliable nature of photovoltaics (Beatley, 2000).

Morra Park (Drachton) is another case study pilot project, consisting of 125 dwellings. Among the sustainable features incorporated in the houses are south orientation, solar panels and conservatories. The problems encountered in this project were technical, as well as social. For example, there was a noticeable lack of interaction among neighbours, perhaps due to the linear configuration of the houses, facing south (see Figure 1.27) (Beatley, 2000).

Project Eco-dus (in Delft) is also a first-generation sustainable development scheme. Combining 250 dwelling units, the project is built through collaboration between the local housing society, the municipality and a private developer. The main sustainable design features consist of south and south-east orientation, solar panels for hot water usage and high-energy efficiency (high-value insulation and high-insulating glass). In addition to this, blocks of flats along the road are designed to act as noise barriers (having bathrooms and kitchens facing the street and living quarters and bedrooms located near the quieter side; this involves a combination of urban planning and urban design, with repercussions for the internal planning of dwellings). Furthermore, all projects promote reduced dependency on car circulation (i.e. narrow roads, providing an extensive bicycle network). In addition, other features include the use of sustainable and recycled construction materials, free composting bins, water-saving toilets and showers, and other sustainable strategies for living. Such a project, completed in 1992, was viewed as a model for future development in the city. In general, solar and energy-efficient strategies succeeded; however, problems arose when residents' desire to add an additional floor was rejected by the city because it obstructed solar access for other units. This incident illustrates the problem of how to design affordable housing where residents can live throughout the different stages of their lives. The fact that those wishing to live in larger units were forced to move out was seen by some as negating the goals of sustainability (Beatley, 2000).

Several housing projects in the UK have taken environmental consideration into account, such as the Pennyland housing scheme at Milton Keynes, where micro-scale spatial structure was used to enhance energy efficiency (Owens, 1986). Another example is the Basildon housing scheme by Ahrends, Burton and Koralek, where – through careful design and environmental considerations – microclimate and passive solar energy were harnessed. North–south access roads, with short east–west access cul-de-sacs, made houses benefit from south exposure; spacing between buildings was studied to minimize overshadowing; and landscaping was designed to reduce wind speed. In this case, there was a conscious effort to keep energy considerations controlled so as not to dominate social, technical, aesthetic and economic factors (Owens, 1986).

The success of certain urban environments is not necessarily based on the faithful application of planning regulations. In Barcelona, the average density of 400 dwellings per hectare, compared with a high density of 100 to 200 dwellings per hectare in some of the lively inner-city areas in English cities (such as London), exceeds what is allowed by current planning regulations. This aspect is actually one of the main features that secured its success as a city, providing a positive urban experience on several levels (e.g. environmental, social and economic). It was categorized as the most vibrant and compact city in Europe (Urban Task Force, 1999). The example of Barcelona illustrates the fact that planning regulations should be based on relevant precedents and actual examples so that amendments of them are likely to meet the sustainability goals.

Energy-efficient urban planning and design versus amenity, equity and aesthetics

Would an energy-efficient environment mean great sacrifices in terms of amenity, equity or aesthetics? Hypothetically, urban sustainability encompasses environmental concerns and economic viability, as well as liveability and social equity. It has been demonstrated in examples of environmental pilot projects that it is difficult to create new settlement quarters that encapsulate the human qualities that older cities have, the latter having grown and developed organically and changed incrementally over many years (Beatley, 2000).

Environment versus human behaviour

The environment plays a major and crucial role in the behavioural pattern of people. People respond to good design and desirable environments by choosing to frequent them. 'Good design (should) provide a stimulus to the senses through choice of materials, architectural form and landscaping' (Urban Task Force, 1999). Authors agree that:

> ... *design and policy and decisions should be related – where possible – to the behavioural sciences... Through interdisciplinary collaboration, designers and planners have the possibility of becoming more socially responsible. Reciprocally, behavioural scientists and environmental sociologists will have a greater opportunity to apply their research findings in more practical ways* (Mänty, 1988).

Correlations have been established between excess heat in cities and incidences of rising criminal activity (Landsberg, 1981).

Owens notes that the success of energy-efficient planning considerations relies, ultimately, on their social acceptance, cultural integration and economical feasibility, and not only on their technical energetic performance (Owens, 1986). An example is the frequent failure of centralized shopping in moderate climates despite the cost benefits of centralization. In mild climates, shopping is perceived and better experienced as an outdoor street-shopping experience, rather than a confined experience. It is therefore virtually impossible to devise general guidelines on the ideal energy-efficient city, bearing in mind the contextual climatic, socio-economic and cultural disparities.

Designing for human comfort: Physiological and psychological

There was harsh criticism of completely mitigating environmental stresses, such as noise, radiation, temperature extremes and wind: some degree of stimulation is required for the average healthy person (Fezer, 1982). There is charm and character in the buzz of the city, involving all senses: audible (e.g. traffic noise, rumbling machinery, street cries and the rustle of foliage), olfactory (e.g. odours, plants, exhaust fumes and food) and visual (e.g. signage, people and changing scenery). As Simmonds (1994) points out: 'Many distinctive and familiar sounds add much to the pleasure of city living'.

A set of preliminary recommendations for a pleasant urban environment was put forward by several authors (Simmonds, 1994; Barton, 1995; Urban Task Force, 1999), coinciding with recommendations that relate to a liveable, positive and environmentally pleasing urban context. These include interconnectivity of streets; a clear hierarchy of street networks; and open spaces that are properly designed in such a way that they promote circulation and the gathering of people, as opposed to there being leftover or negative space between buildings (Urban Task Force, 1999). Some authors relied on experience, intuition and practical, anecdotal evidence through observation to draw upon these recommendations.

Tibbalds (1988) went as far as to set out the ten commandments of urban design, which are primarily and explicitly related to the human dimension:

1. Thou shalt consider places before buildings.
2. Thou shalt have the humility to learn from the past and respect thy context.
3. Thou shalt encourage the mixing of uses in towns and cities.
4. Thou shalt design on a human scale.
5. Thou shalt encourage freedom to walk about.
6. Thou shalt cater for all sections of the community and consult with them.
7. Thou shalt build legible environments.
8. Thou shalt build to last and adapt.
9. Thou shalt avoid change on too great a scale at the same time.
10. Thou shalt, with all the means available, promote intricacy, joy and visual delight in the built environment.

Energy-efficient planning versus reality: Observations and examples

Socio-economic and cultural realities are sometimes overlooked in building design, and there can be resistance towards policy implementation and the translation of research into action, as previously discussed. Several computer models were devised to simulate city compactness. The efficient relationships between scale, capacity and containment were computed to a point where human values faded into the background. Although city compactness might seem appealing to planners in terms of energy efficiency, the concept is still regarded with fear for its risk of causing town cramming. A middle ground should be sought, where concern would focus on 'liveability, attractiveness and urban quality, whilst trying to fit them into a policy framework which stresses sustainability and compactness' (Crookston et al in Jenks et al, 1996). For this purpose, several interrelated aspects of urban planning should be addressed simultaneously:

- housing density: not wasting space, but not cramming either;
- transport: playing to the city's strengths;
- parks, schools and leisure: quality services and facilities;
- urban management and safety; and
- the housing market: offering range and choice (Crookston et al in Jenks et al, 1996).

Beatley demonstrates that density in the urban context is not desirable on its own. Instead, it is the configuration and design according to which this density is expressed that is important, and what comes with it as urban design effort, greening, and accessibility to transit and amenities (Beatley, 2000). Several cities clearly illustrate the visual and experiential benefits of fine-grain street patterns and urban texture. In Dutch cities such as Groningen, Delft or Leiden, for example, the dense network of streets provides a great variety of routes. In turn, there is a diversity of sights and sounds in moving through the city as a pedestrian or cyclist. In addition to enhancing enjoyment, there is a sense of safety among people when a choice exists between several different routes and accesses (Beatley, 2000). This has been referred to as 'permeability' of places, both visual and physical (Bentley, 1995). The conditions of ensuring permeability rely upon eliminating cul-de-sac and hierarchical street layouts and providing finer-grain city configuration (Barton, 1995).

On promoting renewable energy in the urban context, there is so much that can be done while preserving the quality of outdoor space (i.e. streets). As highlighted by Beatley (2000), 'the architecture along the street has much to do with its positive feeling. Buildings could ensure an attractive organic feeling to the street by providing a diversity of colours, height and detail'. A geometrically articulated street frontage is more pleasing to pedestrians than a flat monotonous frontage; although each might have a different impact on the environmental performance of the street, the implications of urban geometry on the qualitative perception of the street is of prime importance (Simmonds, 1994).

Occasionally, however, reality defeats common sense. Town-planning experiments in Greenland demonstrated that locals have a strong preference for open urban planning, with an interest in living with nature – although common sense would dictate that Greenlandic towns should be compact. This phenomenon seems to prevail despite numerous socio-economic facts, such as the more economic aspect of compact and high-density planning, as well as environmental arguments, such as the detriments of wind-swept open, loose planning (Mänty, 1988).

As has been demonstrated, it seems that there is no simple environmental determinism for urban form. The Western city has been shown to consist of a certain number of 'types' or 'elements', with small variations: street and square; monument and palace; city block (buildings around courtyards). The organization of these items was a result of defensive needs, powerful structure and technology. 'The urban pattern seems to be determined in spite of, rather than because of, the climate… Climate should therefore be regarded as a modifier of urban form, rather than a determinant'(Mänty, 1988).

Overview

> *Fast progress [towards a sustainable urban form] will not be made in reversing the spread of the city, which appears to embody social aspirations and meet the economic imperatives of the time. Perhaps the only rational course is to treat the urbanized body as we find it, and not to lament that it was not born to be more beautiful in the first place. Logic might indicate what that form might be; but rational thought indicates that all we can do is improve it'* (Welbank in Jenks et al, 1996).

The nature of this reasoning implies that, faced with the virtual impossibility of revolutionizing the whole built environment and rebuilding entire cities based on sustainability and energy-efficiency, the most viable alternative would be to deal with the existing realities and address them as best as one can. As noted by the UTF, the concrete reality of urban centres is that more than 90 per cent of the existing urban fabric will remain with us in 30 years' time (Urban Task Force, 1999). While much has been said about hypothetical optimal city morphology for efficient energy consumption, few researches thoroughly investigated the quantitative possibility and benefits of remedial urban refurbishment for a more sustainable, energy-efficient use and improved environmental performance. Even major triggers of research in this field do not address the problem-solving aspect thoroughly, specifically the European Commission (EC) Green Paper (CEC, 1990), with its primary concern of reducing the energy-intense character of urban neighbourhoods. Although this influential and revolutionary paper tackles possible alternative ways of dealing with the problem (i.e. densification), the actual practicality and feasibility of increasing the density of cities remains vague and inconclusive. Breheny (1992) highlights the unresolved flaw in the Green Paper in its contradictory recommendations, encouraging simultaneously higher urban densities and an increase of urban space.

At the heart of this document lies the important principle that both identifies problems and targets easily implemented solutions to the current energy issue in cities at a time when most of the research taking place in this field necessitates long-term and time-consuming policy decision-making. The latter encompasses transport policies in relation to pollution emissions and energy consumption; studies of city shape for optimal energy performance; road infrastructure and layout for optimal commuting and, therefore, less energy consumption and pollution emission; and idealistic 'city orientation' for solar radiation access and passive solar heating (Davies and Kelly, 1993).

There have been individual attempts by researchers, planners, architects and academics to investigate and implement this evolutionary approach to dealing with urban refurbishment on a small scale. Nijkamp and Rienstra (Nijkamp et al in Jenks at al, 1996) realize the extent to which the spatial inertia of the built environment and of infrastructure networks can act as a powerful barrier to adopting modern solutions to the energy-intensive urban character, such as new transport technologies:

> *Artefacts following from land use, such as housing blocks, industrial estates and transport infrastructure, have a long life cycle in relation to the capital investment involved. As a result, different types of land use are fixed for a number of decades. So, once the infrastructure is built, it will be there for a long period (especially in historical city areas). As a consequence, technologies which imply step by step (incremental) or small-scale change may have a better chance of*

> *adoption in the urban territory than technologies implying radical change of infrastructure and land use* (Nijkamp et al in Jenks at al, 1996).

Ignasi de Solà-Morales, a Spanish architect and professor of history and theory at the Barcelona School of Architecture, introduced the notion of urban micro-surgery,[5] strongly advocating urban rejuvenation that preserves and respects the equilibrium and texture of historic urban centres (such as his recently completed project in the historic centre of Barcelona). Others strongly acknowledge the importance of rehabilitation and regeneration in the urban context as opposed to comprehensive planning that proved to destroy many vital communities in the process of renewing the physical infrastructure. The organic model for city planning devised by McKei in 1974, and called Cellular Renewal, depends upon on-site survey, with each housing unit representing a cell. Although this organic concept of the neighbourhood caters for slow renewal or rehabilitation, piecemeal development did not disturb the community. The organic city is just like an organism though; it dies once unhealthy, and therefore has an optimum size (Moughtin, 1996).

Much of the current work and approaches to urban planning and design rely heavily upon the use of computers, thus reducing vast quantities of information to manageable form. In addition to mapping, storing data and comparative analysis, very important creative contributions are being made by computer technology, such as visual modelling and presentation of schemes in visually understandable formats. However, despite the wide application of the computer in city planning, caution should be taken so that it remains as an aid and not a substitute for the experienced planner and intuitive designer (see Figure 1.28). In general, 'planners need to adopt and devise a set of specifically intra-urban sustainability principles, concerning issues such as public transport, private car restraint, density targets, urban greening, development of transport modes, mixed uses, etc' (Breheny, 1992). Answers to the environmental problems in the urban context should range from recommendations about short-term changes, which are a must for solving pressing problems, to target long-term development and implementation of processes.

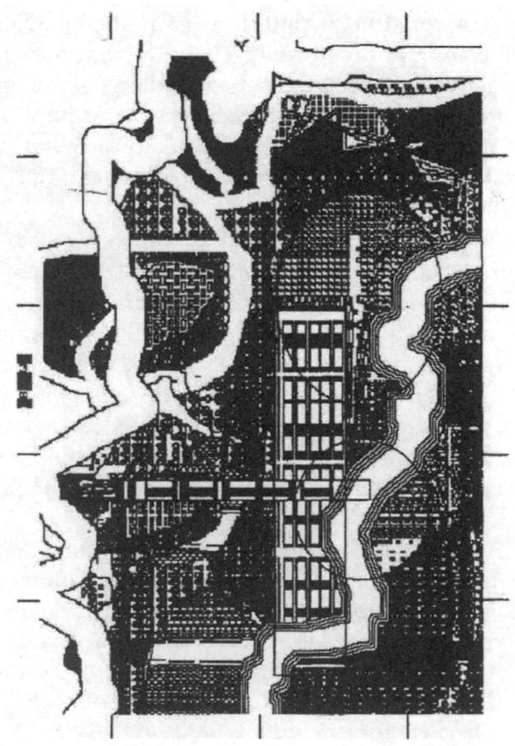

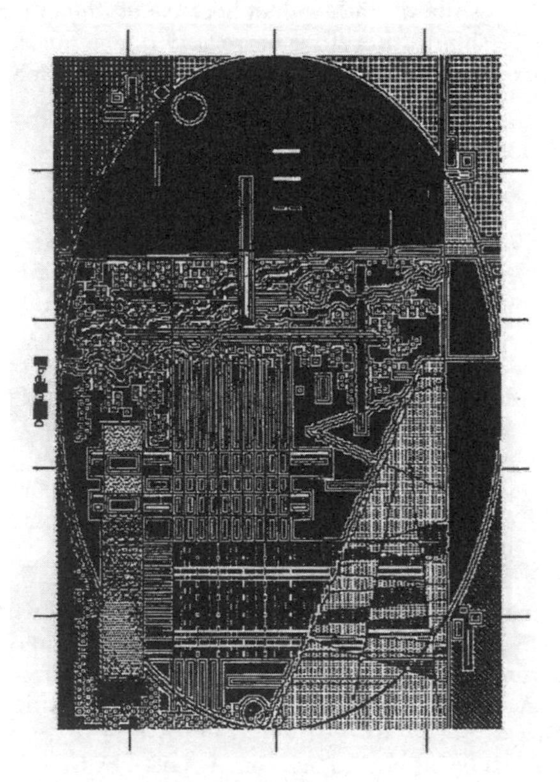

Source: Simmonds, 1994

Figure 1.28 *Experimental work that examines symbiotic relationships between the natural and built form in the urban landscape; the givens are natural features of the site and the outcome is planning proposals*

Synopsis

This chapter started by providing an overview of vernacular urban planning as a background to the knowledge that exists in relation to a climate-responsive urban design. This was followed by a series of sections focusing on current research in urban climatology, energy use, renewable energy and environmental potential related to building form. The final section raised issues related to urban planning, such as amenity, equity and aesthetics, in the context of energy-efficient urban design.

References

Alexander, C. et al (1978) *A Pattern Language: Towns, Buildings, Construction*, Oxford University Press, New York

Alexander, C. (1992) 'A city is not a tree', in G. Bell and J. Tyrwhitt (eds) Humanity Identity in the Urban Environment, Penguin, London, originally published in 1965

Alexandre, A. (1975) Road Traffic Noise, Applied Science Publishers Ltd, Barking

Ashton, J. (ed) (1992) *Healthy Cities*, Open University Press, Milton Keynes

Associated Press (1999) 'Norway: New malls banned', *The Gazette*, Montreal, 9 January, pA16

Attenborough, K. (1998) 'Acoustics of outdoor spaces', *Proceedings of the 16th International Conference on Acoustics*, Acoustical Society of America, Seattle

Barton, H. (1995) *Sustainable Settlements: A Guide for Planners, Designers and Developers*, Government Management Board, University of West of England, Bristol

Beatley, T. (2000) *Green Urbanism: Learning from European Cities*, Island Press, Canada

Bentley, I., Alcock, A., Murrain, P., McGlynn, S. and Smith, G. (1985) *Responsive Environments: A Manual for Designers*, Architectural Press, London

Beranek, L. L. and Embleton, T. F. W. (1982) 'Sound propagation outdoors', *Noise Control Engineering*, vol 18

Blowers, A. (ed) (1993) *Planning for a Sustainable Environment: A Report by the Town and Country Planning Association*, Earthscan Publication, London

Breheny, M. (ed) (1992) *Sustainable Development and Urban Form*, Pion, London

Breheny, M. (1997) 'Urban compaction: Feasible and acceptable?', *Cities*, vol 14, no 4, pp209–217

Broadbent, G. (1990) *Emerging Concepts in Urban Space Design*, Van Nostrand Reinhold (International), London

Calthorpe, P. (1973) *The Next American Metropolis: Ecology, Community and the American Dream*, Princeton Architectural Press, New York

Carley, M. and Christie, I. (1992) *Managing Sustainable Development*, Earthscan Publications, London

CEC (Commission of the European Communities) *Green Paper on the Urban Environment*, EUR 12902, 1990, Brussels

CETAT (Centre d'Etudes Techniques pour l'Amenagement du Territoire) (1986) *Morphologie Urbaine, Indicateurs Quantitatifs de 59 Formes Urbaines Choisies dans les Villes Suisses*, Georg (ed), CETAT, Geneva

Chandler, T. J. (1976) *Urban Climatology and Its Relevance to Urban Design*, WMO Technical Note No 149, World Meteorological Organization, Switzerland

City of Vienna (1993) 'Vienna: Launching into a new era', June 1993, press release

Croxford, B., Hillier, B. and Penn, A. (1995) 'Spatial distribution of urban pollution', *Proceedings of the 5th International Symposium on Highway and Urban Pollution 95*, Copenhagen

Croxford, B. and Penn, A. (1995) *Pedestrian Exposure to Urban Pollution: Exploratory Results*, Air Pollution 95, Porto Carras, Greece

Dantzig, G. B. and Saaty, T. L. (1973) *Compact City: A Plan for a Liveable Urban Environment*, W. H. Freeman, San Francisco, CA

Davies, J. K. and Kelly, M. P. (1993) *Healthy Cities*, Routledge, London

Department of the Environment (1993) *Migration and Business Relocation: The Case of the South East, Executive Summary*, A. Fielding and Prism Research Limited, Planning and Research Programme, HMSO, London

Edwards, B. (1996) *Towards Sustainable Architecture: European Directives and Building Design*, Butterworth Architecture, London

Egan, M. D. (1988) Architectural Acoustics, McGraw–Hill, Hightstown, US

Elkin, R. et al (1991) *Reviving the City: Towards Sustainable Urban Development,* Friends of the Earth, London

ESRC Research Programme (2000) *The Global Environmental Change: Producing Greener, Consuming Smarter*, ESRC, Swindon

Fathy, H. (1973) *The Arab House in the Urban Setting: Past Present and Future*, University of Chicago Press, Chicago

Fezer, F. (1982) 'The influence of building and location on the climate of settlements', *Energy and Buildings*, vol 4, pp91–97

Fothergill, S., Kitson, M. and Monks, S. (1983) *Changes in Industrial Floor Space and Employment in Cities, Towns and Rural Areas*, Industrial Location Research Project Working Paper 4, University of Cambridge, Cambridge, Department of Land Economy, Cambridge

Givoni, B. (1989) *Urban Design in Different Climates*, WHO Technical Note No 346, Geneva

Givoni, B. (1998) *Climate Considerations in Buildings and Urban Design*, Van Nostrand Reinhold, US

Goethert, R. (1978) *Urbanization Primer*, MIT Press, Cambridge, Massachusetts

Golany, G. (1983) *Earth-sheltered Habitat: History, Architecture and Urban Design*, Architectural Press, London

Golany, G. (1995) *Ethics and Urban Design: Culture, Form and Environment*, Wiley, New York

Goodchild, B. (1998) 'Learning the lessons of housing over shops initiatives', *Journal of Urban Design*, vol 3, no 1, pp73–92

Haughton, G. and Hunter, C. (1994) *Sustainable Cities*, Jessica Kingsley Publishers Ltd, London

Holtzclaw, J. (1991) *Automobiles and their Alternatives: An Agenda for the 1990s*, Proceedings of a Conference Sponsored by the Conservation Law Foundation of New England and the Energy Foundation, p50

Jacobs, A. (1993) *Great Streets*, MIT Press, Cambridge, Massachusetts and London

Jenks, M., Burton, E. and Williams, K. (eds) (1996) *The Compact City: A Sustainable Urban Form?*, E. & F. N. Spon, Oxford

Kihlman, T. and Kropp, W. (1988) 'Limits to the noise limits?' *Proceedings of the 16th International Conference on Acoustics*, Acoustical Society of America, Seattle

Landsberg, H. E. (1981) *The Urban Climate*, Academic Press, New York and London

Leinsberger, C. (1996) 'Metropolitan development trends of the latter 1990s: Social and environmental implications', in Diamond, H. L. and Noonan, P. F. (eds) *Land Use in America*, Island Press, Washington, D C

Ludwig, F. L. (1970) *Urban Temperature Fields in Urban Climate*, WMO, Technical Note No 108, Switzerland, pp80–107

Lyle, J. T. (1993) *Regenerative Design for Sustainable Development*, John Wiley and Sons, New York

Lynch, K. (1960) *The Image of the City*, MIT Press, US

Magnan, R. and Mathieu, H. (1975) *Orthopoles, Villes en Iles*, Centre de Recherche d'Urbanisme, Paris

Mahdavi, A. (1998) 'Toward a human ecology of the built environment', *Journal of South East Asian Architecture*, vol 5, no 1, pp23–30

Maldonado, E. and Yannas, S. (1998) 'Environmentally friendly cities', *Proceedings of PLEA 1998 Conference, Lisbon, Portugal*, James & James Ltd, London

Mänty, J. (1988) *Cities Designed for Winter*, Norman Pressman, Building Book Ltd, Helsinki

March, T. A. and Trace, M. (1972) 'The land use performances of selected arrays of built forms', *Land Use and Built Form Studies*, Working Paper No 2, UK

Martin, L. and March, L. (1972) *Urban Space and Structures*, Cambridge University Press, Cambridge

Morris, A. E. J. (1994) *History of Urban Form: Before the Industrial Revolutions*, Longman Scientific and Technical, Harlow

Moughtin, J. C. (1996) *Urban Design: Green Dimensions*, Butterworth Architecture, Oxford

Mumford, L. (1961) *The City in History: Its Origin, Its Transformation and Its Prospects*, Harcourt, Brace and World, New York, p15

Naess, P. (1991) 'Environment protection by urban concentration', *Scandinavian Housing and Planning Research*, vol 8, pp247–252

Naess, P. (1991) 'Environment protection by urban concentration', Paper presented at Conference on Housing Policy as a Strategy for Change, Oslo (copy available from Norwegian Institute for Urban and Regional Research, Oslo)

Newman, P. and Kenworthy, J. (1989) 'Gasoline consumption and cities – a comparison of US cities with a global survey', *Journal of the American Planning Association*, vol 55, pp24–37

Nijkamp, P. and Perrels, A. (1994) *Sustainable Cities in Europe*, Earthscan Publications Ltd, London

Ojima, T. and Moriyama, M. (1982) 'Earth surface heat balance changes caused by urbanization', *Energy and Buildings*, vol 4, pp99–114

Oke, T. R. (1980) 'Climatic impacts of urbanization', in Bach, W., Pankrath, J. and Williams, J. (eds) *Interactions of Energy and Climate*, D. Reidel Publishing Company, pp339–356

Oke, T. R. (1987) *Boundary Layer Climates*, Methuen, London

Oke, T. R. (1988) 'Street design and urban canopy layer climate', *Energy and Buildings*, vol 11, pp103–113

Owens, S. (1986) *Energy, Planning and Urban Form*, Pion Limited, London

Owens, S. (1992) 'Energy, environmental sustainability and land-use planning', in Breheny, M. (ed) *Sustainable Development and Urban Form*, Pion, London

Pearlmutter, D. (1998) 'Street canyon geometry and microclimate', *Proceedings of PLEA 1998, Lisbon, Portugal*, James & James Ltd, London

Rahamimoff, A. and Bornstein, N. (1982) 'Edge conditions – Climatic considerations in the design of buildings and settlements', *Energy and Buildings*, vol 4, pp43–49

Richards, J. M. (1946) *The Castles on the Ground*, Architectural Press, London

Rickaby, P. A. (1987) 'Six settlement patterns compared', *Environment and Planning B: Planning and Design*, vol 14, pp193–223

Rogers, R. (1997) *Cities for a Small Planet*, Faber and Faber, London

Rudofsky, A. (1964) *Architecture without Architects*, Academy Edition, London

Sadik, N. (1991) 'Confronting the challenge of tomorrow's cities – today', *Development Forum*, vol 19(2)

Simmonds, J. O. (1994) *Garden Cities 21: Creating a Livable Urban Environment*, McGraw-Hill, Inc, US

Steemers, K. (1992) *Energy in Buildings: The Urban Context*, PhD thesis, University of Cambridge, Cambridge

Steemers, K. (ed) (2000) 'PRECIS: Assessing the potential for renewable energy in cities', unpublished EU research report, The Martin Centre, University of Cambridge, Cambridge, UK

Steemers, K. and Yannas, S. (eds) (2000) 'Architecture city environment', *Proceedings of PLEA 2000 Conference, Cambridge, England*, James & James Ltd, London

Tibbalds, F. (1988) 'Urban design: Tibbalds offers the prince his ten commandments', *The Planner* (mid-month supplement), vol 74, p1

UNCHS (1996) *An Urbanising World*, Oxford University Press, Oxford

Urban Task Force (1999) *Towards an Urban Renaissance*, Crown Copyright, London

Van Der Brink, A. (1997) 'Urbanization and land use planning: Dutch policy perspectives and experiences', Unpublished paper

Warren, R. (1998) *The Urban Oasis: Guideways and Greenways in the Human Environment*, McGraw-Hill, US

WCED (World Commission on Environment and Development) (1987) *Our Common Future*, Oxford University Press, Oxford

Whyte, W. H. Jr. (1990) *Rediscovering the Center City*, Doubleday (Anchor), New York

World Media Foundation (1999) 'Living Earth', Transcript of interview with Jasper Simonsen, Deputy Minister for the Environment, Norway, 15 January

Wu, S. and Kittinger, E. (1995) 'On the relevance of sound scattering to the prediction of traffic noise in urban streets', *Acoustica*, vol 81, pp36–42

Recommended reading

The list of relevant publications provided in the references demonstrates not only the wide range and depth of existing knowledge, but also the proliferation of interest in the field of urban environmental issues. However, despite this existing expertise, it is hard to identify documents that successfully cover and integrate the key issues. The following publications achieve a degree of success in particular areas of the field:

1. Breheny, M. (ed) (1992) *Sustainable Development and Urban Form*, Pion, London
   This collection of essays has become a key text that relates the wider aspects of sustainability (not only environmental, but also social and economic) directly to urban design. Breheny was a well-respected academic and expert, and was one of the most important researchers in this field. The emphasis of the book is not primarily on the technical aspects.
2. Elkin, R. et al (1991) *Reviving the City: Towards Sustainable Urban Development,* Friends of the Earth, London
   This book established the key environmental concerns and potential strategies for overcoming the perceived environmental problems of urbanization. It covers a wide range of issues, including energy and pollution, but in a fairly broad-brush manner. There is a campaigning emphasis, with an aim of pushing the political agenda forward. As a result, the book perhaps does not have the objectivity one might wish for. Nevertheless, it is stimulating.
3. Rogers, R. (1997) *Cities for a Small Planet*, Faber and Faber, London
   A book written by an architect and aimed at identifying and addressing the urban design challenges. This book is a somewhat emotive exposé from the perspective of one of the world's leading urban architects. It highlights the primary issues and is richly illustrated with design proposals. This is a useful primer for practitioners interested in raising the debate.

Notes

1. Taken from A. V. Ruiz (1998) *Environmental Design for Cities: Traffic Noise Attenuation through Urban Design*, MPhil dissertation, The Martin Centre for Architectural and Urban Studies, University of Cambridge, Cambridge, UK.
2. In free-field conditions (i.e. conditions in which reflecting surfaces are absent).
3. Taken from A. V. Ruiz (1998) *Environmental Design for Cities: Traffic Noise Attenuation through Urban Design*, MPhil dissertation, The Martin Centre for Architectural and Urban Studies, University of Cambridge, Cambridge, UK.
4. The Healthy Cities project was inaugurated by the World Health Organization (WHO) to inform policy throughout the world.
5. Talk given by Professor I. de Solà Morales, Martin Centre for Architectural and Urban Studies, Department of Architecture, University of Cambridge, 24 May, 2000

Activities

Activity 1

Discuss the lessons that today's urban planners may learn from studying vernacular urban plans, drawing on historic examples. Do the pressures on today's urban environment make the study of historical examples irrelevant?

Activity 2

Outline the nature of the urban environment – what are the primary characteristics that distinguish it from a rural climate?

Activity 3

What, in your opinion, are the social and technical advantages and disadvantages associated with both evolutionary and revolutionary urban change?

Answers

Activity 1

The climate-responsive aspect of vernacular settlements has been highlighted by many scholars. For example, Morris (1994) compares the prevalence of courtyard planning in hot-arid climates and the resistance it received in northerly cool climates when Roman conquerors tried to introduce it in urban planning. It was believed that the absence of climatic stresses in moderate northern latitudes did not necessitate the clustering of shelters and introversion around a courtyard in order to filter the climate. The housing type that prevailed was, rather, extroverted and often free standing (Morris, 1994). Examples of such vernacular settlements are the Pueblo Navajos and Anasazis Native American settlements (Golany, 1983), and Middle-Eastern settlements where compact urban planning defied the harshness of hot-arid climates (Rudofsky, 1964) such as in Ur, Olynthus and Cordoba (Morris, 1994). Historical examples are still relevant today despite the modern factors of urban pollution, noise and traffic. In Finland, for example, following the age of functionalism where buildings were free-standing within arteries of infrastructure, the courtyard house re-emerged. The revival of the vernacular south-oriented courtyard type of planning took place for several reasons. In addition to the creation of a pleasant and 'energy-concentrating' sun pocket, formed at the corner of two buildings, the same configuration acts as a buffer against the wind and street pollution (traffic, noise, dirt, particles), in addition to providing efficient land use.

Activity 2

Givoni (1998) acknowledges the effect of urban morphology on the urban microclimate and, therefore, on energy consumption. The physical parameters that he was able to extract were size of city, density of built-up area, land coverage, building height, orientation and width of streets, and building-specific design details that affect the outdoor conditions. There is a subtle trade-off in street design which aims to maximize ventilation, dispersion of pollutants and solar access, while not compromising shelter and urban warmth. On urban warmth and the urban heat island with respect to geometry, Oke (1987) observes that a city with an elevated occurrence of a high H/W, and therefore tending towards compactness, promotes the trapping of solar radiation and urban warmth, especially at night.

Activity 3

A response to this is elaborated briefly in the 'Overview' section.

2

Architectural Design and Passive Environmental and Building Engineering Systems

Spyros Amourgis

Recent examples of green buildings serve to remind architects that sustainability rests on the long-standing basis of smart design (Snoonian and Gould, 2001, p96).

Scope of the chapter

This chapter briefly discusses the building design process and the utilization and integration in this process of the opportunities offered by the natural environment. These opportunities contribute towards inventing a comprehensive building concept which interacts with the elements of the natural environment.

Learning objectives

After reading this chapter you should be able to integrate passive and active environmental systems within building concepts, as well as design processes. These environmental systems help to reduce energy consumption and improve indoor environmental conditions, generated by non-renewable sources.

Key words

Key words include:

- building concept;
- building design process;
- passive and active environmental systems.

Introduction

The emphasis of this chapter is on the building design process and not on specific technical information. It discusses the significance of understanding passive and active systems in making design decisions, as well as the opportunities to conceive interesting and inventive concepts generated by such factors. This approach to design responds to all concerns, including a suitable ambient environment that depends less upon non-renewable energy sources (Fry and Drew, 1956, p23).

In his book *Biology and the History of the Future*, Waddington (1972, p27) characteristically states:

> *I would not settle for something a little less than incorporating the total planet in every analytical small-scale operation. What I think does need to be incorporated is not so much an idea about the planet as a whole, but an idea about the situation as a whole.*

The building concept

Building design is a synthetic process. The key element of this process is the concept. As a process it relies upon the creativity of the designer to produce a central idea, which is the concept. This ought to be generated through the consideration of, and response to, the following factors:

- functional requirements of the users;
- context in which the building is set (natural and man-made environment);
- construction method and materials;
- initial investment; and
- cost of operation and maintenance.

The final design ought to appeal to the user, meet the code requirements and professional standards, and satisfy the designer. This process is not as precise as engineering methodologies, since the preferences of the users or clients, as, indeed, those of the designer, may be influenced by biases or current trends that may reduce (during the decision-making process) the effectiveness of a building design.

Functional requirements of the users

The requirements of the users are:

- appropriate *space* allocation and *layout* of spaces, accommodating and serving all human activities that are anticipated to take place in a given building;
- a *healthy* and *comfortable* environment, as Ruth Sager (cited in Waddington, 1972, p60) appropriately described in her proposal for a 'biological bill of rights for mankind': 'the right to live in an equitable physical environment [that is] aesthetically attractive and physiologically healthful';
- *convenience* of *movement* and use of the building;
- a *pleasing environment* – variable condition since it depends upon individual perceptions of spaces and prevailing social influences and trends.

Context

The context in which the building is set includes consideration of the following:

- *general topography* of the site and the adjacent environment;
- *orientation* of the site;
- *climate* data;
- *land uses* of the adjacent sites and vicinity;
- *environmental conditions*, in general, and specifically the access road (e.g. traffic and noise pollution);
- *historic, cultural* and *social environment* of the area.

Construction method and materials

The choice of construction method and materials is based on the following:

- *safety of the structure*, in general, and specifically addressing local natural hazards (e.g. earthquakes, floods and winds);
- *cost and life expectancy* of building (e.g. limited use as a temporary or a permanent structure);
- *materials available* in the local market and *skills* of local trades people;
- *appropriateness of materials* for certain uses, and with particular regard to the use of renewable materials, recycled materials and materials that least affect the environment through the manufacturing process.

Initial investment cost

The initial cost of construction, as an investment, affects almost all other choices. However, even if funding is not a problem, it is wise to invest in methods that reduce energy consumption since they also increase the effectiveness and quality of the interior environment of the building.

From the environmental point of view in urban areas, when it is feasible, it is preferable to recycle buildings instead of demolishing and rebuilding – a point widely recognized now on both sides of the Atlantic (see Waddington, 1972; Snoonian and Gould, 2001, p94).

Cost of operation and maintenance

Operating costs consist primarily of utilities through the consumption of:

- *water* for kitchen, baths, irrigating plants and general cleaning purposes;
- *energy* sources for lighting, ventilation, heating, cooling, cooking, hot water, the use of electric appliances and mechanical systems for movements.

Maintenance costs include the upkeep of interior and exterior materials that have a limited life span (e.g. paints and replacing or renewing roofing materials).

The building design process

There is not one precise universal method or system for 'designing' buildings that is generally acknowledged. The assimilation and synthesis of information are performed by humans whose thinking processes differ; therefore, as a process, building design varies. For example, some individuals follow a linear thought process and synthesize on the basis of their analysis. Other individuals are intuitive and quickly reach a decision on a concept, based on their experience. While the concept is developed into an actual building design, both approaches should respond to the same issues and criteria.

These two methods of approaching the design process define extremes in the way in which most designers work. What all competent building designers share is that their designs respond to the same list of key issues.

The emphasis placed by some professionals on some of the design issues has not always been equal. Advancements in technology and the low cost of energy, from the early 1950s until the early 1970s, offered opportunities to control the interior environment of buildings with mechanical means, and the emphasis on the natural environment diminished. The result has been that numerous buildings are almost totally dependent upon technology with regard to their interior environment and are excessive energy consumers. This trend has slowly changed since the first major oil crisis of the 1970s.

Environmental approach to building design

The architectural design process is an inventive process. It relies on the ingenuity of the designer to use all key issues of a building problem (and interpret them as spatial elements which are incorporated within the building design). There are numerous fine examples of building designs that illustrate this point, including the development of 'indigenous' housing prototypes in Los Angeles during the first half of the 20th century (Amourgis, 1995, pp121–124).

Indigenous architecture in all parts of the world evolved through time to consider the functional needs of people, the context in which they were built, and locally available materials and construction methods. The cost was measured in human effort and time. The inherited empirical knowledge and wisdom of the local master builders and craftsmen have been replaced by formal education (selective knowledge) and technology (machines and industrial products). The ability, in particular, to control interior environments has lowered the sensitivity of the contemporary building designer towards nature and natural phenomena.

Consumer trends have caught up with buildings, and the search for new products has emphasized the image of a new building, instead of the social and environmental values incorporated within the building concept. Several architectural critics of the last two decades described this formalistic trend as 'the emphasis on the package', meaning that the concept is focused primarily on the external image, that is, 'envelope' of the building.

The emphasis on the environmental design of buildings and bioclimatic architecture is a step in the right direction and not a new invention. It simply has gained ground because of increasing awareness that the global environment is in danger, natural resources are not abundant, and consumption of non-renewable energy sources ought to be reduced or replaced by alternate means.

Today, most of the traditional and empirical knowledge used in the past to improve the environmental condition of buildings has been substantiated through science. The alternative to current building design practices is the use of passive systems. Such systems have become more effective through the invention and utilization of new materials and technological aids.

Buildings as elements of the urban fabric

Urban buildings:

- externally define open and public spaces;
- internally provide sheltered space(s) from the natural elements.

The way in which buildings are located and the form of buildings affect the immediate external environment. The organization of the interior plan and cross-section of a building determine the relationship of the interior spaces to the exterior environment. Furthermore, the way in which building elements, such as openings, are designed controls the visual and functional relationship of the interior to the exterior. Finally, the choice of materials and method of construction are equally important in achieving the best results.

In order to be functional, the interiors of buildings need light, air for ventilation, and heating or cooling when temperatures drop or rise to uncomfortable levels. Humidity can also cause discomfort and in extreme conditions may harm or endanger human life. Ambient conditions are very important as they affect the physiology and the well-being of humans. It is important, therefore, to remember that the effectiveness of the interior environment also depends upon the way in which it has resolved its relationship to the external environment.

Passive systems in buildings

The term 'passive' implies those methods employed in building design that use renewable energy sources and 'simple integrated technologies' (Santamouris and Asimakopoulos, 1996, p75) for heating and cooling, as well as methods that maximize the use of natural light and ventilation for buildings.

There are several alternative applications or variations of passive systems that have been used in buildings during the last 30 years. Some building designers have successfully used passive systems to generate a design concept,

Source: Amourgis archive

Figure 2.1 *Row housing in Munich*

Source: by permission of Professor J. Lang

Figure 2.2 *Example of house incorporating passive systems*

and others have simply applied elements of passive systems within the building design only, or have compromised other aspects of a building design to ensure environmental benefits. An early example of a successful design concept is the row housing in Munich designed by Tomas Herzog and Bernard Schilling in 1976 (see Figure 2.1).

Exploring the benefits of solar energy, Herzog and Schilling designed a building section that is environmentally sustainable, with interesting and pleasant interior spaces. It also offers good views and privacy, and the plan of the complex relates well to the urban fabric.

Another innovative design using the same principle of incorporating passive systems is a house in San Fernando, near Los Angeles, designed by Jurg Lang (Amourgis, 1991, pp87–92) (see Figure 2.2).

Building activities undoubtedly alter the natural environment as they are an intrusion on nature. Therefore, the aim is to minimize the negative impact of buildings in the environment at all levels, from the way in which they relate to the topography, to the use of natural resources, and to the manufacturing method of building materials. The accumulative effects of some manufacturing processes and some urban activities are the main causes of damage to the global environment.

The environmental approach to building design should address such issues and utilize methods and means that serve human needs with the least damage to nature. Lopez Barnett and Browning of the Rocky Mountain Institute refer to this approach in general terms as 'sustainable design' or 'green development'. 'Although this new architecture is difficult to describe in a sentence or two, its overall goal is to produce buildings that take less from the Earth and give more to people' (Lopez Barnett and Browning, 1995, p2).

Building design can minimize or reduce the use of non-renewable energy sources while still providing comfortable conditions for human life in buildings. Such design employs methods that are passive, and incorporates active environmental building engineering systems. When both are employed, they increase the environmental effectiveness of building design.

Active systems utilize mechanical, electrical and electronic equipment in order to increase the effectiveness of passive systems. For example, the use of thermostats in combination with solar powered electric fans may improve the performance of a passive solar heating system by monitoring and regulating temperature conditions. There are many options available through the use of technology and software in combination with passive systems, as described in Chapter 5.

Passive systems can be employed to provide:

- natural light to all interior spaces during daytime;
- natural ventilation;
- heating and/or cooling.

The use of renewable energy sources and simple integrated technologies, such as the Trombe wall (see Chapter 10), can produce positive results.

As mentioned above, when passive systems are combined with mechanical and electrical devices, and

Note: Architects, N. Kalogeras and S. Amourgis (1978)

Source: Amourgis Archive

Figure 2.3 *Interior of terminal building of the International Airport of Alexandroupolis*

Note: Architects, N. Kalogeras and S. Amourgis (1978)

Source: Amourgis Archive

Figure 2.4 *Close up of ceiling of terminal building of the International Airport of Alexandroupolis*

with electronic equipment that coordinate and monitor the functions of the various applications, efficiency is further increased. Such technological enhancements can be powered by converted solar and/or wind energy into electricity.

A building concept that is designed to respond to passive systems may utilize all systems or part of the know-how available. The choices depend upon the environmental opportunities and constraints offered by the location, the complexity and function of a building and the available budget.

Natural light

The effective use of natural light directly influences the choice of a building design concept. The configuration of the plan (as well as the size of the rooms) has to adapt to the maximum distance that natural light can travel within an interior space, at a particular site and at a given orientation. The building sections also have to be conceived of in a way that facilitates maximum penetration of natural light. Figures 2.3 and 2.4 show how a structural module was designed for the terminal building of the International Airport of Alexandroupolis in order to provide natural light and ventilation in the interior spaces.[1]

Different space functions also pose different requirements for intensity or quality of light. For example, domestic spaces require different light conditions than an art gallery does. The Kimball Art Museum at Forth Worth, Texas, by Louis I. Kahn (1966–1972) is an excellent design where the building concept responds to several key issues, including the specific natural lighting requirements for gallery spaces (see Figure 2.5).

For each space, the design concept must take into consideration the orientation, position, size and shape of a window opening facing the exterior (walls or roofs), as well as the intensity and type of light (direct or reflected) required to enter the space.

Of course, the issue of choices is not only a matter of calculations; other factors must be considered, such as desirability of views, privacy, appearance of openings and ventilation. Figure 2.6 shows an apartment building designed on the south coast of Athens,[2] where the optimum resolution of the previously mentioned requirements generated the architectural organization and character of the building (Amourgis, 1966).

The source of daylight is the sun. The properties of sunlight are beneficial to human life in many ways, and it is desirable to have the sun's rays enter the interior of buildings for health reasons. Sunlight consists of rays of various wavelengths. In the short wave range, it is the infrared rays that heat surfaces and dry the air; while ultraviolet rays also have the ability to destroy bacteria. In northern climates, a 'habitable' room is defined by the amount of time during which direct sunlight enters a room on the shortest day of the year (in December). There is a legal minimum of one hour. The minimum area of window openings is also defined in this manner.

Source: by permission of Professor P. Helmle

Figure 2.5 *Interior section of the gallery module*

Source: Kalogenes-Amourgis archive

Figure 2.6 *Apartment building on the south coast of Athens*

Light bounces off reflective materials and highly polished surfaces in the opposite direction at the same angle in which the rays hit the surface. On non-reflective or rough surfaces, light bounces off in all directions; this is called 'diffused reflectivity' (Watson, 1992, p83). Reflective properties can be utilized by the designer through appropriate use of horizontal or vertical surfaces in order to direct light further inside an interior space, to block it or create special effects as necessary, such as indirect and diffused natural light. The texture and colour of materials are also important. Indirect and diffused natural light are particularly desirable in museums, art galleries and spaces with special indirect lighting requirements.

The objective, therefore, is to use natural light to its maximum potential inside a building during daytime, and to reduce or totally eliminate the use of artificial lighting during the day.

Natural ventilation, heating and cooling of buildings

As for natural lighting, the effectiveness of natural ventilation and appropriate temperature levels in the interior of a building are, to a large extent, determined by the building design. Choices made at the design stage for siting, plan layout, sections, building form, wall openings, and methods of construction and materials may increase or decrease the 'comfort conditions' of the internal environment for the occupants of a building.

Siting

In urban areas, zoning dictates the building system. The systems most frequently prescribed by planning and zoning codes are continuous, detached or free standing, and semi-detached. The building lines are normally parallel to the street and the requirements are that all or part of the front elevation is on, or parallel to, the building line. The siting of a building and positioning of the front elevation in the continuous system is dictated by the city plan, as is the orientation. The only options are partial set-backs, bay windows and other architectural projections, as prescribed by the local planning regulations. These limited options may be used to redirect wind flow from the street to the interior of a building, or, in adverse situations, may be designed to exclude it. The detached or free-standing building system allows options for siting, and the building design could evolve by adjusting to advantageous orientations and harnessing local winds. The position and type of external plants and trees can be designed to contribute positively to the interior environment of buildings.

Plan layout

The interior layout of buildings can be designed to benefit from cross-ventilation and air movement, resulting from differences in temperature between internal and external areas. Different room functions require different orientations with regard to solar exposure. Various sources and activities that generate heat (e.g. boilers, fireplaces and chimneys) can be placed in key locations in the plan so that the excessive heat benefits other areas inside the building.

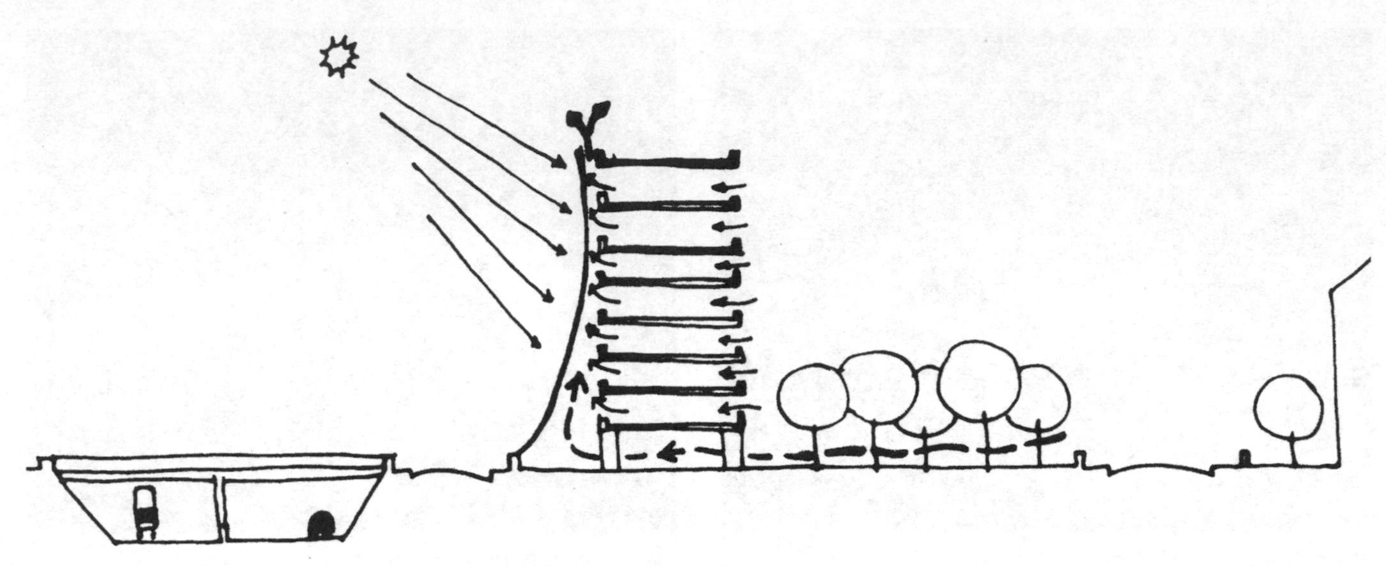

Source: Architectural competition design by S. Amourgis, N. Kalogeras and M. Blake

Figure 2.7 *Section of freeway and building*

Sections

The heights of rooms in relation to the size, position and shape of openings can trap or facilitate air movement. Air stacks or the use of double height-volume spaces – for example, rooms with mezzanines – allow air to move. Buildings on pilotis (sitting on columns or stilts at ground level), in the continuing building system, allow air to circulate between the street and the rear yards. Inventive sections may utilize fresh air in-take of buildings from the rear yard if there are trees and plants, in situations where the front street is very noisy or the air is more polluted.

Figure 2.7 shows the section of a building design case study for high-density commercial offices, next to a busy freeway in Athens (Amourgis, 1993). The curved curtain glass wall facing south and west acts as a 'shield' and as an 'air stack', venting the hot air as it rises to the top. Cleaner air enters from the rear of the building, from the north and further away from the freeway, and exits through the top of the 'air stack'. The designed section provides protection from the freeway pollution to the building interior. Pollution is considerably higher during busy hours up to 67m from the edge of a freeway, beyond which it decreases to the average levels of a built-up area (Proctor, 1989, p27).

Building form

The form of the building can be designed to direct local wind movement in and around the building in order to achieve positive effects. A compact building plan and section has less external wall and roof surface. Therefore, it suffers less heat gains or losses. In warm climates, extensive roofs and eaves protect from heat gains and create different air temperature zones.

Wall openings

The size and shape of fenestration is also critical for air movement. Windows and doors appropriately positioned could improve natural ventilation and internal air movement. High-level windows in warm weather allow hot air to escape; the absence of high-level openings traps the hot air in a space and internal temperatures rise.

Methods of construction and building materials

Methods of construction and building materials can increase the quality of the ambient environment of a building's interior. Different materials have different insulation or heat absorption and storage properties. Another important factor to consider in the choice of materials is the embodied energy (energy consumed in the process, e.g. in manufacture, transportation, etc), as well as the pollution generated through the manufacturing process of the materials (see Chapter 4).

Insulation of external building surfaces helps to maintain differences of temperatures between the exterior and interior of a building. If insulation is uneven between the different materials used for a building's exterior, the insulation effectiveness decreases through 'cold or hot bridges'. Metal frame windows, for example, act as such bridges if the profiles are not properly filled with insulating materials. Detailing is also important as gaps in windows or doors allow undesirable drafts. In dry construction walls, even the gap between the electric switch's plate and the wall allow air droughts, a condition that is remedied with insulating flanges placed between the plastic cover plate of a switch and the wall fitting.

Heat absorption and storage by building materials can be advantageous or disadvantageous. For example, materials that store heat, such as stone, offer advantages when used as interior flooring that is exposed to the winter sun. On the other hand, if the same area is not protected from the summer sun, it becomes disadvantageous, for obvious reasons.

Synopsis

A building's design is a product of a synthetic process. The synthesis is based on the analysis of many factors and considerations regarding the user's requirements, context, technology and the economics of a building. Each one of those factors must be assessed against qualitative and quantitative environmental constraints and possibilities, such as passive and active environmental building systems.

The early consideration of environmental constraints and possibilities will help the creative designer to conceive a building whose design draws upon these factors. There are several contemporary examples of impressive and original architectural designs, where the building concept was generated by ingenious utilization of environmental concerns.

Finally, it seems appropriate to conclude with Donald Watson's summary that a designer ought to 'think of a building in four different ways, each successively more complex, but following a logical progression of development'. A building is a:

1 natural heat exchanger;
2 microclimate;
3 biological system;
4 ecological niche (Watson, 1991, p103).

In conclusion, according to Watson (1991, p102), 'the paucity of environmentally sound design stems from a lack of integration between ecological and architectural principles'.

Notes

1 Kalogeras N. and Amourgis S., Architects, 1978, Alexandroupolis Terminal Building.

2 Amourgis S. and N. Kalogeras, Architects, Vouliagmeni Apartments, 1966.

References

Amourgis, S. (1991) *Critical Regionalism*, CSPU, Pomona, pp87–92

Amourgis, S. (1993) *Competition Design of a Commercial Building Adjacent to Freeways in Athens*, CSU, Pomona

Amourgis, S. (1995) *Design of Amenity*, Kyushu Institute of Design, University Press, Fukuoka, pp121–124

Fry, M. and Drew, J. (1956) *Tropical Architecture in the Humid Zone*, B.T. Batsford Ltd, London, p23

Lopez Barnett, D. and Browning, D. (1995) *A Primer on Sustainable Building*, Rocky Mountain Institute, Colorado, p2

Proctor, G. (1989) 'Building on the edge of a freeway: Case study, Glendale, CA', Research Paper, Urban Design Studio, CSU, p27

Santamouris, M. and Asimakopoulos, D. (1996) *Design Source Book on Passive Solar Architecture*, CIENE, NKUA, Athens

Snoonian, D. and Gould, P. E. (2001) 'Architecture rediscovers being green', *The Architectural Record*, June, pp94, 96

Waddington, C. H. (1972) *Biology and the History of the Future*, Edinburgh University Press, Edinburgh, pp27, 60

Watson, D. (1991) 'Commentary: Environmental architecture', *Progressive Architecture*, March, pp102, 103

Watson, D. and Labs, K. (1992) *Climatic Building Design*, McGraw-Hill, New York, p83

Recommended reading

The following publications are recommended for further reading as an enhancement to the learning process:

1 Herzog, T. (ed) (1996) *Solar Energy in Architecture and Urban Planning*, Prestel Verlag, Munich and New York
This volume comprises very useful examples of various energy-conserving design projects. Readers will appreciate the range and variety of concepts generated by considering environmental concerns (the text is in English, German and Italian).

2 Dimoudi, A., Panno, G., Santamouris, M., Sciuto, S. and Argiriou, A. (1996) *Design Source Book on Passive Solar Architecture*, CIENE (Central Institution for Energy Efficiency Education – National and Kapodistrian University of Athens), Athens
This is a systematic presentation of the specifics of energy conservation in buildings for architects, engineers, physicists and other professionals involved in the design, constructions operation and maintenance of buildings.

Activities

Activity 1

In order to assess your understanding of the terms 'building concept' and 'building design', write a brief description of each and their significance to the design process. When you complete this task, compare your notes with the answer at the end of this chapter.

Activity 2

During the early stages of developing a passive systems building concept, one usually considers key environmental conditions that will influence design ideas. List the main environmental factors that you will consider when developing different concepts for the same type of building – for example, a private house – to be located in different climatic zones, and the reasons why in:

- a temperate climate with winters that are not too severe and warm summers;
- a cold climate with long, cold seasons and short summers?

Answers

Activity 1

'The building concept' section deals with these issues, and explains the key factors that must be considered in developing a concept. 'The building design process' section outlines the development stages of a concept to the building design.

Activity 2

In a temperate climate:

- The sun must be excluded during the warm months and protection from glare and heat gains must be provided.
- During the other seasons, advantage must be taken of the sun to substantially reduce energy for heating.
- During all seasons, natural light may be used to exclude almost totally artificial light during daytime.
- Natural ventilation during the summer may reduce or exclude the use of mechanical cooling.
- Outdoor plants and outdoor materials must be considered for cooling.

In a cold climate:

- The sun is less of a problem during the warm months.
- During the cold months, solar heat may be used to substantially enhance (but not replace) the heating unless active systems are also used.
- Natural light may be used as adjunct to artificial light during the cold seasons.
- Outdoor plants perform a similar role as in warm climates, allowing the winter sun into the building and protecting it during the summer.
- Building form ought to be compact in order to minimize heat losses during cold periods.

3

Environmental Issues of Building Design

Koen Steemers

Scope of the chapter

This chapter provides a step-by-step overview of environmental design issues, from site analysis and planning to detailed decisions about building fabric and services. The chapter is complemented by Chapter 14, which addresses issues of integration and interrelations that are not dealt with here.

Learning objectives

When you complete the study of the chapter, you will be able to:

- understand how the environmental issues impact upon design, from initial site explorations to final construction detailing;
- be strategically aware of individual issues.

Key words

Key words include:

- design phases;
- environmental design checklist.

Introduction

The purpose of this chapter is to suggest the environmental issues that should be considered during the various stages of design. It is structured in a way that broadly follows the design process: from overall site considerations, built form and orientation, to detailed environmental aspects of the façade design, and, finally, to integration of services. However, this should not be taken to assume that there is a deterministic design approach to environmentally responsive design. It is evident that the design process is an iterative development of ideas; thus, interaction between various stages of design development should be taken into account. Integrated design is a theme discussed in more detail in Chapter 14.

Central to the philosophy of this chapter is the aim of minimizing environmental impact, while ensuring 'comfort conditions'. The energy use of buildings in cities is the key to sustainable urban development (WCED, 1987), but also to comfort and well-being (Bordass et al, 1995), in terms of how it affects global, local and interior environmental conditions. The use of energy in buildings raises concern not only for the consumption in use, but also for the embodied energy in materials.[1] This introduces detailed issues about the choice of materials in construction, as discussed previously in this book.

An aim of this chapter is to encourage, through design, opportunities to exploit climatic conditions in order to maintain comfort, minimizing the need for artificial control that relies upon the consumption of energy. During the last century, particularly since the advent of air conditioning, designers have tended to focus upon the *problems* associated with climate (e.g. counteracting the effects of hot or cold weather and high solar radiation). This resulted in strategies to minimize interaction with the exterior climate and increased artificial environmental control. Buildings typically became deep plan and air conditioned, with high energy use and sometimes low user satisfaction.

The degree to which a building can selectively exploit the climate depends upon the very first strategic considerations in its design. The step-by-step environmental discussion that follows highlights the issues that may be considered at the key design stages.

Context

Synoptic climate data

In order to be able to exploit the climatic context, it is critical to analyse the climate type within which the site is located, and to collate relevant data that will inform the appropriate strategic design. The world's climate regions are commonly defined in terms of their thermal and seasonal characteristics (e.g. hot-dry, warm-humid, composite, moderate and cold). Each requires a distinctive design response, which can be frequently found in the vernacular architectural traditions of a region (Rapoport, 1969).

It is important to note that within a climate zone, a wide range of climatic characteristics can be found as a result of, for example, topography, altitude and urban density. In order to define a local climate more precisely than simply according to the generic typologies, more detailed information is required about the local air temperatures, wind patterns and humidity.

It is possible to obtain large amounts of very detailed information from weather stations, based on hour-by-hour monitoring over periods of decades. However, not all of the information will be important or relevant. The detail required depends upon the potential design implication and the level of environmental analysis to be performed. For example, the large diurnal temperature variations in hot-dry climates are as important as the average daily temperatures because they will influence the design strategy for maintaining comfort (e.g. exploiting the time-lag characteristics of thermal mass). Conversely, in warm-humid climates, the diurnal swings are much smaller and air movement is essential to define comfort. As a result, it is important to know wind speeds and directions.

Local climatic conditions

The synoptic climatic context is the primary consideration; but local conditions can differ significantly and will have implications for design. However, it is equally important to consider that a proposed building will influence the microclimate. The microclimate is affected by characteristics of local topography, urbanization and vegetation.

The topography of the land will result in variations in microclimate. For example, the orientation of a steeply sloping site will affect the amount of solar radiation and the hours of sunshine. The direction of a valley may cause a funnelling of the wind if it comes up or down the valley, but may provide a sheltered environment if the wind is perpendicular to the valley. In a location where land meets water, certain local diurnal wind patterns are created: during the day, the solar radiation increases the temperature of the land above that of water, which (through thermal buoyancy of hot air rising up from the land) creates air flow from water to land. At night this effect is reversed.

Table 3.1 *Urban microclimate compared with the rural environs*

Climatic factor	Compared with rural environs
Temperature (annual mean)	0.5–3.0° Celsius more
Radiation (total horizontal)	0–20% less
Wind speed (annual mean)	20–30% less
Relative humidity (annual mean)	6% less

Source: Landsberg, 1981

Figure 3.1 *The urban environment (in this example, Chicago) is typified by hard surfaces, a lack of vegetation and complex patterns of overshadowing and wind. As a result, it is significantly different from a rural climate*

Figure 3.2 *Vegetation provides a variety of microclimates, including the presence of shade and potentially evaporative cooling, as shown in this courtyard in Alhambra, Granada*

The key effect of urbanization on air temperatures is referred to as the 'heat island' effect, where temperatures in cities can be several degrees Celsius higher than in the neighbouring rural area (Landsberg, 1981). This has been discussed in detail in previous chapters. The heat island is caused by a range of factors, such as the lack of moisture (fast run-off and little vegetation), the production of heat, the absorption of solar radiation and lower wind speeds in cities (see Table 3.1 and Figure 3.1).

Another microclimatic consideration of urbanization is air and noise pollution (see 'Air and noise pollution' on p49). Both may have significant consequences on the options for simple natural ventilation strategies through windows that open, and may, as a result, influence design strategies (see 'Building plan and section' on p50).

The use of vegetation to improve microclimatic characteristics is well understood – for example, the use of trees as a shelterbelt to protect a site from cold winds. Other benefits may include using vegetation to provide evaporative cooling, shading, filtering of dust particles, some noise attenuation and carbon dioxide absorption, as well as psychological advantages for humans and ecological benefits to encourage flora and fauna (Figure 3.2).

Solar geometry

The path of the sun is a key factor in determining energy-efficient building form, orientation and façade design. Absorption of solar radiation through building surfaces with different orientations, particularly transparent elements, will significantly influence comfort conditions and energy performance. A relatively simple technique to assess the availability of sunshine on a building façade is to use graphic tools, such as the sun path diagram and shading mask (Goulding, et al, 1992) (Figure 3.3). This allows the designer to determine the effect of topography, vegetation and other obstructions (such as buildings or shading devices) on sunlight availability. The intensity of solar radiation, in relation to its position in the sky, is useful information that allows the designer to judge where to focus the shading or, conversely, the passive solar elements. In cold climates, the aim may be to maximize winter solar gains, while in hot climates it is to minimize gains, particularly in the summer.

Wind and air movement

Wind conditions will have implications for natural ventilation, which, in turn, has an effect on thermal comfort and on energy use. Natural ventilation can be part of a passive cooling strategy in hot seasons, whereas air infiltration (unwanted) and ventilation (controlled) accounts for a significant fraction of the heating load. In all of the above cases, it is important to know the prevailing wind conditions, either to maximize the advantages or to minimize the disadvantages by manipulating the building design. There are instances where wind may be welcomed from certain directions if it is particularly cool, such as prevailing summer breezes. At other times, the need is for shelter from cold winds. It is thus necessary not only to be aware of prevailing wind conditions, but also of seasonal wind and the temperature of wind from different directions (as well as any potential pollution, dust or sand content).

Available data will normally include information about wind conditions measured at specific weather stations and at a prescribed height (usually 10m). Such data are useful in order to provide the overall conditions; but closer site analysis may be necessary, particularly for urban areas where the wind pattern is complex and turbulent.

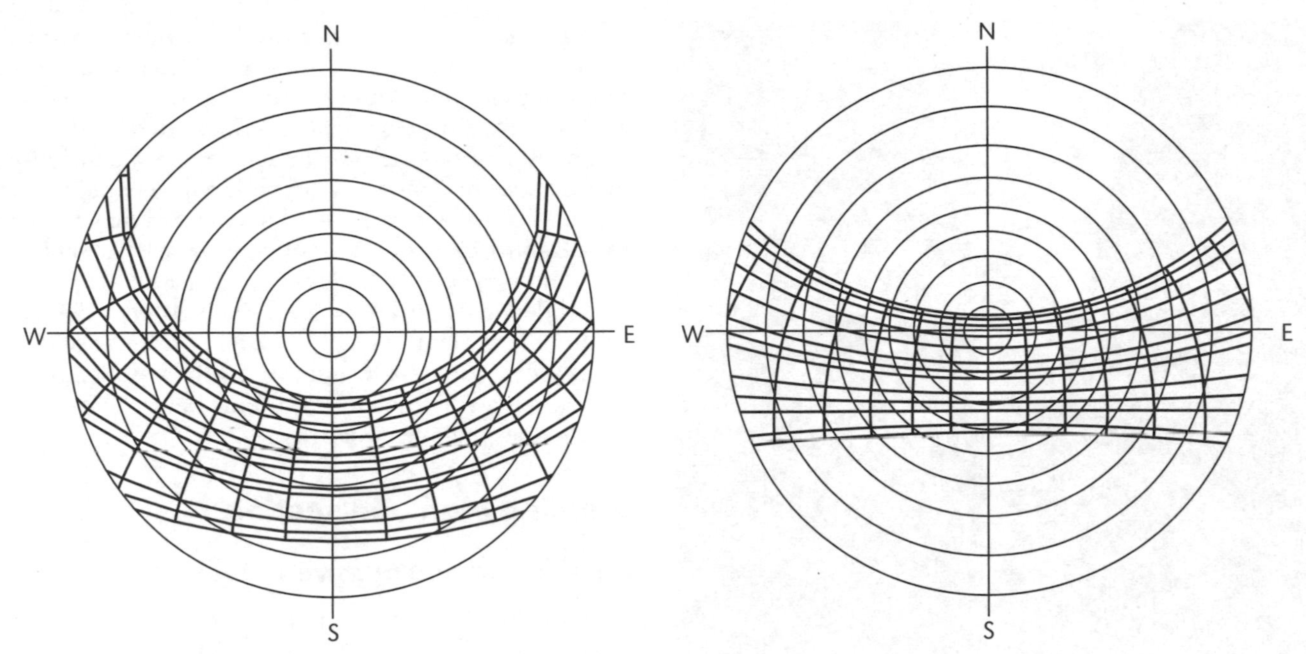

Figure 3.3 *Sun path diagrams for high (53° north left) and low (17° north right) latitude*

Air and noise pollution

Air and noise pollution are factors that are of particular interest with respect to the urban environment, notably in relation to natural ventilation strategies (Figure 3.4). Where possible, it will be useful to monitor (or predict) such pollution levels near to, or on, the site under consideration in order to establish appropriate building plans and sections, as well as more detailed ventilation design solutions.

Site planning

The arrangement of buildings on a site should respond to climatic factors such as solar angles and wind. Where solar access is beneficial to minimizing the heating loads, then the spacing and orientation of buildings may be informed by solar geometry. For example, if the aim is to limit the obstruction to winter sun, then the low angle will result in more widely spaced buildings or reduced building heights. Sunshine is not only of potential benefit to energy savings in buildings, but is also of importance with respect to the quality of the spaces created between buildings. The typically dense urban fabric of vernacular southern towns reflects the concern of reducing solar radiation in outdoor spaces (Figure 3.5).

An analysis of wind speeds, direction and related temperatures can indicate which winds are welcome (e.g. cool summer breezes) and which may be detrimental (e.g. cold winter wind) – they are often not mutually exclusive. The planning of a site can exploit such conditions by the positioning of buildings and vegetation. Wind effects, particularly around tall buildings, will need careful consideration as wind shadows may cause a lack of air movement or turbulence on the leeward side of buildings.

Building design is influenced by microclimate considerations, as outlined above; but the design proposed will also influence the resulting microclimate. Thus, the site plan should contribute to improving the context for the

Figure 3.4 *A haze of pollution covers many urban areas*

Figure 3.5 *Example of urban density from southern Spain (Granada), where narrow urban spaces provide shading from summer sun, but also provide some protection from cold winter winds*

benefit of the overall environment and, in particular, for the benefit of the building itself. For example, the planning of a building site in a noisy, polluted environment may aim to create a quiet, protected public space so that parts of the building facing such a space can be naturally ventilated. This not only reduces the energy use and emissions, but also creates an amenable environment (Figure 3.6).

Mixed uses and movement of people

Other non-climatic issues will have an effect upon the resultant energy consumption levels. The location of uses on large sites will influence the need for transport between activities such as housing and offices, or housing and retail. Without mixed uses it is inevitable that the inhabitants will need to travel (by car or otherwise) outside of the development for retail, employment and entertainment, and will, as a result, consume energy for such journeys. A dispersed urban plan will typically result in greater use of private vehicular traffic, compared with a dense, mixed development (Sherlock, 1991).

Another potential advantage of dense, mixed-use programmes is that energy supply can be centralized and more efficient. A heating load in one area may be offset by a cooling demand in another. Similarly, a combined heat and power (CHP) plant can efficiently supply the mixed demands for hot water and electricity. The more continuous 24-hour energy demand of mixed-use developments is more easily and efficiently supplied locally, compared with the intermittent, short-term peaks required in single-use zones.

Building plan and section

Passive and non-passive zones

Building form is influential in determining the potential interactions between building environment and climate. In order to improve energy performance, building form can be manipulated to exploit those climatic characteristics that are favourable for human comfort. A building that has a high surface-to-volume ratio interacts more with the climate, both potentially positively and negatively. A compact form has less contact with the climate; therefore, the internal environment will need to be controlled artificially. In extreme climates it is often perceived that the less contact with the climate the better; but in terms of energy efficiency, a number of opportunities are lost. A

Figure 3.6 *An example of a courtyard building, where the courtyard potentially provides protection from the noisier and more polluted street environment, creating a quiet haven*

simple example of this is the use of controlled natural light to displace the need for artificial lighting energy. Perimeter zones can benefit from access to daylight – as well as natural ventilation and possibly solar gains – and thus contribute to reducing electric lighting loads. Such zones can be called 'potentially passive zones', and the aim in low-energy buildings is to maximize passive zones and minimize non-passive zones (Baker and Steemers, 2000).

A building with a high percentage of passive zones will have significantly different form characteristics and will appear elongated, linear, courtyard-like or as a finger plan, as opposed to cuboid. It is clear here that selective environmental design tends to enrich the range of formal possibilities, rather than constrain them.

Figure 3.7 *Vernacular architecture from a temperate climate region clearly expresses the selective potential of a simple loggia or overhang to provide seasonal solar control*

Orientation

Building orientation with respect to sun path and wind direction is relevant whether considering hot climates, where minimizing solar gains and maximizing air movement may be the priorities, or cold climates, where the reverse effects may be desirable.

In near-equatorial regions, where the sun path is predominantly overhead (Figure 3.3), radiation on south and north façades can be easily shaded with simple overhangs. However, east and west façades will receive significant direct solar radiation, which is difficult to shade due to the low sun angles. The most significant aspect is the west elevation as this will receive afternoon sun when air temperatures are at their highest of the day. This combination of sun and heat will quickly cause overheating problems if not anticipated. A linear building with a west–east axis, where west and east façades are minimized and have minimal openings, is optimum in terms of solar orientation. Wind direction is a secondary issue as façades can be designed and detailed so as to divert airflow through a building. It is not necessary for the wind direction to be perpendicular, or even near perpendicular, to a façade in order to encourage airflow to the interior.

In cold regions, solar gains, particularly during the heating season, will be welcome. A linear building with a west–east axis is thus also beneficial when solar gains are collected through a solar façade. Shading from high-altitude summer gains may need to be reduced, particularly in temperate climates, by the use of appropriate shading devices – an overhang being traditional and effective (Figure 3.7).

Internal planning

The internal planning of buildings will have implications for energy and comfort performance. For example, the zoning of service areas (e.g. circulation, toilets and plant rooms) as thermal buffer spaces will help to reduce heat losses from the non-solar side and possibly protect the inhabited accommodation from cold winds. Conversely, thermal buffer spaces can be used to protect an interior from excessive solar gains so that, for example, the west

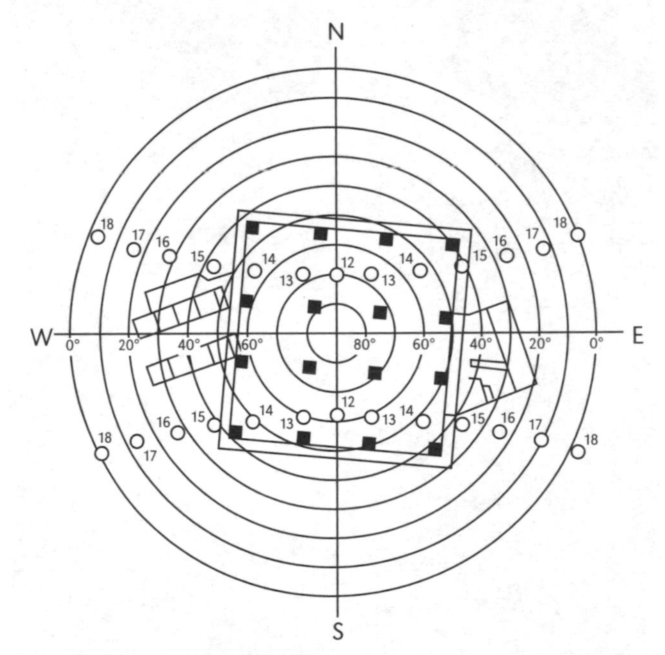

Source: Architect, Ken Yeang, 1994

Figure 3.8 *The plan of this office building (IBM Plaza) has service spaces located on the west and east façades to protect the accommodation from low afternoon and morning sunlight penetration*

or east façades of buildings can be protected from low angle afternoon or morning sun (Figure 3.8).

Noise-generating areas of the accommodation (e.g. plant rooms, machine rooms and workshops) act as acoustic buffer spaces and allow the rest of the building to be protected from a noisy urban context, as well as being potentially naturally ventilated.

Courtyard and atrium spaces

Before the advent of air conditioning and deep-plan buildings, courtyards and light wells were widely used to provide both ventilation and natural light to otherwise deep-plan urban buildings. The benefits in terms of energy use of reducing the need for fan power and artificial light are self-evident. Further advantages may include improved occupant well-being and productivity and reduced costs by avoiding air-conditioning systems. Typological studies of built form and site densities have also indicated that courts can improve the plot ratios of developments.

The renewed interest in courts, and particularly in glazed courts or atria, provides opportunities for low-energy strategies, briefly outlined below.

Figure 3.9 *An atrium environment can provide a range of environmental benefits for a building, particularly in colder northern climates*

Thermal buffer

The first advantage of a glazed atrium, over an open court, is that heat losses from the building are reduced. This is because the extent of the external envelope directly in contact with the outside is reduced. Heat losses from the building, and solar gains into the atrium, will ensure that the atrium temperature is likely to be always higher than the outside temperature. This means that fabric losses from the building into the atrium are reduced. This is known as the thermal buffer effect of atria.

Daylight

By puncturing an atrium through the plan of a deep building, natural light can reach more of the accommodation. The reflectance of the surfaces are under the control of the designer, unlike adjacent street façades, and can therefore be light coloured to maximize daylight penetration. In the case where a court is glazed over, the structure of the glazing system and the glass itself will reduce the amount of light transmitted. However, in such a case the atrium temperatures are higher than outside, and therefore heat loss from the building will be reduced. This, in turn, means that glazing ratios can afford to be larger to improve daylight without a heat loss penalty.

Ventilation

The advantage of court plans is that the cross-section can be reduced down to a distance where cross-ventilation becomes possible. Typically, this distance is taken to be in the order of 12m. Cross-ventilation requires a pressure difference to exist between the façades, normally provided by wind. However, in still wind conditions air movement in buildings is governed by thermal buoyancy. Where a court is glazed over, the wind effects are likely to be reduced; but the potential for stack effect (thermal buoyancy) is increased where stack heights become significant (e.g. a six-storey or taller atrium). The height of the atrium assists the stack-driven ventilation, assuming that openings are provided at the top of the atrium for warm air to escape. This effect induces fresh air to be introduced from the perimeter of the building, across the accommodation and into the atrium. The necessary summer ventilation rate can thus be maintained even under worst-case conditions of no wind. Opening location and sizes need to be carefully designed to ensure adequate and controllable ventilation rates.

Where the main energy use in a building is for space heating, then an alternative ventilation strategy can be adopted to minimize winter heating loads. The warm air in the atrium can be usefully employed as preheated ventilation air for the adjacent accommodation. The atrium is kept sealed to the outside, and the warmed air is encouraged to enter the building. In this way, thermal losses through ventilation, which in a well-insulated design account for the largest fraction of heat loss, can be reduced. Any solar gains made into the atrium space during the winter become very useful. It is important, when adopting such a strategy, to consider the ventilation paths carefully, both in terms of location and sizing.

Building-use patterns

The way in which buildings are used by occupants has a profound impact upon energy use. It is not simply a direct relationship between hours of occupancy and resultant energy use, but relates to the behavioural patterns of occupants and the thermal mass characteristics of the building fabric. Occupants can change the energy performance of identical buildings by a factor of two as a result of opening windows, misuse of heating or air-conditioning controls, and switching off artificial lighting. Buildings with a continuous occupancy pattern (i.e. 24-hour use) may benefit from high-level thermal mass, depending upon climatic variations. Short-term, intermittent occupancy will demand rapid control of the environmental conditions, which suggests a lightweight structure and a fast response system.

In the design of passive control systems (e.g. windows that open, moveable shading and light switching), consideration must be taken of occupant behaviour. A robust strategy, where the building performance is supported by likely occupant response, is an important notion. Simple, effective and direct manual control will provide the user with confidence and increased satisfaction. If control of the environment is obscured, the occupant is likely to feel less comfortable and is also more likely to interact inappropriately. In such a case, fully automated control will provide better performance. For example, consider ventilation. In the case where ventilation is provided through windows that open, occupants will tend to open windows in order to increase ventilation as necessary (although they may not turn the heating off, or may leave the window open longer than necessary). If, however, fresh air is delivered through a grille in the ceiling and controlled via a thermostat, then users may resort to turning controls excessively, opening doors or switching on fans to increase air movement. The result may be noise and privacy problems and a perceived reduced comfort level. An integrated system where temperature sensors open and close windows automatically, but can be overridden manually, may prove the most effective.

Figure 3.10 *A central function of buildings is to provide an appropriate, comfortable and healthy environment that the occupants can control*

The environmental requirements of various uses need to be established and can have a significanct influence on design strategy. In buildings with mixed uses, this is particularly important. A simple example of this may be an office building containing computer rooms, administrative offices, drawing studios and social spaces. Each of the above has differing requirements in terms of light. Computer rooms do not need high levels of natural light, but do require careful glare consideration. Drawing studios benefit from higher levels of daylight, and social spaces may benefit from sunlight. The plan organization of such spaces will thus relate to their environmental requirements: computer suites on the ground floor, drawing office top, administration in between, and social spaces on the sunny side.

Similarly, thermal criteria and noise issues will inform appropriate planning decisions and the use of buffer spaces to protect the more environmentally sensitive areas. The notion of selective design thus has a bearing not only on early design decisions of built form, but transcends all of the design stages and levels of detail.

Different spaces and activities will require different levels of environmental control, ranging from aiming to maintain temperatures at 20° Celsius ± 1° Celsius for sensitive equipment and materials, to allowing temperatures to swing between 18° Celsius and 27° Celsius for human comfort. Similarly, humidity levels and the need for cleaning and filtering air will vary according to building use. Such considerations will influence decisions about the level of mechanical services and their specifications.

The level of internal gains relates directly to building use and is a function of equipment, occupancy and lighting systems. For example, in an office where computer use is intense, internal gains will be significant. Similarly, in spaces that are designed for dense occupancy, such as classrooms, lecture rooms and auditoria, internal gains are high for certain periods, and such spaces will require high levels of ventilation air for health and comfort. In spaces where high light levels are required, heat gains will need to be assessed. Clearly, internal gains will depend upon the efficiency of the lighting system, and whether daylight can be used to minimize the load.

Construction detail

Insulation and U-values

Thermal insulation is a primary way of avoiding heat loss from buildings. It is thus essential to consider levels of insulation where the heating load is a major percentage of the energy bill. In cases where internal gains are high, it is likely that the heating load is not significant. Therefore, the emphasis of an energy-efficient strategy will be on avoiding the need for cooling rather than minimizing heat loss. Similarly, in hot climates this will also be the case, although thermal insulation can play an important role in reducing solar gains through the fabric.

For example, building regulations recommend maximum U-values (standard measure of the rate of heat loss conducted through a building component) and maximum glazing ratios in order to minimize fabric heat losses. This may be appropriate for domestic buildings; but the conditions in non-domestic buildings require a more sophisticated response to energy efficiency, including considering daylighting, shading and natural ventilation. These issues will clearly affect the nature and composition of the building fabric.

Thermal mass

The amount and distribution of thermal mass in the building fabric will affect the thermal response and performance of the spaces. A heavy-weight building will respond slowly to heat gains, either from the heating system or from other sources, such as solar and internal gains. This can be advantageous in both delaying and reducing the peak temperatures caused by heat gains. However, some building types with short occupancy patterns will benefit from a lightweight fabric. If well insulated, such a building will have a fast response and thus will not require a long lead-in time to warm up.

Thermal mass can also be of use to increase the temperature usefully during occupancy. For example, in housing in cold climates, solar gains can be absorbed into the mass during the day and be released in the evening. Thus, comfort conditions can be maintained for a longer part of the day before resorting to artificial heating sources.

In terms of energy use, thermal mass only has potential benefits if the services are designed for it. For example, a fully air-conditioned building may not benefit significantly in energy terms from thermal mass as much of the energy is required for fan power rather than cooling. However, in mixed-mode buildings, where passive design is integrated with systems, cooling can be avoided until peak temperatures rise. Thermal mass will delay the need for cooling, shortening the cooling season.

Where thermal mass is used, the thickness and the surface area are significant characteristics. The thicker the mass, the longer the time lag. If a large surface area is available, then the mass will be more effective.

Embodied energy and toxicity of materials

Environmental issues raised with respect to the choice of the building materials include energy use in their manufacture and transport, as well as health issues of materials and their treatment (e.g. paints and varnishes) after construction.

The materials used in construction have gone through various processes before being integrated on site. Such processes consume various amounts of energy that are 'embodied' in the final building material. High process elements, such as aluminium, steel and glass, have more embodied energy than, for instance, concrete or timber, per cubic metre. Another aspect that may contribute to the choice of materials or elements in buildings is the transport energy required to deliver the product. Clearly, locally made products will require less transport energy than those products imported from abroad. The embodied energy in a building in Europe or North America is equivalent to typically five times the amount of annual energy required to operate the building.

The choice of materials may also relate to health considerations, both in terms of the fabrication processes of products and when installed in buildings. For example, the 'off-gassing' of volatile organic compounds into a space can increase internal pollution levels to ten times that of outside levels. The lower ventilation rates caused by modern airtight construction techniques and adopted to minimize energy loss exacerbate the problem. Dust collected in carpets and soft furnishings will increase the population of dust mites, whose faeces can cause irritation, allergic reactions and generally reduce the well-being of occupants.

Natural lighting

Lighting goes to the heart of the architectural enterprise. Electric lighting can also account for the largest single primary energy load in buildings; thus any reduced reliance on artificial light can make significant energy savings. This can be achieved by displacing the need for artificial light with daylight. Daylight factors of 2 per cent or more can reduce energy use for lighting by in the order of 60 per cent, compared with a totally artificially lit space. However, to achieve daylight levels throughout a building requires careful planning. Clearly, deep-plan designs will not achieve daylighting deep into the building. Typically, a perimeter depth of 6m, or twice the floor-to-ceiling height, can be potentially day lit. Thus, buildings that are deeper than 12m may require more artificial light. However, the darker central areas are often assigned as circulation space, requiring only low levels of light. Other darker plan areas can be occupied by secondary spaces, such as services, toilets and storage areas.

Innovative devices such as light shelves, light ducts, reflectors, holographic films and fibre optics can be adopted to increase the penetration of natural light.

For daylight strategies to be effective in terms of energy savings, it is essential that automatic switching is adopted to ensure that artificial lights are off when sufficient daylight is available (see 'Artificial lighting systems' p59).

The aim is not only to achieve good levels of light deep in the plan, but the light distribution is also significant in determining visual comfort and how occupants are likely to switch lights. An even distribution and high level of light will tend to make the whole space well lit. However, this is often difficult to achieve as light levels tend to drop off quickly further away from windows. If the back part of a room appears significantly darker (e.g. by a factor of five) than the front part of a room, then it will appear gloomy even if light levels are high. Thus, it is the relative brightness that is important. In conditions of poor light distribution, occupants are likely to turn lights on at the back of the room. Light distribution can be improved by devices such as light shelves, which (although they do not increase light levels at the back) do reduce levels near the window, resulting in a more even distribution. Shading devices often tend to worsen the light distribution, unless carefully designed (e.g. louvered and light coloured).

Figure 3.11 *Daylight availability and distribution will significantly affect the success of a space, both in technical and visual terms*

It is not only the amount of glazing that determines daylighting conditions in a space. The positioning of windows will influence light distribution. High-level windows will give a better daylight penetration compared with low-level windows. Similarly, tall windows will throw light deeper into the plan than wide windows of the same area. As a simple rule of thumb, this effect can be demonstrated by determining the 'no-sky line'. If from any point the sky cannot be seen, it is likely that daylight levels will drop quickly. This 'no-sky line' is thus a good indication of depth of daylight penetration.

Window design clearly has to perform a number of tasks, including those mentioned here. Resolving all of these issues is complex; therefore, priorities have to be decided on to arrive at a satisfactory compromise. In terms of energy use, buildings occupied during the day and requiring good lighting will be dominated by the need for daylight. However, in housing, for example, thermal issues (solar gains versus heat loss), as well as views, will be more important.

Designing for passive solar gains

Where buildings have heating requirements, the thermal load can be reduced by designing for solar gains. The design of passive solar architecture requires an understanding of solar geometry and seasonal heat loads. It is, for example, most likely that peak loads coincide with low sun angles. Thus, solar-oriented, vertically glazed openings may be particularly effective at collecting gains in the winter, while horizontal openings will collect more gains when solar altitudes are high (i.e. in the summer) and air temperatures are high. Passive solar buildings are typically characterized by large solar-oriented windows and small openings to the non-solar façade.

On a daily cycle, it becomes important to consider the sun's position in relation to external temperatures. Morning, easterly sun may be very useful as ambient air temperatures are low. However, afternoon westerly sun should be controlled as it will otherwise contribute to internal gains when outside temperatures are at their highest. The direct solar gain strategies mentioned above require careful control. The issue of solar control is discussed below.

Glazed spaces in the form of conservatories or atria can be used to collect gains indirectly before those gains are transferred into the building or ventilated directly out if unwanted. Thus, indirect solar strategies offer greater control.

The planning of buildings in relation to orientation is necessary. The main habitable areas should be oriented to the solar side, with perhaps the kitchen oriented to receive morning sun and the living space facing the afternoon and early evening sun. Secondary spaces such as circulation space, bathrooms and garages can be situated on the non-solar side, requiring only small windows and lower temperatures. Such spaces would thus act as a thermal buffer.

Once solar gains have been collected in the building, either by direct or indirect gains, it is important to distribute them efficiently to other spaces to get the greatest benefit (or solar utilization) from those gains. In a direct gain design, the distribution of gains from the solar to the non-solar side can be assisted by ventilation strategies. For example, the design of a stairwell as a thermal flue on the non-solar side will draw warm air across the plan from the solar side. Similarly, where mechanical extract is required in kitchens and bathrooms, the negative pressure can be used to encourage air movement in certain directions.

Where solar gains are made indirectly into a sunspace, then it is important to link this space to as many of the other spaces in the building via ventilation openings. Double height conservatories are often used to enable this direct link to other floors. Thus, the ventilation air for the spaces is preheated in the sunspace by solar gains and drawn directly from the sunspace.

Figure 3.12 *Shading and glazing design need to be closely coordinated to optimize control with views*

Solar control is essential where large areas of glazing are adopted. In terms of fixed devices, on solar façades a simple overhang will intercept the high-angle summer sun; but on east and west façades, the sun angle is lower and control becomes more important. A more elaborate device, such as egg crates, is appropriate, or in lower latitudes the avoidance of glazing on such orientations may be necessary to minimize overheating risks. On non-solar façades, shading is often minimal in the form of vertical fins to intercept oblique sunlight in summer. Often it will be unnecessary to shade non-solar façades, depending upon the level of internal gains and summer peak temperatures.

It is recommended that moveable devices should be used to allow for greater control. The climate is not totally predictable and great fluctuations in temperature and solar radiation are likely. To ensure that 'comfort conditions' are maintained, louvered systems or retractable blinds are recommended. A degree of occupant control over their own environmental conditions will help to improve their perception of comfort, and their acceptance of a wider comfort range.

Strategies for natural ventilation

There are three basic levels of requirement with respect to natural ventilation:

1 To provide fresh air (health).
2 To provide air movement for convective and evaporative cooling from the human body (comfort).
3 To dissipate heat from a building without the need for air conditioning (energy efficiency).

The first demands only low levels of air infiltration. The second requires noticeable air movement, carefully designed to pass across the occupied space. The third suggests the need for high ventilation rates to remove accumulated heat and cool the thermal mass of a building.

The two available mechanisms of providing air movement are wind and stack induced. Wind is often unpredictable (see the section on 'Wind and air movement' on p48), although designs should consider prevailing directions and opening positions. Wind pressures are likely to be larger than stack pressures and thus can contribute to providing high ventilation rates. However, during periods of low wind, the stack effect may provide the only source of air movement through the effect of thermal buoyancy. Warm air will rise, and if stacks are correctly designed, this effect can be exploited to generate sufficient air movement for comfort cooling. The stack effect is determined by the stack height between top and lower openings (the higher the stack height, the greater the pressure difference), by the opening size (the larger the opening size, the greater the air flow) and by the temperature difference between inside and outside (the larger the temperature difference, the greater the pressure difference).

In order to maximize the benefit of thermal mass, night time cooling will 'purge' the heat from the structure, cooling it down in preparation for the next day's occupation. For night time cooling to be effective, the ventilation air must have maximum contact with thermal mass. Any obstructions – for example, to an exposed mass ceiling by down stands or light fittings – will redirect the air flow away from the mass, thus reducing its effectiveness. The location of openings with respect to thermal mass is relevant for the same reason. The stack effect can often be relied upon to evacuate heat, if night time temperatures are lower than internal temperatures; and in other situations wind may be the main driving force for ventilation. Finally, openings should be protected to avoid security risks.

Figure 3.13 *Ventilation stacks are an increasingly common feature in low-energy, naturally ventilated buildings*

The aim of ventilation to provide fresh air for health and comfort can be jeopardized if the source of air is polluted or noisy. These are often quoted as the reasons for using air conditioning, particularly in urban environments. However, although noise and air pollution impose constraints on possible design solutions, there are techniques to minimize the problems.

With respect to noise, the way in which air enters a building can be manipulated to minimize noise transfer

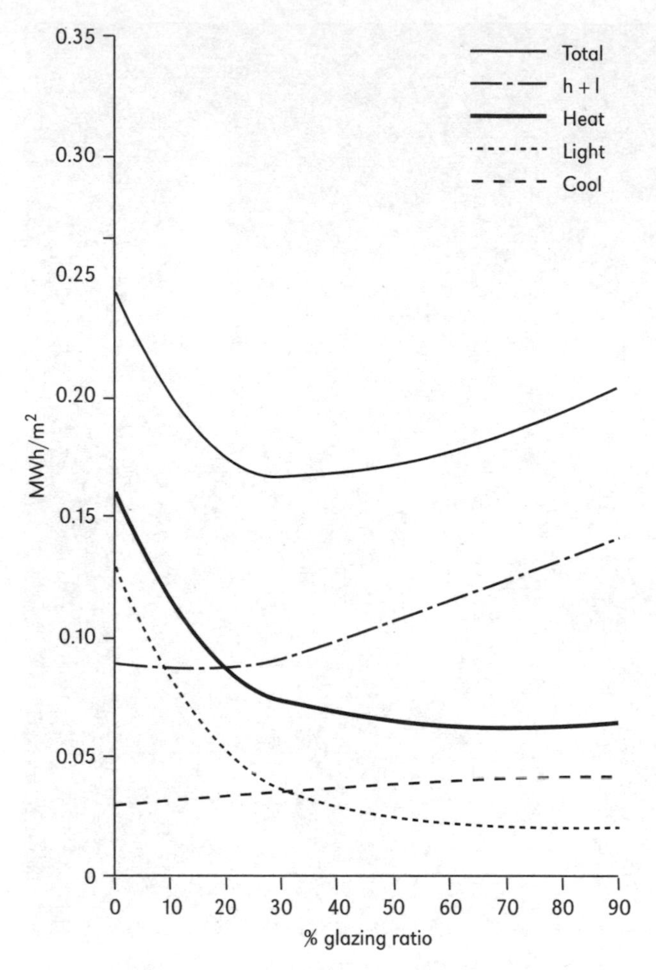

Figure 3.14 *Glazing ratios (i.e. the percentage of glazing in a façade) has a significant impact on the energy use, as shown for a south-facing façade on an office building in the UK*

(e.g. increasing the path length, acoustically lined ducts, acoustic shelves outside windows, and the positioning of acoustic absorbent panels on surfaces inside). It may be that certain areas of a site require particularly careful acoustic consideration; but the site can be protected by the design of the building form, allowing other areas to be naturally ventilated. Those areas located on the noisy edge may be ones that require mechanical ventilation in any case, and thus serve to act as a buffer for the rest of the building.

The problem of air pollution can be tackled by building form in a similar way as described above. The creation of protected courtyards and planting to help settle out and filter dust particles may be considered. The source of incoming air should be removed from the polluted areas, perhaps raised up to the top of the building where any pollution will be more diluted.

Avoiding overheating and increasing comfort

In order to minimize the risk of overheating from solar gains, the area and orientation of glazing needs to respond to the amount and timing of radiation. In temperate climates, the aim is to control summer gains, particularly in buildings with high internal casual gains. Overheating in such a case is most likely to occur in the early afternoon, when external temperatures are highest. If a building has west-facing windows, then low-angle afternoon solar gains will enter the building and add to internal gains. East-facing glazing is not such a problem since air temperatures are generally lower and solar gains may usefully contribute to warming the building up in the morning. On solar-oriented windows, only a fraction of the solar radiation available will be transmitted to the interior because of the high solar altitude and the oblique angle of incidence. The amount of solar radiation on such façades peaks during spring and autumn months.

Shading devices are an essential technique of avoiding unwanted solar gains, but need to respond in design to orientation. Fixed devices are rarely sufficient to control sunshine and moveable devices are often required not only for thermal control, but also for glare control.

In order to provide comfort, solar shading must be integrated with a ventilation strategy. The design of the window to perform the functions of solar shading, glare control, provision of natural light and source of ventilation air needs careful consideration. Shading systems must not interfere with ventilation flow or increase air temperatures locally. An example of this may be a brise soleil, which, although it intercepts unwanted solar radiation, ironically heats up incoming air as it flows over the solar warmed device.

The provision of high- and low-level openings will ensure that the stack effect will induce ventilation when there is no wind. Furthermore, in a situation where cross-ventilation is achieved, then low- and high-level openings provide control of ventilation for the three levels discussed earlier (i.e. health, comfort and energy efficiency). A single opening ensures top-up fresh air for health; low-level openings help to provide air movement across the occupied zone; and high-level windows are appropriately located for high rates of ventilation (above the occupied zone) for dissipation of heat and for night time cooling. Ventilation strategies are described in more detail under 'Strategies for natural ventilation' on p57.

Figure 3.15 *The exposure of thermal mass is essential if it is to assist in providing a stable internal environment*

Thermal mass, as discussed earlier in 'Building plan and section' on p50, assists in reducing peak temperatures and is therefore useful in reducing the likely risk of overheating. Another mechanism at work, apart from reducing air temperature, which is more important with respect to comfort, is that the mean radiant temperature (MRT) is potentially reduced. Comfort depends upon the MRT, and the lower surface temperature of thermal mass will reduce the MRT and make higher air temperature more tolerable. However, in order for the effect of thermal mass to be noticeable, it must be exposed.

Artificial lighting systems

The use of natural light to displace the need for artificial light can have significant potential energy benefits (see 'Natural lighting' on p55), but depends upon lamps being switched off when not required. Manual switching cannot be relied upon, particularly in open-plan buildings; but there are alternative switching controls that range in complexity, including the following:

- time switching off/manual on;
- photoelectric switching off/manual on;
- photoelectric switching on/off;
- photoelectric dimming;
- occupancy sensors (movement or noise).

Each has significant energy saving potential and is appropriate in different circumstances. Economic appraisals have shown that automatic switching can be cost effective in both new and existing buildings.

A simple strategy for reducing lighting loads is to adopt low-energy lamps, with higher efficacies. This generally results in the use of fluorescent lamps, both in the tubular form or as compact lamps suitable for a wider range of fittings. The choice of fittings will also play a role in how effectively the light is distributed, and therefore has implications for energy use.

The choice of efficient lamps and luminaries means that more light can be obtained for less heat output. The amount of fittings can thus be reduced, which, in turn, reduces the level of internal gains. If daylight is relied upon for a substantial time of the occupied period, internal gains from lights will be minimal. A consequence will be a reduced cooling load, or even the avoidance of air conditioning.

Providing heat

Fuel choice is often limited to what is available, but should be considered in terms of environmental impact. Thus, in the UK, the use of gas for heating is typically three times more energy efficient (both in terms of cost and pollutants) than electricity. However, where electricity is generated by, for example, hydropower, the balance of this simple comparison will greatly change. The use of renewable sources of energy, such as solar and wind, should be assessed early on in the design to allow full architectural integration.

Once the source of energy has been determined, the nature of the heating plant will play an important role in

Figure 3.16 *A light fitting that has acoustic absorption integrated within the design, and which enables the avoidance of ceiling tiles and, thus, the exposure of thermal mass*

terms of energy efficiency. Although this guide will not provide detailed comparisons of plant, the location, zoning and distribution of heat is important. The planning of a heating plant should reflect the use and occupancy patterns of a building. The first step is to zone the heating system according to use. This will involve considering whether the plant should be centralized or decentralized. Furthermore, boilers should not be operating at low efficiencies when only a small part of the building is occupied, but should be sized to run at near full capacity. If the load reduces, then banks of boilers can be turned off so that even when loads are small, heat can be provided efficiently.

The choice of heat emitters is important, both in terms of their efficiency (minimal energy for maximum comfort) and in terms of their integration with the building fabric. Comfort is determined by a combination of radiant and convective conditions. Any strong asymmetry in these conditions will reduce comfort. Thus, if heat emitters rely solely on transferring heat by convection, an improvement of the radiant environment should be considered (e.g. use of an emitter with a radiant heat output component, and use of thermal mass or solar energy). Most heat emitters combine heat output via convection and radiation. For example, a 'radiator' will radiate 40 per cent of heat, but will cause convection currents across it which transfer 60 per cent of the heat to the room air.

The distribution of thermal energy is again an issue that will affect energy efficiency and design integration. It may be that a totally decentralized plant is appropriate, where heat is generated and provided at the location where needed. However, if systems are centralized, efficient control and maintenance become more effective. Heat reclamation may also require a more centralized plant for maximum efficiency so that excess heat from one area can be reclaimed for an area with high loads.

In the same way that spaces are planned and linked by circulation routes, so a heating system and its emitters will need to be planned. However, the planning of the system, and the location of plant, distribution runs and emitters, need to be integrated and sympathetic to the overall architectural and environmental aims of the design. There are some simple rules. For example, heat emitters should be positioned under windows to minimize discomfort from downdrafts and to counterbalance radiative losses through glazing.

Services

The avoidance of air conditioning can make very significant energy savings; thus, there is a need to address the question of whether air conditioning is required. Can comfort be achieved by passive means? Often, air conditioning is seen to be necessary for reasons such as guaranteeing temperature conditions of between 19° Celsius and 21° Celsius. It is known that comfort conditions range from about 19° Celsius to 27° Celsius, so that if such criteria are used, cooling may be unnecessary for all or large parts of the building. The use of shading devices, thermal mass, shallow plans for daylight and natural ventilation, and planning for noise are all techniques that reduce and potentially eliminate the need for air conditioning.

It may be possible to avoid air conditioning by largely passive means, and by only using mechanical ventilation (i.e. no cooling) to provide fresh air for heat dissipation and evaporative cooling.

Although there may be spaces in the design where internal loads, or other considerations, demand the need for air conditioning, it is unnecessary to air condition the whole building. Certain areas may be ideal for natural ventilation; others will only require some mechanical ventilation. Thus, the zoning of such areas and the notion of mixed-mode systems design can save significant amounts of energy use.

The integration of services, whether hidden or expressed, is important for successful design. The choice of air conditioning system will affect the complexity of integration. For example, an all-air system requires larger volumes for the integration of ductwork compared with a refrigerant system with its chilled water pipes and local air handling. Similarly, a centralized plant will require larger and longer pipe and duct runs, which will need to be carefully planned in conjunction with the structure and fabric of the building.

Synopsis

In this chapter it has been demonstrated that environmental issues can inform the full range of building design decisions, from site planning to detailed design. At each stage of the design process such concerns can play a role in developing an appropriate environmental strategy. Importantly, the interrelationships between the stages and the development of a consequential strategy lie at the root of successful environmental architecture, and are discussed in more detail in Chapter 14.

Note

1 Embodied energy is not easily quantifiable, but can be loosely defined as the amount of energy that has been used to extract and transport raw materials, and to manufacture and transport building components.

References

Baker, N. and Steemers, K. (2000) *Energy and Environment in Architecture*, E. & F. N. Spon, London

Bordass, W. T., Bromley, A. K. R. and Leaman, A. J. (1995) *Comfort, Control and Energy Efficiency in Offices*, BRE Information Paper, IP3/95, February

Goulding, J. R., Lewis, J. O. and Steemers, T. C. (1992) *Energy in Architecture: The European Passive Solar Handbook*, Batsford, London

Landsberg, H. E. (1981) *The Urban Climate*, Academic Press, London

Rapoport, A. (1969) *House Form and Culture*, Prentice Hall, New York

Sherlock, H. (1991) *Cities Are Good for Us*, Paladin, London

WCED (World Commission on Environment and Development) (1987) *Our Common Future*, Oxford University Press, Oxford

Recommended reading

1 Baker, N. and Steemers, K. (2000) *Energy and Environment in Architecture*, Spon Press, London
This fundamental text provides an overview of the key environmental strategies and issues that impinge upon the design process. It also includes a simplified energy assessment tool to link the key design parameters – such as glazing ration and building form – to the energy performance.

2 Goulding, J. R., Lewis, J. O. and Steemers, T. C. (1992) *Energy in Architecture: The European Passive Solar Handbook*, Batsford, London
This text comprises essentially informative design guidelines, based on detailed explanations of energy-related parameters. It thus provides a breadth of fundamental building science with a wealth of visual material that is readily accessible to students and practitioners alikc.

Activities

Activity 1

Outline the potential energy advantages of an atrium: Why and in which climate type is it most appropriate?

Activity 2

Describe how and what aspects of the urban context will impact upon the energy performance of a window?

Activity 3

How does orientation influence the energy performance of a building?

Answers

Activity 1

Energy advantages are related to thermal-buffer effect reducing heat loss; improved reflected daylight reducing reliance on artificial light; and passive ventilation strategies (stack ventilation in summer and ventilation preheating in winter).

Atria are most appropriate to cold climates where the thermal advantages can be exploited fully and will show the greatest benefits.

Activity 2

Obstructions will reduce the availability of solar gains (useful or unwanted) and daylight (increasing reliance on electric lighting).

The urban heat island effect, urban noise and pollution will reduce or negate the potential for direct natural ventilation and thus increase the reliance on mechanical ventilation and cooling.

Activity 3

Orientation will influence solar availability, which could reduce winter heating or summer cooling loads.

Prevailing wind directions affect the potential for passive cooling strategies.

4

Sustainable Design, Construction and Operation

Evangelos Evangelinos and Elias Zacharopoulos

Scope of the chapter

This chapter briefly discusses the building construction process, isolating its effects on the natural environment. Sustainable construction techniques and materials are suggested, as well as the ways in which they can be evaluated according to their environmental qualities.

Learning objectives

This chapter will enable readers to better appreciate how to integrate construction techniques and materials within designs, according to their environmental qualities. Furthermore, readers will be able to make basic design decisions, taking into consideration environmental criteria.

Key words

Key words include:

- environmental consumption;
- environmental deterioration;
- the global environment;
- the local environment;
- the indoor environment.

Introduction

Contemporary building activities, as most other human activities, affect the natural environment. This chapter looks primarily at the environmental effect of the building process. It discusses the significance of understanding how the building process affects the natural environment and suggests environmentally friendly techniques.

Sustainability and building

Buildings are major consumers of energy and resources for their construction, maintenance and operation. The resources are natural or manufactured building materials that have also consumed energy during processing and transportation. The principle construction techniques use materials that require variable amounts of processing. For example, stone may be used for building foundations, walls and paving, or may be crushed to produce gravel or sand, to be added in concrete mix. Stone is also used as the basic material for manufacturing cement.

Energy is consumed at all stages of the construction process, from the extraction of materials from the natural environment, to processing and transportation to the building site, as well as during the construction phase itself. Large amounts of energy are consumed during the lifetime of the building. Energy consumption ends with the demolition and disposal of building materials back to nature.

Buildings, therefore, consume materials and energy during three distinct periods of their life:

- The first is the *manufacturing-construction period*, during which materials are produced from the natural environment, processed or manufactured (energy used for this process is termed 'embodied energy'), and transported to the building site (using 'grey energy'). This period ends with the construction stage of the building (energy used for this process is termed 'induced energy').

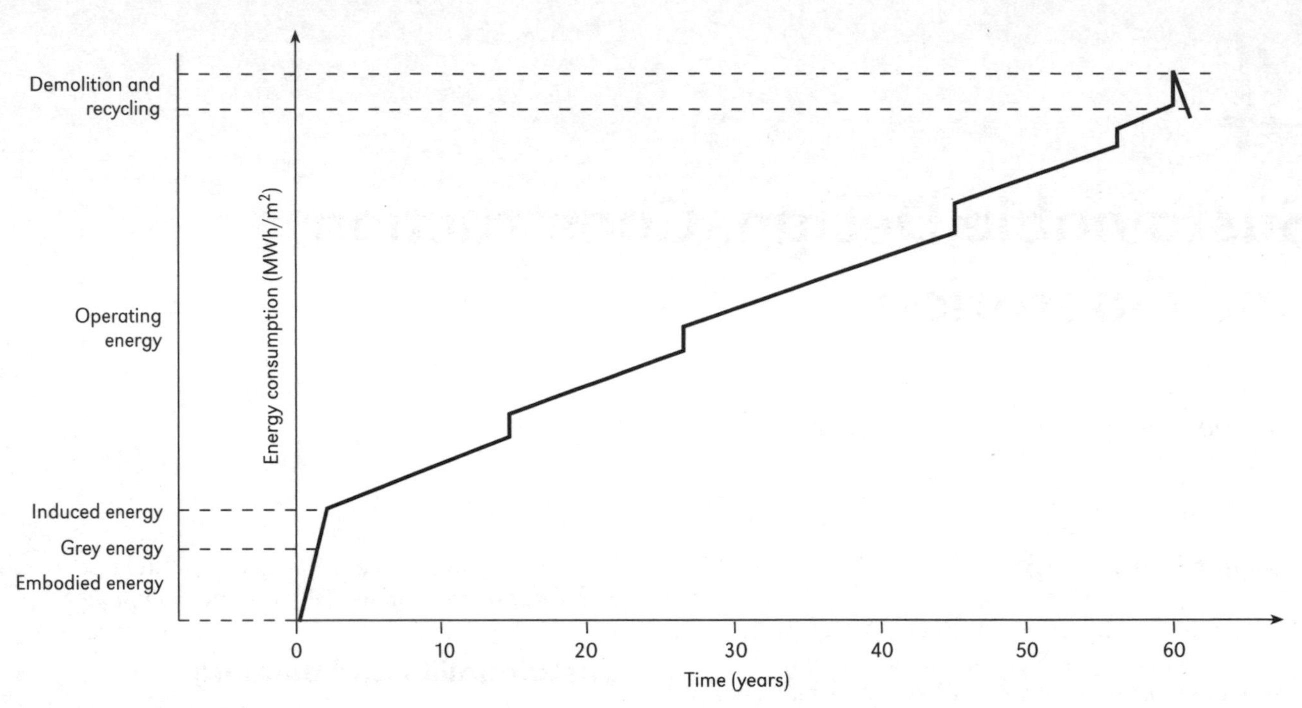

Figure 4.1 *Energy consumed in the life of a building*

- The second period is the useful life of the building, during which it uses energy ('operating energy') for its operation. During this period, energy and resources are also used for the maintenance of the building. It should be noted that this is the most significant period of a building's life with regard to its consumption of energy, and it should equally be a period of improving energy consumption.
- The third period is the demolition and recycling, during which a building has completed its useful life, and energy will be used for its demolition and recycling or disposal of its materials.

Environmental consequences of buildings

The brief description of a building's life cycle demonstrates that the natural environment is affected in two ways.

The first is the fact that all buildings need and, indeed, use natural resources in the form of building materials and energy. This, as an environmental consequence, may be described as the effect of *environmental consumption* and should be the basic concern of sustainable design. For the construction process, environmental consumption comprises building materials on the one hand and non-renewable energy resources on the other.

The second environmental consequence relates to the use of non-renewable energy in buildings and may be described as *environmental deterioration*. All use of non-renewable (or conventional) energy will affect the environment as a result of pollution. Greenhouse gases, (principally carbon dioxide, or CO_2) are produced as a result of burning hydrocarbons. Environmental degradation also occurs as a result of the manufacturing process, which in this case relates to the production of building materials and the disposal of demolition products.

The global, local and indoor environment

In order to control the environmental deterioration that results from the building process, we have to consider the environmental consequences in more detail.

Building materials are defined by their environmental behaviour. A material used in a building may directly affect the health of its inhabitants. The same material during its manufacture may have contributed in a number of ways towards the degradation of the environment in the area where it was manufactured. Finally, this material may be responsible for climatic changes due to the large amounts of greenhouse gases emitted to the atmosphere during its manufacture.

In order to control all environmental consequences that result from the building cycle, we have to measure environmental impacts according to three different scales.

During the life cycle of buildings, solid waste and air pollution are produced. The production of pollution such as CO_2 augments the atmospheric content of CO_2, contributing to the global greenhouse effect. The same happens with the use of chlorofluorocarbons (CFCs) and other gases that escape national borders with the air movement around the globe, affect the global climate by depleting the ozone layer and, consequently, cause the *global* environment to deteriorate.

The use of materials and energy contributes to local environmental deterioration, with solid waste and air pollution, which affects the local atmosphere of an area, as well as the natural environment. The second scale of control is the *local* environment.

The construction of modern buildings often includes new techniques and materials that have not been adequately tested. The extended use of buildings, combined with inadequate ventilation, brings the inhabitants into contact with an atmosphere of dubious quality, which may contain, besides pathogenic micro-organisms, carcinogenic and toxic substances in high concentrations. The exposure of inhabitants to such an internal atmosphere may affect their health. It is imperative, therefore, to control the *indoor* environment of a building.

As a result, the control of all environmental consequences of the building cycle should be achieved according to three environmental scales:

1 The global environment, for environmental consequences on a global scale.
2 The local environment, for environmental consequences on a local scale.
3 The indoor environment, for environmental consequences on an indoor scale.

Sustainable construction techniques and materials

The aim of this section is to define what sustainable construction techniques and materials are and to set forth criteria for their environmental evaluation.

A construction technique is the entire procedure of using one or several building materials. In this sense, stonewall masonry is a technique that uses stone as the principle material during construction; but it also uses many kinds of mortars as binding agents and for different types of joint finishing. A construction technique, therefore, consists of the materials used, their joining together and finishing, and the environmental consequences of extracting them from the natural environment, processing them and transporting them to the site. The natural resources, the method used to extract them from the environment, their processing and the way in which they were used during construction define the environmental consequences of the technique. Obviously, to evaluate this, we should assess materials and processes that constitute a technique, and try to minimize detrimental environmental impacts. A construction technique, in order to be sustainable, should minimize *environmental consumption* and *deterioration*, as discussed earlier. In addition, its environmental deterioration should be tested according to its *global, local and indoor impacts* upon the environment. The method of evaluating the techniques and materials available to construction technicians and engineers at a given place is quite difficult to perform. Several references deal with materials and construction methods for specific countries within the European Union (EU). In particular, the *Handbook of Sustainable Building: An Environmental Preference Method for Selection of Materials for Use in Construction and Refurbishment* (Anink et al, 1996) covers techniques and materials (encountered primarily in The Netherlands).

In order to overcome the lack of ready-made evaluation methods, one can use simple rules for selecting environmentally friendly or sustainable building materials. Materials that conform to the majority of the following rules are preferable.

Use local materials

It is possible to minimize transport energy costs by selecting local building materials. As most building materials are heavy to transport and handle, the energy savings are considerable (conversely, the associated pollution generation of transporting materials long distances is significant).

Use materials in abundance

Although this rule sounds like common sense, it is important to determine the extent of what is considered to be abundant. The use of non-renewable materials has to be related to the degradation of the extraction site. Problems usually arise due to large-scale extraction to satisfy global demand, such as in the case of exports.

Use naturally renewable materials

The use of materials according to the pace in which they are naturally renewed is an important rule for supporting sustainability.

Use materials with low embodied energy

Energy is consumed during all stages of building material production. The amount of energy spent is embodied within the material. Consequently, reduced overall energy consumption can be achieved by selecting materials with low embodied energy.

Use materials that are proven not to create health problems

Sick building syndrome is, to some extent, caused by materials that emit odours, gases, chemicals or fibres. Considering that reduced ventilation rates are needed to save energy, it is imperative to select building materials that do not degrade the indoor environment.

Reuse building materials

Reusing building materials provides a multitude of benefits. Degradation of the material extraction site is reduced, less landfill volume is occupied, and energy for the production of new materials is saved. The designer can facilitate the future reuse of a building's materials by considering this target during the design stage of the building.

The most important environmental factor in the above rules is energy:

- By using local materials, transportation energy costs are reduced: when building materials are heavy, the transportation factor becomes very important.
- The use of natural materials contains, as a criterion, the energy factor. In general, natural materials require less energy for processing than do man-made materials. Renewable natural materials are the most sustainable because they are renewed by natural processes. In doing so, they can be employed in quantities proportional to their natural yield.
- The use of materials with small amounts of embodied energy reduces the problem of atmospheric pollution produced during their manufacture, and cuts down on the burning of hydrocarbons or electrical energy from non-renewable sources.
- The recycling of building materials after the demolition of a building closes the cycle of their use and diminishes energy use in the building process.

Besides the energy factor that was obvious in the above rules, concern for health and safety is also vital. Certain materials are either suspected or proven to create health problems for those individuals who are exposed to an internal environment that is affected by them. Therefore, the highest priority must be placed on the use of materials and techniques that cannot adversely affect the health of a building's inhabitants.

Following on from this discussion, it is clear that for the environmental evaluation and selection of construction techniques and materials, several other elements that have already been mentioned must be considered.

Renewable materials

Natural renewable materials are beyond doubt the ones with the best environmental 'behaviour'. They form a family that, until recently, constituted the main construction materials. They exist in nature and are used in their natural state, or very near to it. Wood and wooden products, such as bark, branches and leaves, are the principle renewable products that are extracted from forests. The wood industry produces many products that entail various degrees of processing and exhibit diverse characteristics. Frequently, the objective is to use most of the available forest resource with as little waste as possible. This attitude can lead to the exploitation of the available forest in ways that obstruct it from regenerating naturally. The rate of growth of a forest is defined according to local conditions, such as ground composition, local climate and type of forest cultivation. For the forest to remain sustainable, the rate of its exploitation should not be greater than that of its growth.

Recycling materials

The life cycle of a building material starts from its extraction from nature and ends with its return, at the end of the building's life, to nature. With the reuse of materials in more than one building, this cycle acquires longer time scales and the relative 'environmental consumption' diminishes. The history of architecture cites many cases of buildings that are constructed with the materials of older structures. In vernacular architecture, the recycling of building materials is a common practice. Stone walls were built with old cornerstones and, indeed, with most available stones from nearby ruins. That was made easy because mortar was very weak and easily removed, facilitating recycling.

The use of a building material after recycling may not be the same as its original function (i.e. a brick as a brick in the construction of a wall), but according to the imagination of the user may be recycled for a different use altogether (i.e. brick as paving material). Such is the case with a retaining wall in Pelion, Greece, which is made out of fragments of a demolished concrete slab (see Figure 4.2).

Figure 4.2 *Retaining wall made out of fragments of concrete slab*

Modern building construction uses strong binding mortars, usually cement based, that remain on the construction element after demolition, making their reuse difficult, if not impossible. The same applies to binding practices in the majority of construction techniques that render recycling practically impossible. We could give many examples of this practice, from wall construction with bricks or cement blocks, to paving using ceramic tiles that are universally glued together.

To facilitate the recycling of building materials, we have to rethink construction techniques with respect to the useful life of a material and the possibility of reuse. Timber is a construction material that may be used in many applications in a building: as a structural element in the form of a post, beam or truss; as a material for making frames and panels for windows and doors; or as material for flooring. The life of wooden elements is generally limited, depending upon many parameters – but mainly that of maintenance. Recycling wood is possible, after evaluating its strength with simple methods of inspection and testing. Metallic materials (namely, steel beams, tubes or sheets) – while they theoretically have a longer expected life – usually have reduced structural strength due to wear and tear. As they cannot be easily evaluated, it is not advisable to use them structurally again. Natural stone building materials have a very long useful life, while cement or ceramic ones have considerable shorter lives. The recycling possibility depends upon the way in which they are joined together. Stone or ceramic tile material can be recycled if they can be 'unstuck' from the original site. The construction technique in this case needs drastic change in relation to the binding process.

During the recycling process, we should try to reuse as many original pieces of a building as possible. Reusing an entire wooden window is preferable to using its wood in chipboard production. The same applies to reusing entire metallic elements instead of recycling the metal.

Finally, in order to facilitate recycling, the design of the building's construction should be rethought so that buildings are assembled rather than constructed; and instead of demolishing, we should be dismantling them.

Manufacturing materials

Energy is used in the form of hydrocarbons and electricity during the manufacture of building materials. Depending upon the method of production, electrical energy may be renewable or conventional. Renewable energy is produced from renewable sources such as the sun, the wind, hydro plants, geothermy and biomass. Conventional or non-renewable energy is produced primarily from burning hydrocarbons and from atomic reactors. The difference between renewable and non-renewable energy is pointed out here to highlight the environmental consequences of using electrical non-renewable or conventional energy.

The energy intensity of every building material is the amount of energy that has been used in the entire cycle of the building process and that is 'embodied' within it. The evaluation of embodied energy does not end with its calculation (in kilowatt hours, or kWh), but may proceed to its separation into renewable or non-renewable energy, and the estimation of CO_2 or other greenhouse gases released into the atmosphere during its manufacture. Apart from the gases related to energy use, each material may be responsible for the production of polluting substances and gases during its manufacture. Evaluating the environmental consequences according to the three mentioned scales (global, local and indoor) will evaluate the material's overall environmental performance.

The method of estimating the embodied energy of building materials is based on the logistics of energy cost. For any given industrial method, the sum of the energy cost of all the inputs (materials and energy) equals the cost of the output (Stein and Serber, 1979). In other words, to calculate the embodied energy of a material we use the statistical records of the manufacturer, where materials and energy used for manufacturing are recorded. The total energy expended in a time period is considered to be the amount used for the production of however much material is yielded in that time period.

Table 4.1 *Embodied energy of building materials in kilowatt hours per kilogram (kWh/kg)*

	Author				
Material	**Szokolay, 1980**	**Wright, 1974**	**Chapman, 1973**	**Various researchers**	**Stein, 1977**
Sand, gravel	0.01	–	–	–	–
Stone	–	0.85	–	–	–
Lime	1.50	–	1.30	–	–
Cement	2.20	1.58	2.30	–	–
Concrete	0.20	–	–	300kg * 0.305 200kg * 0.199	0.26
Wood	0.10	–	–	0.40	–
Plywood	–	12,90 kWh/m2	–	–	–
Brickwork	1.20	1.74/brick	–	–	–
Steel	10.00	6.60	13.20	3.78	12.07
Copper	16.00	19.08	20.00	–	–
Aluminium	56.00	24.40	85.00	20.16	59.40
Zinc	15.00	10.50	–	16.20	–
Lead	14.00	7.14	–	–	–
Gypsum	–	–	–	0.30 McKillop	–
Glass	6.00	–	7.20	–	–
Plastics	10.00	–	–	–	–
PVC	–	–	–	19.27 (Smith)	–
Polyethylene	–	–	–	12.19 (Smith)	–
Glass wool	3.90	–	–	–	–

For the environmental evaluation and the selection of building materials, embodied energy is an important criterion. Table 4.1 gives values in kilowatt hours per kilogram (kWh/kg) of the basic building materials according to various authors. The estimation of CO_2 according to the kind of energy used is given in Table 4.2.

Healthy materials

In our era, people spend a considerable part of their time indoors, arguably longer than their predecessors. Emphasis on the creation of a mechanically controllable, closed thermal environment is widespread, and frequently modern materials and appliances are used for designing indoor surroundings.

Table 4.2 *Kilograms carbon dioxide per kilowatt hours (kg CO_2/kWh) of embodied energy*

	Bibliographic source	
Type of energy	**kg CO_2/kWh[1]**	**(2) kg CO_2/kWh[2]**
All types of energy	0.24	–
Electrical energy	0.22	0.832
Natural gas	0.19	0.198
Coal	0.31	0.331
Oil	0.28	0.302

Note: 1 Baker, N. V. and Steemers, K. (1994) 'The LT 3.0 Method', The European Commission, Brussels
2 British Research Establishment (BRE)

Alarmed by cases of people who felt sick indoors, health professionals investigated the cause to find that some incidents can be attributed to the quality of the indoor air. The so-called sick building syndrome is credited to a multitude of factors, such as the following.

Chemical factors

The most common and therefore the most significant health hazard factors in the internal atmosphere of buildings related to building materials, furniture and cleaning practice are chemical. Chemical factors that influence the internal atmosphere of buildings can take the form of gas, steam, particles of dust or fibres. Some of the chemical substances that have been traced in the internal atmosphere of buildings are toxic, carcinogenic, are suspected of mutative action, and are irritating or allergy inducing, whereas others just smell bad. Important categories are the volatile organic compounds (VOCs) and fibres.

VOCs are usually diluting agents for various substances, such as paints, varnishes and cleaning agents, or may be included in glues, plasterboards, particleboards and foam insulation. VOCs are emitted from these materials in diminishing rates from the time of application, and their concentration in the atmosphere is inversely proportional to the available ventilation. As a result, care must be taken during the application of paints and varnishes to protect workers and to allow time for fumes to dissipate before the space is offered for use.

VOCs are also found in substantial quantities in synthetic carpets that are used for floor coverings. These emissions are from the glues that adhere carpet fibres to their bedding or that fix the carpet to the floor. Research has shown that the most persistent concentrations of VOCs are those emitted from carpets, and a period of 61 to 98 months is needed for indoor air quality to fall within the accepted limits.

An equally important hazard is presented by the emissions of formaldehyde, which is present in glues used for the production of wood products, such as particleboard, plywood and block board, as well as thermal insulating boards that are produced from resins of urea formaldehyde. Formaldehyde is a proven carcinogen for workers in the chemical and timber industries.

Another category of materials that may pose risks for human health are those capable of releasing fibres and dust into the internal atmosphere of a building. An example points to fibrous insulating materials without appropriate surface sealing, which can release dust and fibres into the air through simple friction or usual wear and tear. Extensive research has been carried out on the damage to lungs caused by fibres, and the danger factors (Brownie, 1992, p45).

Radiation factors

Natural factors with long-term impacts on the health of inhabitants exist in buildings, such as the non-ionizing electromagnetic radiation from electric and communication appliances and ionizing radiation from radioactive materials or gases – most commonly, radon. Radon is a gas that is generated by the decomposition of uranium, present in the ground in various concentrations. It is also present in building materials and is related to their origin and composition. The radon traced in the atmosphere of a building comes primarily from the ground due to infiltration through cracks in the walls or the floor of the basement, or is exhaled from building materials used in its construction. Radon affects the respiratory system and, consequently, increases the risk of lung cancer.

Bacterial contamination

The concept of a sealed building, which aimed to minimize heat losses, led to a mechanically ventilated internal environment that had to be continuously controlled for temperature and humidity. Unfortunately, controlling the quality of the air is not an easy task, as was proven by the most serious problem associated with this practice: the growth of pathogenic micro-organisms in the air-conditioning system and their subsequent introduction to the indoor environment (Legionnaire's disease).

In order to protect inhabitants from dangerous materials, directives have been drawn setting accepted limits for the concentration values of various substances in the internal atmosphere of buildings. The verification of those limits in real conditions is a complex task demanding specialized personnel and equipment. To overcome this difficulty, specialized organizations (such as the American Society for the Testing of Materials, the British Research Establishment and the German Institute for Quality Assurance and Certification) have come forward with quality certification programmes, along with the control of VOCs and other noxious emissions, and the placement of a quality tab on the materials that comply.

The International Commission on Radiological Protection and the European Council has set limiting values for radon emissions in buildings, as well as values for radiation emissions of building materials.

Besides the various efforts for certification, manufacturers voluntarily try to replace potentially hazardous materials with healthy ones. An interesting case is that of many paint industries that have substituted VOCs with water as a diluting agent, creating products with high-quality characteristics and competitive prices.

Recycling buildings

This section discusses the environmental benefits of recycling entire buildings. The building shell requires the largest amount of building materials. By using an existing building, or at least its shell, one can economize on natural resources in the form of materials and energy. Obviously, in order to reuse an old building, a lot of work has to be done to bring it up to date. The economic burden of its budget sometimes reaches or, in some cases, surpasses new building costs. This is because repair techniques are more costly, mainly because they are labour intensive and need higher skills.

There are two main categories of recycling buildings. The first is that of retrofitting old buildings without altering the use for which they were originally designed. In the process of retrofitting, we are extending their life and bringing them up to date. The second is that of bringing old buildings up to date while altering the function for which they were originally designed. In this category a major part of the success of the endeavour lies in the choice of function that will be housed in the shell of the old building.

Retrofitting buildings

The aim in retrofitting buildings is to bring them up to date in all respects, saving them from demolition. The environmental profit from such an undertaking is substantial, not only because of the natural resources saved, but because of avoiding the trauma of demolition. Demolition is a process that expends large amounts of energy in tearing down a building, at the same time creating great environmental disturbance. Transporting the products of a demolition is an additional expenditure of energy and causes disturbance in the vicinity, while disposing of disused products within the natural environment is responsible for substantial environmental degradation.

In the process of retrofitting, great care should be taken in trying to upgrade buildings in matters of structural adequacy, safety and energy saving. In matters of energy saving, a complete evaluation of the energy systems and performance of the building in terms of heat losses and solar gains, heat gains from interior sources and lighting should be made in order to decide on the upgrading of its building fabric and electromechanical installations. A successful retrofitting should result in a building with a smaller consumption of energy; as in the upgrading process, all energy-saving techniques should be used.

New functions in old building shells

The previous section established the environmental benefits of reusing or recycling buildings. The question that remains is what happens to a building that cannot keep its original function, either because it is obsolete or because it is no longer needed. In such cases, the old buildings can house new or different functions provided that they can adequately fit into the old shell. We have many examples of such cases, not only of isolated buildings but also of whole areas that have changed function, creating new and very interesting focal points that have proved successful.

Sustainable construction processes

The construction process is the first period in the life of a building during which materials are produced from the natural environment, processed or manufactured, and transported to the building site where the actual construction of the building takes place. In order to ensure that this process is sustainable, we should try to minimize all non-renewable energy used, in all of its stages. Additionally, we should try to use as much renewable natural material as possible, trying to rely less upon manufactured or processed materials. The general idea is to build with as little non-renewable energy as possible, whether this energy was expended in manufacturing or building methods. A sustainable construction process should, in all its stages, be sustainable. Such an undertaking is not an easy task. It should be set as a target for a gradual and progressive attainment.

From the above description it is obvious that we have to consider sustainability according to:

- the materials used;
- construction energy.

Earlier, we covered the topic of sustainable materials, in general. This section provides some information on the conditions of use of natural materials, either renewable or non-renewable. Regarding construction energy, it is obvious that sustainability is labour intensive. We do not propose here to return to the sole use of manpower, but suggest the use of environmentally friendly machinery for the tasks required. In general, labour-intensive construction techniques are more sustainable than machine-intensive ones, and it is true that the most sustainable of all energies is the energy of the human mind.

Using renewable natural materials

The use of natural materials is paramount to sustainable design. This section elaborates upon the conditions of processing that natural materials should meet in order to be sustainable.

By definition, natural as opposed to manufactured materials have the least embodied energy. Nevertheless, we do not disregard the fact that most natural materials require a variable amount of processing in order to reach the conditions acceptable in a modern building. This processing is the factor that weighs against their sustainability by the embodied energy and other added substances (e.g. glue in plywood). The first condition, therefore, is to minimize processing or use the material as *close to its natural state* as possible.

In order to qualify as being renewable, natural materials should be certified to indicate that the speed of their exploitation equals the speed of their cultivation. This is a condition that is very hard to certify unless certification is made by a reliable international organization. If there is no such organization, the question of being renewable lies with the reliability and knowledge of the supplier. The second condition, therefore, is to use *certified renewable materials*.

Conditions for using non-renewable natural materials

Non-renewable natural materials should be used sparingly so that they continue to exist. Materials that exist in large quantities in the natural environment may be harnessed under the condition that their extraction does not alter and distort to a large extent the landscape's image. In cases of organized mining, the natural environment should be restored and the native fauna and flora of the area should be preserved.

As in most cases of environmental abuse, the exploitation of natural resources is a question of scale. The use of local materials usually does not create environmental problems of a large magnitude. In contrast, a resource which is to be exported in large unspecified quantities usually creates environmental problems that are difficult to manage. Such is the case with cement factories that exploit huge amounts of rock, altering the natural environment of a site.

In sum, the conditions for using non-renewable natural materials are as follows.

Use as little as possible

All non-renewable natural resources are finite. This means that in order to continue to exist, they should be used sparingly. In this sense, techniques that use large amounts of resources should be discarded in favour of less wasteful ones. An example may be the use of terrazzo flooring instead of stone or marble. The amount of stone used in the terrazzo technique is a small fraction of that for marble flooring. Additionally, the quality of the terrazzo gravel may be produced from lower-quality marble than marble slabs.

Recycle

The technique of recycling has been discussed elsewhere in this chapter. Here, we isolate the conditions that will facilitate the use of non-renewable natural materials. As discussed above, natural resources are finite. This means that by reusing them, we economize on the actual finite resource. Recycling is a technique that has been practised extensively in the past. We could easily adapt our present building practices to facilitate the use of recycling in the future.

Use local materials

The use of local building materials economizes on transport energy, which for heavy objects is considerable. Moreover, using locally quarried material creates a smaller excavation site and, therefore, lessens environmental degradation, provided that local needs are small. A very important factor is linking architecture aesthetically to the local environment by using local substrates. Frank Lloyd Wright terms this aesthetic continuity as 'plasticity'. Finally, the support of the local economy through the maintenance of skills and jobs that are essential for small communities is also important.

Manage quarry sites

Managing quarry sites is a difficult task, not only because their physical extent is usually not defined from the start (with precise planning and responsibilities), but primarily because the task of restoring the natural environment begins after full exploitation of the site. In this way the picture of environmental degradation lingers for a long time during the working years of a quarry.

Introduce environmental taxes

The discussion of environmental economics is beyond the scope of this chapter. Nevertheless, according to the 'polluter pays' principle, environmental taxation is a possibility. The proposed tax should be proportional to the environmental consequence of each material. Two materials of the same economic value but with different environmental consequences will appear priced differently because they are taxed differently, according to their environmental behaviour. Such a tax should give incentives to use environmental friendly materials and should signify that nature is not a free commodity. Finally, the income from such taxation should be spent on managing and restoring the environment so that the long-term effects of human intrusions are minimized.

Synopsis

In this chapter we have discussed the issues that define sustainability in the building process. Our discussion began with the question of sustainability and building, and defined the environmental consequences of the building process on a global, local and indoor basis. Next, we briefly investigated sustainable construction techniques and materials, giving general information on the use of renewable materials, ways of recycling materials, and details on the process of manufacturing materials and the related energy content. The practice of recycling buildings, by retrofitting them, or allocating new uses for old shells was highlighted. The question of how to achieve sustainable construction processes was attempted, with the proposal that managed natural, renewable materials should be used and that stringent conditions should apply to the use of non-renewable materials. Finally, based on the issues discussed here, an attempt was made to isolate a few principles that are based on solid environmental reasoning and are easily supported. These six sustainable design axioms are as follows:

1 *Use an existing building*: using an existing building shell, with the necessary alterations and improvements of its structure and installations, can be a very costly undertaking. Nevertheless, the environmental benefit is substantial due to the saving of natural resources from the building shell and avoiding energy-intensive demolition, and consequent environmental deterioration.
2 *Optimize needs in a building's design brief*: by reducing the size of a building environmental impacts are also reduced. A smaller building will use less energy in its operation and fewer resources in its construction. A designer should try throughout the design period to reduce the size of a building by rationalizing and optimizing the original requirements.
3 *Reduce energy-intensive mechanical movement*: often the designer can choose whether a building will be high or low rise. It is possible for a designer to choose a compact design that reduces movement needs. By reducing the need for movement (especially mechanical movement), considerable energy can be saved.
4 *Use bioclimatic design*: when designing a building it is necessary to use bioclimatic design principles, which highlight the use of renewable energies, such as the sun and wind, for heating and cooling.
5 *Design for longevity*: a building should be designed and constructed with longevity in mind. Materials that last longer are not always more expensive. It is important to note that occasionally, design or aesthetic choices may contribute to the sense of a building's ephemeral nature (what in industrial design is termed 'planned obsolescence').
6 *Use environmentally friendly or sustainable construction techniques*: this chapter has highlighted the importance of selecting appropriate and environmentally friendly construction techniques. We believe that when selecting a material for use, three scales of environmental control should be considered. In addition, techniques should comply with as many other criteria put forward as possible.

References

Anink, D., Boonstra, C. and Mak, J. (1996) *Handbook of Sustainable Building: An Environmental Preference Method for Selection of Materials for Use in Construction and Refurbishment,* James & James, London

Boonstra, C. (1995) 'Choice of building materials: The environmental preference method' in Lewis, O. and Goulding, J. (eds) *European Directory of Sustainable and Energy Efficient Building*, James & James, London

Boonstra, C. (1996) 'Sustainable choice of building materials', in Lewis, O. and Goulding, J. (eds) *European Directory of Sustainable and Energy Efficient Building*, James & James, London

Brownie, K. (1992) 'Health check: Fibers in the lungs', *The Architects Journal*, 19 February, pp45–48

Chapman, P. F. (1973) *The Energy Costs of Producing Copper and Aluminium from Primary Sources*, Open University Report

Curwell, S. (1996) 'Specifying for greener buildings', *The Architects Journal,* 1 November, pp38–40

Fox, A. and Murrell, R. (1989) *Green Design: A Guide to the Environmental Impact of Building Materials,* Architecture Design and Technology Press, London

Holliman, J. (1974) *Consumers' Guide to the Protection of the Environment*, Pan/Ballantine, London

Kwisthout, H. (1996) 'Choosing the right timber', in Lewis, O. and Goulding, J. (eds) *European Directory of Sustainable and Energy Efficient Building*, James & James, London

Lopez Barnett, D. and Browning, W. (1995) *A Primer on Sustainable Building*, Rocky Mountain Institute, Colorado

Lloyd Jones, D. (1998) *Architecture and the Environment: Bioclimatic Building Design,* Laurence King Publishing, London

Marshal, H. and Ruegg, R., (1979) 'Life-cycle costing guide for energy conservation in buildings', in Watson, D. (ed) *Energy Conservation through Building Design*, McGraw-Hill, New York

Stein, R. (1977) *Architecture and Energy*, Anchor Press, New York

Stein, R. and Serber, D. (1979) 'Energy required for building construction', in Watson, D. (ed) *Energy Conservation through Building Design*, McGraw-Hill, New York

Szokolay, S. (1980) *Environmental Science Handbook*, The Construction Press, London

Vale, B. and Vale, R. (1975) *The Autonomous House*, Thames and Hudson, London

Vale, R. (1995) 'Selecting materials for construction', in Lewis, O. and Goulding, J. (eds) *European Directory of Sustainable and Energy Efficient Building*, James & James, London

Wright, D. (1974) 'Goods and services: An input–output analysis'. *Energy Policy*, December, pp307–315

Yates, A., Prior, J. and Bartlett, P. (1995) 'Environmental assessment of industrial buildings using BREEAM', in Lewis, O. and Goulding, J. (eds) *European Directory of Sustainable and Energy Efficient Building*, James & James, London

Recommended reading

1. Anink, D., Boonstra, C. and Mak, J. (1996) *Handbook of Sustainable Building: An Environmental Preference Method for Selection of Materials for Use in Construction and Refurbishment*, James & James, London
   This volume highlights the 'environmental preference method', which was developed as a tool for selecting building materials according to their environmental performance. It also covers building techniques that are used for the construction of dwellings in The Netherlands.
   This handbook is considered essential reading for anyone concerned with sustainable design and construction.
2. Fox, A. and Murrell, R. (1989) *Green Design: A Guide to the Environmental Impact of Building Materials*, Architecture Design and Technology Press, London
   This is an A to Z guide of building materials, with comments on their environmental performance. The introduction touches on issues of environmental degradation and factors that influence this. As a guide, it is considered useful for any investigation into building sustainability.

Activities

Activity 1

Classify, in order of magnitude, the possible environmental effects of constructing the frame of a small building that is made alternatively of:

- concrete;
- steel;
- timber.

Comment (in no more than 150 words) on the possible environmental consequences of each technique according to the three environmental scales of:

- the global environment;
- the local environment;
- the indoor environment.

Consider that all material was produced locally and that the aggregate of concrete is granite with a high radon content.

Activity 2

In order to appreciate the benefits of retrofitting buildings instead of demolishing them and constructing new ones, we propose a rough calculation of the amount of embodied energy and corresponding CO_2 released into the atmosphere with the new construction. In order to simplify the calculation, consider the benefits from the construction of the concrete frame alone, supposing that the area of the one-storey building is 100 square metres, the amount of concrete needed is 50 cubic metres and the energy used for its manufacture is 50 per cent coal and 50 per cent oil. Provide a short answer with no more than 50 words and your calculations.

Activity 3

Using the simple rules discussed in 'Conditions for using non-renewable natural materials' on p70, consider (in less than 50 words) the alternative improvements resulting from specifying environmentally friendlier materials for a housing project in the construction of:

- walls (clay bricks);
- flooring (PVC tiles);
- thermal insulation (polystyrene).

Answers

Activity 1

Steel is the material with the highest embodied energy, with concrete second and timber (by far) third on the list.

The possible environmental consequences of each technique according to the three environmental scales are as follows:

- *The global environment*: since steel has the highest embodied energy, it will contribute the most to CO_2 emissions, while concrete will be second and timber third.
- *The local environment*: steel (being manufactured locally) will degrade the local environment the most. Concrete also will contribute to environmental degradation due to emissions from its manufacture. Timber is by far the least polluting and most environmentally beneficial material, contributing to the 'cleaning' of the local atmosphere during its growth.
- *The indoor environment*: timber is a material that has been used for years with no ill effects on the quality of the indoor environment. Nevertheless, the treatment of wood to withstand infestations, either from insects or moulds, may be noxious for humans and animals, at least during the early stages of its application. Steel is a material that, by itself, does not pose any problems; but like wood its treatment can cause the indoor environment to deteriorate. Concrete, on the other hand, relies upon the aggregate being used, as well as the qualities of cement. Radon content should be considered seriously. If used, additional building ventilation should be specified.

Activity 2

According to Table 4.1, 50 cubic metres of concrete of 300kg cement content will have an energy content of $50 \times 0.305 \times 2200 = 33{,}550$kWh.

To calculate the CO_2 released into the atmosphere:

$$50\% \times 33{,}550 \times 0.31 = 5200.25\text{kg}$$
$$50\% \times 33{,}550 \times 0.28 = 4697.00\text{kg}$$

The total sum being:

$$5200.25 + 4697.00 = 9897.25\text{kg}$$

Activity 3

In terms of specifying environmentally friendlier materials for a housing project:

- *Walls*: instead of clay bricks, which are energy intensive, use stone, if available, or cement bricks.
- *Flooring*: instead of PVC tiles, use natural, renewable material, such as cork tiles or linoleum tiles.
- *Thermal insulation*: instead of polystyrene, use a natural renewable material, such as cellulose or cork.

5

Intelligent Controls and Advanced Building Management Systems

Sašo Medved

Scope of the chapter

The main task of architects is to design a building that provides safety, comfort, pleasure and an optimal living environment for its occupants. These demands should be met with the least possible amount of energy consumed, and should affect the environment as little as possible. This is why all of the building's technological systems must be well controlled and synchronized. In contemporary buildings this rather complicated task can be fulfilled by using microelectronic-based building management systems. The purpose of this chapter is to introduce the fundamentals of microelectronic control systems, and to explain what building management systems are and how they operate. The chapter is divided into two parts. The first part deals with the basics of control systems; the possibilities, advantages and configurations of building management systems are presented in the second part.

Learning objectives

At the end of the chapter, readers will be able to:

- understand the basic principles of microelectronic controls;
- identify the concepts of building management systems;
- compare different building management system standards.

Key words

Key words include:

- control systems;
- control algorithms;
- hardware;
- software;
- building management systems;
- building management system standards.

Introduction

Conditions in nature and in buildings are continuously changing because of variable meteorological conditions, air pollution, the occupancy behaviour and appliances used within them. Therefore, building indoor conditions are dynamic and quite unpredictable. This is why energy and material flows in buildings change constantly. In theory, sustainable buildings should regulate energy flows. In practice, we need building service systems to control energy and materials flows.

The efficient operation of building service systems can be achieved only with efficient controls and management systems. Such systems reduce energy and material flows, while improving the indoor environmental quality. Because different occupants perceive the same indoor environment very differently, intelligent controls should provide individuals with the possibility of adjusting indoor parameters according to their needs.

In modern society, information exchange is as important as energy or materials supply. Up-to-date building management systems transfer information flows in order to provide supervision and a safe indoor environment.

Only a decade ago these intelligent controls and building management systems were only economical for large buildings. However, over the last few years these systems have become more cost effective for individual residential premises due to a broader number of producers and new technology solutions, which have given rise to a growing number of operating functions.

The aim of this chapter is to describe intelligent controls and building management systems, to explain their functioning and to provide some examples.

Intelligent buildings

The term 'intelligent building' arose during the beginning of the 1980s. In the beginning, definitions of intelligent buildings were linked to innovative construction technologies and to automation of mechanical systems, with an emphasis on greater energy efficiency. At present, 'intelligent buildings' are defined as those that have been built using the latest techniques and technologies in order to optimize their service systems and improve the efficiency of their maintenance and management. Intelligent buildings provide a high degree of comfort, safety and economy to their owners, managers and users. By considering that the true cost of the edifice comprises more than its construction cost, the overall value of the building during its life cycle must be taken into account.

Intelligent buildings of the future will continuously and independently respond to the changes within them and in their surrounding environment by using information and communication technologies. Intelligent materials – such as glazing with variable optical properties, materials with temperature recollection, dynamic thermal insulation, intelligent devices with microchips that communicate with users, as well as intelligent control, supervision and communication systems – will be installed in buildings. The first steps towards intelligent buildings of the future, however, involve the digitally managed control and supervision processes of technological systems, which are discussed later in this chapter.

Fundamentals of control systems

Control algorithms

Control systems ensure that building service systems will automatically adapt to internal and external environments without the intervention of users. They function in such a way that the actual value of a physical quantity, which is measured with a sensor (for example, the room temperature), is compared with the expected or desired value (e.g. 20° Celsius in the winter). This is termed set-point temperature. If there is a difference between actual and desired physical quantities, a controller mediates the corresponding information to the controlled device in the system (for instance, to the heater switch).

A controller uses different kinds of control actions, using the feedback from the sensor, and sending information to the controlled device in order to equalize actual and desired values. There are various control actions; however, the best known are:

- two positions (on/off) that regulate action;
- proportional (P);
- proportional plus integral (PI);
- proportional plus integral plus differential (PID);
- artificial intelligence (AI).

Two-positions or *on/off* control is the most simple. The controlled device is either turned on or shut off. Oscillations of the actual value of the controlled variable are therefore periodic and large. For example, the room is heated by an electric heater to 20° Celsius. This temperature is the so-called set-point temperature. The gap between the temperature that would cause the controller to transmit switch-on information (e.g. 19° Celsius) and the temperature that causes the heater to receive switch-off information (e.g.) 21° Celsius) is called the control differential or hysteresis. The actual temperature in the room differs more than hysteresis due to the heat accumulation in the heater, which is emitted into the room after switching off the heater. Likewise, when switched on, the heater first warms up only after it starts to emit the heat into the room. The temperature interval between the lower and upper temperatures in the room is known as the operating differential. Lower thermal comfort and greater energy consumption is the result of the larger temperature oscillation in this case.

The heat flow that is emitted from the electric heater when turned on is most likely constant no matter what the actual temperature difference between the actual and desired temperature in the room. A more detailed analysis of the operation would also show that the operating

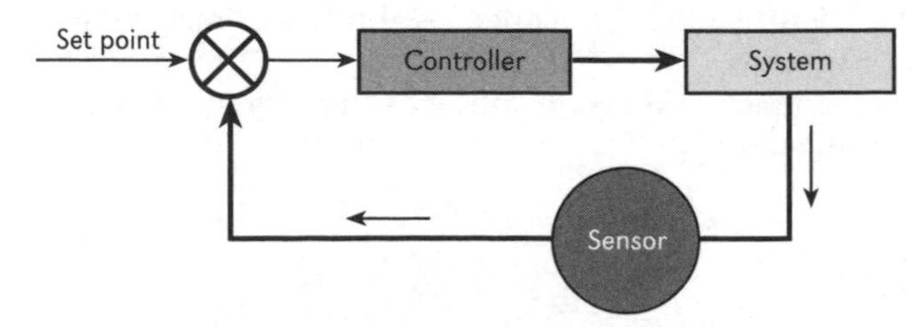

Figure 5.1 *Illustration of a loop-control system*

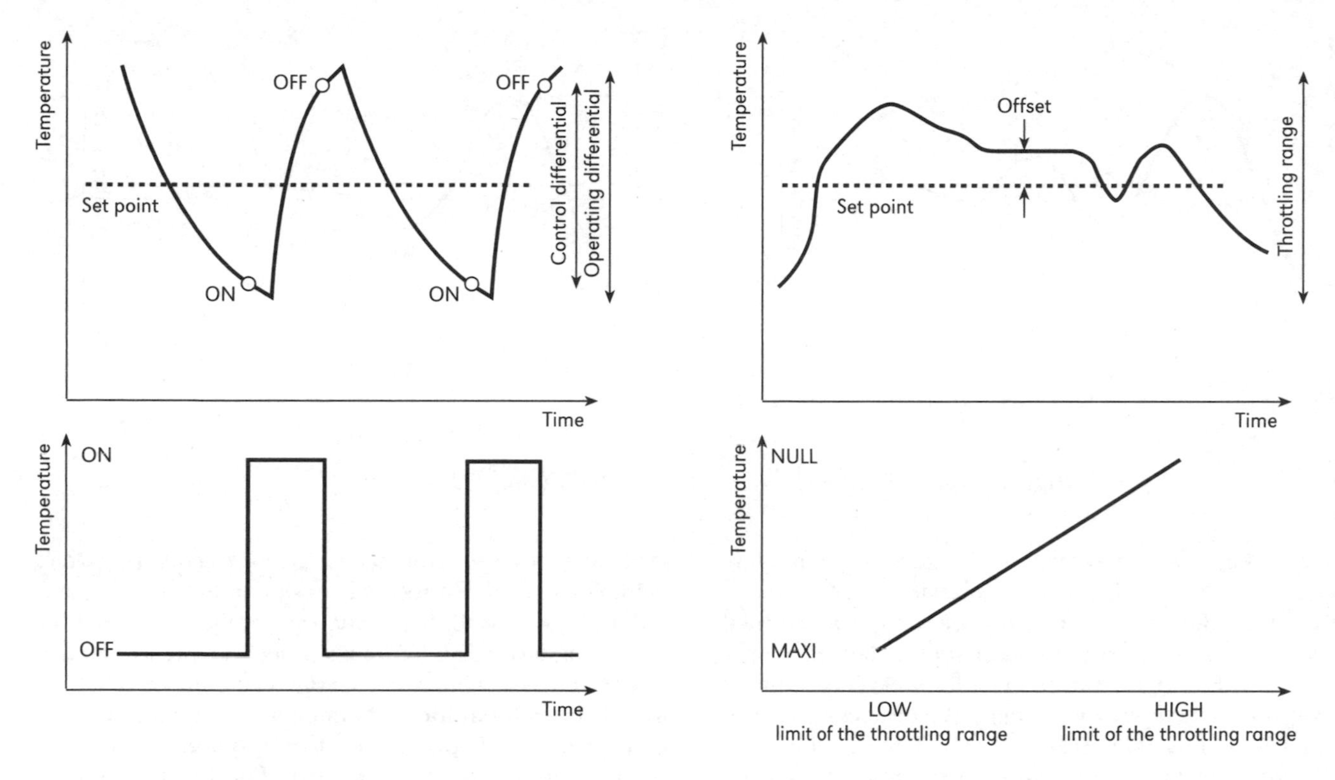

Figure 5.2 *Illustration of the controller operation that uses two-position control* (left) *and P control* (right)

differential is greater when a smaller heat flow for heating a room is required or in the case of lower system loading. This disadvantage of the two-position control action can be improved if the controller uses *proportional control action*. In such a case, the controller sends the proportional action signal to the controlled system. In this case, the signal will change the electric power of the heater proportionally (linearly) in the range between minimum and maximum value. This range is called the throttling range. The proportional action can be mathematically described with the factor of *proportional gain* K_p. The difference between the actual and desired temperature within the monitored time interval is much smaller than it is in the case when an on/off controller is used. The difference between actual and desired value is called offset.

Proportional controllers can be improved by adjusting proportional gain factor, K_p, to the established differences, between measured and desired temperature. The differences can be integrated and averaged over a pre-selected period of time. This process is called reset; such control action is known as *proportional plus integral action (PI)*. This correction can be mathematically described with the factor of *integral gain* K_i. We choose the integration time or the time between two resets when setting the controller. It remains constant during the whole operation period of the controller. Controllers of this type are usually used for regulating heating and air-conditioning systems in the buildings.

With an additional function that changes the intervals amidst individual resets, PI controllers can be upgraded to *proportional plus integral plus differential* (or *PID*) controllers. Mathematical correction of the reset intervals is described with the factor of derivative gain, K_d. With the common controllers, the factors K_p, K_i, K_d are the constant characteristics of the controller; with the *adaptive or self-adaptable controllers*, on the other hand, the constants K_p, K_i, and K_d are changing automatically and constantly in relation to the characteristics of the system.

The functioning of adaptive controllers is based upon *artificial intelligence* (*AI*). These are special algorithms that imitate people's thinking and decisions through their operation. From the sytems operations' past experiences, they learn to anticipate the course of events in the future in the same manner as people learn. For instance, we have learned from past experience that on a hot summer's day it is much cooler under a tree's shade than it is in an adjacent open space. We have learned this without actually 'measuring' the temperature in each individual case.

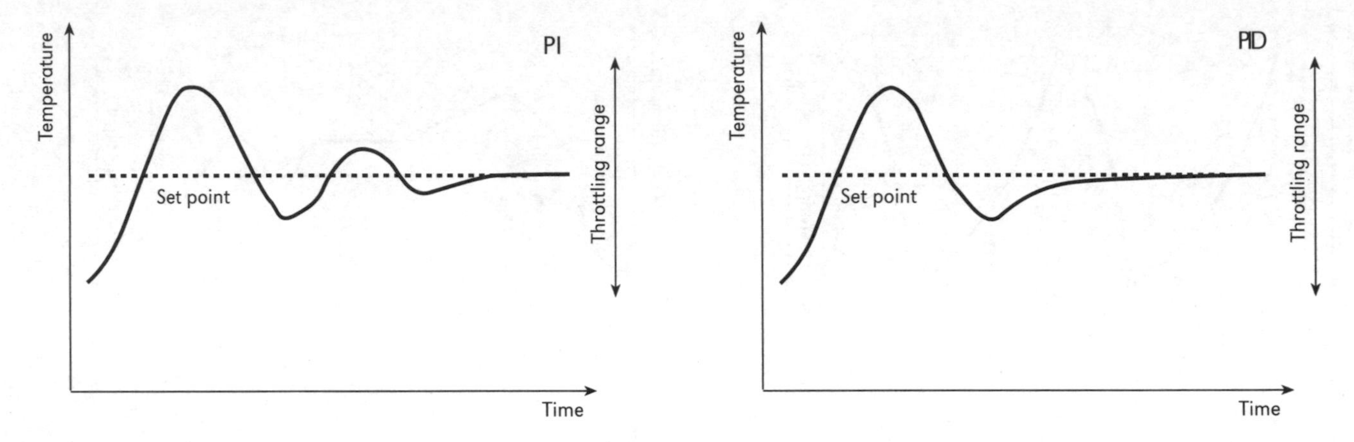

Figure 5.3 *Illustration of the operation of the PI and PID controller*

When making decisions, we analyse several parameters simultaneously, which are then classified according to their future importance. Finally, we compare the final effect of each selected combination of the relative parameter's value. The algorithm by which modern controllers imitate this human process is called the *neuron network*.

In case of multiple parameters the 'yes' or 'no' answer is not easy and fuzzy logic can be used for approximate reasoning. In the case of fuzzy logic operating controllers, linguistic instead of physical variables (for example, room temperature) are introduced as shown in Figure 5.5. The linguistic variables (e.g. cold or cool) refer to the overlapping values. These triangular values are called membership functions. Therefore, the value of each linguistic variable is between 0 and 1. The fuzzy controller calculates output information – for example, for a room heater – according to the degree of truth of each fuzzy sets value. Using the so-called centre of gravity method, the power of a room heater is adjusted continuously.

The design of control systems

Control systems capture data from the 'outer world' with sensors, (e.g. temperature sensors, photosensors, CO_2 concentration sensors and occupancy sensors) in the form of analogue signals. Table 5.1 depicts various sensors that are used in different applications as they appear in building management systems. The sensors can be chosen, for example, according to their sensitivity characteristics, integration compatibility, geometrical dimensions and price.

The signal is then transformed to a digital one using the converters. Digital signals can be analysed by hardware devised from *direct digital control* (*DDC*) technology. A DDC process is designed on digital signals that are processed by a microprocessor or a central processing unit (CPU). A microprocessor exchanges data with the outside world through the input/output (I/O) unit or/and stores them into data memory. The operation of a microprocessor is prescribed through a code, which is saved in the programme memory. The listed elements, also named microchips (CPU, I/O device, clock, data and programme memory) comprise a microcomputer. The signals are translated from microcomputer to *switches* and *driving mechanisms*, which are installed in the systems (actuators, valves, drive on heaters, switches, etc.) in order to maintain set values in the indoor environment.

Control systems can be integrated within specific intelligent devices. On the other hand, one microcomputer can be connected to several devices and act as a remote control. In this case, the devices must be connected via a central microcomputer. The data exchange between the devices and the microcomputer can be organized using different protocols. The selection of the communication protocol is based on the quantity of data that can be transferred in a unit of time.

Software in the control systems enables the functioning of individual controllers to be monitored, facilitates information flow through the network connecting the controllers and optimizes operation. In contrast to classic electric, mechanical or pneumatic controllers, the operation in *building management systems* (BMS) can be changed with the installation of a new programme algorithm, which is sent to the controller via the network. Software is also designed for the communication between the BMS and users.

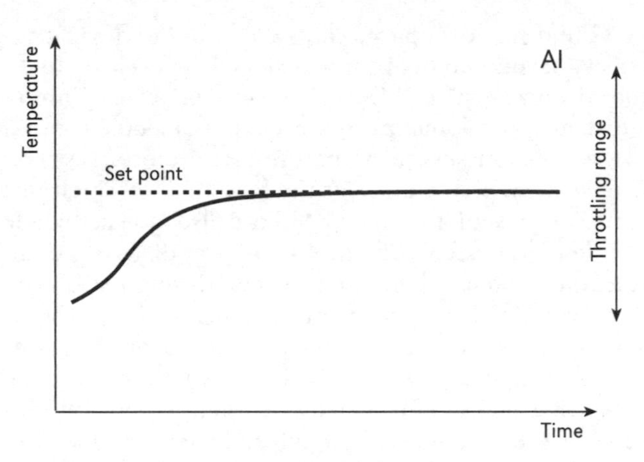

Figure 5.4 *Illustration of the operation of the AI controller*

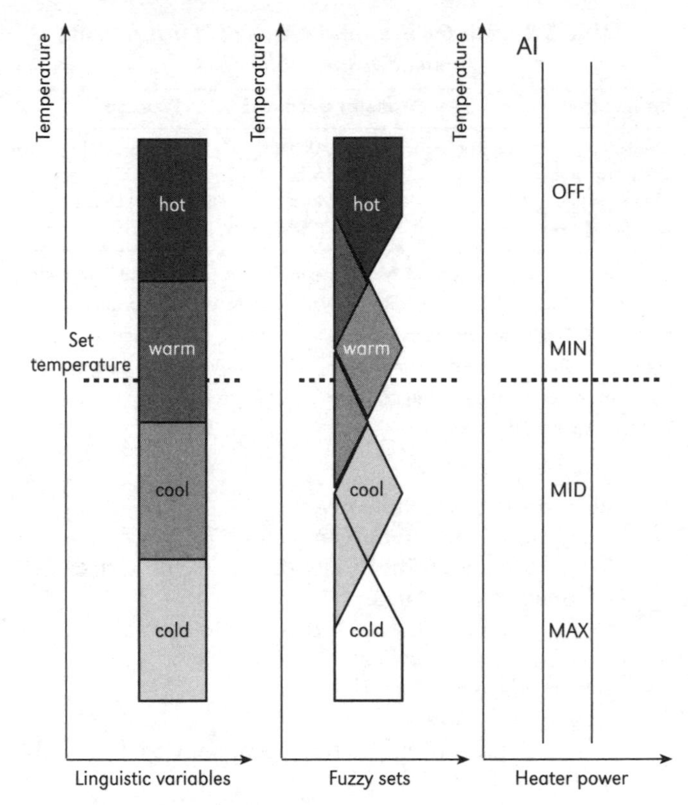

Figure 5.5 *Illustration of the linguistic variables and fuzzy sets (actual room temperature is characterized by 75 per cent warm and 25 per cent cool)*

Building management systems

A characteristic of big structures and building complexes is a very branched-out system of installations. These installations include heating, ventilating and air-conditioning systems, as well as electric grids, lighting, sanitary and transport installations, information and communication systems, safety systems and others. These systems have to be constantly controlled and managed. At first, monitoring the functioning of these systems was only limited to monitoring operating mistakes and problems; however, the supervisory systems have grown into management systems. Furthermore, the supervision of energy consumption has been installed within the operation, resulting in cost benefits. This is how modern *building management systems* (*BMS*) or *building energy management systems* (*BEMS*) came into existence. Beside surveillance, these systems take care of the lowest possible energy consumption in the building. BMS tasks can be divided into the following:

- Managing and supervising energy consumption and other resources:
  - switching on/off of devices according to time and occupancy;
  - limiting electricity demand peaks;
  - ensuring the optimal operation of the heating, ventilating and air-conditioning systems;
  - regulating shading and electric lighting.

Table 5.1 *Review of sensors used in different application in buildings*

Application	Lighting burglary	HVAC Presence detection	Gas recognition	Fire Supervision	Device
Sensors for:					
Temperature		✓		✓	✓
Humidity		✓		✓✓	
Brightness	✓				
Infrared radiation			✓		✓
Ultrasound waves		✓			
Gases				✓	✓
Microwaves				✓	
Electricity					✓

Note: HVAC, heating, ventilation and air conditioning system.

Table 5.2 *Data transfer and protocols for different applications in buildings*

Application	Transfer data rate	Protocol
Measure, control, define	1kb/s–10kB/s	EIB, EHS, LON
Voice transfer	up to 1Mb/s	ISDN, DECT
Video transfer	> 10Mb/s	Fire Wire/IEEE1394
Computer net	1Mb/s to 100Mb/s	TCP/IP

Note: Kb/s, kilobits per second; Mb/s, megabits per second; EIB, European Installation Bus; EHS European Home System; LON, Local Operating Network; ISDN, Integrated Services Digital Network; DECT, Digital Enhanced Cordless Telecommunications; IEEE 1394, Institute of Electrical and Electronics Engineers Standard 1394; TCP/IP, Transmission Control Protocol/Internet Protocol.

- Safeguarding, which involves:
  - diminishing the human factor;
  - personal identification with electronic cards;
  - image surveillance;
  - hierarchically restricted access to rooms;
  - anti-burglar alarms;
  - fire alarms;
  - gas detection;
  - simulations of the virtual occupancy of the buildings.
- Managing informatics:
  - internal phone and video connection;
  - video conferencing;
  - satellite communications;
  - electronic mail;
  - access to the internet.
- Providing automation of working places:
  - central data processing;
  - electronic documents transfer;
  - data transfer through computer-aided design among experts;
  - notifying and informing.

Advantages of building management systems

A system that involves the complete surveillance of the systems operation within a building enables constant *monitoring*. All the measured values can be monitored on line or stored. This allows permanent analyses of the system function and optimization resulting in better energy efficiency in real conditions. The meteorological conditions, the quality and realization of the systems and the habits of the users can all significantly contribute to a discrepancy between the predicted and the actual system response. Information that is gathered by a BMS can be followed and controlled from a distance, thus making visits to each building unnecessary. In smaller BMSs (e.g. in a building), occupants can switch on the devices and check on their status from a distance. Therefore, one of the advantages of a BMS is also *communication*. Through a distant control, one person can simultaneously operate the systems in several structures. Therefore, BMS can enable *manpower savings* or the formation of 'single seat' surveillance stations. Since BMS can also indicate when a problem has occurred, problems are discovered and remedied sooner than would occur through site visits. However, through permanent monitoring of the discrepancies, the possible problem can be discovered before it actually happens. The same is also true with the energetic flows in BEMS. This is why *maintenance* of building services is better, more thorough and cheaper. One of the possible applications of a BMS is also *commissioning*. In big structures, the inspections conducted by the designer/installer and the commissioner, are long and expensive. Many of the appliances and systems have to be checked out and regulated (for example, there is a need to hydraulically balance the pipes, to set the inflow grid and to set the control valves).

Designing building management systems

Hierarchy and *compatibility* have to be ensured in order to make the operation of the BMS possible. In relation to today's technological development, hierarchy is achieved through the three operation levels:

1. management level;
2. control and automation level;
3. field level.

The field level is designed for capturing and transmitting (input/output) data from the single systems within a building. This level is therefore intended for managing the systems in the room (e.g. control over heating and cooling, lighting, position of the shading devices, etc.).

Information between the sensors and BMS (inputs from the building service system) interchange in the form of signals (binary numbers), measured values of the quantities (analogical values) or impulses (sensors transmit an impulse according to the physical unit of the measured quantity – for instance, 1 impulse = kilowatt hours (kWh), 1 impulse = kilograms per second (kg/s). BMSs, however, output switching or control commands to the building services system or to the control mechanisms installed in the system. Systems on this level also have the hand-control option. Communication among the BMS elements on the field level runs either between the individual sensor and controller or through the communication network. This is termed a field-level network (FLN).

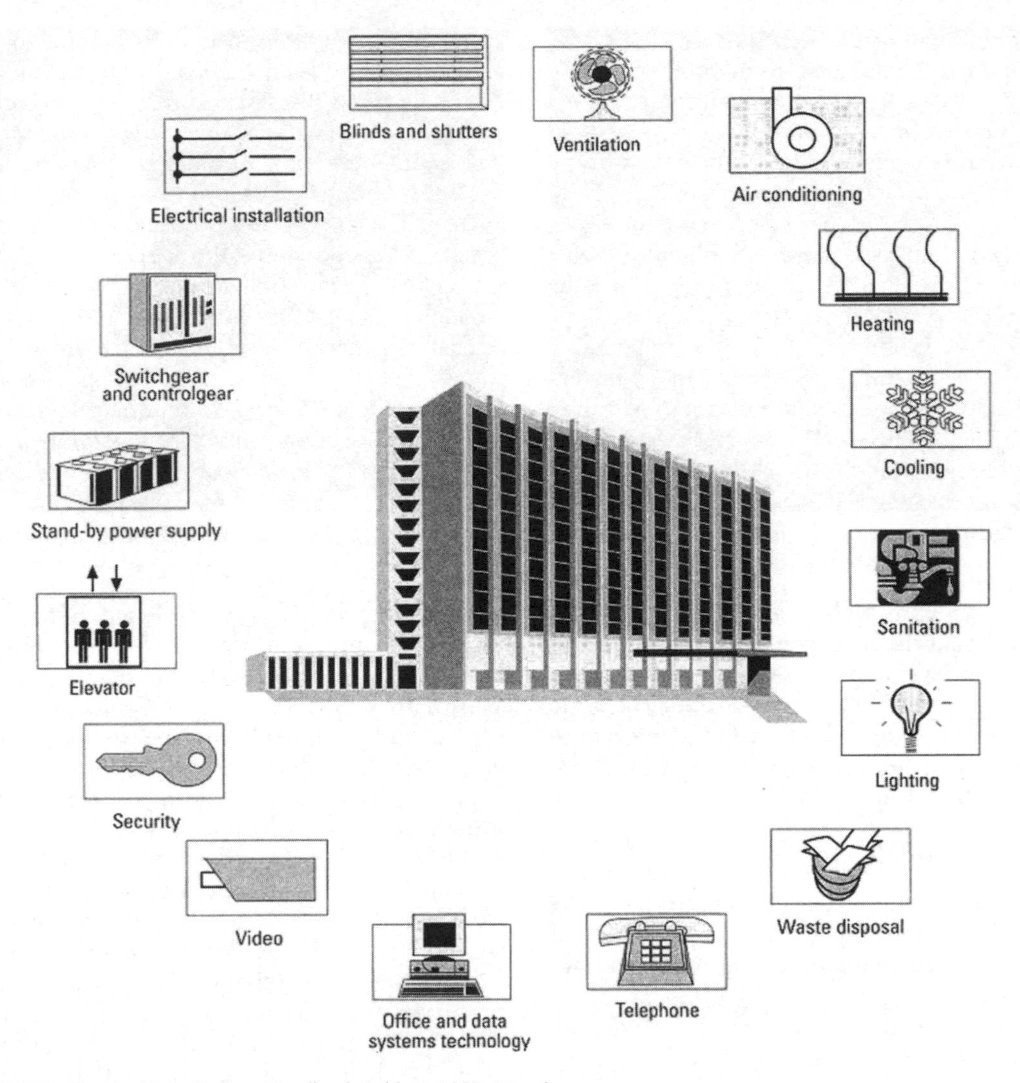

Source: Gunter G. Seip, *Electrical Installations Handbook*, Publocis MCD, Munchen

Figure 5.6 *Buildings service systems that can be monitored, controlled and optimized by BEMS*

On the control and automation level, BMS monitor, control and optimize the building services systems. *Application-specific* controllers (ASC) or *modular controllers* are used for these tasks. On this level, building management systems become aware of the problems in the building system's operation, display the measured values and ensure that they stay within the allowed margins. The continual monitoring of the appliances and their functioning ensures that timely maintenance is carried out, thus increasing the life span of the system and devices. Furthermore, they take care of the information transfer about the system in digital form between levels. Controllers are interconnected with the control and automation-level network (CLN). Due to the great volume of data that are processed and transferred between the field and management levels, CLN has to be capable of faster data transfer than is the case with the network in the field level.

Management is at the top of the hierarchy in a BMS. It comprises all the data needed for the statistical data processing from the field, as well as control and automation levels, together with the displayed and out-printed quantities' values and events. Moreover, graphic review of the building, systems condition and the values of the measured quantities in the single systems and rooms of the building are included. Simultaneous monitoring of the data at the management level allows system operation mistakes to be detected and provides an integral survey

Figure 5.7 *Luxmeter* (left), *movement sensor* (right) *and valves with the control drive on as an actuator*

of energy consumption, as well as an estimation of the costs. BMS elements at the management level are linked up in a management-level network (MLN), which makes communication with outside systems possible. The wide dissemination of information and access to the internet as a global network opens up new possibilities of surveillance and the control of distant systems.

Networks and protocols

Information transfer on the different levels of the BMS has to be managed through protocols. Interconnection of the different BMS components from different manufacturers occurs through the use of the same protocol. Therefore, the protocols have to be standardized. The following protocols have been suggested as draft standards by the European Committee for Standardization, Technical Committee 247 (CEN TC247), Working group 4 group experts:

- Building Automation and Control Network (BACnet) and Firm Neutral Datatransmission (FND) at the management level;
- BACnet (Local Operating Network, or LON), European Installation Bus Network (EIBnet), Process Field Bus (PROFIBUS) and WorldFIP at the control and automation level;
- BatiBUS, European Home System (EHS), European Installation Bus (EIB) and LONTalk at the field level.

BACnet was designed in the US and is the most widespread protocol at the management level. It is accepted as an American standard and is proposed for a European one. The FND protocol is especially common in Germany, where it was developed. EIBnet is a widened field-level protocol for EIB. Siemens developed PROFIBUS for the automation of systems in buildings. WorldFIP was developed in France, above all for the control and automation of processes. France also developed BatiBUS, which was devised for the automation of systems in residential buildings. EHS was made on the initiative of the European Commission within the ESPRIT programme, and its purpose is the automation of systems in residential buildings. It could connect various electric and electronic appliances based on plug-and-play technology. EIB was developed by Siemens and is a widespread protocol that is used by many of the manufacturers of the various electric appliances. LON was developed in the US and is a very common system for the control and survey of heating, ventilating and air-conditioning systems (HVACs). The EIB and LON systems are presented in more detail further on.

The appliances on the single BMS levels are linked with networks. Smaller networks are called local area networks (LANs). Through a LAN, a central processing unit (e.g. a personal computer) communicates with the individual 'intelligent controllers' (termed outstations), which are equipped with their own microprocessors and which communicate with sensors and control devices. The number of the in-LAN linked controllers can be quite large. The communication time between the central unit and the single controller is usually short due to the fact that most of the jobs are already executed by the outstation. However, smart sensors and actuators are also installed within the microprocessors, which allow them to communicate independently with the central unit through a connection to the LAN.

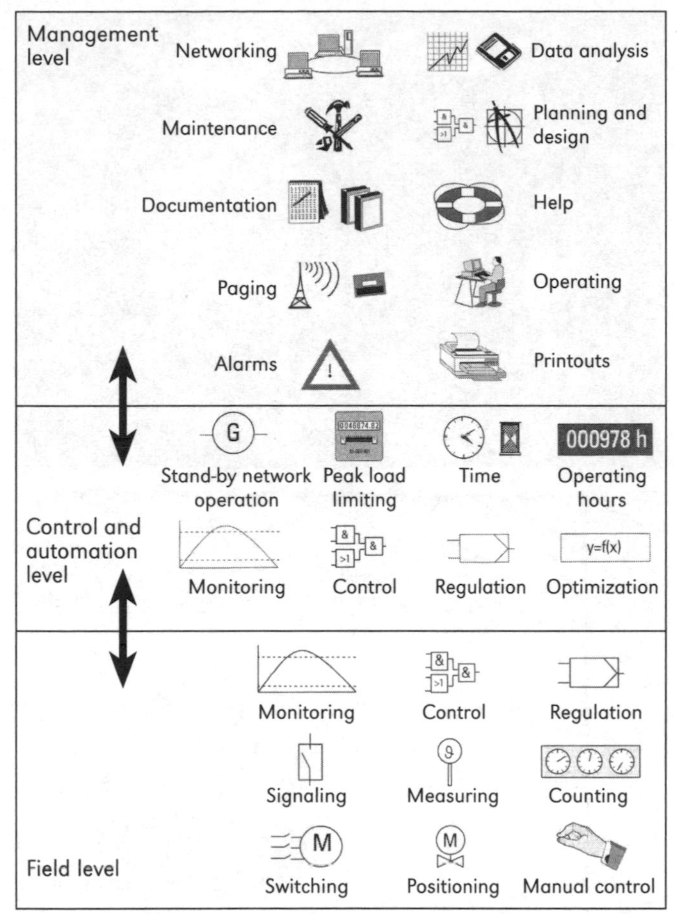

Source: Gunter G. Seip, *Electrical Installations Handbook*, Publocis MCD, Munchen

Figure 5.8 *Review of the building management system's functions at different levels*

The number of bits sent per second, which is also termed baud rate, measures the speed of the data transfer into the LAN. Through the modern DDC systems, which are connected into a LAN, the characteristic speed of the data transfer is between 300 and 1,250,000 bauds or, to put it differently, 300 bits per second (bps) and 1.25 megabits per second (Mbps).

Networks differ in topology based upon the way in which a central unit is linked to the other devices. The most commonly used topologies are as follows:

- Point-to-point topology is the simplest, and directly connects a central computer with only one outstation.
- Star topology is similar to point-to-point topology, but more outstations are connected in the same manner.
- Bus topology: outstations can communicate independently among themselves and with a central unit.

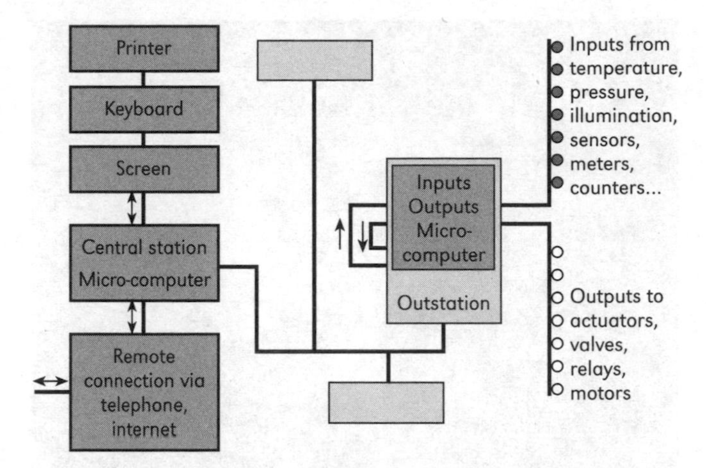

Figure 5.9 *Illustration of the LAN with outstations*

Every outstation has its own identifying mark; the spread of a network is executed in a simple way by annexing new outstations to the LAN. Communications are transmitted in both directions; therefore, the protocol must also contain a part that takes care of the correct order of precedence of the received and sent information.

- Ring topology: here, outstations are linked so that information travels very fast and in one direction only. The protocol has a part added through which an outstation recognizes whether the information is meant for it; if not, the information is sent to the next outstation.
- Tree or hierarchical topology: the LAN connects outstations which exchange information in the vertical direction of influence without a central unit.

Data transfer in a LAN is carried out through the conductors. The selection of a conductor is conditional upon the distance of data transfer and upon its capacity, which increases along the length of a conductor. In most cases, twisted-pair wiring and co-axial and fibre-optic cables are used as conductors. Twisted-pair wiring is combined with two cables in a bandage. They are interlaced in order to reduce the electromagnetic induction. They represent the least expensive choice of conductor; however, they must not be put in the vicinity of high-voltage conductors. The electromagnetic induction in a conductor is reduced by metal armour in the co-axial cable. In the fibre-optic conductors, information is transferred by a coherent light, which reflects within a flexible fibreglass. These are the most efficient conductors; however, they are also the most expensive ones. LANs in which different protocols or BMSs of different manufacturers are used can be interconnected through a protocol compiler, termed a gateway.

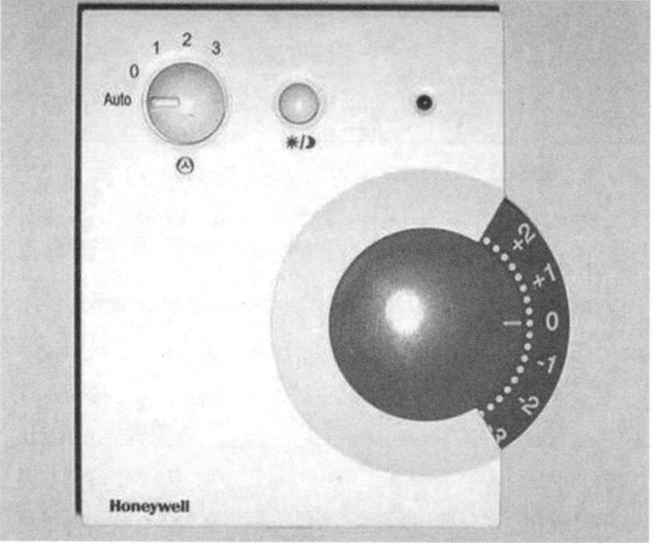

Figure 5.10 *Outstation* (left) *for fan-coil* (top right) *control with temperature sensor and window switch as input devices and valves with control drives on the hot and cool pipeline as output devices* (bottom right)

Selecting a building management system

The advantages of BMSs have already been outlined; however, there are also shortcomings. It should be emphasized that these systems are rather expensive and for this reason they are economical, as a rule, only in larger buildings, settlements or towns. If a decade ago it held true that BMSs were used mostly as controlling systems within a building (a questionnaire among 50 energy managers has shown that buildings, which are being monitored, are – in 82 per cent of cases – equipped with a BMS; however, only 1 per cent of them compare gathered data with anticipated outcomes), then today the role of a BMS is to maximize building performance. With the development of microelectronics, telecommunication and more user-friendly applications, however, BMS are also moving into smaller buildings as a first step towards realizing intelligent buildings. One obstacle in implementing BMSs is the relatively large number of manufacturers, whose systems, as a rule, are not compatible with others. The choice of a system therefore dictates the choice of all components by the same producer, which means additional expense. That is why a better way to choose the right BMS is to use and follow these methodological steps:

- Become aware of the advantages that are offered by BMSs in terms of energy-consumption reduction, improved comfort of living, building protection, fire alarming and monitoring of the building from a distance. However, not even the perfect BMS enables effective control over the operation of badly maintained or even broken systems within a building.
- Choose the tasks that the BMS in your building should carry out; carefully investigate what service systems are already installed in the structure, and

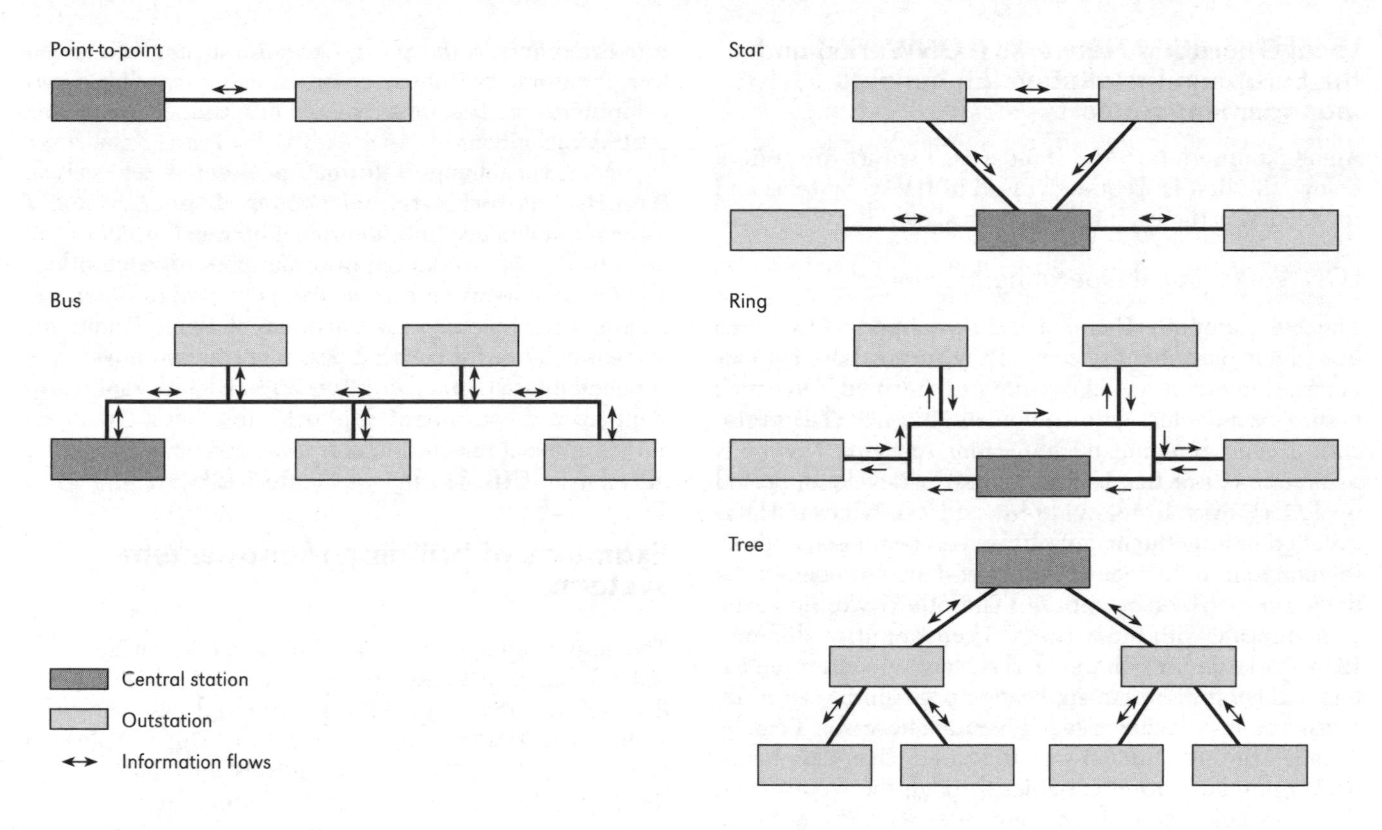

Figure 5.11 *LAN topologies*

how you can reduce the operation costs through their management. Furthermore, determine whether it is appropriate to install the additional service systems through which building performance can be improved. A small building management system cannot be upgraded later on; a large system may not be fully exploited, may be complicated to manage and is expensive to run.

- Study the tenders of more manufacturers; check on the professional reputation and references of the tendering firms or BMS manufacturers; find out how many systems the bidder has already implemented. Your starting choice will affect the success of the operation of the whole life cycle of the system. Within this time, the system will require maintenance and upgrading of software. In the future, you will most probably expand the systems in your building; therefore, check out the expansion capabilities for the doubled number of control points.
- Go through the entire functioning of the BMS; incomplete testing after the system has been installed is one of the most frequent errors. With the designer, check out the functioning of every sensor, every command and the entire software; after the test, sign the notes about the commissioning of the system.
- Ensure that the people who will handle the BMS on all levels are well trained. Use of a BMS calls for additional education, in order to ensure that managers understand the numerous functions and jobs which can be performed by the system.
- During the operation of the system, take care of the regular monitoring of the operation and the simultaneous registration of the quantities' values, such as temperature conditions in rooms and energy consumption. This makes maintenance and problem-solving substantially easier.

Local Operating Networks (LONWorks) and the European Installation (EI) building management systems

Among numerous BMS standards, two are presented below: the first is frequently used in HVAC systems and the second is the leading European standard.

LONWorks (Local Operating Networks)

The US Company Echelon has developed LONWorks building management systems. They are composed of four main elements: a microprocessor termed 'Neuron'; network conductors with equipment; the LONTalk protocol and management; and application software. Neuron is a microprocessor that is built into every device supported by LONWorks BMS. A 48-bit address Neuron ID is installed in it by the manufacturer. It is composed of three integrated circuits; one circuit is designed to operate the device into which it is built. The other two, however, communicate with the network. There are many different LONWorks devices on the market, such as sensors, actuators and controllers. An application programme is already registered in a device installed Neuron; however, it can be changed through the network if desired. Every appliance with input and output variables through the Neuron and the LONTalk protocol communicates with the network (LAN), which is called a channel. The channel typology enables information exchange between devices, therefore all output information can be used as input information for other devices. The typical speed of data transfer is 78 kbauds (78kbps). Appliances from different manufacturers that can be built into LONWorks have a LONMark label that indicates compatibility. Manufacturers of LONMark appliances (there are more than 200 worldwide) are members of the LONWorks Association.

Devices are connected with channels into groups. Each group can include up to 64 appliances. 256 groups are linked into a domain. A domain is then linked into the control and automation level through a network driver, the LONWorks/RS232 interface and a personal computer, which is equipped with a Supervisory Control and Data Acquisition (SCADA) software package.

EIBA (European Installation Bus Association)

Many of the European manufacturers of electric equipment take part in the European Installation Bus Association (EIBA). Their products are compatible with one another and with the EIB system. This system is divided into two constituent complexes: power supply and control part. Energy is supplied to the consumers or to a group of consumers through the EIB system. Devices are connected to both the power supply line and the control line. Sensors, switches, communication modules and computers, on the other hand, are connected to the control line only.

Devices exchange information, termed 'telegrams', through a network with bus topology. Every device has its own intelligence and identification number. All of the bus devices can exchange information with each other. The distance between two devices is limited to 700m; the length of the bus line is a maximum of 100m. Parameter entry is achieved through a personal computer, which is connected to EIB bus with the ETS (EIB tool software) standardized equipment. Up to 15 lines with 256 appliances can be connected to one area. Fifteen areas can be linked to an EIB system (see Figure 5.12).

Examples of building management systems

The design and operation of a BMS in a modern commercial building is presented below. The VO-KA building is a three-storey structure with a south and east wing (its architect is Mlakar&Berg and the investor is Vodovod kanalizacija Ltd). The BMS controls air conditioning, heating and ventilation, lighting, water supply, vehicle access heating in winter, security video surveillance, and fire surveillance.

System settings

Description of the BMS system

Systems on the field level are controlled with LON controllers (see Chapter 4), which operate as an individ-

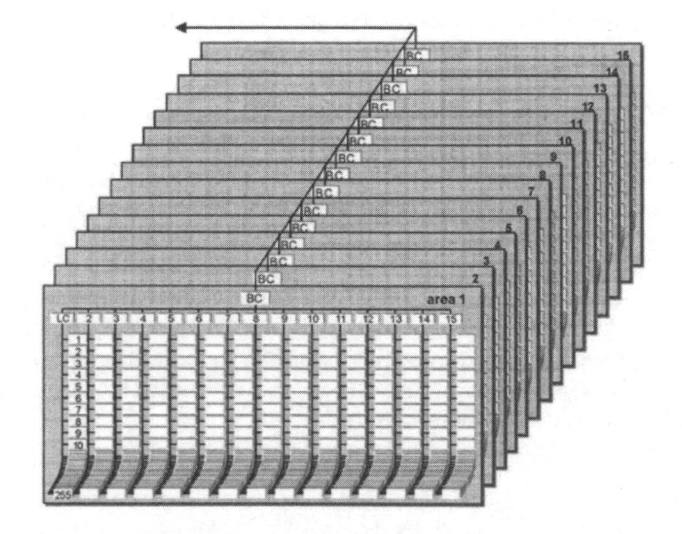

Figure 5.12 *Instabus EIB control lines*

Figure 5.13 *Main entrance, daylit atrium and east façade with movable shading devices*

ual unit and control appliance or sub-system, for example, the central heating sub-station or each of the four-pipe fan-coil units. All of the devices or sub-systems are equipped with LON controllers input/output modules and the modules are connected via a local network. The controllers are connected via a local network. They communicate with the LONTalk protocol.

The building is heated via a district heating system. Heat is supplied through a heat exchanger, which connects the district heating system with building installations. This heat is used for air conditioning, two-zone (south and east wing) office heating with fan-coils and sanitary hot-water preparation. There is additional heat storage in a sanitary water pipelines system.

Besides managing and controlling heating, air conditioning and ventilating systems, the BMS also controls illumination in offices by adjusting electrical lighting (see Figure 5.18a) and the position of shading devices (see Figure 5.13) in relation to daylight. The BMS also controls safety lighting (see Figure 5.18b) and notifies the users in case of potential danger.

After the system has been installed, the entire functioning of the BMS is commissioned. Representatives of the designers and owners test the functioning of sensors, commands and the entire software on the management level. After testing is complete, the certification is signed. Designers also train staff who handle the BMS on all levels.

Figure 5.14 *Central heat sub-station LON controller*

Figure 5.15 *Fan-coil LON controller*

Operation of the system

The LON network is configured as an individual system that connects independent LON controllers. The LON network is connected with management level by communication cards and hardware lines. The management-level computer is equipped by iFix software (a product of Intellution) to depict all processes and their conditions in graphical form. The operators can control the status of building systems and building services systems. In this way, current and past systems operation states and values of heat flow, energy use, temperatures, etc. are available.

During operation of the system, regular monitoring occurs, as well as simultaneous registration of the quantity values. Therefore, maintenance and problem-solving are more efficient.

Possibilities for the individual user

Individual users can adjust their local indoor environment parameters according to different levels:

- by manually switching on/off the fan in heating/cooling units as well as the lights, and by adjusting shading devices;
- by setting point temperature correction (–3 to +3; see Figure 5.20) on the room thermostat;
- by changing the parameters settings with authorized access on the management level; previous set-point values can be changed in a user friendly way (see Figure 5.20);
- the user can also monitor the office from a distance via the 'pcAnywhere' system.

Synopsis

The maintenance of a high quality of living in buildings and the rational use of energy are tasks that can be fulfilled in the process of planning and operating buildings. In modern buildings, where indoor environmental quality and a secure residence are provided by numerous installations, devices and systems must be controlled simultaneously and with a quick response to outside conditions and occupancy behaviour. BMSs can be very effective for this purpose.

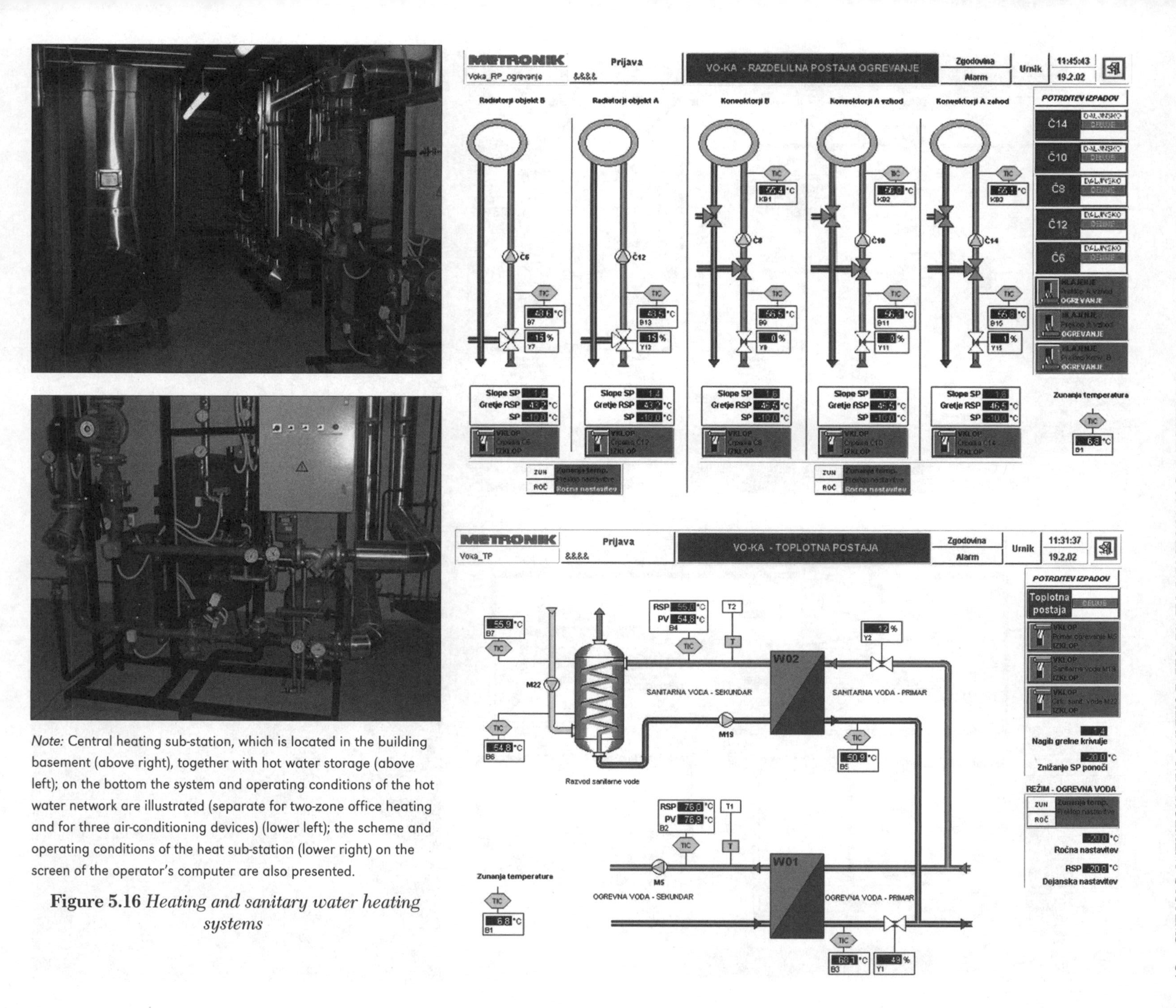

Note: Central heating sub-station, which is located in the building basement (above right), together with hot water storage (above left); on the bottom the system and operating conditions of the hot water network are illustrated (separate for two-zone office heating and for three air-conditioning devices) (lower left); the scheme and operating conditions of the heat sub-station (lower right) on the screen of the operator's computer are also presented.

Figure 5.16 *Heating and sanitary water heating systems*

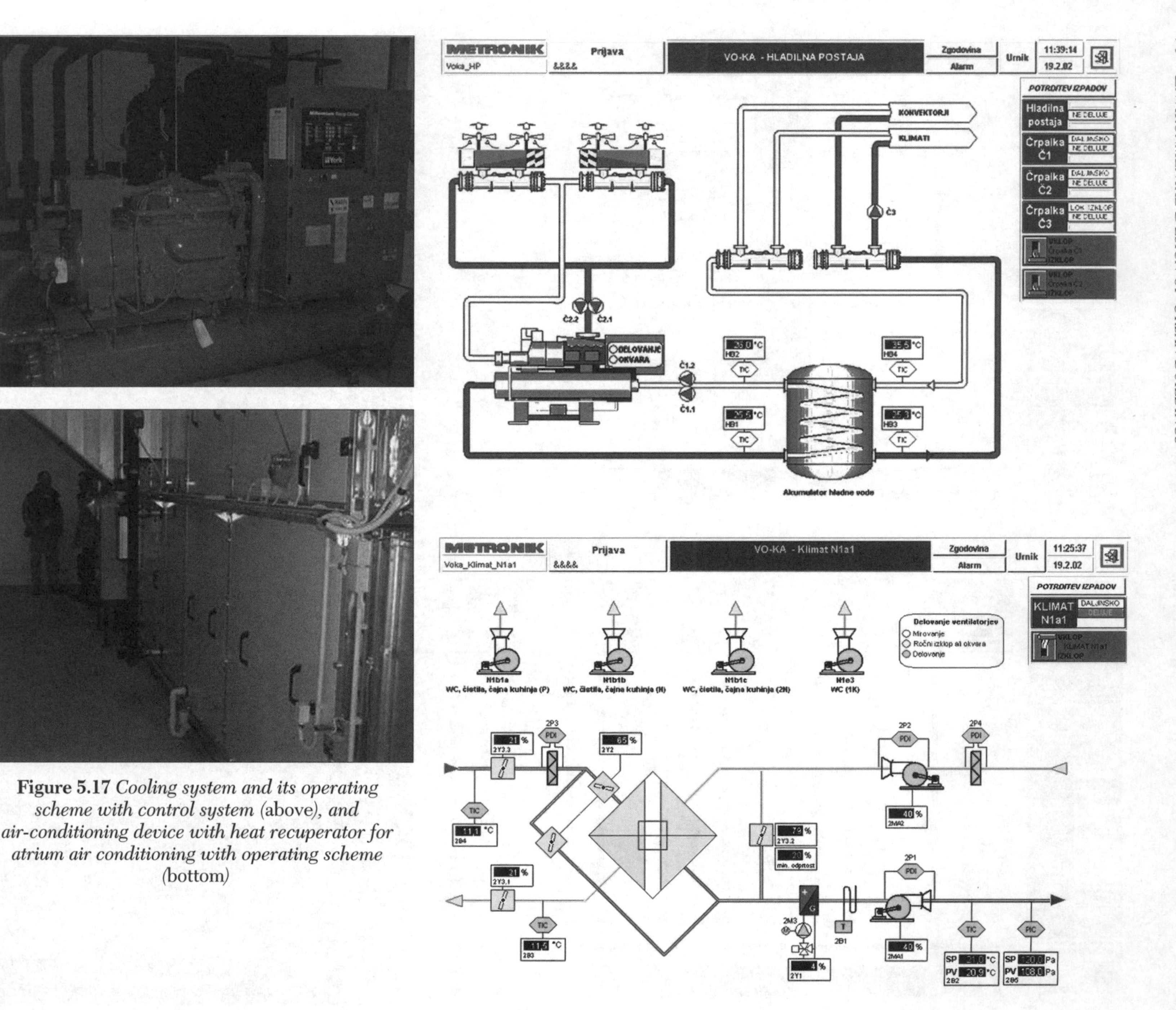

Figure 5.17 *Cooling system and its operating scheme with control system* (above)*, and air-conditioning device with heat recuperator for atrium air conditioning with operating scheme* (bottom)

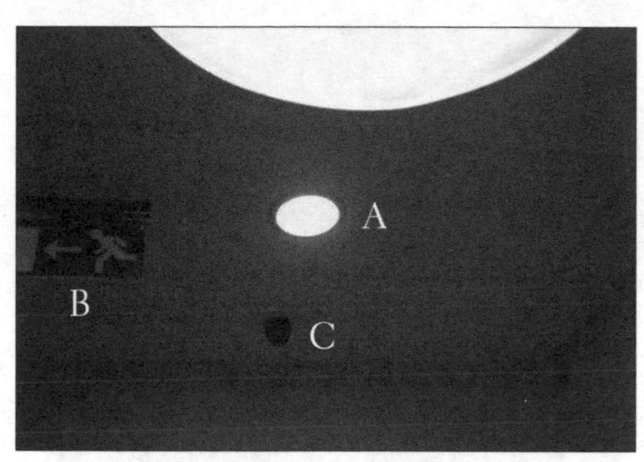

Figure 5.18 (above) *Lighting control is based on illumination level and room presence (sensor C)*

Note: A comparison of the set-point value of the supply water temperature with actual value is also shown. The proportional plus integral control action (PI) algorithm is used for controlling all heating sand air-conditioning systems.

Figure 5.19 (right) *Office condition control* (top) *and scheme of past values of hot water temperatures in the central heating sub-station in case room temperature decreases during the night* (bottom)

Figure 5.20 (below) *The room thermostat gives users the possibility of correcting the set-point temperature according to individual indoor environmental requirements* (left)*; user interface window for changing set-point values of indoor environment parameters* (right)

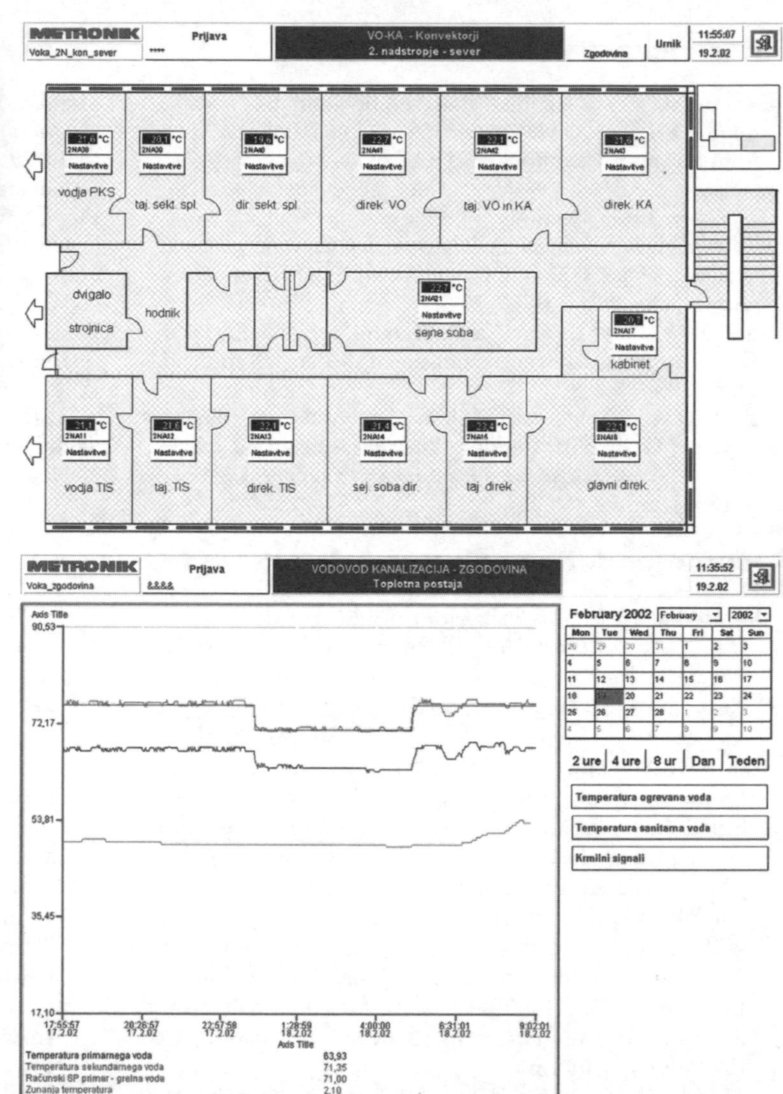

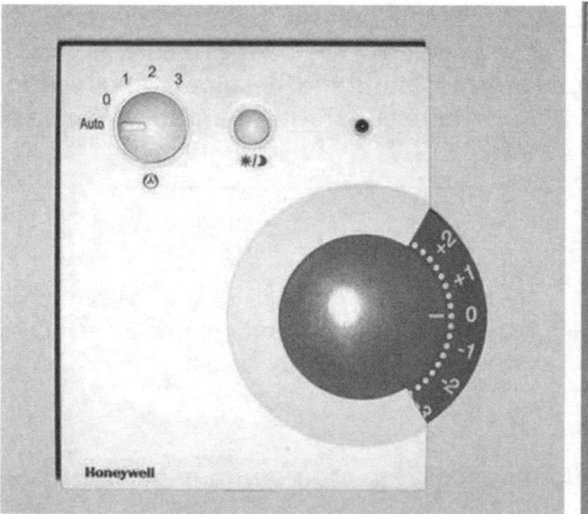

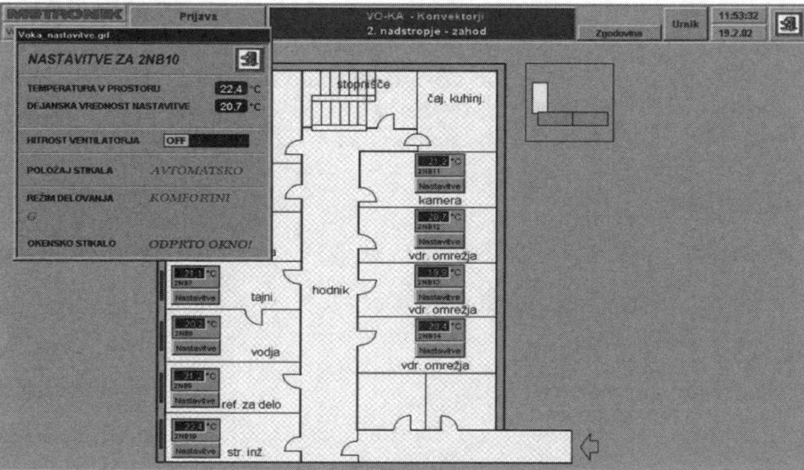

References

American Society of Heating, Refrigeration and Air-conditioning Engineering (1995) *ASHRAE Handbook*, HVAC Applications, SI Edition, ASHRAE, Atlanta

Avtomatika (2001) *Metronikova izdaja revije o avtomatizaciji procesov*, June, Metronik, Ljubljana

Brambley, M. R., Chassin, D. P., Gowri, K., Kammers, B. and Branson, D. J. (2000) 'DDC and the web', *ASHRAE Journal*, December, pp38–50

Coffin, M. J. (1998) *Direct Dogotal Control for Building HVAC Systems*, Kluwer Academic Publishers, Dordrecht, The Netherlands

Coggan, D. A. (2002) 'Smart buildings', www.coggan.com

Levermore, G. J. (1992) *Building Energy Management Systems: An Application to Heating and Cooling*, E& FN SPON, London

Mandas, D. (1995) *A Manual for Conscious Design and Operation of A/C Systems*, Save Publication, Atene

Moult, R. (2000) 'Fundamentals of DDC', *ASHRAE Journal*, November, pp19–23

Piper, J. (2002) 'Riding hard on energy costs', www.facilities.com

Seip, G. G. (2000) *Electrical Installations Handbook*, John Wiley and Sons, Munchen

Trankler, H. R. and Schneider, F. (2001) *Das Intelligente Hause*, Richard Pflaum Verlag GmbH & Co, Munchen

Wilkinson, R. J. (2001) 'Commissioning inoperable system', *ASHRAE Journal*, March, pp44–53

www.europa.eu.int/comm/energy_transport/atlas, accessed November 2002

Recommended reading

1 Trankler, H. R. and Schneider, F. (2001) *Das Intelligente Hause*, Richard Pflaum Verlag GmbH & Co, Munchen

This book starts by describing the technological aspects of the intelligent building. The introduction is followed by a detailed description of micro-system technology and integration of this system within buildings. The most interesting chapters provide descriptions of different sensor technologies, such as passive and active sensors, gas sensors, microwave and ultrasound sensors, as well as multi-chip modules with integrated sensors. Readers will find descriptions of research projects in Europe, Japan and the US.

2 Levermore, G. J. (1992) *Building Energy Management Systems: An Application to Heating and Cooling*, E& FN SPON, London

This book is intended as both a student text on the control of a building services plant and a practitioner's guide to the basics of practical control. The book starts by detailing the development of building management systems, and outlines the advantages and disadvantages of BMS from case studies. The following chapters are devoted to a description of the outstations, central units and basics of control algorithms. A wide range of analytical solutions are presented, dealing with sensors and their responses, dead time and distance velocity lag, preheated time and optimizer control. A separate chapter is devoted to unsteady building heat loss and heating. This is a useful source of knowledge for all of those interested in numerically modelling the heat transfer in building and in controlling HVAC systems, including BMS.

3 Seip, G. G. (2000) *Electrical Installations Handbook*, John Wiley and Sons, Munchen

This handbook offers a basic introduction to the construction and dimensioning of electrical distribution systems, with particular reference to building services automation and building system engineering for residential and functional buildings. One of the focal points of the book is communication installation equipment, particularly networks with Instabus EIB. All topics are presented on the basis of international and European standards. The book provides an excellent foundation for buildings designers involved in planning, erecting or operating buildings management systems.

Activities

Activity 1

Describe the term 'intelligent buildings'.

Activity 2

Explain loop-control systems.

Activity 3

Compare control algorithms that can be used in building management systems for controlling HVAC systems.

Activity 4

Analyse the possibilities and advantages of a building management system in a low-energy house.

Activity 5

Describe the functional levels in building management systems.

Answers

Activity 1

The term 'intelligent building' arose during the beginning of the 1980s. In the beginning, definitions of intelligent buildings were linked to innovative construction technologies and to automation of mechanical systems, with an emphasis on greater energy efficiency. At present, the term 'intelligent building' is connected with techniques and technologies for optimizing the operation, maintenance and management of building services systems. Intelligent buildings provide a high degree of comfort, safety and economy to their owners, managers and users. By considering that the true cost of the edifice comprises more than its construction, and encompasses its operation and maintenance over its lifespan, the complete value of the building is put into the foreground during the design phase.

Intelligent buildings of the future will continuously and independently respond to the changes within them and in their surrounding environment by using information and communication technologies. Intelligent materials – such as glazing with adaptive optical properties, materials with temperature recollection, dynamic thermal insulation, intelligent devices with microchips that communicate with users, as well as intelligent control, supervision and communication systems – will be installed in buildings. The first steps towards intelligent buildings of the future, however, involve the digitally managed control and supervision processes of technological systems.

Activity 2

Control systems allow the building's operation to automatically adapt to internal and external environments without user intervention. They function in such a way that the actual value of a physical quantity, which is measured with a sensor (e.g. the room temperature), is compared with the expected or desired value (e.g. 20° Celsius in the winter). This is termed set-point temperature. If there is a difference between actual and desired physical quantities, a controller mediates the corresponding information to the controlled device in the system (for instance, a heat exchanger's valve).

A controller uses different kinds of control actions, which use information based on the feedback of the sensor, and sends information to the controlled device in the system in order to achieve the quickest possible equalization between actual and desired values.

Activity 3

There are various control actions; however, the best known include:

- two positions (on/off) that regulate action;
- proportional (P);
- proportional plus integral (PI);
- proportional plus integral plus differential (PID);
- artificial intelligence (AI).

For a complete answer, one should describe the basics of each control action.

Activity 4

In your new building, the building management system can undertake the following tasks:

- Managing and supervising energy consumption and other resources:
  - switching on/off of devices according to time and occupancy;
  - limiting electricity peaks;
  - ensuring the optimal operation of the heating, ventilating and air-conditioning systems;
  - regulating shading and electric lighting.
- Safeguarding, which involves:
  - diminishing the human factor;
  - personal identification with electronic cards;
  - image surveillance;
  - hierarchically restricted access to rooms;
  - anti-burglar alarms;
  - fire alarms;
  - gas detection;
  - simulations of the virtual occupancy of the buildings.
- Managing informatics:
  - internal phone and video connection;
  - video conferencing;
  - satellite communications;
  - electronic mail;
  - access to the internet.
- Providing automation of working places:
  - central data processing
  - electronic documents transfer
  - data transfer through computer-aided design among experts;
  - notifying and informing.

Activity 5

Hierarchy and compatibility have to be ensured in order to make the operation of the BMS possible. In relation to today's technological development, hierarchy is achieved through the three operation levels:

1 management level;
2 control and automation level;
3 field level.

Information between the sensors and BMS (inputs from the building service system) interchange in the form of signals (binary numbers), measured values of the quantities (analogical values) or impulses (sensors transmit an impulse according to the physical unit of the measured quantity). Systems on this level also have the hand-control option. Communication among the BMS elements on the field level runs either between the individual sensor and controller or through the communication network. This is termed field-level network (FLN).

On the control and automation level, BMSs monitor, control and optimize the building services systems. *Application-specific* controllers (ASC) or *modular controllers* are used for these tasks. On this level, BMSs become aware of the problems in the building system's operation, display the measured values and ensure that they stay within the allowed margins. Furthermore, they take care of the information transfer around the system in digital form amidst single levels. Controllers are interconnected with the control and automation-level network (CLN). Due to the large amounts of data processed and transferred between the field and management levels, CLN has to be capable of faster data transfer than is the case with the network in the field level.

Management is at the top of the hierarchy in a BMS. It comprises all of the data needed for the statistical data processing from the field, as well as control and automation levels, together with the displayed and out-printed quantities' values and events. Moreover, graphic review of the building, systems condition and the values of the measured quantities in the single systems and rooms of the building are included. Simultaneous monitoring of the data at the management level allows system operation mistakes to be detected and provides an integral survey of energy consumption, as well as an estimation of the costs.

6

Urban Building Climatology

Stavroula Karatasou, Mat Santamouris and Vassilios Geros

Scope of the chapter

The purpose of this chapter is to study urban climatic environments in order to make use of them in evaluating design options and determining design strategies. Urban areas are characterized by complex urban microclimates, modulated by a complex set of meteorological, morphological, topographical and other factors. The scope of this chapter is to clarify how and why the urban climatic conditions are modified compared with the surrounding rural areas.

Learning objectives

Upon finishing this chapter, readers will be able to describe:

- how and why the urban climate differs from the climatic conditions of the surrounding rural areas;
- the main characteristics of the heat island and the canyon effect, and their impact upon the urban climate;
- the role of construction materials and the effect of green spaces.

Key words

Key words include:

- heat island effect;
- canyon effect;
- microclimate;
- urban climate;
- wind profile;
- air flow;
- green space;
- construction materials.

Introduction

The urban environment is dynamically related to urbanization and industrialization. In particular, urban and industrial growth and their implied environmental changes have caused the urban environment to deteriorate and have modified the urban climate.

This modification is highly variable and depends upon the local climate, particular topography, regional wind speeds, urban morphology, human activity and other factors. However, in general, urban climates are warmer and less windy than rural areas. All inadvertent climatic changes are briefed by the concept of 'urban heat island effect' and 'urban canyon effect'.

Consequently, urban areas use more energy for air conditioning in summer, less energy for heating during winter and even more electricity for lighting. Moreover, discomfort and inconvenience to the urban population due to high temperatures, wind tunnel effects in streets and unusual wind turbulence due to badly designed high-rise buildings are very common.

This chapter is divided into four main parts. The first three parts look at the analytical presentation of the heat island effect, the urban wind field and the canyon effect, while the last part looks at the impact of construction materials and the green effect (the impact of green spaces), and their potential to improve the urban climate.

The urban temperature

Heat island effect

The diurnal temperature in almost every city in the world today is warmer than in the surrounding open (rural) countryside. In most cases, the highest differences between urban and rural temperatures occur during clear nights with light winds, and temperature elevations are

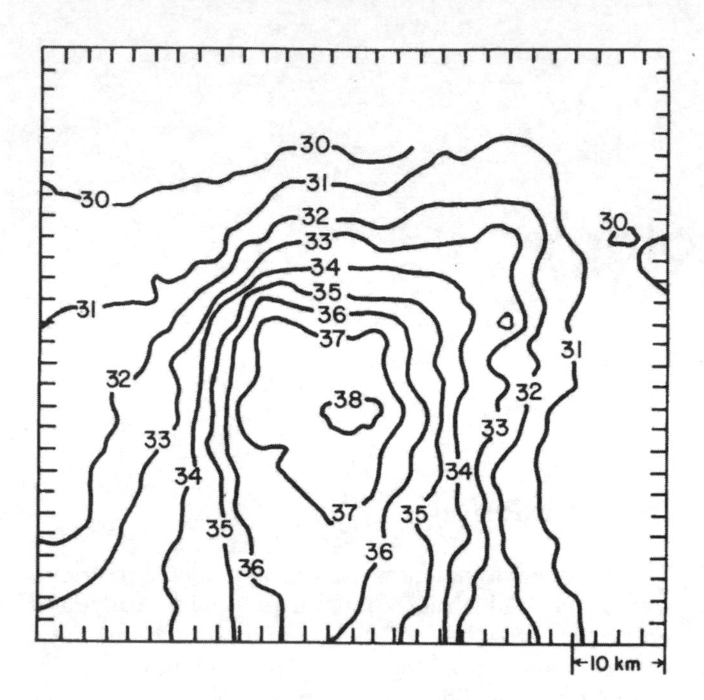

Source: adapted from Byun, 1987

Figure 6.1 *Surface isotherms showing the heat island phenomenon over the St Louis metropolitan area*

commonly about 1–4° Celsius, although elevations of 8–10° Celsius are also observed. This difference between urban and rural temperatures is called the 'urban heat-island effect'. Drawing the isotherms for an urban area and the surrounding rural area, one can observe that the closed isotherms separate the urban area like the contour of elevation for small, isolated islands in the ocean – hence, the use of the term 'heat island'. Figure 6.1 shows the surface isotherm curves over St Louis City during a summer evening under clear sky conditions, which indicate the heat island phenomenon over the city.

The urban temperature is affected by several independent factors, especially near the ground, which contribute to the development of the urban heat island. Oke (1982) lists a number of factors, including altered energy balance terms that lead to a positive thermal anomaly:

- Increased incoming long-wave radiation ($R_{L\downarrow}$) due to air pollution: the outgoing long-wave radiation is absorbed and then re-emitted by the polluted urban atmosphere (urban green house effect).
- Decreased outgoing long-wave radiation loss ($R_{L\uparrow}$) from street canyons: as long-wave radiation is emitted from the various buildings and streets surfaces within the canyon, their sky view factor is reduced and much warmer surfaces replace the cold sky hemisphere. These surfaces receive a high proportion of the infrared radiation emitted from the ground and radiate back an even greater amount (canyon radiative geometry).
- Greater daytime storage of perceptible heat (ΔH_s) due to the thermal properties of urban materials and heat release at night time.
- Addition of anthropogenic heat (H_a) in the urban area by the combustion of fuels from both mobile and stationary sources (transportation, heating/cooling, industrial operations).
- Decreased evaporation and, hence, latent heat flux (H_L): the reduction of evaporating surfaces and the surface waterproofing of the city puts more energy into perceptible heat and less into latent heat.

As shown in Figure 6.2, ambient temperature varies with the distance between the rural area and the city centre. For a large city, during a clear day with light winds, just after sunset, the boundary between rural and urban areas presents a steep temperature gradient to the urban heat island, while the rest of the urban area is characterized by a weak gradient of increasing temperatures, with a final peak at the city centre where the urban maximum temperature is found. The temperature difference between the maximum urban temperature and the background rural temperature is defined as the urban *heat island intensity* ($\Delta T_{u\text{-}r}$) (Oke, 1987).

The heat island intensity depends upon meteorological factors, such as the cloud cover, the humidity and the wind speed. Furthermore, many aspects of the urban structure, such as the size of cities, the density of the built-up areas and the ratio of buildings' heights to the distances between them can have a strong effect on the magnitude of the urban heat island. Therefore, morphology is strongly affected by the particular character of each city and presents an important spatial and temporal variation.

Over large urban areas, and under clear and calm conditions, the heat island near the surface is likely to display a complex spatial structure and isotherms that follow the built form of the city: the sharp urban–rural boundaries exhibit a steep temperature gradient, while in the greater urban area one can observe many small-scale variations in response to distinct intra-urban land uses, such as parks or recreation areas and industrial units, as well as topographical characteristics such as hills, lakes or rivers. Figure 6.3 shows a small part of the Athens heat island, where the geographic centre is occupied by a cool 'area', about 2° Celcius cooler than the surrounding temperature due to the presence of a large park.

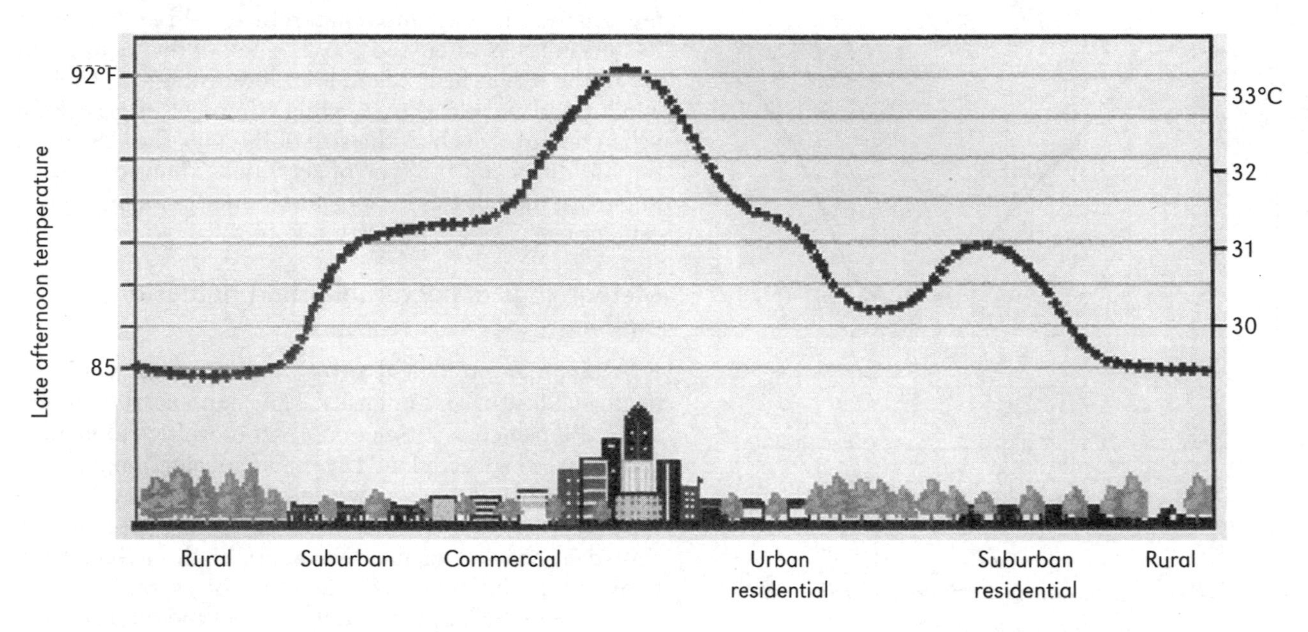

Source: Heat Island Group, http://eetd.lbl.gov/heatisland/

Figure 6.2 *Representation of variation in air temperature from a rural to an urban area*

With regard to the temporal variation of the heat island, a simplified diurnal picture arises for constant weather conditions. The heat island phenomenon may occur during the day and/or the night (see Figure 6.4). In cold climates during winter, the greatest temperature differences are observed at night since the heat island is attributed mainly to urban–rural cooling, rather than to heating differences, especially during the period around sunset. Hence $\Delta T_{u\text{-}r}$ grows rapidly around, and just after, sunset, reaching its maximum three to five hours later, while during the rest of the night it declines slightly. Changes in weather conditions can considerably modify this diurnal picture, as $\Delta T_{u\text{-}r}$ is inversely related to wind speed and cloud cover.

The heat island phenomenon has been intensifying throughout this century. Scientific data from many cities shows that July's maximum temperatures during the last 30 to 80 years have been steadily increasing, ranging from 0.1 to 0.5° Celsius per decade.

Data from various cities have been compiled by the Intergovernmental Panel on Climate Change (IPCC, 1990) in order to assess the impact of the heat island. The data show that the effect is quite strong in large cities. The temperature increase due to heat island varies between 1.1 and 6.5° Celsius (see Table 6.1).

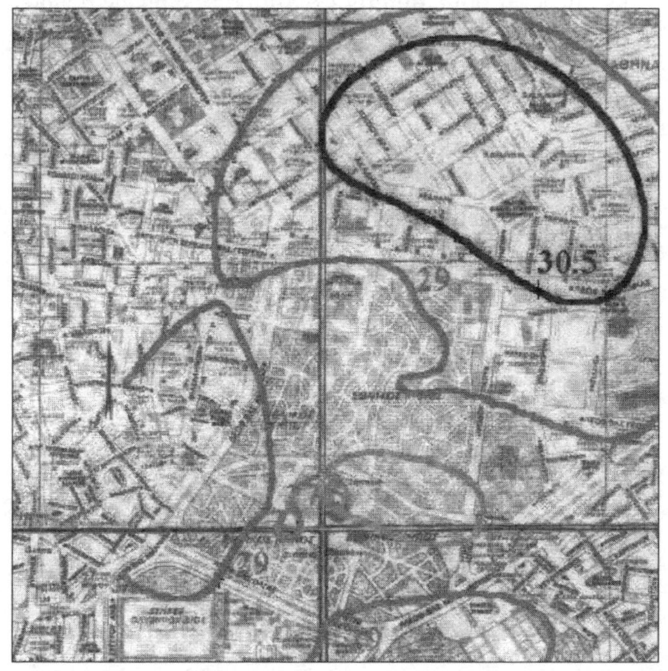

Source: Santamouris, 2001

Figure 6.3 *Temperature distribution in and around a park in Athens, Greece*

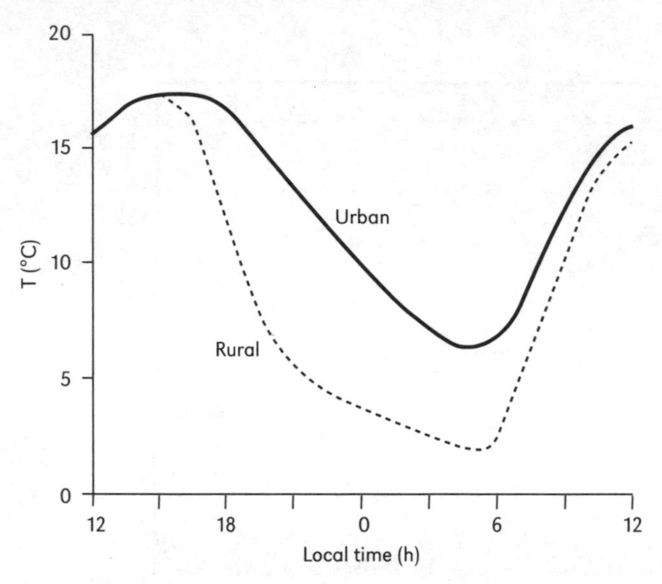

Source: Oke, 1982

Figure 6.4 *Typical temporal variation of urban and rural air temperature*

Higher urban temperatures have a serious impact upon energy consumption for heating or cooling buildings. This impact is quite different in different climatic regions, and it is also different during different seasons for a given region. In cold climates where winters are cold and summers are comfortable, the effect of higher urban temperatures is beneficial. Of course, in summer the phenomenon of the heat island always increases the energy consumption and aggravates thermal discomfort. Beyond this, heat island increases smog production, while it contributes to an increase in emissions of pollutants from power plants, including sulphur dioxide, carbon monoxide, nitrous oxides and suspended particulates. Thus, the heat island phenomenon has a negative connotation.

Heat island models

Numerous studies have been carried out to analyse and understand the heat island. Most of the studies concentrate on night heat islands during the winter period, and few analyse the daytime temperature field and summer heat islands. As discussed above, some of the factors that affect the urban heat island are meteorological, such as wind speed and cloudiness, while other factors arise from urban features, such as the size of the city, the density of the buildings and the type of activities. Therefore, existing urban models can be separated into the following two categories.

Table 6.1 *Heat island effects in some cities*

City	Temperature increase (degrees Celsius)
30 US cities	1.1
New York	2.9
Moscow	3–3.5
Tokyo	3.0
Shanghai	6.5

Source: IPCC, 1990

Meteorological nocturnal urban heat-island models

Existing meteorological urban models deal with the nocturnal heat-island intensity. They express the temperature difference as a function of meteorological factors, such as wind speed, cloud cover and specific humidity.

Ludwig (1970) has suggested a formula that predicts the heat island as a function of the lapse rate, based on the statistical analysis of measurements of the urban–rural temperature differences (dT) and the corresponding lapse rate (in degrees Celsius per millibar) over the rural area (Y):

$$dT = 1.85 - 7.4\,Y \qquad (1)$$

Note that the lapse rate is negative: temperature decreases with height. The lapse rate is very sensitive to the cloudiness conditions; thus, the model expresses indirectly the effect of cloudiness on the heat island.

Different statistical models relating various meteorological parameters, which vary according to location, have also been suggested.

Sundborg (1950) has suggested a model that relates the nocturnal heat island of Uppsala, Sweden, with the following meteorological parameters: cloudiness (N), wind speed (V), temperature (T) and specific humidity (q). The equation, developed by Sundborg, is:

$$dT = 2.8 - 0.1\mathrm{N} - 0.38V - 0.02T + 0.03q \qquad (2)$$

Summers (1964), using data from Montreal, has correlated wind speed with the heat island intensity and proposed the following equation:

$$DT = \frac{2r\dfrac{\partial T}{\partial z}Qu}{\rho c_p u} \qquad (3)$$

where r is the upwind edge of the city to the centre, $\partial T/\partial z$ is the potential temperature increase with height z, Qu is the urban excessive heat per unit area, ρ is the air density, c_p is the specific heat and u is the wind speed.

All of these equations are useful in predicting the variations of the heat island intensity for various meteorological conditions. As meteorological models do not deal with factors that are influenced by urban design, they are of limited interest to urban designers. Furthermore, since they primarily deal with the maximum urban temperature elevation on a given night, these models cannot be applied when estimating the heat island effect on energy use for heating or cooling, which is related to the diurnal average temperature instead of nocturnal conditions. Estimations for summer cooling energy consumption and peak load demand knowledge of the daytime average and maximum temperatures, rather than the extreme conditions during the night.

Thus, meteorological models are primarily of interest in order to understand how these factors affect the heat island. To have an applicable value to urban design, the urban heat island should also be expressed as a function of urban design factors. A brief analysis of urban heat island models follows.

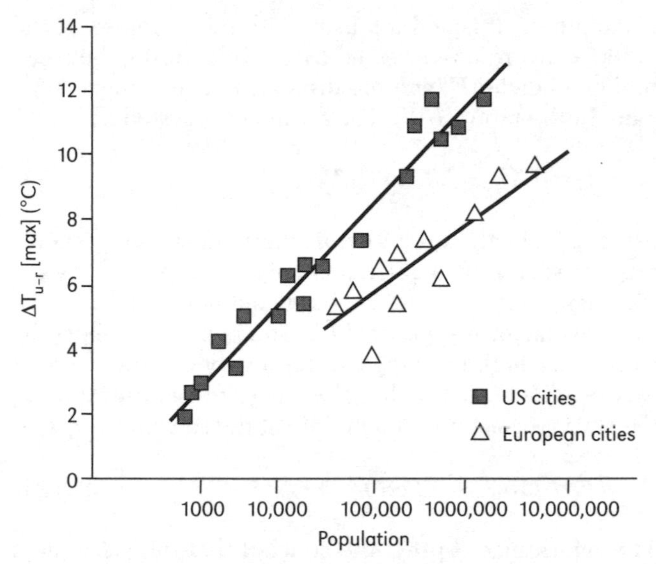

Source: adapted from Oke, 1982

Figure 6.5 *Relation between maximum heat island intensity and population for North American and European cities*

Urban design-oriented heat-island models

There are few heat island models that correlate the heat island with a number of characteristics of the urban structure. Usually, such models include only very general urban characteristics.

Oke (1982) has correlated urban heat island to the size of the urban population (P). The heat island intensity is found to be proportional to log$P$, and under calm winds and clear sky it is very well related to log$P$ for many North American and European cities (see Figure 6.5). As shown, the expected heat island intensity for a city of 1 million inhabitants is close to 8 and 12° Celsius in Europe and the US, respectively. Oke has developed two different regression lines for the two sets of data. He attributed this discrepancy to the fact that the centres of North American cities have taller buildings and higher densities than typical European cities.

Furthermore, by taking into account the wind effect, Oke suggests the following for the calculation of the heat island intensity near sunset and under cloudless skies:

$$dT = P^{0.25}/(4 \times V)^{0.5} \qquad (4)$$

where dT is the heat island intensity in degrees Celsius, P is the population and V is the regional non-urban wind speed at a height of 10m in metres per second.

Jauregui (1986) added a number of cities located in low latitudes in South America and India to Oke's data (Figure 6.6). As can be seen from this figure, the heat island in these cities is weaker even than in the European cities. Jauregui suggests that this phenomenon can be attributed, in part, to the difference in morphology (physical structure) between the South American and European cities.

Another model of Oke's (1981) correlates the

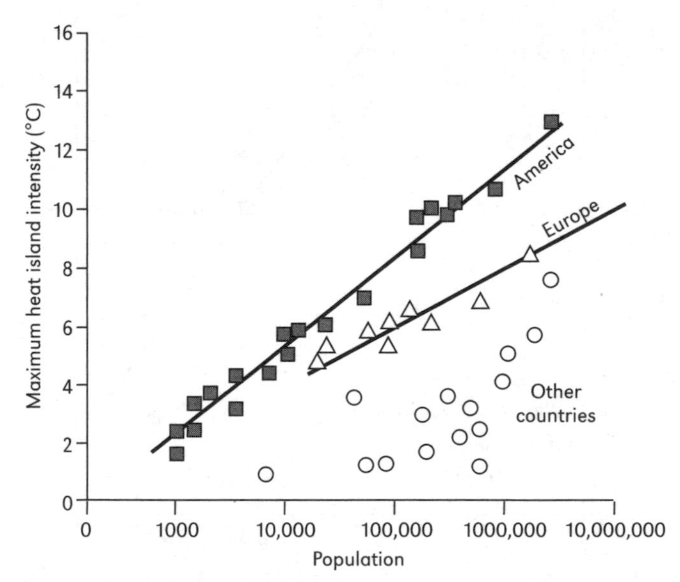

Source: adapted from Jauregui, 1986)

Figure 6.6 *Relation between maximum heat island intensity and population for North American, European and South American cities*

maximum heat island intensity with the geometry of the 'urban canyon', as expressed by the relationship between building height (H) and the distance between them (W) – namely, the ratio (H/W). The formula suggested is:

$$dT = 7.54 + 3.97 \ln (H/W) \quad (5)$$

Alternatively, the urban hemispheric height-to-distance ratio, as seen from a given point, can be expressed by the 'sky view factor'. For an unobstructed horizontal area the sky view factor is equal to 1.0. For a point surrounded by close, very high buildings, or for a very narrow street, it may be about 0.1. Oke has also suggested a formula using the sky view factor of the middle of the canyon floor, Y_{sky}:

$$dT = 115.27 - 13.88\, Y_{sky} \quad (6)$$

These formulae express the concept that the urban heat island is caused by reduced radiant heat loss to the sky from the ground level of densely built urban centres due to the restricted view of the sky. Figure 6.7 shows the relationship between the sky view factor and the urban heat island intensity for North America, Europe and Australasia.

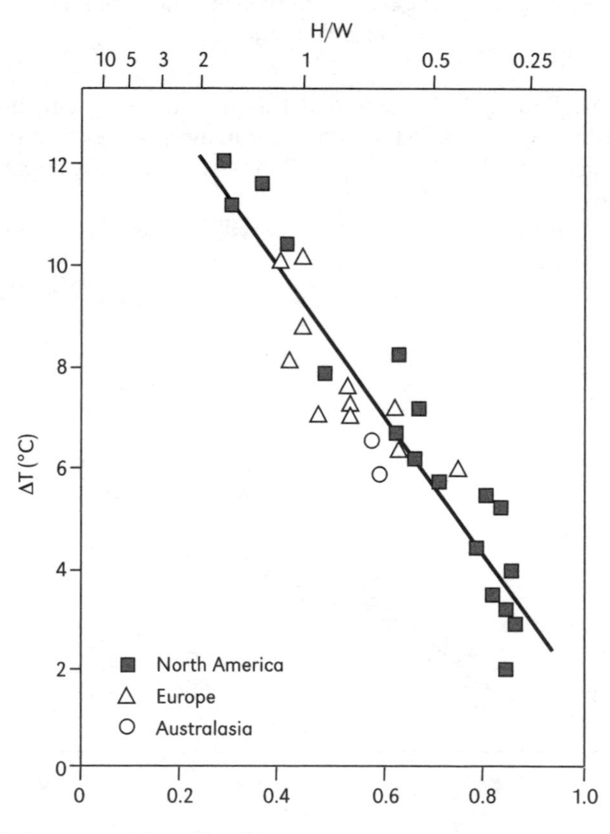

Source: adapted from Oke, 1981

Figure 6.7 *Relation between maximum heat island intensity and population for North American, European and Australasian cities*

Urban wind field

The process of urbanization has a notable effect upon the speed and direction of near-surface winds. This effect is mainly attributed to the change of surface roughness and the heat island, and this results in a complicated wind field in urban areas. Even under the least complicated synoptic conditions (e.g. with a clear sky and light winds in the centre of an extended area), irregular air flows can be brought about by some of the many local factors that influence wind distribution.

Wind distribution in the boundary layer is influenced by many factors, such as horizontal pressure and temperature gradients, the diurnal cycle of heating and cooling of the surface (which determines the thermal stratification of the boundary layer) and surface topographical features (which can provoke local or meso-scale circulation); but it is mainly controlled by the frictional drag imposed on the flow by the underlying rigid surface. The air, flowing from the rural to the urban environment, must adjust to the new and totally different set of boundary conditions defined by cities. Thus, an internal boundary layer develops downwind from the leading edge of the city (see Figure 6.8). According to Oke (1976), the air space above a city may be divided into the urban boundary layer and the urban canopy, which is the space beneath the roof level and is produced by micro-scale processes operating in the street canyons between the buildings.

Some general characteristics of airflow in the urban boundary layer are presented in the following section. Particular airflow patterns in urban streets within urban canyons are discussed in detail in the section on 'Urban canyon effect'.

Wind profile in urban areas

As has already been mentioned, surface roughness influences wind speed. The drag decelerates motion close to the ground, and thus the mean horizontal wind speed ($\overline{u}$) is decreased as the surface is approached (see Figure 6.9).

The profiles in Equation 7 are based on measurements in strong winds (i.e. in the absence of strong thermal effects). In this case, the depth of the frictional influence depends only upon the roughness of the surface. The height that defines the top of the boundary layer (e.g. the height above which the drag is negligible and the

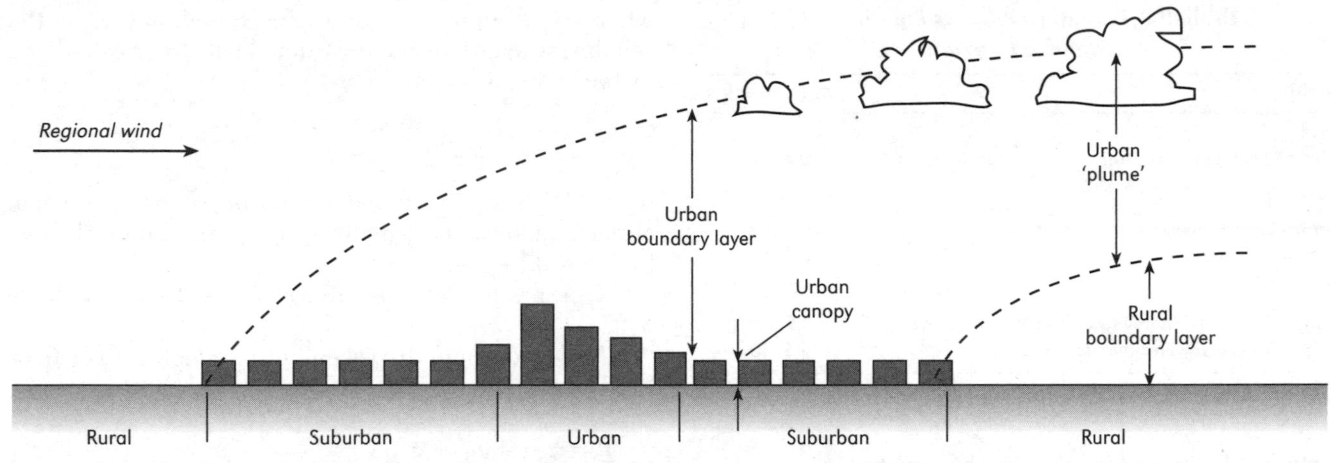

Source: adapted from Oke, 1976

Figure 6.8 *Schematic representation of a two-layer classification of an urban modification*

mean wind velocity is constant) increases with roughness. In light winds the depth of the boundary layer also depends upon the amount of thermal convection generated at the surface. With strong surface heating, this height is greater than in Equation 7, and with surface cooling it is less.

Summarizing the previous characteristics, the wind variation with height depends upon the surface roughness and the atmospheric stability conditions. Under neutral stability (e.g. with strong winds to homogenize the temperature structure), in the free surface layer above roof tops, the vertical wind structure is described by a logarithmic decay curve, which is known as the logarithmic wind profile:

$$\overline{u}_z = \frac{u_*}{k} \ln \frac{z + d + z_0}{z_0} \tag{7}$$

where $\overline{u}_z$ is the mean wind speed at the height z, k is the von Karman constant ($\cong$ 0.40), d is the zero-plane displacement, z_o is the roughness length and u_* is the

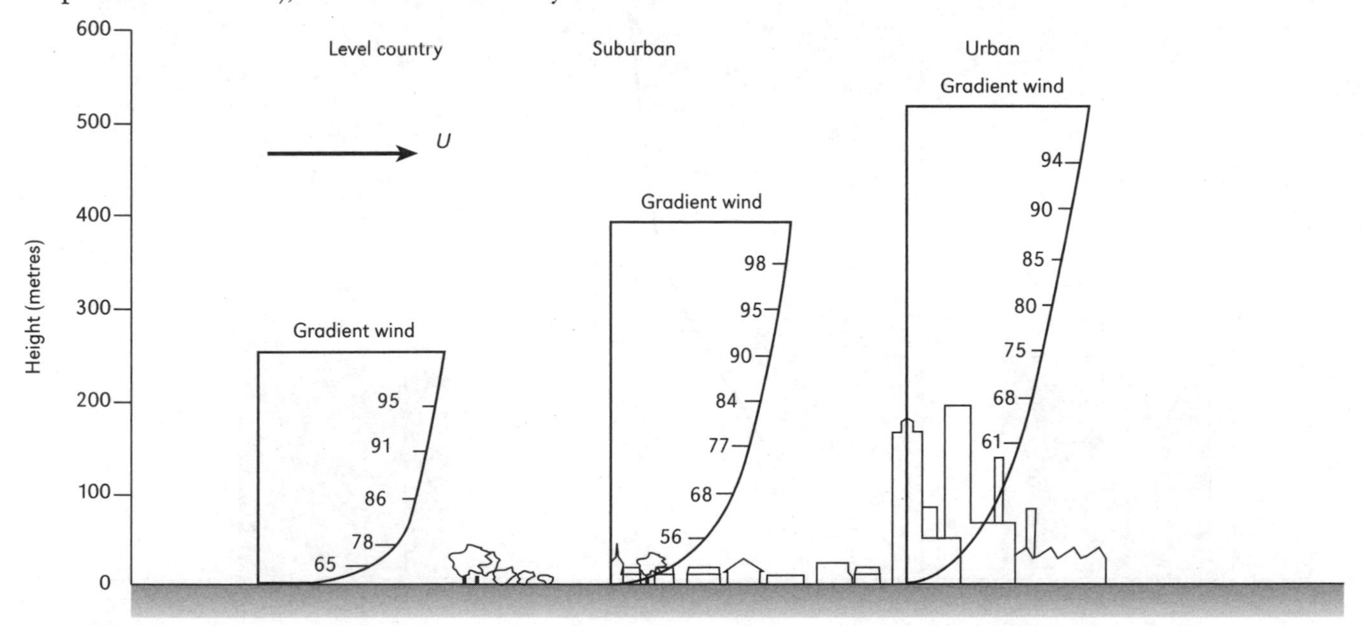

Source: adapted from Davenport, 1965)

Figure 6.9 *Vertical wind speeds (percentage of the gradient wind at various heights) over terrain of different roughness*

Table 6.2 *Typical roughness length z_o of urbanized terrain*

Terrain	z_o (metres)
Rural	
Scattered settlement (farms, villages, trees, hedges)	0.2–0.6
Suburban	
Low-density residences and gardens	0.4–1.2
High density	0.8–1.8
Urban	
High density, < five-storey row and block buildings	1.5–2.5
Urban high-density plus multistorey blocks	2.5–10

Source: adapted from Oke, 1987

friction velocity and is given by the equation:

$$u_* = \frac{\tau}{\rho} \qquad (8)$$

where τ is the surface shearing stress and ρ is the atmospheric density.

In the obstructed sub-layer, the variation of wind with height is described by the exponential low used to describe airflow beneath forest canopies (Cionco, 1965, 1971; Inoue, 1963):

$$u = U_0 e^{z/Z_0} \qquad (9)$$

where U_0 is a constant reference speed, and Z_0 is the roughness length of the obstructed sub-layer calculated by the expression:

$$Z_o = h_b D^*/z_o \qquad (10)$$

where D^* is an effective diameter of air space between obstacles and can be tentatively approximated for the city by $D^* = 0.1\, h_b$.

Typical values of z_o are given by Oke (1987) (see Table 6.2).

The zero-plane displacement parameter d is calculated by the following expression:

$$d = z_o x - (h_b + z_o) \qquad (11)$$

where:

$$x \ln x = 0.1\, (h_b)^2/(z_o)^2 \qquad (12)$$

Both d in equation (11) and U_o in equation (9) are determined by the requirement that the log law and the exponential law must give the same u and $\bar{u}_z$ at height $z = h_b$.

Both the logarithmic and exponential profiles are mathematical idealizations and do not hold at a particular point in the city. Instead, they represent an average over the entire city or city sector (see Figure 6.10).

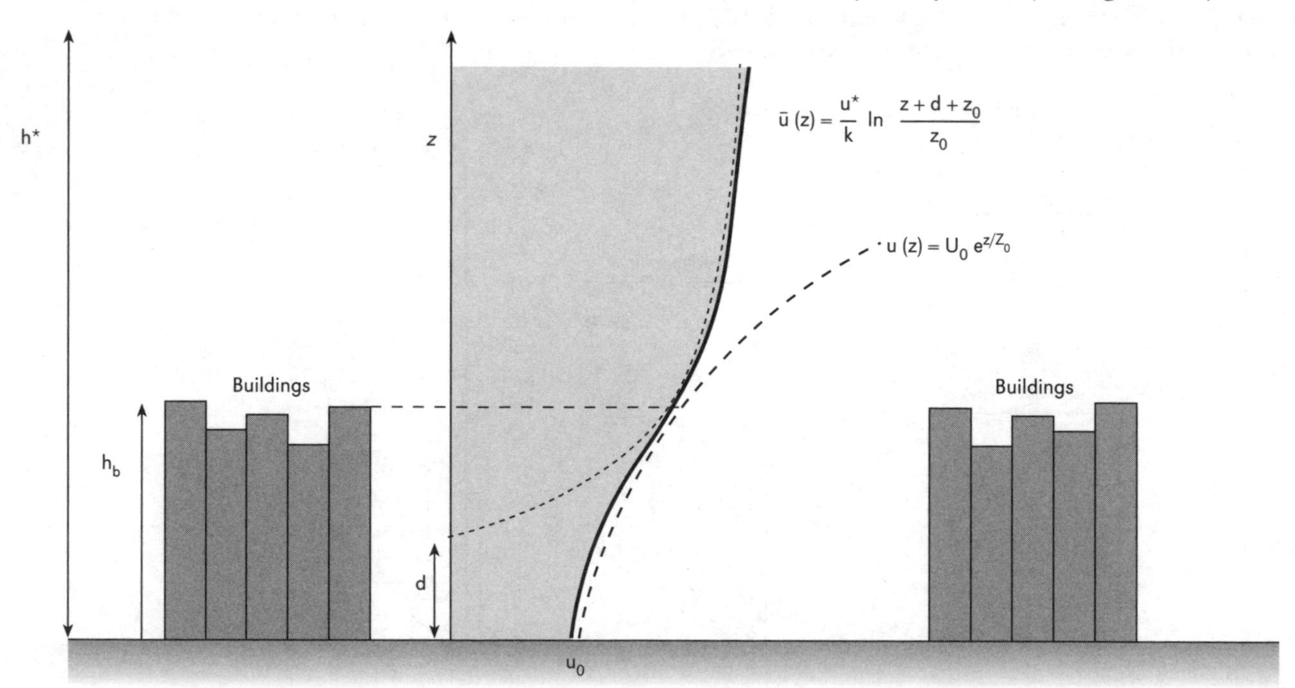

Source: adapted from Nicholson, 1975

Figure 6.10 *Logarithmic and exponential wind profiles in the surface layer, above and below height*

Estimations of the wind speed in a city are of vital importance for passive cooling applications, especially in the design of naturally ventilated buildings. This section has indicated that wind speeds measured above buildings or at airports differ considerably from the speed at an urban monitoring site. Roughness length, z_o is greater in an urban area than in the surrounding countryside, and the wind speed u at any height z is lower in the urban area, and much lower within the obstructed area.

Urban canyon effect

Urban canyon climate

The air space above a city, according to Oke (1987), may be divided into the boundary layer over the city space, 'the urban air dome' and the urban air 'canopy'. The urban air canopy is the space bounded by the urban buildings up to their roofs. The air dome layer is the portion of the planetary boundary layer whose characteristics are affected by the presence of an urban area at its lower boundary, and is more homogeneous in its properties over the urban area at large. Beneath roof level, micro-scale processes operate in the streets between the buildings (i.e. the street canyons produce the urban canopy layer).

Various urban configurations result in an unlimited number of microclimates. The specific climatic conditions at any given point within the canyon are determined by the nature of the immediate surroundings and, in particular, by the landscape's geometry, the materials and their properties.

In the preceding sections, urban effects on the urban dome have been discussed. However, the local environment that encompasses the streets around buildings canyons is more important as temperature distribution and air circulation within urban canyons directly affect the energy consumption of buildings, as well as pollutant dispersion and human comfort. Thus, it is important to understand the thermal and airflow conditions that predominate within such urban structural forms.

Thermal conditions

The energy balance of an urban canyon is very important as it determines the temperature distribution in its elements (i.e. the buildings and street surfaces and the air). Surface temperatures are very important, primarily because they dominate the thermal exchanges with the air. Urban surfaces absorb the incident solar radiation, which is then transformed to sensible heat, and emit long-wave radiation to the sky and other surfaces. A large amount of solar radiation impinges on roofs and the vertical walls of buildings, while only a relatively small part reaches ground level. Furthermore, the intensity of the emitted radiation depends upon the view factor of the surface regarding the sky. The geometry of the urban environment reduces the sky view factor of the vertical (e.g. building walls), horizontal (e.g. street surface) or other declination surfaces, and thus the long-wave radiant exchange does not really result in significant losses.

The net balance between solar gains and heat loss as a result of emitted long-wave radiation determines the thermal balance of urban areas. Because the radiant heat loss is slower in urban areas, the net balance is more positive than in the surrounding rural areas; thus, higher temperatures occur.

Surface temperature

The thermal balance of a surface in a canyon determines its temperature. It can be expressed as follows (Mills, 1993):

$$Q^* = Q_H + Q_G \tag{13}$$

where Q^* is the net radiation, Q_H comprises the convective heat exchanges and Q_G represents the conductive heat exchanges with the substrate. Net radiation is the balance of the received beam, diffuse and reflected solar radiation (K), as well as the received and emitted long-wave radiation (L):

$$Q^* = K{\downarrow}_S + K{\downarrow}_T + K{\uparrow}_r + L{\downarrow}_S + L{\downarrow}_T + L{\uparrow}_e \tag{14}$$

where the arrows represent directions to (↓) and from (↑) the surface, and subscriptions T and S represents the sky and surrounding terrain sources, respectively, r the reflected radiation and e the emitted radiation.

The urban canyon surfaces (i.e. walls, roofs and streets) absorb solar radiation as a function of their absorptivity and their exposure to solar radiation. They also absorb long-wave radiation emitted by surrounding surfaces and emit long-wave radiation to the sky as a function of their temperature, emissivity and view factor. Finally, they transfer heat to or from the surrounding air and exchange heat via conduction procedures with the lower material layers.

Within an urban canyon, two categories of surfaces are usually considered: the surfaces of buildings and the surfaces of streets. The optical and thermal characteristics of materials used in the construction of these elements, especially the albedo-to-solar radiation and the emissivity-to-long-wave radiation, are the factors that determine their thermal condition.

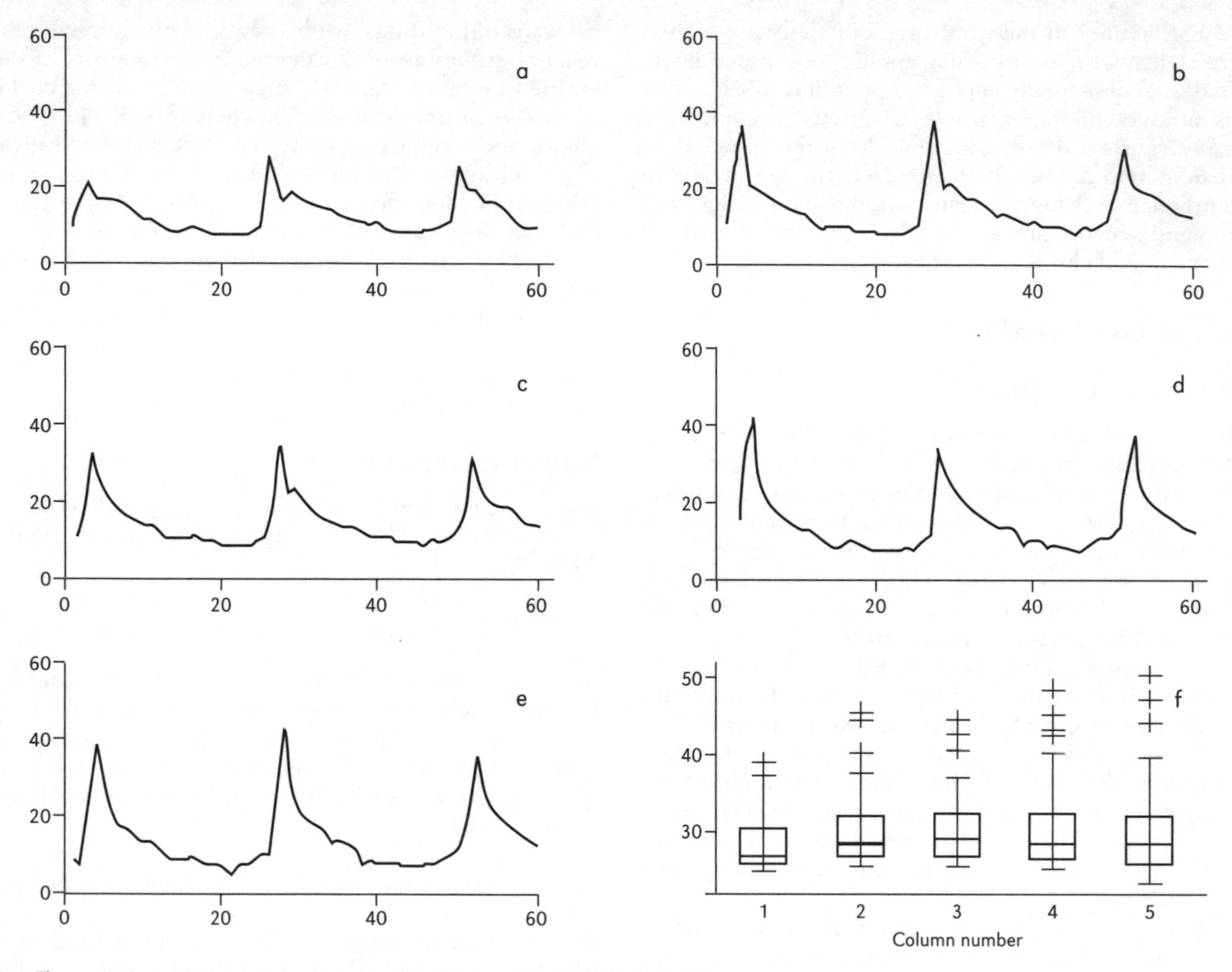

Note: The measurement points are close together. For all points, the material is dark slab.

Source: Santamouris, 2001

Figure 6.11 *Canyon measurements; surface temperature is measured at five different points in the street: (a) on the south-west façade and (e) on the north-east façade of the street, with (b), (c) and (d) in between*

Surface temperature measurements have been performed in different canyons all over the world, for both the surfaces of buildings and of streets. For the surfaces of streets, experiments indicate that construction materials that are ordinarily used have a significant impact. During the day, the variation of surface temperature depends principally upon the solar radiation that reaches the ground. Thus, temperature distribution along the street is determined by the accessibility of solar radiation. During the night, street surface temperature is determined by the heat transfer via radiation and convection, as well as via conduction procedures with lower layers.

Extensive measurements of the surface temperature distribution of pavements and roads have been performed in seven different canyons, during the summer period, in Athens, Greece, on an hourly basis and for a period of two to three days (Santamouris et al, 1997). Asphalt temperatures during the day reach peak temperatures up to 57° Celsius (see Figure 6.11), while the corresponding maximum temperatures for white and dark slab pavements are close to 45° and 52° Celsius, respectively. The mean temperature of all the materials during the night is close to 23–25° Celsius.

In addition to the type of materials used, the

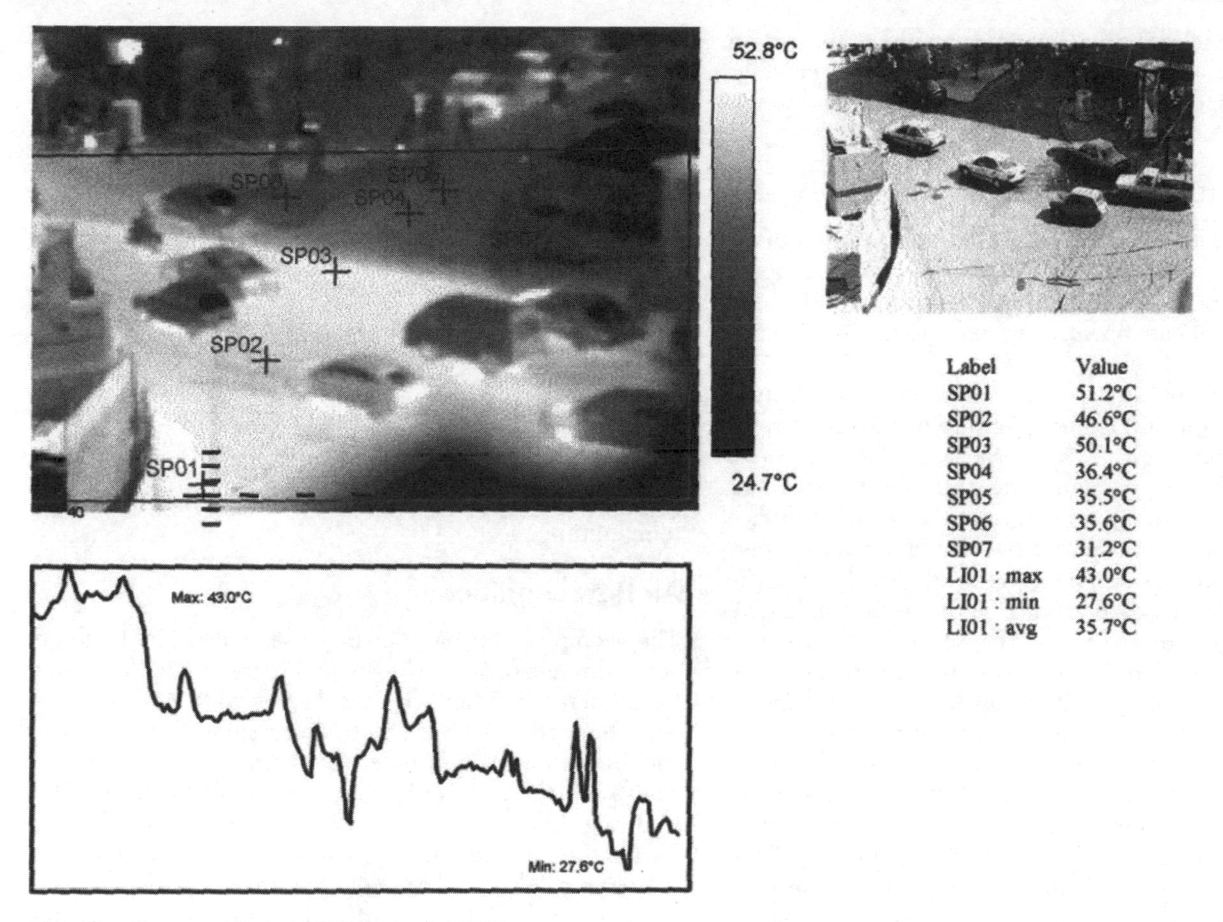

Note: The picture was taken at 16:00, September 1998

Source: Santamouris et al, 1997

Figure 6.12 *Infrared thermography of a section of Omonia Square in Athens, Greece*

orientation of the streets as well as the H/W (height-to-width ratio of the canyon), directly affect the surface temperature of the materials. The impact of the absorbed solar radiation on the temperature increase of the materials used in pavements and roads is found to be very important.

Furthermore, Santamouris et al (1997), using infrared thermography, assessed the temperature of the materials used in pavements and streets in Omonia Square in Athens during the summer. A typical picture is shown in Figure 6.12, where the temperature of non-shaded and shaded asphalt was close to 52° Celsius and 35° Celsius, respectively. The temperature of shaded white slabs used in pavements was between 28° and 31° Celsius. During the experiment the ambient temperature was around 31° Celsius.

Similarly, the surface temperature of building façades is determined by the absorbed solar radiation and the emitted thermal radiation. Factors that also have an important impact are building orientation, relative position of façades and the view factor.

During the day, south-facing façades reach higher temperatures than north-facing façades. The temperature profile indicates that the temperature increases with height and that this is because the lower façades receive much less radiation than the upper ones. However, the maximum temperature is not always observed on the top of the canyon. This is due to the geometrical characteristics of the canyon; lower parts of building façades can absorb the same amounts of solar radiation as higher parts, but due to urban canyon geometry the former parts absorb a greater amount of thermal radiation than the more elevated areas. At night, the temperature of façades is governed by the radiative balance. The temperature decreases with height, as lower surfaces have lower sky view factors and higher view factors for the other canyon buildings. Under normal conditions, maximum temperatures are observed during the day, while minimum temperatures occur at night.

Air temperature

The mechanisms that determine the distribution of air temperature in the urban canyon are complex. In general, the air temperature in urban canyons is influenced by the temperatures of canyon surfaces and the flux divergence per unit volume of air.

The air temperature near construction materials is influenced by the surface temperature as energy is transferred through convective processes. It has been observed that close to the façade of the buildings an air film governed by the temperature of the building surface and the vertical air transport is developed. Lower temperatures are measured near ground level and the temperature increases as a function of the canyon height. The air temperature close to the south, south-west or south-east façades of a canyon is usually higher. The difference between opposite façades varies according to the canyon layout and the surface characteristics.

In the middle of the canyon and at ground level, air temperature depends more upon the flux divergence per unit volume of air, including the effects of the horizontal transport. Thus, air temperature at the middle of the canyon is very different from the average temperatures of the two air films that have developed close to the façades of the buildings since they are, in most cases, lower than the corresponding air-film temperature.

Measurements of surface and air temperatures during the summer period show clearly that, in most cases, surface temperatures are higher. As would be expected, the temperature differences are up to 13° Celsius higher for south, south-west or south-east façades, while the greatest differences that have been observed for north, north-west or north-east façades were up to 10° Celsius. In all cases, the air temperature inside the canyon was higher than the undisturbed temperature measured above the canyon.

The temperature distribution in a canyon during the night is low. During the summer period, the maximum temperature difference between the different canyon levels never exceeds 1.5° Celsius. (Santamouris et al, 1997). In agreement with the distribution of the surface temperature in the canyon during the night period, higher temperatures are measured at ground level, and temperature is found to decrease as a function of height. The temperature of the air in the middle of the canyon is higher than that of the air film close to the façades of the canyon, while significant air temperature differences are not observed between the air temperature close to the south, south-west or south-east façades and the north, north-west or north-east façades. In general, the south, south-west or south-east façades featured a higher air temperature; but the temperature difference rarely exceeded 0.5° Celsius, (Santamouris et al, 1997). The temperature of the air in the middle of the canyon is higher than that of the air film close to the façades of the canyon.

For some canyons, higher air temperatures were recorded than surface temperatures. In these canyons, the temperature of the asphalt on the ground was always higher by approximately 1° Celsius than the air temperature. Thus, there was a convective flow from the street surface to the adjacent air that contributed to increased temperature.

Air flow conditions

The geometry of urban canyons is characterized by three main dimensions, as shown in Figure 6.13: the mean height of the buildings (*H*); the canyon width (*W*); and the canyon length (*L*). Given these dimensions, the geometrical description is determined by three simple parameters. These are the ratio *H/W*, the aspect ratio *L/H* and the building density $j = A_r/A_1$, where A_r is the plan of roof area of the average building and A_1 is the 'lot' area or unit ground area occupied by each building.

Due to the inherent difficulty in field experiments within cities, much of our knowledge and understanding of wind flows within and around urban canyons comes from numerical wind tunnels simulations. Most of the existing studies focus on pollution characteristics within the canyon and emphasize situations where the ambient flow is perpendicular to the canyon long axis, when the highest pollutant concentration occurs in the canyon.

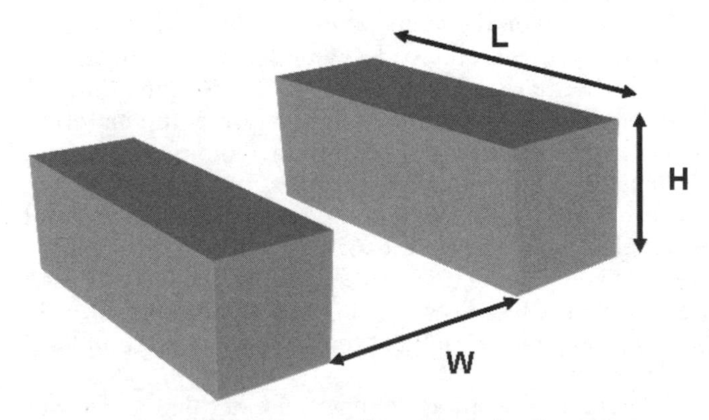

Figure 6.13 *Height, width and length of a canyon*

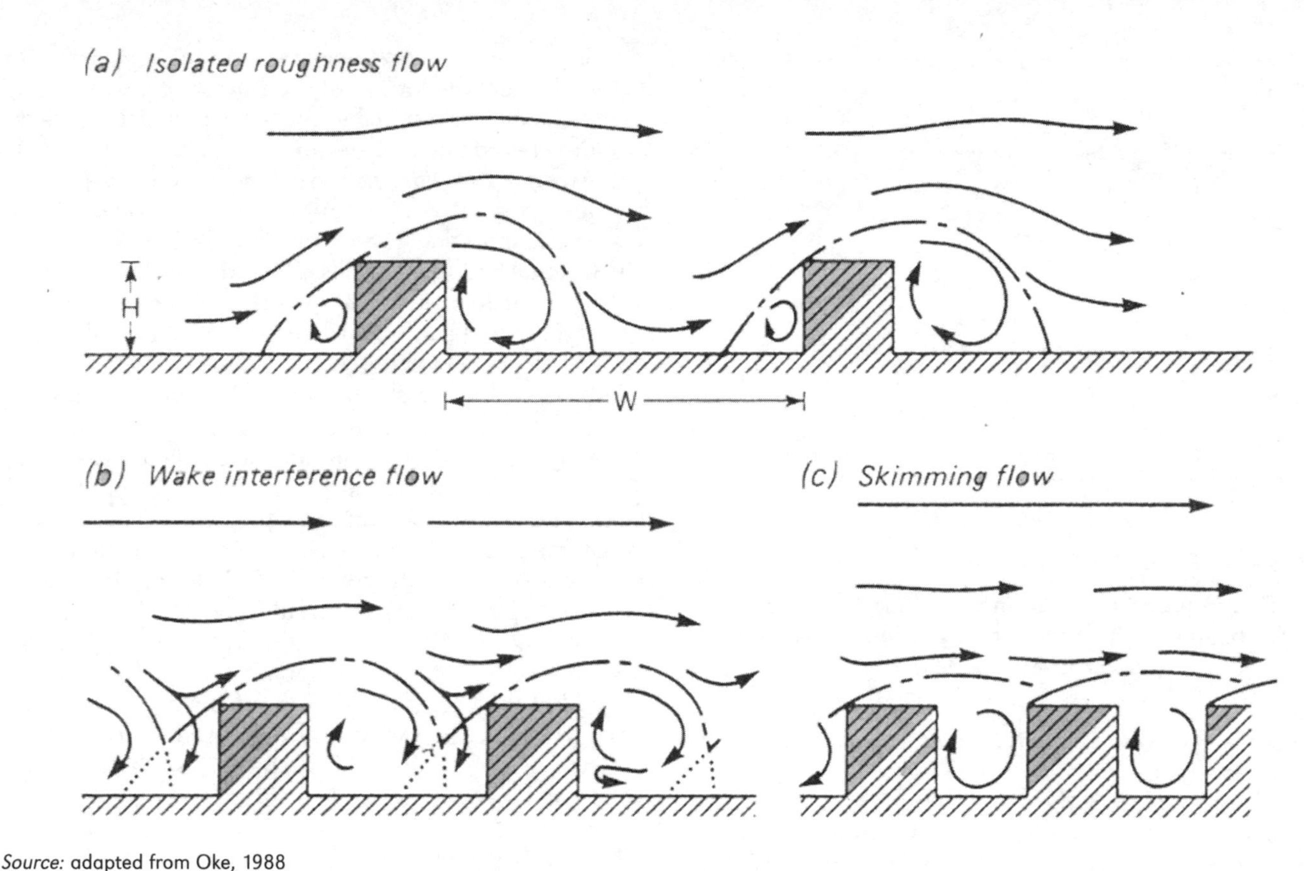

Source: adapted from Oke, 1988

Figure 6.14 *The flow regime associated with airflow over building arrays of increasing height-to-width ratio* (H/W)

The following sections outline current knowledge of airflow in urban canyons when the flow is perpendicular, parallel or oblique to the canyon axis.

Perpendicular wind speed

The flow over arrays of buildings depends upon the geometry of the array and, in particular, upon the canyon height-to-width ratio (*H/W*). When the predominant direction of the airflow is approximately normal (±30°) to the long axis of the street canyon, three type of airflow regimes are observed as a function of the this ratio (*H/W*) (Oke, 1988; Hussain and Lee, 1980) (see Figure 6.14).

When the buildings are well apart (*H/W*>0.05), their flow fields do not interact.

When the buildings are relatively widely spaced (*H/W*<0.4 for cubic and <0.3 for row buildings), their flow pattern appears almost the same and the flow regime is known as 'isolated roughness flow' (see Figure 14a). At closer spacing (*H/W* of up to about 0.7 for cubic and 0.65 for row buildings), the flow pattern appears more complicated as a result of the interference of building waves (see Figure 6.14b). This flow regime is referred to as 'wake interference flow' and is characterized by secondary flows in the canyon space, where the downward flow of the cavity eddy is reinforced by the deflection down the windward face of the next building downstream. At even closer spacing (i.e. greater *H/W* and density), a stable circulatory vortex is established in the canyon because of the transfer of momentum across a shear layer of roof height, and transition to a 'skimming' flow regime occurs where the bulk of the flow does not enter the canyon (see Figure 14c). The existence of the vortex flow within the urban canyon was first measured by Albrecht (1933), and has been confirmed through numerous wind tunnel and field studies.

The transitions between these three regimes occur at critical combinations of *H/W* and length-to-width ratio (*L/W*). Oke (1988) has proposed threshold lines that divide flow into three regimes as a function of the building (*L/H*) and canyon (*H/W*) geometry. The proposed threshold lines are given in Figure 6.15.

In most cities, building arrays are commonly characterized by high *H/W* ratios; thus, skimming airflow regime

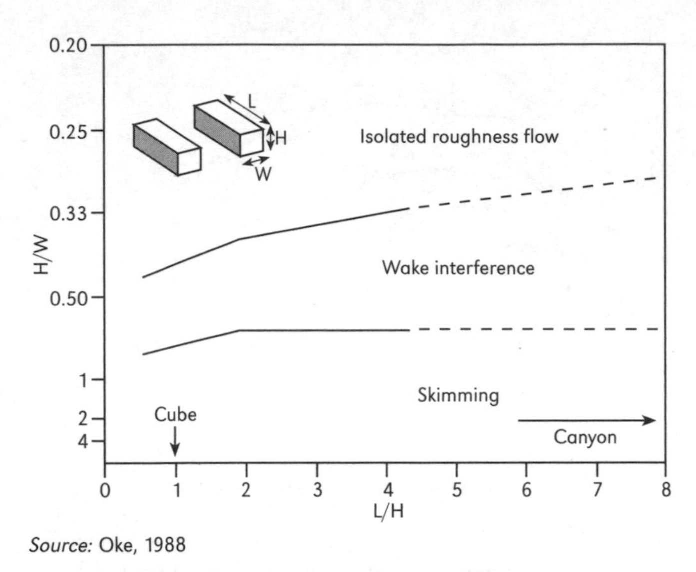

Source: Oke, 1988

Figure 6.15 *Threshold lines dividing flow into three regimes as functions of the building* (L/H) *and canyon* (H/W) *geometries*

has been extensively studied, both with wind tunnel and field experiments.

Since the airflow in the canyon is a secondary circulation driven by the above roof-imposed flow, the wind speed out of the canyon is a crucial parameter. If that speed is below some threshold value, the coupling between the upper and secondary flow is lost (Nakamura and Oke, 1988), and the relation between wind speeds above the roof and within the canyon is characterized by considerable scatter. If the wind speed is higher than the threshold value, a stable vortex is produced within the canyon. The threshold value for symmetrical canyons with a *H/W* ratio of between 1.0 and 1.5 is from 1.5–2 metres per second.

The vortex velocity increases with the speed of the cross-canyon flow. Within symmetrical canyons, and for wind speeds normal to the canyon axis and up to 5 metres per second, the relation between velocity in and out of the canyon appears to be linear, $u_{in} = pu_{out}$ (Nakamura and Oke, 1989), where p varies between 0.66 and 0.75 under the condition that winds in and out are measured at $0.06H$ and $1.2H$, respectively. Furthermore, it has been reported that the transverse vortex speed inside the canyon is proportional to the above-roof transverse component and independent of the longitudinal component. But no evidence of such dependence between the vortex speed and the horizontal or total wind velocity has been found in asymmetric canyons with a *H/W* ratio close to 1.52.

The vortex direction near the ground is opposite to the wind direction outside of the canyon. This is what is expected by a flow normal to the canyon, which is driven by a downward transfer of momentum across the roof-level shear zone. For the case of a symmetrical and a step-up canyon configuration, this is verified by Nakamura and Oke (1988), and Hoydysh and Dabbert (1988), and Arnfield and Mills (1994) report that in a step-up asymmetric canyon, the vortex direction is also consistent with the mechanism described above, although a reversed vortex was detected in some cases with wind speeds below the threshold value of 2 metres per second.

Furthermore, the vertical and horizontal components of wind speed in the canyon have been examined. Downdraft vertical velocities are a strong function of height and reach a maximum close to 95 per cent of the ambient horizontal velocity, at heights of near three-quarters, while at the centre of the canyon they have been measured close to zero. Updraft vertical velocities are relatively independent of height and had a maximum close to 55 per cent of the ambient velocity at a height of one half of the height of the upwind field. Respectively, horizontal velocities varied from 0 up to 55 per cent of the free-stream wind speed and reach their maximum at the bottom and higher parts of the canyon.

Flow along the canyon

It has become clear that the airflow in the canyon has to be seen as a secondary circulation, driven by the above roof-imposed flow. As in the case of perpendicular wind speeds, there is a threshold value below which a considerable scatter characterizes the relation between the two wind speeds. For wind speeds higher than the threshold, the parallel ambient flow generates a mean wind along the canyon axis (Wedding et al, 1977; Nakamura and Oke, 1988).

The flow is characterized by a velocity along the canyon that is almost always parallel to the axis of the canyon and that has a downward angle of incidence relative to the canyon floor of between 0 and 30°. Retardation of the airflow due to the friction from the building walls and the street surface can cause uplift along the canyon walls (see Figure 6.16).

This is verified by the fact that with no winds along the canyon, the mean vertical canyon velocity is close to 0m sec^{-1}.

Regarding the relationship between the above-roof free-stream wind speed out of the canyon, U, and the corresponding component of the wind velocity within the canyon, v, Yamartino and Wiegand (1986) reported that they are directly proportional. The constant state of

Source: Santamouris, 2001

Figure 6.16 *Airflow characteristics along the canyon*

proportionality is a function of the approach-flow azimuth, and the proposed relation, at least to first order, is $v = U\cos\theta$, where θ is the incidence angle. Nakamura and Oke (1988) report that for wind speeds of up to 5 metres per second, the general relation between the two wind speeds appear to be linear: $v = p\ U$. The value of p varies between 0.37 and 0.68 under the condition that the winds inside and outside the canyon are measured at about $0.06H$ and $1.2H$, respectively, of asymmetrical canyon with $H/W = 1$. The lower p values are ascribed to the deflection of the flow by a side canyon.

The vertical wind speeds at the canyon top and the along-canyon free-stream wind speed are also related. Arnfield and Mills (1994) report that the vertical wind increases with the along-canyon free-stream velocity. The relationship between the two wind speeds is almost linear, while the free-stream wind travels down only a short section of the canyon and is still actively decelerating in response to the sudden imposition with the canyon facets. When a partial equilibrium with the frictional effect on the surfaces is attained, a positive association between the two wind speeds still exists but is characterized as much more scattered.

The mean vertical velocity at the canyon top, w, resulting from mass convergence or divergence in the along-canyon component of flow, can be expressed as follows (Arnfield and Mills, 1994):

$$w = -H\ \partial v/\partial x \qquad (15)$$

where H is the height of the lower canyon wall, x is the along-canyon coordinate, and v is the along canyon component of motion within the canyon averaged over time and the canyon cross-section. A linear relationship between the in-canyon wind gradient $\partial v/\partial x$ and the along-canyon wind speed has also be found by Arnfield and Mills (1994), as well as by Nunez and Oke (1976) According to Arnfield and Mills, the value of $\partial v/\partial x$ varies between $-6.8\ 10^{-2}$ and $1.7\ 10^{-2}\ sec^{-1}$, while according to Nunez and Oke, $\partial v/\partial x$ varies between $-7.1\ 10^{-2}$ and $0\ sec^{-1}$.

In deep canyons, the situation is more complicated. Measurements performed in a canyon of $H/W = 2.5$ (Santamouris et al, 1999) have not shown any clear threshold value where coupling is lost. Furthermore, the correlation between the ambient wind speed and the along-canyon wind speed inside the canyon was not clear, primarily because most of the data corresponded to ambient wind speeds lower than 4 metres per second, where the relation between the two wind speeds is not clear. However, statistical analysis of the data has shown that there is a statistical correlation. Figure 6.17 shows the variation of the median, upper and lower quartile, as well as the outliers of the along-canyon wind speed inside the canyon for four classes of increasing ambient wind speed and, in particular:

- $0<Vx<1$ metres per second;
- $1<Vx<2$ metres per second;
- $2<Vx<3$ metres per second;
- $3<Vx<6$ metres per second.

As shown, both the median and the quartiles increase as ambient wind speed increases. The existence of high wind speed outliers inside the canyon, but low ambient wind speeds, does not permit such a conclusion to be derived from Figure 6.17.

With regard to the vertical wind speed within the same deep canyon, an important downward flow has been observed. Downward air movement close to the canyon walls could be the result of finite length canyon effects associated with intermittent vortices. These vortices shed on the building corners and are responsible for the mechanism of a downward advection flow from the building corners to the mid-block position in the canyon (Yamartino and Wiegand, 1986; Hoydysh and Dabberdt, 1988).

Flow at an angle to the canyon axis

The preceding flow types are found when the wind is normal or parallel to the long axis of the canyon street,

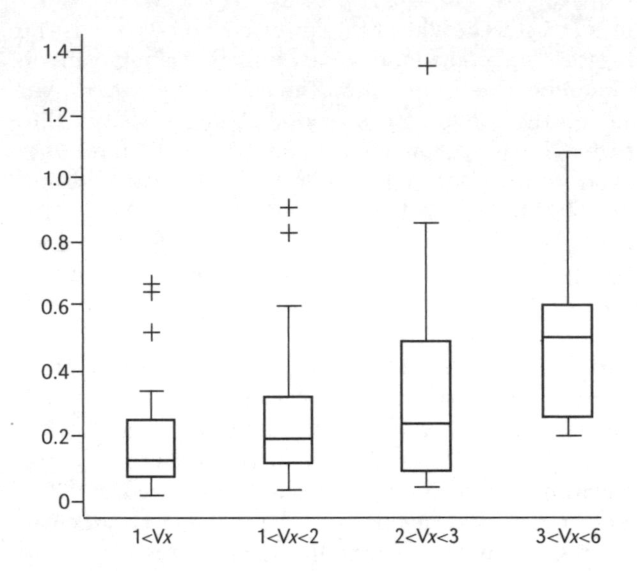

Source: Santamouris et al, 1999

Figure 6.17 *Box plot of wind speed inside the canyon for various clusters of ambient wind speed for wind directions parallel to the canyon*

although the most common case is when airflows at some other relative angle. Unfortunately, research on this type of flow is considerably limited compared with the perpendicular and along-canyon flows, and the actual level of knowledge arises primarily from wind-tunnel experiments and numerical calculations.

When the flow above the roof is oriented at some other angle to the canyon axis, a spiral vortex is induced along the length of the canyon – a cork-screw motion within an elongation to the street (see Figure, 6.18) (Nakamura and Oke, 1988).

Results from wind tunnel experiments have also shown that a helical flow pattern develops in the canyon. For intermediate angles of incidence to the canyon long axis, the canyon airflow is the product of both the transverse and parallel components of the ambient wind, where the former drives the canyon vortex and the later determines the along-canyon stretching of the vortex. As far as the direction of this flow is concerned, it has been reported that, in a first approximation, the angle of incidence on the windward wall is the same as the angle of reflection of the wall, which forms the return flow of the spiral vortex across the canyon floor (Nakamura and Oke, 1988). However, some evidence was found that the angle of incidence is greater than that of reflection, and this could be caused by the along-wind entrainment in the canyon.

In case studies (Lee et al, 1994), the wind speed inside the canyon, in a canyon with $H/W = 1$ and a free-stream wind speed equal to 5 metres per second, flowing at 45 degrees relative to the long axis of the canyon, was less than the wind speed above the roof level by about an order of magnitude. Maximum across-canyon air speed inside the canyon was 0.6 metres per second and occurred at the highest part of the canyon. The vortex was centered at the upper middle part of the cavity and, in particular, to about $0.65H$. The maximum along-canyon wind speeds were close to 0.8 metres per second. Much higher along-canyon wind speeds are reported for the downward façade (0.6 to 0.8 metres per second) than for the upward façade (0.2 metres per second). The maximum vertical wind speed inside the canyon was close to 1.0 metre per second. Much higher vertical velocities are reported for the downward façade, (0.8 to 1.0 metres per second) than for the upward façade (0.6 metres per second). Results by Santamouris et al (1999) have shown that an increase of the ambient wind speed corresponds almost always to an increase of the along-canyon wind speed, for both the median and the lower and upper quartiles of the speed.

Wind tunnel experiments have also been performed by Hoydysh and Daberdt (1988) for symmetrical, even, step-down and step-up canyons, in order to study the distribution of pollutant concentration. The wind angle for which the minimum of the concentration occurs has been found for different configurations. For the step-down configuration, the minimum of the concentration occurs for along-canyon winds (the incidence angle equals 90°). For the symmetrical, even configuration, the minimum on the leeward façade occurs for an incidence angle of 30°, while on the windward the minimum is achieved for angles of between 20 and 70°. Finally, for

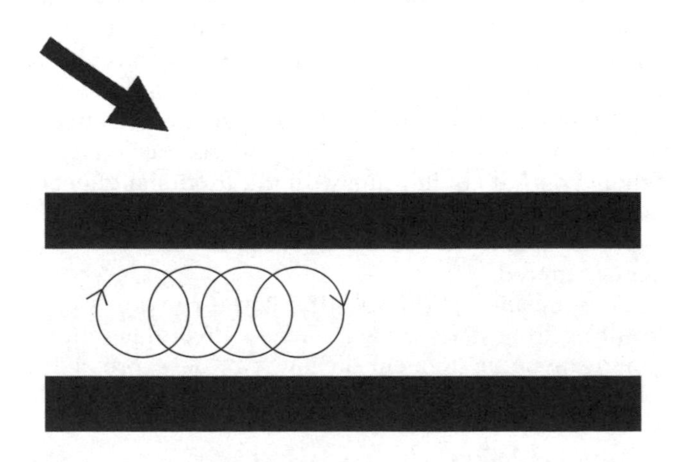

Source: Nakamura and Oke, 1988

Figure 6.18 *A cork-screw type of flow*

step-up canyon configurations, the minimum on the leeward façade occurs at incidence angles of between 0 to 40°, while for the windward façade the minimum is found for incidence angles of between 0 to 60°.

How to improve the urban climate

It is evident that urbanization, due to its increased thermal capacity, lack of water for evapotranspiration and the 'canyon effect', has aggravated the negative effects of climate. The question is how can these negative effects be sufficiently reduced in order to improve the urban climate?

Urban heat islands increase cooling energy use and contribute to the formation of urban smog. Among the factors that have already been mentioned, those that primarily determine summer heat islands are high solar radiation absorption by urban surfaces, lack of vegetation and lower urban wind speeds. Thus, the main methods of reducing the urban heat island comprise the surface albedo and local winds; urban trees and high albedo surfaces can modify the urban environment by altering energy balance and cooling requirements.

The impact of materials

In order to reduce heat island intensity, the first method is to design the external environment in order to minimize solar heat gains. In general, this can be achieved by increasing the solar reflectivity of construction material surfaces and reducing the insulation of susceptible structures. During this procedure, the albedo of the various materials and surfaces is a crucial factor.

The albedo of a surface is defined as the reflected solar radiation divided by the incident solar radiation. On a spherical coordinate system, the albedo is given by the following equation:

$$a = \frac{\int_{\lambda_1}^{\lambda_2}\int_{0}^{2\pi} I \uparrow \cos\theta \, d\bar{\omega} \, d\lambda}{\int_{\lambda_1}^{\lambda_2}\int_{0}^{2\pi} I \downarrow \cos\theta \, d\bar{\omega} \, d\lambda} \qquad (16)$$

where I is the radiant intensity (W/m^2), and θ is the zenith angle defined as the angle between the normal to a surface and the incident beam. Furthermore, λ is the wavelength and ω is the solid angle defined as the ratio of a partial spherical area of interest to the square of the sphere radius. The upward and downward pointing arrows indicate reflected and incident radiation, respectively. In order to find an average albedo, an additional integral time should be added to Equation 16.

Urban areas and cities are characterized, in general, by a relatively reduced effective albedo. This is due to the greater absorption of solar radiation by urban construction materials and multiple reflections inside urban canyons. Thus, the optical characteristics of used materials have an important impact – their albedo-to-solar radiation and emissivity of long-wave radiation being the key parameters.

High albedo materials reduce the amount of solar radiation absorbed through building envelopes and urban structures, and in this way keep their surfaces cooler. Respectively, materials with high emissivities are good emitters of long-wave energy and release the energy that has been absorbed as short-wave radiation. In this way their surface temperature is lower and contributes to the reduction of ambient temperature through the mechanism of heat convection.

Tables 6.3 and 6.4 give the albedo of various typical urban materials and areas, their emissivity, and the albedo for selected materials.

The use of appropriate materials to reduce the heat island and improve the urban environment has gained increasing interest during the last few years. Much research has been carried out to identify the possible energy and environmental gains when light-coloured surfaces are used.

The impact of green spaces

Trees and green spaces contribute significantly to cooling our cities and saving energy. Trees can provide solar protection to individual houses during the summer period, while evapotranspiration from trees can reduce urban temperatures. In parallel, trees absorb sound and block erosion-causing rainfall, filter dangerous pollutants, reduce wind speed, and stabilize soil and prevent erosion.

The major mechanism through which trees contribute to decrease urban temperatures is evapotranspiration, which is defined as the combined loss of water to the atmosphere by evaporation and transpiration. Evaporation is the process by which a liquid is transformed into gas (Oke, 1987), and in the atmosphere usually water changes to water vapour. Transpiration is the process by which water in plants is transferred as water vapour in the atmosphere. The energy transfer to the latent heat from plants is very high, at 2324 kilojoules (KJ) per kilogram of water evaporated (Montgomery, 1987). Moffat and Schiller (1981) report that an average tree evaporates 1460kg of water during a sunny summer day, which consumes about 860 megajoules (MJ) of

Table 6.3 *Albedo of typical urban materials and areas*

Surface	Albedo
Streets	
Asphalt (fresh 0.05, aged 0.2)	0.05–0.2
Walls	
Concrete	0.10–0.35
Brick/stone	0.20–0.40
Whitewashed stone	0.80
White marble chips	0.55
Light-coloured brick	0.30–0.50
Red brick	0.20–0.30
Dark brick and slate	0.20
Limestone	0.30–0.45
Roofs	
Smooth-surface asphalt (weathered)	0.07
Asphalt	0.10–0.15
Tar and gravel	0.08–0.18
Tile	0.10–0.35
Slate	0.10
Thatch	0.15–0.20
Corrugated iron	0.10–0.16
Highly reflective roof after weathering	0.6–0.7
Paints	
White, whitewash	0.50–0.90
Red, brown, green	0.20–0.35
Black	0.02–0.15
Urban areas	
Range	0.10–0.27
Average	0.15
Other	
Light-coloured sand	0.40–0.60
Dry grass	0.30
Average soil	0.30
Dry sand	0.20–0.30
Deciduous plants	0.20–0.30
Deciduous forests	0.15–0.20
Cultivated soil	0.20
Wet sand	0.10–0.20
Coniferous forests	0.10–0.15
Wood (oak)	0.10
Dark cultivated soils	0.07–0.10
Artificial turf	0.05–0.10
Grass and leaf mulch	0.05

Source: Bretz et al, 1992; Baker, 1980; Oke, 1987; Martin et al, 1989

energy, a cooling effect outside a home 'equal to five average air conditioners'. The same authors report that the latent heat transfer from wet grass can result in temperatures of 6 to 8° Celsius cooler than over exposed soil, and that $1m^2$ of grass can transfer more than 12MJ on a sunny day.

Evapotranspiration contributes to lower temperature spaces in an urban environment, a phenomenon known

Table 6.4 *Albedo and emissivity for selected surfaces*

Material	Albedo	Emissivity
Concrete	0.3	0.94
Red Brick	0.3	0.90
Building brick	–	0.45
Concrete tiles	–	0.63
Wood (freshly planed)	0.4	0.90
White paper	0.75	0.95
Tar paper	0.05	0.93
White plaster	0.93	0.91
Bright galvanized iron	0.35	0.13
Bright aluminium foil	0.85	0.04
White pigment	0.85	0.96
Grey pigment	0.03	0.87
Green pigment	0.73	0.95
White paint on aluminium	0.80	0.91
Black paint on aluminium	0.04	0.88
Aluminium paint	0.80	0.27–0.67
Gravel	0.72	0.28
Sand	0.24	0.76

Source: Bretz et al, 1992; Edwards, 1981

as 'the oasis phenomenon'. The magnitude of the temperature reduction is related to the overall energy balance of the area; but, in general, such oases are characterized by the Bowen ratio, which is the ratio of the sensible-to-latent heat fluxes. Taha (1997) reports that, under average oasis conditions, Bowen ratios in vegetative canopies are within 0.5 to 2.0. A Bowen ratio equal to 2.0 was measured in a pine forest in England in July during noontime, (Gay and Stewart, 1974), and corresponded to a 400 and 200 Watts (W) square metres sensible and latent flux, respectively. Furthermore, in a Douglas fir stand in British Columbia, the Bowen ratio was equal to 0.66, with a 200W and 300W per square metres sensible and latent flux, respectively (McNaughton and Black, 1973). Slso reported by Taha (1997) the Bowen ratio is typically around 5 in urban areas; in a desert it is close to 110; and over tropical oceans it is about 0.1.

The positive impact of trees on energy savings has gained an increasing acceptance during the last few years. For example, the Sacramento Municipal Utility District is supporting and funding tree-planting to offset the projected need for a new power plant in the next decade (Summit and Sommer, 1998). As reported by Akbari et al (1992):

> *Field measurements have shown that through shading, trees and shrubs strategically planted next to buildings can reduce summer air-conditioning costs typically by 15 to 35 per cent, and by as much as 50 per cent or more in certain specific situations.*

Simply shading the air conditioner – by using shrubs or a vine – a covered trellis can save up to 10 per cent in annual cooling energy costs.

As a result, several communities and non-profit organizations, primarily in the US, have initiated and coordinated tree-planting efforts and encourage residents to plant trees.

Trees also help to mitigate the greenhouse effect, filter pollutants, mask noise, prevent erosion and calm human observers. As pointed out by Akbari et al (1992): 'The effectiveness of vegetation depends on its intensity, shape, dimensions and placement. But, in general, any tree, even one bereft of leaves, can have a noticeable impact on energy use.' Trees in paved urban areas intercept both the advected sensible heat and the long-wave radiation from high-temperature paved materials, such as asphalt (Halvorson and Potts, 1981; Heilman et al, 1989).

Trees absorb gaseous pollutants through leaf stomata and can dissolve or bind water-soluble pollutants onto moist leaf surfaces, while tree canopies intercept particulates. Trees also reduce ambient air ozone concentrations, either by absorbing them or other pollutants as nitrogen dioxide (NO_2), directly, or by reducing air temperatures, which reduces hydrocarbon emission and ozone formation rates (Cardelino and Chameides, 1990; McPherson et al, 1998). Bernatsky (1978) reports that a street lined with healthy trees can reduce airborne dust particles by as much as 7000 particles per litre of air. Trees also remove atmospheric carbon dioxide (CO_2) and store it as a woody biomass. However, biogenic hydrocarbon emissions from trees may play a role in ozone formation (Winer at al, 1983; Chameides et al, 1988).

In addition, trees reduce and filter urban noise. As pointed out by Akbari et al (1992), leaves, twigs and branches absorb high-frequency sounds that are more bothersome to humans. The same authors report that a belt of trees 33m wide and 15m tall reduces highway noise by 6 to 10 decibels (dB) – a sound reduction of close to 50 per cent. Data on the attenuation effect of vegetation as a function of vegetation are given by Broban (1990). He reports that in order to obtain significant reductions in noise level, in the order of 110dB, at least 100m-deep dense vegetation is required.

Reduction of wind speeds by trees is important and may contribute to important energy savings. Heisler (1989) reports that an increase by 10 per cent of the tree cover in a residential area can reduce wind speeds by 10 to 20 per cent, while an added 30 per cent cover can reduce it by 15 to 35 per cent. Huang et al (1990) simulated the effect of wind shielding by 30 per cent tree cover on the heating and cooling energy use of existing houses, for various US cities, and found that for all locations the heating load was significantly reduced. Studies by DeWalle (1983), dealing with the role of windbreaks of height H on the infiltration and heating energy needs of a mobile house, have shown that at distances of 1 to $4H$ from the wind break the air speed was reduced by 40 to 50 per cent of the undisturbed wind. There was a 55 per cent reduction of the infiltration at $1H$, and a 30 per cent reduction at $4H$ and $8H$. The corresponding heating energy was reduced by about 20 per cent at $1H$ to about 10 per cent at $4H$.

In summary, the use of appropriate materials in combination with planting new trees can reduce the heat island and improve the urban environment. Since measuring the indirect energy savings from large-scale changes in the urban albedo is almost impossible, the possible change of urban climate conditions is usually evaluated using computer simulations. Many research studies have been carried out in this field. Taha et al (1988) have shown that localized afternoon air temperatures on summer days can be lowered by as much as 4° Celsius by changing the surface albedo from 0.25 to 0.40 in a typical mid-latitude warm climate. Taha (1988) reports simulation results for Davis, California, using the URBMET PBL model. Results of the simulations show that a vegetative cover of 30 per cent could produce a noontime oasis of up to 6° Celsius in favourable conditions, and a night-time heat island of 2° Celsius. Huang et al (1987), report that computer simulations predict that increasing the tree cover by 25 per cent in Sacramento and Phoenix, US, would decrease air temperatures at 14.00 in July by 6° to 10° Fahrenheit. Rosenfeld et al (1998) show that white roofs and shade trees in Los Angeles, US, would lower the need for air conditioning by 18 per cent or 1.04 billion kilowatt-hours, equivalent to a financial gain close US$100 million per year. These results are valid for southern climates with significant cooling problems.

Synopsis

Urban areas have particular climatic conditions as a result of complex urban microclimates, where many factors come into play. It is important to describe how and why the urban climate differs from the climatic conditions of the surrounding rural areas.

Urban and industrial growth has caused significant changes in the radiant balance of the urban space, the convective heat exchange between the ground and its buildings, the air flowing above the urban area and the heat generation within the city. The main consequence is differences in values of air temperature and in wind conditions between urban and rural areas. Furthermore, as the thermal structure over cities changes, many other climatological parameters are affected.

Air temperatures in dense urban areas are higher than temperatures in the surrounding countryside. This difference is known as the 'urban heat-island effect' and characterizes almost every city or town in the world. The heat that is absorbed during the day by all urban constructions (buildings, roads, etc.) is re-emitted after sunset, and the greatest temperature differences are observed during the night. Furthermore, additional heat is given by all human heat sources.

Wind speeds, in general, usually decrease in urban areas compared with rural winds at the same height. This is mainly due to changes in surface roughness, but also to the heat island effect and the channelling effects through canyons. However, the urban wind field can be very complicated as it is affected by many local parameters. Such parameters include topography, building geometry and dimensions, streets or trees. Small differences in any of these parameters continuously modify the boundary conditions in which the flowing air must be adjusted, and therefore can cause irregular flows.

Global solar radiation is seriously reduced because of increased scattering and absorption by urban pollutants, especially by particulates. Sunshine duration in many industrial cities can be reduced by 10 to 20 per cent in comparison with the surrounding countryside, and similar losses are observed in energy received.

Precipitation and cloud cover are also affected by urbanization. There are a number of factors that could cause an increment of cloud cover and precipitation in urban areas, such as the heat island, the obstacle effect and polluting elements, which contribute to cloud formation and change the drop-size spectra. The exact effect of these factors is not always clear; furthermore, it is complicated by the topographic influence and the relative place of a specific city in relation to the general atmospheric circulation Yet, there is enough evidence to suggest that precipitation is higher over urban areas when compared with rural environments near cities. For example, Escourrou (1991b) mentions that urbanization causes a proportional increase in precipitation in cities, which, because of their geographic location, are more often in a perturbation zone.

Thus, in winter, most urban microclimates are more moderate than those found in suburban or rural areas. They are characterized by slightly higher temperatures and, away from tall buildings, weaker winds. During the day, wide streets, squares and non-planted areas are the warmest parts of a town. At night, the narrow streets have higher temperatures than the rest of the city. In summer, during the hot months the heat island creates considerable discomfort and stress, with waves of blistering heat emanating from roads and dark buildings. During nightfall, the streets are still radiating heat, while surrounding rural areas are rapidly cooling.

Evidently, the two most effective methods of reducing the inadvertent effects of urbanization are by increasing the amount of vegetation in cities and using light-coloured façades and appropriate construction materials. These techniques can be implemented in whole urban areas or in individual locations such as pedestrian streets.

References

Akbari, H., Davis, S., Dorsano, S., Huang, J. and Winett, S. (1992) *Cooling Our Communities: A Guidebook on Tree Planting and Light Colored Surfacing*, US Environmental Protection Agency, Office of Policy Analysis, Climate Change Division, January 1992, Pittsburgh

Arnfield, A. J. and Mills, G. (1994) 'An analysis of the circulation characteristics and energy budget of a dry, asymmetric, east, west urban canyon: I. Circulation characteristics', *International Journal of Climatology*, vol 14, pp119–134

Barring, L., Mattsson, J. O. and Lindovist, S. (1985) 'Canyon geometry, street temperatures and urban heat island in Malmo, Sweden', *Journal of Climatology*, vol 5, pp433–444

Baker, M. C. (1980) *Roofs: Designs, Application and Maintenance*, Multi-science Publications, Montreal, Canada

Bernatsky, A. (1978) *Tree Ecology and Preservation*, Elsevier Scientific Publishers, New York

Bornstein, R. D. (1986) 'Urban climate models: Nature, limitations and applications', in *Urban Climatology and Its Applications with Special Regard to Tropical Areas*, WMO, Geneva, pp237–276

Bretz, S., Akbari, H., Rosenfeld, A. and Taha, H. (1992) *Implementation of Solar Reflective Surfaces: Materials and Utility Programs*, LBL Report 32467, University of California, Berkeley

Broban, H. W. (1967) 'Stadebauliche Grundlagen des Schallschutzes', *Deutsche Bauzeit*, vol 5

Byun, D.-W. (1987) 'A two-dimensional mesoscale numerical model of St Louis urban mixed layer', Department of MEAS, North Carolina State University, PhD Thesis

Cardelino, C. A. and Chameides, W. L. (1990) 'Natural hydrocarbons, urbanization and urban ozone', *Journal of Geophysical Research*, vol 95, no 13, pp971–979

Chameides, W. L., Lindsay, R. W., Richardson, J. and Kiang,

C. S. (1998) 'The role of biogenic hydrocarbons in urban photochemical smog: Atlanta as a case study', *Science*, vol 241, pp1473–1475

Chandler, T. J. (1965) 'City growth and urban climates', *Weather*, vol 19, pp170–171

Cionco R. (1965) 'A mathematical model for air flow in a vegetative canopy', *Journal of Applied Meteorology*, vol 4, pp517–522

Cionco, R. (1971) 'Application of the ideal canopy flow concept to natural and artificial roughness elements', Technical report ECOM – 5372, US Army Electronics Command, Fort Monmouth, NJ

Davenport, A. G. (1965) 'The relationship of wind structure to wind loading', National Physics Laboratory Symposium, no 16, *Wind Effects on Buildings and Structures*. Her Majesty's Stationery Office, London, pp54–112

De Paul, F. T. and Shieh, C. M. (1986) 'Measurements of wind velocities in a street canyon', *Atmospheric Environment*, vol 20, pp455–459

DeWall, D. R. (1983) 'Windbreak effects on air infiltration and space heating in a mobile house', *Energy and Buildings*, vol 5, pp279–288

Edwards, D. K. (1981) *Radiation Heat Transfer Notes*, Hemisphere Publishing Corporation, Washington, DC

Eliasson, I. (1996) 'Urban nocturnal temperatures, street geometry and land use', *Atmospheric Environment*, vol 30, no 3, pp379–392

Ershad, M. H. and Nooruddin, M. (1994) 'Some aspects of urban climates of Dhaka City', in *Report of the Technical Conference on Tropical Urban Climates*, WMO, Dhaka

Escourrou, G. (1991a) 'Climate and pollution in Paris', *Energy and Buildings*, vol 15–16, pp673–676

Escourrou, G. (1991b) *Le Climat et la Ville*, Nathan University Editions, Paris, France

Gay, L. W. and Stewart, J. B. (1974) *Energy Balance Studies in Coniferous Forests*, Report No 23, Institute of Hydrology, Natural Environment Research Council, Wallingford, Berks, UK

Halvorson, H. and Potts D. (1981) 'Water requirements of honeylocust (*Gleditsia triacanthos f. inermis*) in the urban forest', USDA Forest Service Research Paper, NE-487

Heilman, J., Brittin, C. and Zajicek, J. (1989) 'Water use by shrubs as affected by energy exchange with building walls', *Agricultural and Forest Meteorology*, vol 48, pp345–357

Heisler, G. M. (1989) *Site Design and Microclimate Research*, Final outcome to the Argonne National Laboratory, University Park, PA, US Department of Agriculture Forest Service, Northeast Forest Experimental Station, Morgantown

Howard, L. (1833) *The Climate of London*, vols I to III, Harvey and Darton, London

Hoydysh, W. and Dabbert, W. F. (1988) 'Kinematics and dispersion characteristics of flows in asymmetric street canyons', *Atmospheric Environment*, vol 22, no 12, pp2677–2689

Huang, Y. J., Akbari, H., Taha, H. and Rosenfeld, A. H. (1987) 'The potential of vegetation in reducing cooling loads in residential buildings', *Journal of Climate and Applied Meteorology*, vol 26, pp1103–1116

Huang, Y. J., Akbari, H. and Taha, H. G. (1990) 'The wind shielding and shading effects of trees on residential heating and cooling requirements', 1990 ASHRAE Transactions, American Society of Heating, Refrigeration and Air-conditioning Engineers, Atlanta, January

Hussain, M. and Lee, B. E. (1980) 'An investigation of wind forces on three – Dimensional roughness elements in a simulated atmospheric boundary layer flow. Part II. Flow over large arrays of identical roughness elements and the effect of frontal and side aspect ratio variations'. Report No BS 56. Department of Building Sciences, University of Sheffield, Sheffield

IPCC (Intergovernmental Panel on Climate Change) Working Group II (1990) *Climate Change: The IPCC Impacts Assessment*, IPCC, Australian Government Publication Service, Canberra, pp3–5

Jauregui, E. (1986) 'The urban climate of Mexico City', in *Urban Climatology and Its Applications with Special Regard to Tropical Areas*, WMO, Geneva, pp63–86

Kawamura, T. (1979) *Urban Atmospheric Environment*, Tokyo University Press, Tokyo, Japan

Landsberg, H. (1981) *The Urban Climate*, Academic Press, New York

Lawrence Berkeley Laboratory (1998) *Mitigation of Heat Islands*, www.lbl.gov/HeatIsland

Lee, I. Y., Shannon, J. D. and Park, H. M. (1994) 'Evaluation of parameterizations for pollutant transport and dispersion in an urban street canyon using a three-dimensional dynamic flow model', *Proceedings of the 87th Annual Meeting and Exhibition*, Cincinnati, Ohio, 19–24 June 1994

Ludwig, F. L. (1970) 'Urban temperature fields in urban climates', WMO Technical Note No 108, pp80–107

McNaughton, K. and Black, T. A. (1973) 'A study of evapotranspiration from a Douglas forest using the energy balance approach', *Water Resources Research*, vol 9, pp1579

McPherson, E. G., Scott, K. I. and Simpson, J. R. (1998) 'Estimating cost effectiveness of residential yard trees for improving air quality in Sacramento, California, using existing models', *Atmospheric Environment*, vol 32, no 1, pp75–84

Martin, P., Akbari, H. and Rosenfeld, A. (1989) 'Light colored surfaces to reduce summertime urban temperatures: Benefits, costs, and implementation issues', presented at the 9th Miami International Congress on Energy and Environment, 11–13 December, Miami Beach

Mills, G. M. (1993) 'Simulation of the energy budget of an urban canyon: Model structure and sensitivity test', *Atmospheric Environment*, vol 27B, no 2, pp157–170

Moffat, A. and Schiller M. (1981) *Landscape Design that Saves Energy*, William Morrow and Company, New York

Monteiro, C. A. F. (1986) 'Some aspects of the urban climates of tropical South America: The Brazilian contribution', in *Urban Climatology and Its Applications with Special Regard to Tropical Areas*, WMO, Geneva, pp166–198

Montgomery, D. (1987) 'Landscaping as a passive solar strategy', *Passive Solar Journal*, vol 4, no 1, pp79–108

Nakamoura, Y. and Oke, T. R. (1988) 'Wind, temperature and stability conditions in an E-W oriented urban canyon', *Atmospheric Environment*, vol 22, no 12, pp2691–2700

NASA Climate News and Research (1998), www.climatenews.com

Nicholson, S. E. (1975) 'A pollution model for street-level air', *Atmospheric Environments*, vol 9, pp19–31

Nunez, M. and Oke, T. R. (1976) 'Long wave radiative flux divergence and nocturnal cooling of the urban atmosphere. II. Within an urban canyon', *Boundary Layer Meteorology*, vol 10, pp121–135

Oke, T. R. (1976) 'The distance between canopy and boundary layer urban heat island', *Atmosphere*, vol 14, no 4, pp268–277

Oke, T. R. (1981) 'Canyon geometry and the nocturnal urban heat island: Comparison of scale model and field observations', *Journal of Climatology*, vol 1, pp237–254

Oke, T. R. (1982) 'Overview of interactions between settlements and their environments', WMO Expert Meeting on Urban and Building Climatology, WCP-37, WMO, Geneva

Oke, T. R. (1987) *Boundary Layer Climates*, University Press, Cambridge

Oke, T. R. (1988) 'Street design and urban canopy layer climate', *Energy and Buildings*, vol 11, pp103–113

Oke, T. R. and East, C. (1971) 'The urban boundary layer in Montreal', *Boundary Layer Meteorology*, vol 1, pp411–437

Oke, T. R., Johnson, G. T., Steyn, D. G. and Watson, I. D. (1991) 'Simulation of surface urban heat islands under "ideal" conditions at night – Part 2: Diagnosis and causation', *Boundary Layer Meteorology*, vol 56, pp339–358

Park, H. S. (1987) 'City size and urban heat island intensity for Japanese and Korean cities', *Geographical Review Japan*, September A, vol 60, pp238–250

Rosenfeld, A., Romm, J., Akbari, H. and Lloyd, A. (1998) 'Painting the town white and green', paper available through the website of Lawrence Berkeley Laboratory, www.lbl.gov

Sham, S. (1990/1991) 'Urban climatology in Malaysia: An overview', *Energy and Buildings*, vol 15–16, pp105–117

Santamouris, M. (ed) (2001) *Energy and Climate in the Urban Built Environment*, James & James, London

Santamouris, M. and Assimakopoulos, D. (eds) (1996) *Passive Cooling of Buildings*, James & James Science Publishers, London

Santamouris, M., Argiriou, A. and Papanikolaou, N. (1996) *Meteorological Stations for Microclimatic Measurements*, Report to the POLIS Project, Commission of the European Commission, Directorate General for Science Research and Technology, September (available through the authors, msantam@cc.uoa.gr)

Santamouris, M., Papanikolaou, N. and Koronaki, I. (1997) *Urban Canyon Experiments in Athens: Part A – Temperature Distribution*, Internal Report to the POLIS Research Project, European Commission, Directorate General for Science, Research and Technology, Athens, Greece

Santamouris, M., Papanikolaou, I., Koronakis, I., Livada and Assimakopoulos, D. N. (1999) 'Thermal and air flow characteristics in a deep pedestrian canyon under hot weather conditions', *Atmospheric Environment*, vol 33, no 27, pp4503–4521

Summers, P. W. (1964) *An Urban Ventilation Model Applied to Montreal*, PhD thesis, McGill University, Montreal

Summit, J. and Sommer, R. (1998) 'Urban tree planting programs – A model for encouraging environmentally protective behavior', *Atmospheric Environment*, vol 32, pp1–5

Sundborg, A. (1950) 'Local climatological studies of the temperature conditions in an urban area', *Tellus*, vol 2, pp222–232

Swaid, H. and Hoffman, M. E. (1990) 'Climatic impacts of urban design features for high and mid-latitude cities', *Energy and Buildings*, vol 14, pp325–336

Taha, H. (1988) 'Site specific heat island simulations: Model development and application to microclimate conditions', LBL Report No 26105, University of California, Berkeley

Taha, H. (1997) 'Urban climates and heat islands: Albedo, evapotranspiration and anthropogenic heat', *Energy and Buildings*, vol 25, pp99–103

Taha, H., Douglas, S. and Haney, J. (1997) 'Mesoscale meteorological and air quality impacts of increased urban albedo and vegetation', *Energy and Buildings*, vol 25, pp169–177

Taha, H., Sailor, D. and Akbari, H. (1992) *High Albedo Materials for Reducing Cooling Energy Use*, Lawrence Berkeley Lab Report 31721, UC-350, Berkeley, CA

Wanner, H. and Hertig, J. A. (1983) *Temperature and Ventilation of Small Cities in Complex Terrain (Switzerland)*, Study supported by Swiss National Science Foundation, Berne

Wedding, J. B., Lombardi, D. J. and Cermak, J. E. (1977) 'A wind tunnel study of gaseous pollutants in city street canyons', *Journal of Air Pollution Control Assessment*, vol 27, pp557–566

Winer, A. M., Fitz, D. R. and Miller P. R. (1983) *Investigation of the Role of Natural Hydrocarbons in Photochemical Smog Formation in California*, Final Report, Contract No AO-056-32, California Air Resources Board. Sacramento, CA

WMO (World Meteorological Organization) (1986) *Urban Climatology and its Applications with Special Regard to Tropical Areas*, Proceedings of the Technical Conference in Mexico City, WMO, No 652, Mexico City

WMO (1994) *Report of the Technical Conference on Tropical Urban Climates*, WMO, Dhaka

Yamartino, R. J. and Wiegand G. (1986) 'Development and evaluation of simple models for the flow, turbulence and pollution concentration fields within an urban street canyon', *Atmospheric Environment*, vol 20, pp2137–2156

Yamashita S., Sekine, K., Shoda, M., Yamashita, K. and Hara, Y. 'On the relationships between heat island and sky view factor in the cities of Tama River Basin, Japan', *Atmospheric Environment*, vol 20, no 4, pp681–686

Recommended Reading

1 Oke, T. R. (1987) *Boundary Layer Climates*, Cambridge University Press, Cambridge
This is a classic book, which covers (at a comprehensive level for non-meteorological specialists) important topics in geography, agriculture, forestry, ecology, engineering and planning. It is recommended for its discussion of the atmospheric layers.

2 Landsberg, H. (1981) *The Urban Climate*, Academic Press, New York
This book analyses the topic of urban climatology and features detailed illustrations and figures. It is recommended for a quick check of each topic of interest and its historical review.

3 Santamouris, M. (ed) (2001) *Energy and Climate in the Urban Built Environment*, James & James, London
This is a book that treats the built environment and urban influence in an integrated way, covering all of the important theoretical, experimental and practical aspects of urban design.

Activities

Activity 1

Heat island effect

Give a description of the heat island effect and mention some of the main factors that contribute to this phenomenon.

Activity 2

Heat island effect

Provide a short description of some of the heat island models and discuss the main factors that should be taken into account.

Activity 3

Urban canyon effect

Give a short description of the thermal phenomena that occur inside urban canyons during the day and at night.

Activity 4

Airflow conditions

Describe the three main airflow patterns inside a canyon according to the angle of incidence of the wind to the canyon axis.

Activity 5

Improving the urban climate

Some of the improvements that reduce the heat island effect concern the materials that form urban exterior surfaces and green spaces. Give a short summary of how these elements affect the urban microclimate.

Answers

Activity 1

Heat island effect is related to the fact that the centre of cities is warmer than surrounding rural areas. This phenomenon is observed in diurnal temperature profiles and primarily during the night period. In most situations there is a difference of between 1 to 4° Celsius (in extreme situations a difference of between 8 to 10° Celsius can be observed). Heat island intensity depends upon meteorological factors, as well as upon many aspects of the urban structure (e.g. size of cities, the density of built-up areas and urban geometry); therefore, this phenomenon is characterized by the character of each city.

Some of the main factors that contribute to the heat island effect can be summarized as follows:

- increased incoming long-wave radiation due to air pollution;
- decreased outgoing long-wave radiation loss from street canyons;
- greater daytime storage of sensible heat due to the thermal properties of urban materials and heat release at night time;
- anthropogenic heat in urban areas;
- decreased evaporation and, hence, latent heat flux.

Activity 2

There are three main models that concern the heat island effect during the night period. In general, these models take into account meteorological factors such as wind speed, cloud cover and specific humidity. Furthermore, these models do not use factors that are influenced by urban design.

The Ludwig model is based on statistical analysis of urban–rural temperature differences and takes into account the lapse rate that describes how temperature decreases with height.

The Sundborg model concerns a specific location in Uppsala, Sweden, and calculates the temperature difference between urban and rural areas by using the following meteorological parameters: cloudiness, wind speed, temperature and specific humidity.

Finally, the Summers model is based on data collected in Montreal and correlates wind speed with the heat island intensity.

Another category of heat island models concerns the urban design-oriented model, which correlates the heat island with a number of characteristics of the urban structure.

Oke (1973) suggests a simple equation that calculates the heat island intensity near sunset and under a cloudless sky by taking into account the population and the regional non-urban wind speed at the height of 10m. Other models of Oke correlate the temperature difference between urban and rural areas by using the H/W ratio (H = height of the canyon; W = width of the canyon) or take into account the sky view factor in the middle of the canyon.

Activity 3

The geometry and the orientation of an urban canyon play an important role in its energy balance. This balance determines the temperature distribution of the canyon's elements, which can be divided into three component categories: the canyon vertical surfaces (building façades), the canyon horizontal surfaces (street surfaces) and the air. During the daytime period, the urban surfaces absorb the incident solar radiation and emit long-wave radiation to the sky and to other surfaces. The absorption of solar radiation is transformed to sensible heat that can affect the cooling and heating requirements of the canyon buildings. There is a greater frequency of solar radiation incidents on building roofs and vertical walls, and only a relatively small part reaches ground level. Part of the long-wave radiation that is emitted by the opaque elements of the canyon is lost in the sky according to the sky view factor, while another part is emitted to other surfaces and remains 'trapped' inside the canyon. Therefore, the long-wave radiant exchange does not really result in significant losses, especially when the canyon is narrow and high. Additionally, the high temperature canyon surfaces heat the air close to the surfaces by convection and increase its temperature. Furthermore, during the daytime, the temperature inside the canyon increases with height and this is because the lower façades receive much less radiation than the upper ones. However the maximum temperature is not always observed on the top of the canyon, as the geometrical characteristics of the canyon determine this maximum.

During the night-time period, the missing energy source is the sun; therefore, solar radiation does not contribute to the energy balance of the canyon. This gives a dominant role, from the heat transfer point of view, to the radiative exchange between the canyon surfaces, as well as the surfaces and the sky (the radiation that is emitted to the sky according to the sky view factor of each surface). The temperature of the façades decreases with height, as lower surfaces have lower sky view factors and higher view factors for the other canyon buildings. Under normal conditions, maximum temperatures are observed during the day, while minimum temperatures are observed during the night.

Activity 4

In general, there are three situations that concern the airflow patterns inside the canyons.

The first situation assumes that the wind is perpendicular to the long axis of the urban canyon (30° Celsius). For this condition, four airflow regimes are observed (see Figure 6.14):

1 When the buildings are well apart (*H/W*>0.05), their flow fields do not interact.
2 When the buildings are relatively widely spaced (*H/W*<0.4 for cubic and <0.3 for row buildings), their flow pattern appears almost the same and the flow regime is known as 'isolated roughness flow'.
3 At closer spacing (*H/W* up to about 0.7 for cubic and 0.65 for rows building), the flow pattern appears more complicated as a result of the interference of building waves (wake interference flow), and this flow is characterized by secondary flows in the canyon space.
4 At even closer spacing (greater *H/W* and density), a stable circulatory vortex is established in the canyon, presenting a 'skimming' flow regime that occurs where the bulk of the flow does not enter the canyon.

The second situation concerns the airflow patterns when the wind flows parallel to the urban canyon long axis (angle of between 0° and 30°). In this case, the retardation of the airflow due to the friction along the building walls and the street surface can cause a vertical motion of the air along the canyon walls. It is also important to underline that the relationship between the above-roof free-stream wind speed out of the canyon and the corresponding component of the wind velocity within the canyon, is reported to be directly proportional under certain conditions.

The third situation concerns the airflow patterns when the wind direction outside the canyon forms an angle with the canyon long axis; this is the most common case. In general, when the flow above the roof is oriented at some other angle to the canyon axis, a spiral vortex is induced along the length of the canyon – a cork-screw motion within an elongation along the street.

Activity 5

During the daytime, the absorption of solar radiation by the urban exterior's surface materials is one of the major reasons for increased temperatures in the urban areas. This also affects the nocturnal time period due to the thermal storage of the solar radiation in urban materials. If one can decrease the amount of the absorbed solar radiation, the surface temperature of the exterior construction elements can be reduced; and heat storage can be also be minimized. The simplest solution to reduce the absorption of the solar radiation is to use high albedo materials in order to keep their surfaces cool. Furthermore, as the long-wave radiation changes are very high inside the canyons, it is important to use materials with high emissivities. As a result, these materials are good emitters of long-wave energy and release the energy that has been absorbed as short-wave radiation. In this way their surface temperatures are lower and contribute to the reduction of ambient temperature through the mechanism of heat convection.

The contribution of trees and green spaces can be significant in cooling urban areas and therefore can save on energy (e.g. by decreasing the cooling requirements of urban buildings). Trees can provide solar protection to individual houses during the summer period, while evapotranspiration from trees can reduce urban temperatures (the major mechanism related to trees that decreases urban temperatures, otherwise known as the oasis phenomenon). In addition, trees absorb sound and block erosion-causing rainfall, filter dangerous pollutants, reduce wind speed, stabilize the soil and prevent erosion. For example, the latent heat transfer from wet grass can result in temperatures 6° to 8° Celsius cooler than over exposed soil. In addition, trees and shrubs strategically planted next to buildings can reduce summer air-conditioning costs by 15 to 35 per cent and, in certain situations, by 50 per cent. Simply shading the air conditioner – by using shrubs or a vine-covered trellis – can save up to 10 per cent in annual cooling energy costs.

7

Heat and Mass Transfer Phenomena in Urban Buildings

Samuel Hassid and Vassilios Geros

Scope of the chapter

The scope of this chapter is to understand the main heat transfer processes that take place in buildings, as well as to become familiarized with the notions and the parameters that concern heat transfer procedures. Since these processes are quite complex and of various kinds, the purpose of this chapter is to make clear the differences between the various heat transfer phenomena that contribute to the thermal balance of buildings.

Learning objectives

After completing this chapter, readers will able to understand:

- the principles of the various heat transfer processes that occur in buildings;
- how and why heat transfer processes affect the energy balance of buildings;
- which are the main parameters that influence the heat and mass transfer phenomena in urban buildings;
- how to perform simplified calculations for the various heat transfer processes.

Key words

Key words include:

- heat transfer;
- conduction;
- convection;
- radiation;
- solar radiation;
- heat transfer coefficients;
- time constant of opaque elements;
- effusivity;
- free convection;
- forced convection;
- infiltration;
- ventilation;
- heat gains;
- energy balance.

Introduction

In order to understand heat transfer processes in the urban environment, an understanding of the heat transfer processes in individual buildings is required.

There are many kinds of heat and mass transfer processes in buildings: conduction through walls; solar energy transmission through windows or absorption in the external surfaces of the envelope; radiation from surfaces into the upper atmosphere; heat and moisture transfer by infiltration through cracks; and natural ventilation through openings and mechanical ventilation using fans. All of these processes contribute to the thermal balance of the building and interact with each other. In addition, heat transfer in buildings is affected by thermal storage in the different building elements. It is the purpose of this chapter to apply heat and mass transfer analysis to buildings in order to calculate the different heat transfer rates for the different modes.

This chapter is divided into two main parts. The first part concerns the physics of the heat transfer and rate equations, while the second part presents the principles of heat transfer phenomena in buildings.

Physics of heat transfer and rate equations

An understanding of the physical mechanisms that underlie the different heat transfer modes is important so that rate equations quantifying the amount of energy transferred can be established. The main heat transfer modes (i.e. *conduction*, *convection* and *radiation*) will be considered.

Conduction

Conduction is the mode of heat transfer that is related to activities in the *atomic* or *molecular* level. The processes in the atomic or molecular level depend upon the phase (solid, liquid or gas) of the material and the form of the material (metal, ceramic, etc.); on the microscopic level, they are, indeed, very different.

In liquids and gases, heat transfer by conduction is related to collisions between free molecules: higher temperatures are associated with higher molecular energies. In ceramic materials, the molecules are not free to collide, but are, rather, vibrating nodes of crystal lattices. In metals, heat transfer is related to electrons shared by the metallic lattice being transferred between atoms. In all cases a transfer of energy is taking place from the hotter (more energetic) parts to the colder (less energetic) ones. However, since the heat transfer scientist or engineer is not really interested in the molecular and the atomic level, but in giving a macroscopic description of heat transfer, these differences are not really important.

The rate equation associated with conduction heat transfer is *Fourier's law*:

$$q_x'' = -\lambda \frac{dT}{dx} \tag{1}$$

where: q_x'' is the *heat flux* defined as the heat transfer rate per unit area perpendicular to the direction of heat transfer (W/m^2); dT/dx is the *temperature gradient* in this direction and the constant of proportionality; and λ is a property known as the *thermal conductivity* of the material (W/m.K) (see Table 7.2).

Under steady state conditions and uni-directional heat transfer, the temperature distribution is linear and the heat flux can be expressed as:

$$q_x'' = -\lambda \frac{T_2 - T_1}{L} \tag{2}$$

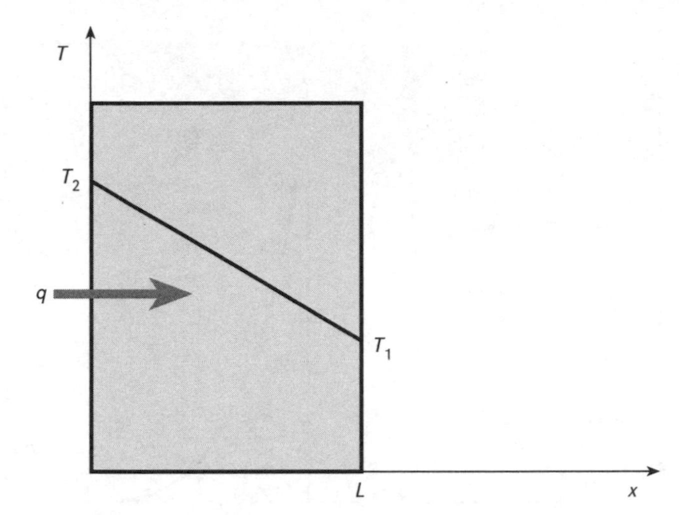

Figure 7.1 *Conduction in a single-layer wall*

where: L is the thickness of a slab; T_2 and T_1 are the temperatures at each side (see Figure 7.1).

The negative sign in both Equations 1 and 2 is a consequence of heat being transferred in the direction of decreasing temperatures.

Convection

Convection is the mechanism whereby energy is transferred by *bulk macroscopic* motion of fluids (liquids or gases), as opposed to the random motion of individual molecules, which characterizes conduction. Convection is associated with large numbers of molecules moving as aggregates.

For heat transfer by convection to occur, two conditions are necessary (see Figure 7.2): fluid motion is taking place and there is a temperature gradient.

When the fluid motion itself is caused by the temperature gradient, then one speaks of *free* or *natural* convection. Free (natural) convection is caused when buoyancy forces result in the upward movement of lighter, hotter fluid, whose place is taken by heavier, cooler fluid.

Convective heat transfer occurs simultaneously with conductive heat transfer. However, in most cases, convection is a much more efficient way of transferring heat than conduction. Fluid flow can be *laminar* (i.e. in a regular pattern constant, or changing only slowly with time), or *turbulent* (i.e. with strong fluctuations of velocity around a mean value). In the first case, heat transfer in the direction of the fluid flow is mainly (in most cases, exclusively) by convection; but heat transfer in direction perpendicular to it is by conduction. In the latter case convection is

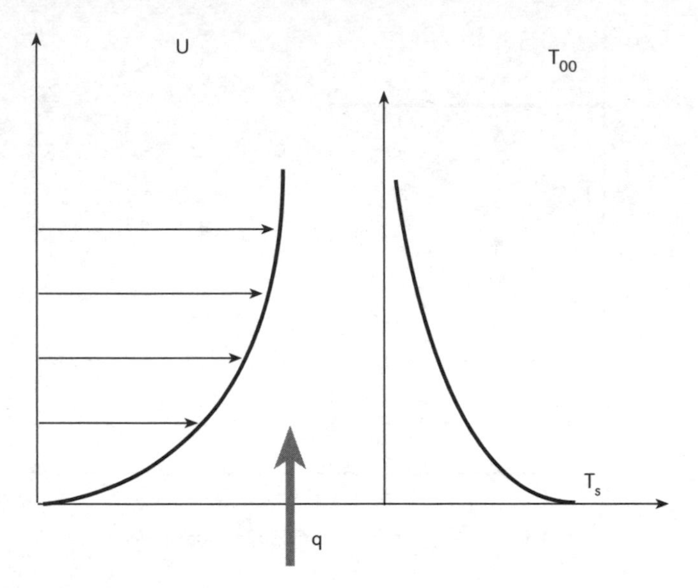

Figure 7.2 *Convection*

the heat transfer mechanism both in the direction of the flow and perpendicular to it.

Regardless of the particular nature of convection heat transfer process (free or forced, laminar or turbulent) the appropriate rate equation is usually of the form:

$$q'' = h_c (T_s - T_\infty) \qquad (3)$$

where: q'' is the convective heat flux (presumed positive when away from the surface and negative when into the surface, although nothing prevents it from being defined the other way round)

T_s is the surface temperature

T_∞ is the fluid temperature

h_c is the convective heat transfer coefficient ($W/m^2.K$).

The h_c coefficient depends upon the fluid velocity or wind speed in the case of forced convection, on the temperature difference itself in the case of free convection, and, of course, on the physical parameters of the fluid. Being proportional to the density, it is much higher for liquids than for gases.

Radiation

Radiation is emitted by matter and transported by electromagnetic waves (although in the alternative description, one considers photons rather than waves). The emission results from changes in the energy levels of electrons in the atoms of matter.

Radiation differs from the other two mechanisms of heat transfer in two ways:

1. Conduction and convection require the presence of matter as a medium. Radiation requires no material medium and, in fact, occurs most efficiently in a vacuum.
2. In conduction and convection, heat transfer is proportional to a temperature difference. In radiation, it is also dependent on the absolute temperature T, measured in Kelvin (temperature in Celsius plus 273.15).

The basic law of radiation is the one governing the rate at which energy is released from a surface per unit area – or emissive power E_b:

$$E_b = \varepsilon \sigma T^4 \qquad (4)$$

where: σ is a universal constant called the Stefan Boltzmann constant

$(= 5.67 \times 10^{-8}\ W/m^2.K)$

ε is a radiative property of the surface called *emissivity* $(0 \leqslant \varepsilon \leqslant 1)$

When the emissivity is equal to 1, emissive power is at maximum and the emitting surface is called a *blackbody*.

Radiation may also be incident on a surface from its surroundings. The radiation may originate from a special source, such as the sun or a lighting device, or it may originate from other surfaces to which the surface of interest is exposed. We designate the rate at which such radiation is incident on a unit surface as the irradiation G.

Since radiation is transmitted by electromagnetic waves, it has a wavelength and a frequency – in fact, a whole range of wavelengths. The wavelength of radiation with which one is concerned in buildings varies between some tenths of a micron (micrometer) to approximately 100 microns. The wavelength of radiated energy depends upon the temperature of the emitting surface: the characteristic wavelength (at which the spectre of emissive power is maximum) is given by Wienn's law:

$$\lambda_{max} T = 2898 \mu m.K \qquad (5)$$

The two particular radiation kinds one is most interested in buildings is *solar* (*short-wave*) radiation, characterized by a wavelength of approximately 0.5μm (*microns* or *micrometers*) and *terrestrial* (*long-wave*) radiation, characterized by a wavelength of approximately 10μm. The part of the spectrum associated with solar radiation is subdivided into visible (wavelengths between 0.4 and 0.76μm accounting for 44 per cent of the energy), *ultra-violet* (wavelengths lower than 0.4m, accounting for 3 per cent of the energy) and *infrared* (wavelengths higher than 0.6μm, accounting for 53 per cent of the energy).

Incident radiation may be *absorbed*, *reflected* or *transmitted*. The coefficients governing these processes are the absorptivity (α) (see Table 7.3), the reflexivity (ρ) and the transmissivity (τ) – all those coefficients varying between 0 and 1. A material with non-zero transmissivity is called *transparent*. Whereas absorption and emission increase and reduce, respectively, the thermal energy of matter, transmission and reflection have no effect upon it. The above properties depend upon the characteristic wavelength of the radiation and are often very much different for solar and for long-wave radiation: glass transparent to solar radiation is usually opaque to long-wave radiation and a wall is white (i.e. reflective) to solar radiation, but almost black (absorptive) to long-wave radiation.

One is usually interested in the *net radiative heat transfer* from a surface, which is equal to the difference between the emitted and absorbed energy:

$$q''_{rad} = \varepsilon\sigma T_s^{\,4} - \alpha G = \varepsilon\sigma(T_s^{\,4} - T^4_{mrt}) \qquad (6)$$

where: T_s is the absolute temperature of a surface (in K); T_{mrt} is the mean radiant absolute temperature of the surroundings (see Figure 7.3).

The mean radiant temperature will be defined later, but in Equation 6 it is shown that net radiative heat transfer is proportional to the difference of the fourth power of absolute temperatures.

In many cases it is convenient to express net radiation exchange terms of a temperature difference:

$$q_{rad}'' = h_r(T_s - T_{mrt}) \qquad (7)$$

where h_r can be derived from Equation 6:

$$h_r = \varepsilon\sigma(T_s + T_{mrt})(T_s^{\,2} + T_{mrt}^{\,2}) \approx 4\varepsilon\sigma[(T_s + T_{mrt})/2]^3 \qquad (8)$$

Here radiation is modelled in a manner similar to convection and made proportional to a *linearized* difference. Note, however, that there are several differences between the two expressions. In Equation 8, the mean radiant temperature T_{mrt} replaces the fluid temperature, and h_r is much more dependent on temperature than h_c.

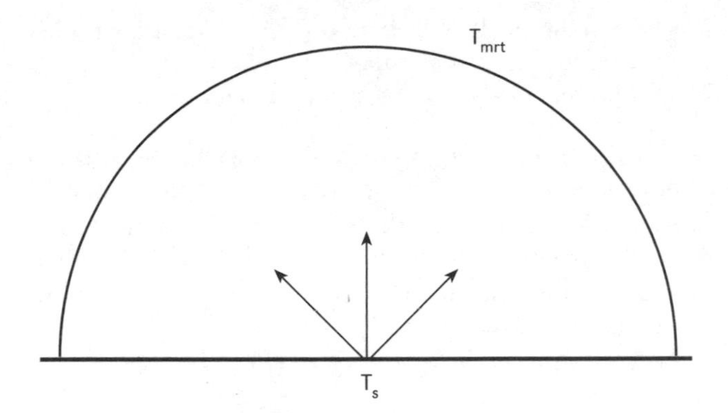

Figure 7.3 *Radiation*

Equations 3 and 7 are often combined to form an equation giving the thermal balance from a surface:

$$q'' = h(T_s - T_o) \qquad (9)$$

where:

$$h = h_c + h_r \qquad (10)$$

$$T_0 = \frac{h_c T_\infty + h_r T_{mrt}}{h} \qquad (11)$$

Principles of heat transfer in buildings

Conduction

Uni-directional steady-state conduction heat transfer

Conduction heat transfer in building elements (walls and roofs) is usually considered uni-directional, since the dimensions of walls in the direction of heat transfer are much larger than the ones perpendicular to it. These temperature gradients are usually much larger in that direction (see Equation 2).

There are some limited regions where two building elements border each other or in the corners of buildings, where gradients in the direction perpendicular to the main flow are appreciable. These are called *thermal bridges* and are usually treated separately, their effect being relatively small, although by no means negligible (and will be summarily treated further on).

Conduction in walls, on the other hand, is *not* usually steady state for the time constants of meteorological phenomena: it involves storing and releasing energy. However, steady-state heat transfer is relevant in two cases:

1 When one deals with long-term, averaged quantities, as opposed to instantaneous quantities.
2 When one deals with light-weight elements.

The reason that conduction steady-state heat transfer is so important is because, in many countries, in the insulation standards and in other regulations connected to energy consumption in buildings much attention is given to the conduction steady-state properties of building elements for the reasons stated above.

Equation 2 can be rewritten in several forms:

$$q_x'' = -\frac{T_2 - T_1}{r} \qquad (12)$$

where: r is the *thermal resistance* of the building element.

This last term is based on the analogy between thermal and electrical resistance. Equation 12 is effectively similar to Ohm's law, with the heat flux replacing the electric and the temperature difference replacing the potential difference (voltage).

Using the electrical analogy, one can use Equation 12 for multilayered walls, as well – and most walls consist of more than onc layer:

$$r = r_1 + r_2 + r_3 + \ldots = \frac{L_1}{\lambda_1} + \frac{L_2}{\lambda_2} + \frac{L_3}{\lambda_3} + \ldots \qquad (13)$$

where: r is the total thermal resistance of the wall (or roof); r_1, r_2, r_3, etc are the resistances of each individual layer (see Figure 7.4).

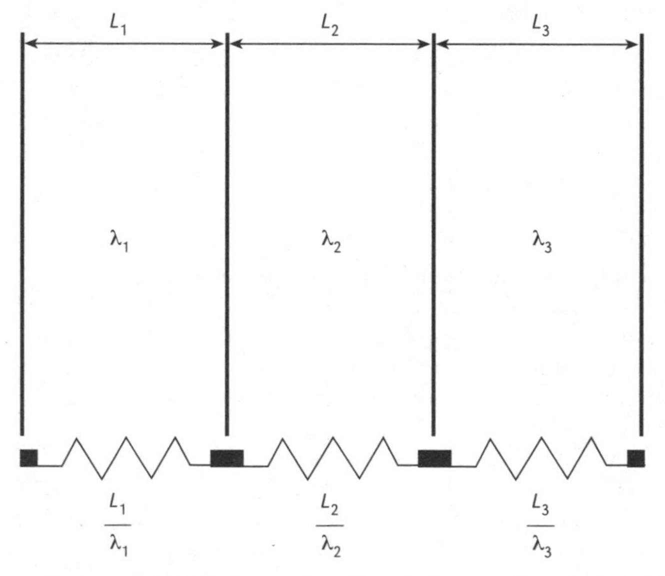

Figure 7.4 *Multilayered wall and electrical analogue*

Equation 13 can be used in conjunction with Equation 12 to relate the steady-state heat flux through a multilayered wall to the temperature difference between the internal and the external *surface* temperatures.

Usually one needs to relate the heat flux through a wall to the difference between the internal and external *air* temperatures (rather than the surface temperatures). To this end, one needs to account for heat transfer between the surface and the air, by convection, and to other surfaces by radiation. This is usually done by using a film resistance for each surface, defining a *total thermal resistance* R (m^2.K/W) and a *total thermal conductance* U (W/m^2.K), also called the U-value in several US or UK references:

$$q_x'' = -\frac{T_o - T_i}{R} = -U\,(T_o - T_i) \qquad (14)$$

$$R = \frac{1}{h_o} + r + \frac{1}{h_i} \qquad (15)$$

where: T_o is the internal air temperature; T_i is the external air temperature; h_o is the external heat film conductance; h_i is the internal heat film conductance.

In Equation 14, the heat flux is positive when it is from inside the house in the external direction.

Equation 14 is often used; but one should clarify that it is correct only in the case that the outside air temperature is equal to the mean radiant temperature and does not account well for radiative heat transfer. This can be done by using the sol-air temperature, to be defined later.

Uni-directional unsteady-state conduction heat transfer

As stated above, conductive heat transfer in building elements is seldom steady state: the time constants involved are of the same order of magnitude as the time constants of climatic change. Therefore, conductive heat transfer in walls is accompanied by heat storage. In this section, we will consider one-dimensional heat transfer in building elements and the ways of accounting for variation of temperature with time – concentrating on heat transfer in buildings.

Heat conduction (diffusion) differential equation

The heat diffusion equation runs:

$$\rho c_p \frac{\partial T}{\partial t} = \frac{\partial}{\partial x}\left(\lambda \frac{\partial T}{\partial x}\right) \qquad (16)$$

Where ρ is the *density* of the material (kg/m³); c_p is the *specific heat* of the material (for gases, the specific heat at constant pressure), measured in J/kg.K (although Wh/kg.K is sometimes used) (see Table 7.2 for values concerning various materials).

Equation 16 is sometimes presented as a diffusion equation, using the *thermal diffusivity* κ (m²/s):

$$\frac{\partial T}{\partial t} = \frac{\partial}{\partial x}\left(\kappa \frac{\partial T}{\partial x}\right) \tag{17}$$

where:

$$\kappa = \lambda / \rho c_p \tag{18}$$

The conduction equation has several kinds of boundary conditions. In building analysis, one usually focuses on simplified solutions that can easily be applied to building applications, involving some kind of *lumping* of the conductance.

Note that since many building elements (especially concrete) have a specific heat of approximately 1kJ/kg.K (0.24kWh/kg.K) thermal diffusivity is dependent on density only.

Lumping of thermal mass in building elements: Thermal time constant

The easiest way of analysing the unsteady-state thermal behaviour of a slab of thickness L is by lumping all its thermal capacitance C in the middle:

$$C = \rho c_p L \tag{19}$$

In the electric analogue, this is equivalent to connecting an earthed capacitance C to the middle of the resistance (see Figure 7.5). Thus, an element is formed whose time constant is:

$$T_c = \frac{L^2}{2\kappa} = \frac{1}{2}\left(\frac{L}{\lambda}\right)\left(L \rho c_p\right) = CR/2 \tag{20}$$

Thus, considering a single layer wall, the response of the internal temperature $T_i(t)$ to a step change ΔT in the outside temperature (the initial temperature in the slab being uniform and equal to the outside temperature $T_o(0)$ before the change) is given by:

$$T_i(t) = \Delta T[1 - \exp(-t/T_c)] + T_o(0) \tag{21}$$

More generally, the response of the internal temperature to changes in external temperature is given by:

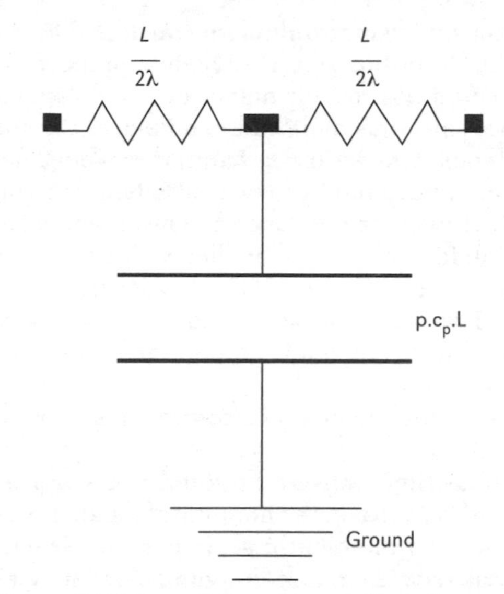

Figure 7.5 *Electrical analogue of one-layer wall (unsteady state)*

$$T_i(t) - T_i(0)\exp(-t/T_c) = \int_0^t T_o(t')\exp[-(t-t')/T_c]dt'/T_c \tag{22}$$

Equation 21 is a special case of Equation 22, in which the initial value of T_i and a continuous variation of T_o with time are taken into account. Both equations being linear, one can measure the temperature above an arbitrary level (say, 0° Celsius or 20° Celsius).

Time constant of multilayer constructions

It is possible to generalize the thermal time constant of Equation 20 for multilayer constructions. For such constructions, the thermal time constant can be defined as the energy stored per unit area in the construction per unit change in heat flux and is equal to:

$$T_c = \sum_{i=1}^{n} \rho_i c_{pi} L_i \left(\frac{1}{h_o} + \sum_{j=1}^{i-1} \frac{L_j}{\lambda_j} + \frac{L_i}{2\lambda_i}\right) \tag{23}$$

In Equation 23, the layers are numbered from the outside inwards. Equation 23, strictly speaking, gives the external thermal time constant and an internal one can be defined by applying the equation with the terms being numbered from the inside outwards; but such a time constant is seldom used.

Equation 23 can be used in conjunction with Equations 21 and 22 when the internal temperature response to external temperature variations is required.

A major characteristic of Equation 23 is the importance of the relative position of the layers. If a concrete slab is insulated using an insulating material, the thermal resistance, describing the steady state properties of the construction, is the same whether the insulating material is placed inside, outside or in the middle of the slab. The time constant, on the other hand, describing the non-steady state properties of the construction, is completely different. It is very large when the insulating material is placed outside and much smaller when the insulating material is placed inside. This shows that two building elements may have the same steady-state heat transfer properties, but a completely different transient behaviour.

Limitations of the thermal time constant method: Effusivity

The thermal time constant method is an approximate method and is based on the lumping of the thermal capacity of a building element in a single node. The thermal storage capacity in real elements is not lumped, but distributed along the element, and the lumping may introduce an error which might become appreciable. It works reasonably well for that time constant, but is not appropriate for describing the thermal behaviour for times shorter than that time constant. In particular, it is not appropriate for describing the thermal response to solar energy incident inside the thermal element.

For such problems, the *effusivity* method is more appropriate.

Consider Equation 16, with an initially uniform temperature distribution and a step change q_0 in the thermal flux. The expression for the temperature for an infinite slab, which can be found using an integral method, is:

$$T(t) = q_o \sqrt{t} / \alpha \qquad (24)$$

where: α is the effusivity, is defined as:

$$\alpha = \sqrt{\lambda \rho c_p} \qquad (25)$$

As already stated, the effusivity method, consistent with a variation with time which is proportional to $\sqrt{t}$, is valid for relatively short times, when the heat wave has not reached the other side of the slab. For longer times, the influence of the boundary conditions on the other side of the slab is felt. Effusivity represents the response of the surface temperature to the heat flow changes on the wall surface. The lower the effusivity, the higher the sensitivity of the surface temperature to the heat flow changes.

Other solutions of the transient uni-directional conduction equation

As already stated, the uni-directional transient conductive heat transfer equations mentioned above are approximate – the reason for the approximation being the desire to keep a physical insight into the interplay of conductive heat transfer and storage.

There are several cases where the above equations fail to adequately describe heat transfer, and alternative methods can be used. These methods are as follows:

- Methods based on a *numerical solution* of a difference equation based on the conduction Equation 16, with appropriate boundary conditions on each side of the wall and at the interface between two elements. Usually, the layers are subdivided into two or three parts for an adequate solution to be obtained.
- Methods based on *response factors*, giving the heat flux in terms of not only the temperature, but also the temperature and heat flux history of the construction. The response factors are a series of numbers calculated by solving Equation 16 using Laplace and Fourier analysis and are mainly used in the US (ASHRAE Fundamentals, 1985).
- The *admittance* method, based on Fourier analysis, used in the UK.

Convective heat transfer

The study of convective heat transfer is linked to that of fluid flow. The equations describing fluid flow are essentially non-linear and their closed solution is possible only in some special cases. Engineering calculations are usually based on a combination of theory, experiment and dimensional analysis.

For heat transfer calculations in buildings, the main formulae used are the ones of *Newton's Law of Cooling*, describing heat transfer between a solid surface and a fluid in contact with it (see Equation 9), as well as the equation describing bulk heat transfer between two fluids with different temperatures, used in calculations of *infiltration* and *ventilation* heat transfer:

$$Q = \rho\, c_p\, U\, (T_u - T_d) \qquad (26)$$

where: Q is the bulk heat transfer; U is the bulk velocity; T_u is the temperature upstream; T_d the temperature downstream

In what follows, the emphasis is on heat transfer to air (density of the order of 1.1 to 1.2 kilograms per cubic metre, specific heat of 1000J/kg.K, viscosity 1.6 x 10^{-5}, volumetric expansion coefficient equal to $1/T$).

Convection coefficients from infinite plates: Free convection

As already stated, in *free* or *natural* convection the motive force of the movement of the fluid is the temperature difference itself and the ensuing buoyancy forces created by the density differences. Thus, one expects the convective film heat transfer coefficient to depend upon the temperature differences.

Of particular importance in building heat transfer are the convective heat transfer coefficients from horizontal plates. These are influenced by two factors:

1 Whether the heat transfer flow is *upwards* (for example, from a hot floor or roof to the air above it or from a cold ceiling to the air below it) or *downwards* (from a hot ceiling to the air below it or from a cold floor or roof to the air above it). In the former case, the heat transfer is more intense than in the latter case because of the direction of the buoyant forces.
2 Whether the heat transfer is *turbulent* or *laminar*, which is the determining criterion for this case.

A characteristic geometric parameter that is important for estimation of the convective heat transfer coefficients is the length (L) of the plate in the direction of the buoyancy-driven flow (e.g. in a vertical wall, L is the height of the wall). For heat flow *upwards*, the heat convection coefficient is given by:

$$h_c = 1.32\ (\Delta T/L)^{1/4} \text{ for laminar flow } (L^3\ \Delta T \leq 1)$$
$$h_c = 1.52\ (\Delta T)^{1/3} \text{ for turbulent flow } (L^3\ \Delta T > 1) \quad (27)$$

For heat flow *downwards*, only laminar flow is possible since the stabilizing buoyant forces dampen turbulence:

$$h_c = 0.59\ (\Delta T/L)^{1/4} \quad (28)$$

Correlations similar to Equation 27, with slightly different coefficients, are valid for *horizontal* heat flow, from *vertical* plates (walls):

$$h_c = 1.42\ (\Delta T/L)^{1/4} \text{ for laminar flow } (L^3\ \Delta T \leq 1)$$
$$h_c = 1.31\ (\Delta T)^{1/3} \text{ for turbulent flow } (L^3\ \Delta T > 1) \quad (29)$$

Equation 29 can be generalized to cover flow from inclined surfaces with tilt β, between 30° and 90° from the horizontal plane:

$$h_c = 1.42\ (\Delta T \sin \beta/L)^{1/4} \text{ for laminar flow } (L^3\ \Delta T \leq 1)$$
$$h_c = 1.31\ (\Delta T \sin \beta)^{1/3} \text{ for turbulent flow } (L^3\ \Delta T > 1) \quad (30)$$

In Equations 27 to 30, the heat transfer coefficient is in W/m^2.K and the other parameters are in SI units.

Convection coefficients from infinite plates: Forced convection

In forced convection, the fluid movement is independent of the thermal gradients and therefore is independent of the orientation of the plane (horizontal or vertical). For laminar flow over plates, the convective heat transfer coefficient in air is approximately:

$$h_c = 2\ (U/L)^{1/2} \quad (31)$$

where: L is the length of the surface in the direction of the flow; U the velocity away from the plate (the wind speed in metres per second).

However, air flow around buildings is very seldom laminar, and Equation 31 applies only if $UL < 1.4$ square metres per second. For turbulent flow (UL ˘ 1.4 square metres per second), the convection coefficient is given by:

$$h_c = 6.2\ (U^4/L)^{1/5} \quad (32)$$

Equations 31 and 32 are applicable for flows in open spaces. Equation 32 is applicable for smooth surfaces (such as glass) and the coefficient can be twice as big for rough ones (such as stucco walls). It has to be modified for buildings in urban spaces.

Convective heat transfer: Infiltration and ventilation of buildings

Heat transfer in buildings may also be due to *infiltration* and *ventilation*. In both cases, heat transfer comes as a result of *airflow* into and out of buildings. In the case of *infiltration*, the movement is uncontrolled and involuntary and is through cracks in the building walls and around openings. In the case of *ventilation* the airflow is desirable in order to maintain indoor air quality and occasionally for thermal comfort reasons, by lowering the internal temperature or increasing wind speed. *Ventilation* may be *natural* (i.e. through opened windows) or *forced* through fans and air-conditioning systems.

In both cases, the heat loss/gain (depending upon whether the internal temperature is higher/lower than the external one) Q_{inf} is given by:

$$Q_{\inf} = \rho c_p \dot{V}(T_{in} - T_o) \qquad (33)$$

where: $\dot{V}$ is the volumetric flow rate of air.

The volumetric flow rate of air is expressed in cubic metres per second; but often one sees it in *air changes per hour* (*ACH*), which are obtained by dividing the volumetric flow rate by the volume V of the building and then multiplying by 3600 (number of seconds in an hour):

$$ACH = 3600\dot{V}/V \qquad (34)$$

Both infiltration and ventilation are governed by two main physical mechanisms:

1. The *stack effect* or *buoyancy forces*: this results from the density difference between the internal and the external air and is proportional to $\Delta\rho \cdot g \cdot H_L$, $\Delta\rho$ being the difference between the internal and external air density and H_L the difference in height between the uppermost and lowermost opening.
2. The *wind forces*, resulting from differences in static pressure over the building envelope due to the wind. These forces are proportional to ρV^2 and depend upon the relative position of the building in relation to the wind direction.

The wind forces are usually the dominant ones, with the stack effect being important only for very low wind speeds (less than 1 metre per second) or in cases of very high multistorey buildings. Although the two forces may theoretically oppose each other, they usually enhance each other. However, their combined effect cannot be described by adding the airflow rate one calculates by each one separately.

The airflow rate across an opening is proportional to ΔP^n, n being an exponent with typical values two-thirds for tight cracks to one-half for larger cracks and for openings.

Estimating infiltration rate

Different approaches are used in estimating infiltration, with varying degrees of complexity. The simplest one is the air exchange method, in which the airflow rate depends only upon the number of openings in each room and the volume of the room (see Table 7.1).

This method is easy to use, but fails to take into account the dynamics of airflow in buildings. It would give the same flow rate in a windy and in a calm region, and for the same house it would imply that the air flow rate is constant independent of the external and internal conditions.

Table 7.1 *Airflow rate due to infiltration according to the number of windows and exterior doors*

Type of room	Air changes per hour
No windows or exterior doors	0.5
Windows or exterior doors on one side	1.0
Windows or exterior doors on two sides	1.5
Windows or exterior doors on three sides	2.0
Entrance halls	2.0

Note: for windows with weather stripping or storm sash, use two-thirds of these values.

A method applicable for detached houses is the *LBL method*, in which the air-tightness of a building is characterized by the *effective leakage area* (*ELA*), which is defined as the area of a perfect orifice that at a pressure difference of 4 Pascales (Pa) between inside and outside the house gives the same airflow rate Q_4 (in cubic metres per second) as in the house. One can picture the ELA as concentrating the area of all the cracks of a house in a single orifice:

$$ELA(m^2) = \sqrt{Q_4 / (8/\rho_{air})} \qquad (35)$$

The effective leakage area is determined experimentally by the blower-door method – although the airflow rate at 4Pa is not actually measured since it is too unreliable. Usually, measurements are performed between 10Pa and 50Pa and the result extrapolated to 4Pa. Alternatively, the ELA can be estimated from the length of the cracks on the surface of the building.

Once the ELA is known, the infiltration can be estimated by calculating first the contribution of the stack effect Q_{st} and then the contribution of the wind forces Q_w (see Figure 7.6):

$$Q_{st} = 0.25\ ELA\sqrt{gH_L|\Delta T|/T} \qquad (36)$$

$$Q_w = C\ ELA\ U\ \frac{\alpha_b H_b^{\gamma_b}}{\alpha_s H_s^{\gamma_s}} \qquad (37)$$

where: ΔT is the difference between the internal and the external temperature; C is a coefficient dependent on the terrain of the house (varying between 0.3 for open terrain to 0.1 for city centres); U is the wind speed (in metres per second); H is the height of the building (in metres); H_L is the difference in height between the highest and the lowest cracks (in metres).

α and γ depend on the surroundings and vary between 1 (for open spaces) to 0.5 (to city centres) for α and between 0.5 (for open spaces) and 0.1 (for city centres) for γ. The subscripts b and s refer to the building and the weather station in which the measurements were taken correspondingly.

Once Q_{st} and Q_w are calculated, the total infiltration may be estimated:

$$Q = \sqrt{Q_{st}^2 + Q_w^2 + Q_{vent}^2} \tag{38}$$

where: Q_{vent} is the forced ventilation rate, if applicable.

In more sophisticated models, a mass flow balance of the flows in each room is solved in conjunction with the equations relating the pressure difference across each opening. The equation is non-linear and therefore the use of an iterative procedure is required. These are network and modal models, such as COMIS, AIRNET, AIOLOS and others.

In addition, the use of more analytical models such as computational fluid dynamics (CFD) models can facilitate a more detailed study of ventilation and infiltration. CFDs are based on the solution of Navier-Stokes equations by using numerical methods. More precisely, these models use the conservation equations of mass, of momentum and of energy, as well as the transport equations for the turbulent velocity. One can use these numerical models in order to determine the air velocity and temperature profiles, as well as pressure, in a limited space (e.g. a building). These models require the geometrical description of the space and of the external environment when natural ventilation is studied. This representation is performed by using a grid that divides the model into basic volumes (cells).

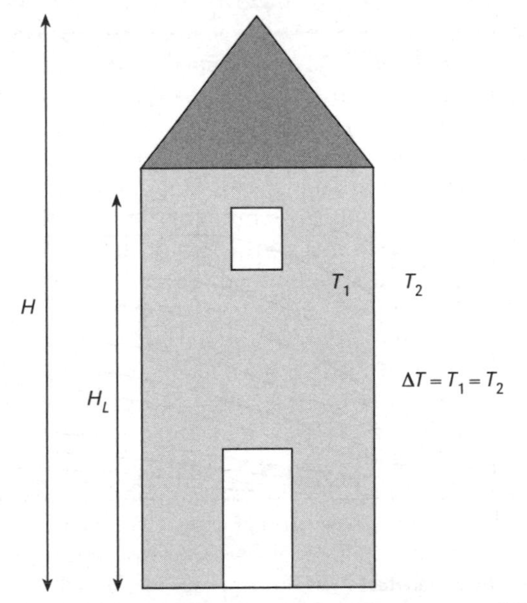

Figure 7.6 *Stack effect in LBL infiltration model*

Finally, a third category of models comprises the zonal models. These models are based on an approach that is intermediate between network models, which give no information about airflow patterns, and CFD models, which give detailed temperature and flow distributions but are computationally intensive. Such intermediate models execute much faster than CFD calculations, yet model heat and mass transfer in greater detail than the network approach and provide temperature, concentration and flow distributions that are detailed enough to evaluate parameters such as infiltration and ventilation.

Radiation heat transfer in buildings

Radiation is a very important mechanism of heat transfer. In buildings, there are two different radiation applications:

1 Radiation between surfaces at terrestrial temperatures (or long-wave radiation), as is the case between two walls or a wall and neighbouring buildings or the sky. This particular form of radiation is usually treated using the *grey surface* methodology.
2 Solar radiation, transmitted, reflected or absorbed by different surfaces.

Grey surfaces

In principle, the radiation properties of matter (emissivity, transmissivity, absorptivity, reflexivity) depend upon the wavelength of incident radiation. For monochromatic (i.e. single wavelength) radiation, it can be proved on the basis of thermodynamic equilibrium considerations that the emissivity is equal to the absorptivity:

$$\varepsilon(A) = \alpha(\lambda) \tag{39}$$

However, given that the radiation spectrum is usually different for the radiation *emitted* by a surface and the one *absorbed* by it, which originates from surfaces with temperatures different from the one of the absorbing temperature, the energy-spectrum weighted-mean emissivity is different from the mean absorptivity:

$$\varepsilon \neq \alpha \tag{40}$$

However, if the temperature difference between the surfaces that exchange energy by radiation is relatively small, one can ignore the dependence of the emissivity (and absorptivity) on the spectrum and use the grey surface theory, in which the emissivity is assumed to be independent of the spectrum. Therefore, Equation 39 gives:

$$\varepsilon = \alpha \tag{41}$$

A corollary of Equation 41 for non-transmissive surfaces is:

$$\varepsilon + \rho = 1 \tag{42}$$

Where ρ is the reflectivity when Equation 42 is valid, the radiosity J of the surface is related to its irradiation G and its emissive power E_b (see Figure 7.7):

$$J = \varepsilon E_b + \rho G = \varepsilon \sigma T^4 + (1 - \varepsilon) G \tag{43}$$

The irradiation G_i is related to the radiosity of surfaces J_j with which surface i exchanges heat by radiation through the radiation *shape factors* F_{ji}, a quantity that is strictly geometric and depends upon the relative position and orientation of surfaces i and j. F_{ji} does not depend upon the emissivity and temperature of the surfaces.

G_i is related to the radiosity of the surfaces through the following linear relation:

$$G_j = \frac{1}{A_j} \sum_{i=1}^{n} A_i F_{ji} J_i \tag{44}$$

A basic property of the shape factor is the reciprocity relation:

$$A_i F_{ij} = A_j F_{ji} \tag{45}$$

which simplifies Equation 44:

$$G_j = \sum_{i=1}^{n} F_{ji} J_i \tag{46}$$

Another relationship that can be derived from the first law of thermodynamics states that the sum of the shape factors for a given surface is equal to unity:

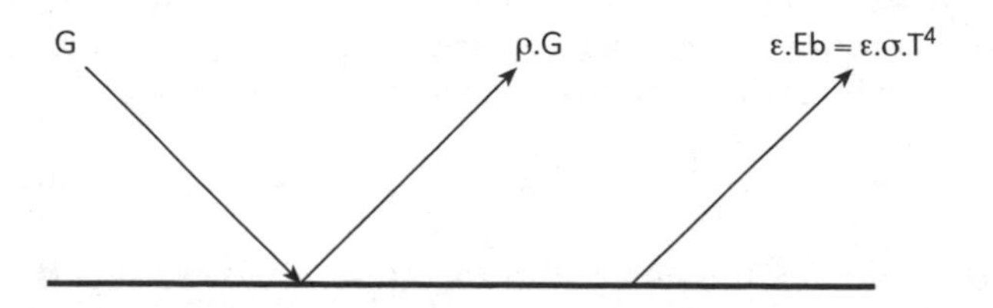

Figure 7.7 *Grey surface radiation*

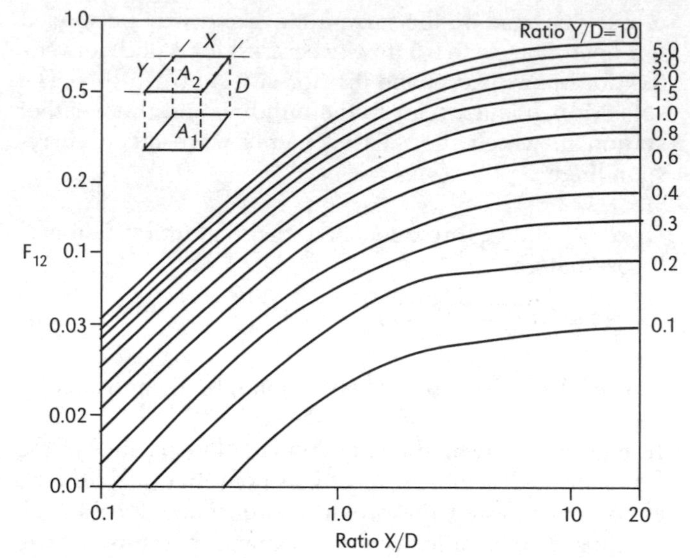

Source: Kreider and Rabl, 1994

Figure 7.8 *View factors between parallel and equal rectangles as a function of geometry*

$$\sum_{i=1}^{n} F_{ji} = 1 \tag{47}$$

F_{ji} is 0 for plane and convex surfaces, but has a finite value for concave surfaces, in which some of the energy radiated by the surface is incident upon it. The shape factor algebra applies for diffusely emitting and reflecting surfaces.

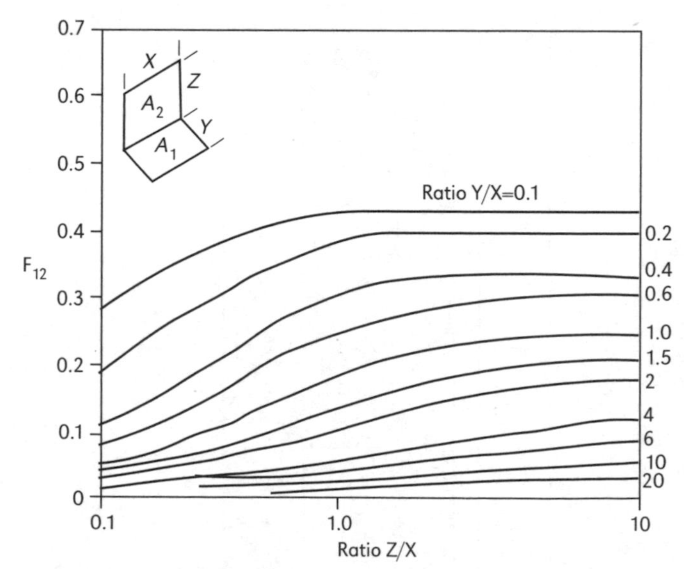

Source: Kreider and Rabl, 1994

Figure 7.9 *View factors between perpendicular rectangles as a function of geometry*

For radiative heat transfer between adjacent rectangular walls (or between vertical walls and ceilings) and heat transfer between opposing planes of a rectangular room, the shape factors as a function of the ratios of the edges (see Figures 7.8 and 7.9). A special case is the faces of a cube, in which both the shape factor between adjacent surfaces and between opposing surfaces is equal to 0.2.

Radiative exchange

The simplest case of radiative heat transfer is between opposing parallel infinite surfaces. Using the methodology developed in the previous section, with the shape factors equal to 1, one can obtain an equation for the heat transfer by radiation between two parallel surfaces, which is used to calculate the heat transfer rate by radiation in gaps, which are parts of walls and double-glazed windows (see Figure 7.10):

$$q_1 = \frac{\sigma (T_2^4 - T_1^4)}{1/\varepsilon_1 + 1/\varepsilon_2 - 1} \tag{48}$$

Equation 48 is a special case of heat transfer between two surfaces, which are configured so that one closes completely on the other:

$$q_1 = \frac{\sigma (T_2^4 - T_1^4)}{(1-\varepsilon_1)/\varepsilon_1 + (1-\varepsilon_2)A_1/\varepsilon_2 A_2 + 1/F_{12}} \tag{49}$$

A special case in Equation 49 is when one of the two surfaces is flat, which can be used to calculate heat transfer between the attic floor and the surfaces of an attic, assuming that they have the same temperature (see Figure 7.10):

$$q_1 = \frac{\sigma (T_2^4 - T_1^4)}{(1-\varepsilon_1)/\varepsilon_1 + (1-\varepsilon_2)A_1/\varepsilon_2 A_2 + 1} \tag{50}$$

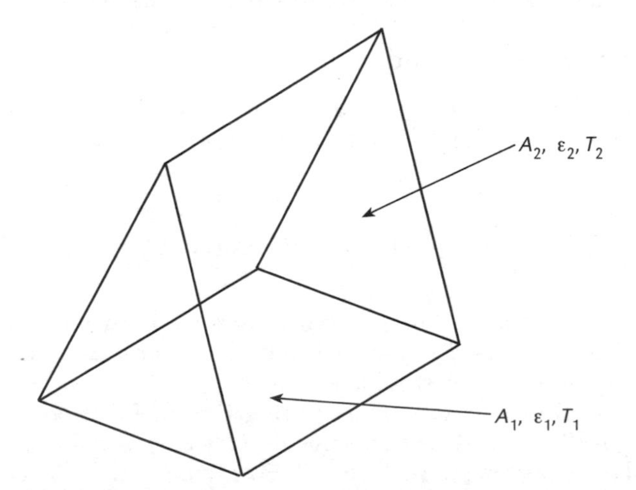

Figure 7.10 *Radiative heat transfer between a flat floor and the surfaces of an attic*

Radiative heat transfer coefficients

Radiative heat transfer is often presented in terms of a *heat transfer coefficient* – similar to the convective heat transfer coefficient. For example, the heat transfer coefficient characterizing the radiative heat transfer between two infinite plates is equal to:

$$h_r \approx \frac{4\sigma [(T_1 + T_2)/2]^3}{1/\varepsilon_1 + 1/\varepsilon_2 - 1} \tag{51}$$

For $\varepsilon_1 = \varepsilon_2 = 0.8$ and mean temperature equal to 300K, the heat transfer coefficient is approximately equal to 4.1W/m^2.K.

Another convenient transformation is to express the heat transfer from a surface in terms of the *mean radiant temperature* T_{mr}, defined as the temperature of a black enclosure exchanging heat by radiation with the given surface at the same rate as the actual surroundings. Thus, the radiative heat exchange with a surface with temperature T_s and emissivity ε_s is equal to:

$$Q/A = \varepsilon_s \sigma (T_s^4 - T_{mr}^4) \approx h_r (T_s - T_{mr}) \tag{52}$$

The mean radiant temperature is a convenient way of lumping all of the temperatures of the surfaces with which radiative exchange occurs into a single temperature – equivalent to a delta-star transformation in electrical networks.

Solar radiation

Solar radiation has a major contribution to the energy balance of buildings, both during the winter, when this contribution is desirable, and during the summer, when it may be a major component of the cooling load.

Solar radiation calculations feature three stages:

1. calculation of the solar coordinates;
2. calculation of radiation intensity at a particular surface;
3. calculation of solar heat gain.

Solar coordinates

The position of the sun in the sky is defined by the two solar coordinates (see Figure 7.11):

1. the solar *elevation* or *solar height* β (i.e. the angle between the sun's rays and the horizontal plane);
2. the solar *azimuth* φ (i.e. the angle between the projection of the solar rays on the horizontal surface and the south, in the Northern hemisphere).

Both angles are functions of:

- the *solar hour angle H*;
- the *latitude L*;
- the *declination angle* δ.

The *solar hour angle H* can be derived from the apparent solar time (AST):

$$H \text{ (in degrees)} = (AST - 12)/15 \quad (53)$$

It is negative before solar noon and positive thereafter.

The apparent solar time can be derived from the local standard time (*LST*):

$$AST = LST + ET - 4(LSM - LON) \quad (54)$$

where *ET* is the equation of time, expressing the non-uniformity of the sun movement in the sky; *LSM* is the Local Standard Meridian characterizing the time of a given place (in degrees east of the Greenwich meridian); *LON* is the Local Longitude of that place.

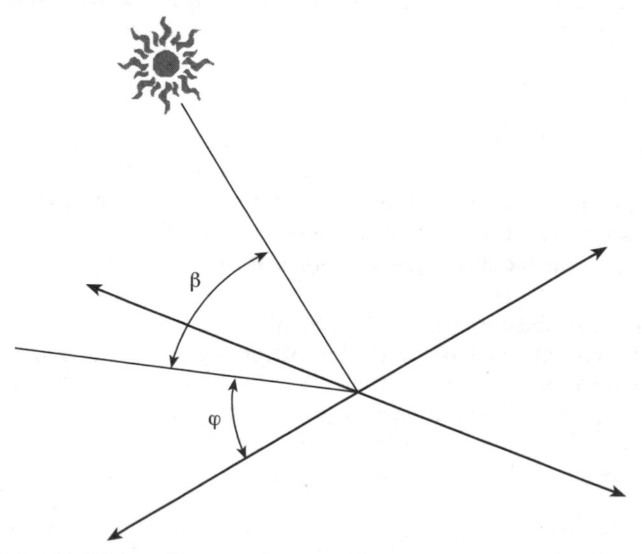

Figure 7.11 *Solar coordinates (elevation β and azimuth φ)*

The factor '4' is used to convert degrees of latitude into minutes of time.

ET is a function of the day of the year *n* (varying from 1 to 365) and is usually given in minutes:

$$ET = 9.87 \sin(4\pi \frac{n-81}{364}) - 7.53 \cos(2\pi \frac{n-81}{364}) - 1.5 \sin(2\pi \frac{n-81}{364}) \quad (55)$$

ET varies from +16 minutes at the end of October to –14 minutes in mid-February.

The local standard meridian (LSM) characterizes the standard time of a certain place and is usually a multiple of 15°: 0° for the UK and Ireland, 15° for Western Europe and 30° for Eastern Europe (Finland and Greece). During daylight savings time, the LSM is pushed 15° eastwards. Note that in textbooks applicable to the US, the sign in front of *LSM – LON* term in Equation 54 is positive, since Western latitudes are considered positive there.

Solar height and azimuth

The solar height and the azimuth are given by:

$$\sin\beta = \cos H \cos L \cos\delta + \sin\delta \sin L \quad (57)$$

$$\cos\phi = \frac{\sin\beta \ \sin L - \sin\delta}{\cos\beta \ \cos L} \quad (58)$$

The equation for the azimuth should be used with the proviso that the azimuth is *negative* (east of south) in the morning and *positive* (west of south) in the afternoon.

Intensity of solar radiation

The *solar constant* I_0 gives the intensity of solar radiation *outside* of the Earth's atmosphere (normal to the solar rays):

$$I_o\,(W/m^2) = 1353\,(1 + 0.033 \cos\left(2\pi \frac{n-1}{365}\right)) \quad (59)$$

There is an almost 7 per cent difference between the highest value, on 2 January, when the sun is nearest to the Earth, and the lowest value in June.

Satellite measurements during the time period of 1978–1998 report a mean value of daily averages for the solar constant equal to 1366.1W/m^2, having a minimum–maximum range of 1363–1368W/m^2. Adjustments yielded 'a solar constant' calculated as 1366.22W/m^2.

Once in the atmosphere, some of the solar radiation reaches the ground as *direct normal radiation* I_{DN} (or *beam radiation*), some is *reflected* back to space by clouds, some is *absorbed* and some is *scattered* (by clouds and particles) and reaches the ground as *diffuse* radiation, with little or no directionality.

Solar radiation on clear days

The solar beam radiation on clear days is related to the solar constant through a correlation dependent upon the solar elevation:

$$I_{DN} = I_o[a_o + a_1 \exp(-k/\sin \beta)] \quad (60)$$

The coefficients a_o, a_1 and k are functions of the *visibility* and the *altitude* A (in kilometres) of the site and are given by:

For visibility 23km
$a_o = r_o[0.4237 - 0.00821\ (6 - A)^2$
$a_1 = r_1[0.5055 - 0.00821\ (6.5 - A)^2]$
$k = r_k[0.2711 - 0.01858\ (2.5 - A)^2]$

For visibility 5km
$r_o[0.2538 - 0.0063\ (6 - A)^2]$
$r_1[0.7678 - 0.0010\ (6.5 - A)^2]$
$r_k[0.2490 - 0.0810\ (2.5 - A)^2]$

Values of r_o, r_1 and r_k

Climate type	r_o		r_1	r_k
	Visibility			
	23km	5km		
Tropical	0.95	0.92	0.98	1.02
Mid-latitude summer	0.97	0.96	0.99	1.02
Sub-arctic summer	0.99	0.98	0.99	1.01
Mid-latitude winter	1.03	1.04	1.01	1.00

Under the same conditions, the diffuse radiation on a horizontal surface is given by:

$$I_{dif} = (0.2711 I_o - 0.2939 I_{DN}) \sin \beta \quad (61)$$

The equations above are based on the Hottel (1976) model accepted by ASHRAE (1985).

Radiation on inclined surfaces

Solar radiation on an inclined surface is given by:

$$\begin{aligned} I &= I_{DN}\cos\theta + I_{dif} + I_{gr} \\ &= I_{DN}\cos\theta + I_{dif,h}\frac{1+\cos\Sigma}{2} + \rho_{gr} I_{hor}\frac{1-\cos\Sigma}{2} \end{aligned} \quad (62)$$

where I_{DN} is the direct normal radiation intensity; I_{dif} is the diffuse radiation intensity; $I_{dif,h}$ is the diffuse radiation intensity on a horizontal plane; I_{hor} is the total radiation on a horizontal plane; I_{gr} is the ground reflected radiation intensity; θ is the angle of incidence of the sun rays on the plane; Σ is the slope of the plane in relation to the horizontal; ρ_{gr} is the reflectivity of the ground in front of the plane (taken usually as 0.2 for Earth, but can reach values of 0.7 for sand and 0.9 for fresh snow).

Equation 62 implies that both the diffuse radiation and the ground reflected radiation is perfectly diffuse. Although there are more sophisticated models for the diffuse radiation, this is still the model most commonly used.

The incidence angle is given by the following equation (see Figure 7.12):

$$\cos\theta = \cos\beta \cos(\varphi - \psi)\sin\Sigma + \sin\beta \cos\Sigma \quad (63)$$

where: ψ is the azimuth of the plane (i.e. the angle subtended between the horizontal projection of the normal to the plane and the south).

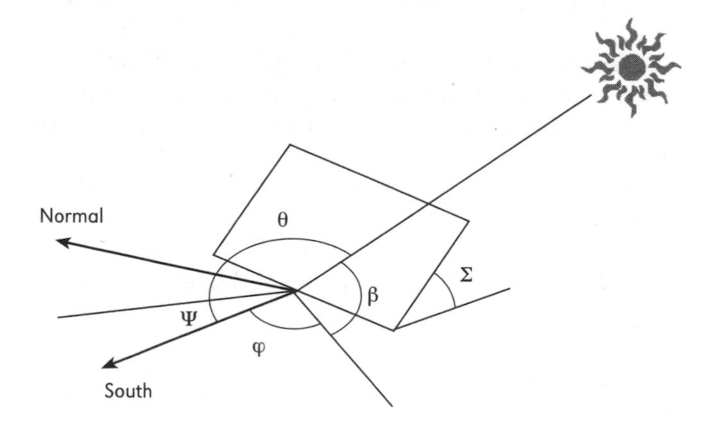

Figure 7.12 *Incidence angle on inclined surfaces*

Two special cases with regard to Equation 63 are the horizontal and vertical planes:

$$\cos\theta_{hor} = \sin\beta \quad (64)$$

$$\cos\theta_{ver} = \cos\beta\cos(\varphi - \psi) \quad (65)$$

The estimation of hourly and daily solar radiation values on inclined or horizontal surfaces can also be performed by using more detailed atmospheric models in order to calculate solar radiation in places where there are no available measurements. These models are capable of estimating solar radiation by using available meteorological data as inputs, such as air temperature, relative humidity, barometric pressure and sunshine duration. Atmospheric models examine the propagation of solar radiation through the atmosphere by taking into account, in most cases, the optical transmittances due to various atmospheric phenomena (e.g. Mie and Rayleigh scattering) and various absorption processes (e.g. due to ozone, water vapour and mixed gases). Most atmospheric models are capable of working efficiently under various sky conditions (clear, overcast, partly cloudy skies).

Solar heat gains through windows

As already stated, solar radiation interaction with transparent surfaces is characterized by the transmissivity τ, the absorptivity α and the reflexivity ρ, all three functions of the incidence angle and the refractive index:

$$\tau(\theta) + \alpha(\theta) + \rho(\theta) = 1 \quad (66)$$

The transmissivity *decreases* with the angle of incidence, whereas the reflexivity and absorptivity *increase*; but their value is nearly constant for angles of incidence of up to 50°.

The solar heat gain (*SHG*) can be divided into two parts: the solar energy *transmitted* through the glass and the solar energy *absorbed* by the glass and *transferred inward* by conduction, convection and long-wave radiation:

$$SHG = (\tau + \alpha\frac{U}{h_o})\,I \quad (67)$$

ASHRAE has prepared tables of the solar heat gain for *double strength glass* (DSG), for which transmissivity and absorptivity at normal incidence are 0.86 and 0.06, respectively. The solar heat gain for double strength glass is the *solar heat gain factor* (SHGF). SHGF is given as a function of latitude, month and solar time for various directions of vertical windows.

For glass whose properties are different from DSG, one has to multiply SHGF by the shading coefficient (SC), which is the ratio of solar heat gain to the corresponding quantity for DSG at normal incidence (contrary to what its name implies, it is not related to shading and is a function of the optical properties of glass):

$$SC = \frac{\tau(0) + \alpha(0)U/h_o}{\tau_{DSG}(0) + \alpha_{DSG}(0)U/h_o} \approx \frac{\tau(0) + \alpha(0)U/h_o}{0.86} \quad (68)$$

In Equation 68, one assumes that although the values of the optical properties of the glass are different from DSG, their variation with the angle of incidence is the same.

Similar expressions can be used for double glazing, characterized by a heat transfer coefficient for the gap between the two glazings h_g:

$$SC = \left(\frac{\tau_i\tau_o}{1-\rho_{io}\rho_{oi}} + (\alpha_{oo} + \frac{\tau_o\rho_{io}\alpha_{oi}}{1-\rho_{io}\rho_{oi}})\frac{U}{h_o} + \right.$$

$$\left. + \frac{\tau_o\alpha_{oi}}{1-\rho_{io}\rho_{oi}})\,U\,(\frac{1}{h_o} + \frac{1}{h_g})\right)\frac{1}{0.86} \quad (69)$$

where the subscripts i and o refer to the inner and the outer glazing, whereas oo refers to the external surface of the external glazing, oi to the internal surface of the external glazing, and io to the external surface of the inner glazing (all the properties are at normal incidence).

In Equations 68 and 69, the resistance of the glazing itself has been neglected.

Heat transfer coefficients used in building calculations

In building heat transfer calculations, Equations 27 and 31 are seldom used. The convective and the radiative heat transfer coefficient are lumped together and are taken as constant. The values usually employed differ from country to country; however, these are approximately equal to:

Outside film conductance:

$$h_o \approx 25\text{W/m}^2.\text{K} \quad (70)$$

Inside film conductance:

$$\text{Vertical surface } h_i \approx 8.3\text{W/m}^2.\text{K} \quad (71)$$

Horizontal surface: heat flow upwards

$$h_i \approx 8.3\text{W/m}^2.\text{K} \quad (72)$$

Horizontal surface: heat flow downwards

$$h_i \approx 6\text{W/m}^2.\text{K} \tag{73}$$

In the US, different values for h_o are used for the summer and the winter, reflecting the different wind speeds between the two periods. A more appropriate differentiation would be between day and night in sites where the wind speed is much smaller during the night than during the day. It might also be appropriate to use different values in urban and rural surroundings.

The approximations involved are valid for most building materials that are characterized by a rather high emissivity ($\varepsilon > 0.8$). This makes the dominant part of the film resistance the radiative part and the dependence of the film resistance on temperature difference might be neglected. Thus, the approximation might be inappropriate when low-emissivity materials (such as aluminium foil or special property films) are used. In this case, closer approximations of the previous sections might be necessary.

Sol-air temperature

Equation 14, which describes heat steady-state conduction heat transfer through a wall, implies that the heat flux is proportional to the temperature difference between the internal and external temperature. This, however, is true only if there are no gains or losses by radiation, whether short wave (solar) or long wave. To correct for radiation, it is customary to replace the external temperature with the *sol-air temperature*. Sol-air temperature is defined as the temperature of air with which (in the absence of solar and terrestrial radiation) the heat transfer is equal as in the given element (or surface) when environmental conditions are normal. The sol-air temperature at a given surface is equal to:

$$T_{\text{sol-air}} = T_{\text{air}} + \frac{\alpha I}{h_r + h_c} + \frac{\Delta q_{ir}}{h_r + h_c} \tag{74}$$

where T_{air} is the air temperature; α is the absorptivity of the surface (for solar radiation); I is the solar radiation intensity incident on the surface; Δq_{ir} is the correction to infrared radiation transfer between surface and environment if sky temperature is different from T_{air}.

In practice, the third term varies from 0 for vertical surfaces to 3.9° K for upward-facing surfaces (the sky overhead is colder than the rest of the environment).

For external surfaces, the second term results air temperature higher than the air temperature the daytime. The third term usually results in lo the sol-air temperature, which becomes lower than the external temperature during night time. The effect of the second term is highest for horizontal roofs during cloudless summer nights. It is lower for urban buildings than for buildings in the countryside since long-wave radiation is shielded by the neighbouring buildings.

Thermal balance of a building

The heat transfer processes that contribute to the thermal balance of a building can be seen in Figure 7.13.

The heat transfer processes involve:

1. heat conduction through walls, roofs and other elements;
2. heat conduction into the ground from slab-on-grade elements;
3. solar radiation gains through glazing elements by transmission through, and/or absorption by, the glass;
4. internal heat gains from people, lighting and equipment;
5. latent internal heat gains (losses) through evaporation (condensation) of water involving the latent heat of humidity;
6. convective heat transfer from the building envelope to the air and long-wave radiative heat transfer to the upper atmosphere (sky) and clouds;
7. convective heat transfer from internal surfaces of building elements to indoor air, and radiative heat

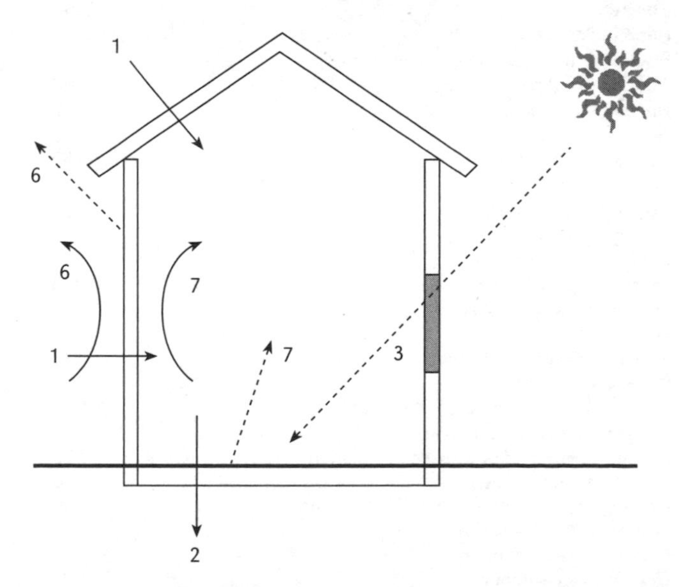

Figure 7.13 *Heat transfer balance of a building*

transfer between the internal surfaces of building elements, or absorption of solar radiation and light radiation by internal surfaces;
8 heat transfer by convection due to infiltration through cracks to natural ventilation through openings and forced ventilation using fans;
9 storage in different building elements, both external (walls, roofs, floors) and internal (structures and internal walls), resulting in the attenuation of internal temperature swing;
10 artificial heating and cooling by HVAC equipment.

Synopsis

The energy needs for cooling and heating in buildings are determined by the energy balance. The heat transfer phenomena that take place in buildings refer to a variety of mechanisms which concern the transfer of heat between the outdoor and the indoor environment, through the envelope, as well as between the various components of the building. Understanding how these mechanisms work allows the designer to have a clear view of how the involved parameters affect the indoor environment and how the control of these parameters can lead to an appropriate building design.

The main purpose of this chapter is to present the principle mechanisms of heat and mass transfer that occur in buildings. Three primary heat transfer processes are described (conduction, convection and radiation), in general, in the first part of this chapter. A more detailed analysis is provided in the second part of the chapter. For each one of these mechanisms, various calculations are presented depending upon the situation and the desired accuracy. In addition to these main mechanisms, emphasis is also placed on the infiltration and ventilation of buildings (both of which are parts of the connective heat-transfer process), as well as on solar radiation and the movement of the sun (which is part of the radioactive heat transfer). At the end of this chapter, the thermal balance of a building is presented in order to explain the contribution of each of the heat transfer processes.

Table 7.2 *Thermophysical properties of various building materials*

Material name	Density (kg/m³)	Conductivity (W/(m.°C))	Specific heat (J/(kg.°C)
Asbestos cement	1501	0.36	1050
Asbestos cement sheet	700	0.36	1050
Asbestos insulation	577	0.16	833
Asphalt	1907	0.62	833
Felt/bitumen layers	1700	0.5	1014
Bitumen composition for floors	2400	0.85	1014
Bitumen impregnated paper	1090	0.06	1014
Roofing felt	960	0.19	833
Concrete	2400	1.75	920
Flooring (Wilton carpets)	186	0.06	1376
Flooring (simulated sheep's wool carpet)	198	0.06	1376
Flooring (wool felt underlay)	160	0.04	1376
Flooring (cellural rubber underlay)	400	0.1	1376
Flooring (cork tiles)	530	0.08	1810
Flooring (rubber tiles)	1600	0.3	2027
Flooring (timber flooring)	650	0.14	1195
Flooring (wood block)	650	0.14	1195
Window glass	2500	1.05	833
Cork insulation	105	0.045	1810
Thatch (reed) insulation	270	0.09	1810
Thatch (straw) insulation	240	0.07	1810
Wood fibre insulation	300	0.06	1014
Bitumen impregnated fibreboard insulation	430	0.07	1014
Woodwool insulation	500	0.1	1014
Expanded polystyrene(EPS) insulation	25	0.034	1014
Polystyrene beads insulation	7	0.04	1014
Extruded expanded polystyrene insulation	32	0.03	1412
Foamed polyurethane (aged) insulation	30	0.026	1014
Foamed urea formaldehyde (UF) insulation	30	0.032	1520
Expanded PVC insulation	55	0.04	1014
Mineral fibres insulation	112	0.042	760
Glass fibres insulation	81	0.04	833
Foamed glass insulation	125	0.045	833
Foamed concrete insulation	550	0.3	1014

Table 7.2 *continued*

Material name	Density (kg/m³)	Conductivity (W/(m.°C))	Specific heat (J/(kg.°C)
Air cavity	1.1	0.026	1000
Brickwork outer leaf (5 per cent mc)	1700	0.84	796
Brickwork outer leaf (5 per cent mc)	1800	0.96	941
Brickwork inner leaf (1 per cent mc)	1700	0.82	796
Brickwork inner leaf (1 per cent mc)	1800	0.71	941
Vermiculite insulating brick	700	0.27	833
Concrete dense	2000	1.13	1014
Concrete dense	2100	1.4	833
Concrete lightweight	1200	0.38	1014
No fines concrete	1800	0.96	833
Foamed slug concrete	1040	0.25	977
Aerated concrete outer leaf (5 per cent mc)	500	0.18	1014
Aerated concrete outer leaf (5 per cent mc)	800	0.26	1014
Aerated concrete inner leaf (3 per cent mc)	500	0.16	1014
Aerated concrete inner leaf (3 per cent mc)	800	0.23	1014
Concrete block inner leaf (3 per cent mc)	2300	1.63	1014
Concrete block inner leaf (3 per cent mc)	1400	0.51	1014
Concrete block inner leaf (3 per cent mc)	600	0.19	1014
Foamed concrete block inner leaf (3 per cent mc)	600	0.16	1014
Foamed concrete block outer leaf (5 per cent mc)	600	0.17	1014
Vermiculite aggregate	450	0.17	833
Glass-reinforced cement	1700	0.5	833
Glass-reinforced cement	2200	1.3	833
Sandstone	2000	1.3	724
Granite	2600	2.5	905
Marble	2500	2	796
Limestone	2180	1.5	724
Slate	2700	2	760
Gravel (general)	1840	0.36	833
Aluminium	2800	160	905
Copper	8900	200	434
Steel	7800	50	507
Dense plaster	1300	0.5	1014
Lightweight plaster	600	0.5	1014
Gypsum plasterboard	950	0.16	833
Perlite plasterboard	800	0.18	833
Gypsum plastering	1200	0.42	833
Perlite plastering	400	0.08	833
Vermiculite plastering	720	0.2	833
Clay roof	1900	0.85	833
Concrete roof	2100	1.1	833
Slate roof	2700	2	760
Asphalt/asbestos roof	1900	0.55	833
PVC/asbestos roof	2000	0.85	833
Tile hanging roof	1900	0.84	796
Concrete screed	2100	1.28	1014
Concrete lightweight	1200	0.41	833
Granolithic	2085	0.87	833
Rendering dry	1300	0.5	1014
Rendering (1 per cent mc)	1431	1.13	1014
Rendering (8 per cent mc)	1329	0.79	1014
Soil density1	1280	0.7	1846
Soil density2	1900	1.4	1701
Timber softwood	630	0.13	2787
Timber hardwood	700	0.15	1412
Timber plywood	530	0.14	2787
Timber hardboard (standard)	900	0.13	2027
Timber hardboard (medium)	600	0.08	2027
Timber chip board	800	0.15	2100
Timber soft board	350	0.55	1014

Note: mc = moisture content

Table 7.3 *Solar absorptivities*

Material	Absorptivity
Asbestos	
Cement	0.60
Concrete	
Heavy/light	0.65
Aerated concrete/block	0.65
Refractory insulation	0.65
Vermiculite aggregate	0.65
Insulation	
Fibreboard	0.50
Woodwool	0.50
Glasswool	0.30
Urea formaldehyde	0.50
Thermalite	0.70
Polyurethane board	0.50
Polystyrene	0.30
Siporex	0.40
Asphalt/Bitumen	
Asphalt/Asphalt mastic	0.90
Bitumen felt	0.90
Roofing felt	0.90
Brick	
Outer/inner leaf	0.70
Insulating	0.70
Carpet	
Wilton	0.60
Felt underlay	0.65
Rubber underlay	0.65
Screeds/Renders	
Light concrete	0.80
Cast concrete	0.65
Granolithic	0.65
White render	0.50
Tiles	
Clay	0.60
Concrete	0.65
Slate	0.85
Plastic	0.40
Rubber	0.82
Cork	0.60
Asphalt/asbestos	0.70
PVC/asbestos	0.60
Wood	
Block	0.65
Flooring	0.65
Hardboard	0.70
Cork board	0.60
Chip board	0.65
Oak (radial)	0.65
Weatherboard	0.65
Fir (20 per cent moist)	0.65
Metal	
Copper	0.65
Aluminium	0.20
Steel	0.20
Plaster	
Dense/light	0.50
Gypsum/gypsum board	0.50
Perlite	0.50
Vermiculite	0.50
Perlite board	0.60
Stone	
Sandstone	0.60
Red granite	0.55
White marble	0.45

References

ASHRAE (American Society of Heating, Refrigeration and Air-Conditioning Engineers) (1985, 1989) *ASHRAE Fundamentals*, ASHRAE, Atlanta, Georgia

Clarke, J. A. (1985) *ESP-r: Energy Simulation in Building Design*, Adam Hilger Ltd, Bristol

Hottel, H. C. (1976) 'A simple model for estimating the transmissivity of the direct solar radiation through clear atmosphere' *Solar Energy*, vol 18, p129

Incropera, F. P. and DeWitt, D. P. (1996) *Fundamentals of Heat and Mass Transfer*, fourth edition, John Wiley and Sons, New York

Kreider, J. F. and Rabl, A. (1994) *Heating and Cooling of Buildings: Design for Efficiency*, McGraw-Hill International, New York

Santamouris, M. and Assimakopoulos, D. (eds) (1996) *Passive Cooling of Buildings*, James & James, London

Santamouris M., Geros V., Klitsikas N. and Argiriou A. (1995) 'SUMMER: A computer tool for passive cooling applications', *Proceedings of the International Symposium on Passive Cooling of Buildings*, Athens, June 1995

Recommended reading

1 Kreider, J. F. and Rabl, A. (1994) *Heating and Cooling of Buildings: Design for Efficiency*, McGraw-Hill International, New York
This book covers technologies (dealing with materials to computers) that exert a profound effect on the design and operation of buildings today. Various examples are presented and are solved in order to reinforce important concepts, and software applications are integrated throughout. The new edition covers the concepts of refrigeration and air conditioning in the context of practical design. The emphasis is on economic and design factors, so one can see the relevance of the technologies covered and the trade-offs that a design engineer must make. Design is further reinforced with the HCB software on a CD-ROM found in the book. It provides a wealth of informational resources for HVAC design.

2 Incropera, F. P. and DeWitt, D. P. (1996) *Fundamentals of Heat and Mass Transfer*, fourth edition, John Wiley and Sons, New York
The *Fundamentals of Heat and Mass Transfer* provides a complete introduction to the physical origins of heat and mass transfer. It gives a clear presentation of the subject and presents various problems that are solved. It may be described as an essential tool for thermal analysis.

3 Santamouris, M. and Assimakopoulos, D. (eds) (1996) *Passive Cooling of Buildings*, James & James, London
This book is compiled under the SAVE European Research Programme and describes the fundamentals of passive cooling, together with the principles and formulae necessary for its successful implementation. This publication will be of interest to building designers, building engineers (including mechanical and electrical engineers), building scientists (especially those involved with building physics) and indoor air specialists.

Activities

Activity 1

R-value of a concrete wall with insulation

The outside wall in a home consists essentially of a 10cm layer of concrete (λ = 1.4W/m.K) with 5cm of polystyrene insulation (λ = 0.03W/m.K). What is the thermal resistance of the wall r (excluding film resistance), the total thermal resistance R, the heat flux per unit area Q, and the internal and external surface temperature, given that the external air temperature is 5° Celsius, the internal one is 20° Celsius and there is no radiation?

Activity 2

Thermal time constant and temperature response to a step change

Find the thermal time constant of a 15cm concrete (λ = 1.4W/m.K, ρ = 2100kg/m^3, c_p = 653J/kg.K) wall, with 5cm polystyrene (λ = 0.03W/m.K, ρ = 25kg/m^3, c_p = 1000J/kg.K) insulation if:

(a) the polystyrene is placed externally;
(b) the polystyrene is placed internally;
(c) the polystyrene is placed in the middle of the concrete.

For each case, find the internal temperature 12 hours after a 10° Celsius rise in external temperature, the initial internal and external temperature being 20° Celsius.

Activity 3

Internal convective heat transfer coefficient

Find the convective heat transfer coefficient of the floor in a room 3m high, when the floor temperature is:

(a) 2K higher than the air above;
(b) 2K lower than the air above;

Find also:

- the heat transfer coefficient of a vertical wall in the same room.

Activity 4

External convective heat transfer coefficient

Find the external convective heat transfer coefficient:

(a) for a wall 2m long, when the wind speed is 0.5 metres per second;
(b) for a wall 10m long, when the wind speed is 5 metres per second.

Activity 5

Effective leakage area

The blower door test in a single house whose volume is 500 cubic metres gives 3 air changes per hour (ACH) for a pressure difference of 50Pa and 1 ACH for a pressure difference of 10Pa. Find the effective leakage area of the building.

Activity 6

Infiltration rate

The building above is 6m high, whereas the height difference between the lowest and the highest crack is 4m. It is situated in the countryside. Calculate the infiltration rate:

(a) when the internal temperature is 20° Celsius, the external one is 30° Celsius and the wind speed (measured in a station 10m high situated in the countryside) is 0.5 metres per second;
(b) when the wind speed is 5 metres per second;
(c) when the wind speed is 5 metres per second and there is an additional artificial ventilation of 2 air changes per hour (ACH).

Activity 7

Shape factors

Find the shape factors in a rectangular room of 5m × 10m × 3m (height), from the floor to the ceiling and from the floor to each wall.

Activity 8

Radiative heat flux

Find the radiative heat flux from a 10m × 10m surface whose emissivity is 0.9 and whose temperature 40° Celsius to a parallel surface whose emissivity is 0.22 (aluminium foil) and whose temperature is 20° Celsius.

Activity 9

Radiative heat flux

Find the radiative heat flux from the 10m × 10m floor of an attic made of equilateral triangles. The floor is covered by aluminium foil and its temperature is 20° Celsius, whereas the ceiling's uniform temperature is 40° Celsius.

Activity 10

Radiative heat transfer coefficient and radiative heat flux

Find the radiative heat transfer coefficient and the radiative flux from a 10m × 10m surface whose emissivity is 0.9 and whose temperature 40° Celsius to a parallel surface whose emissivity is 0.22 (aluminium foil) and whose temperature is 20° Celsius.

Activity 11

Solar time

Find the apparent solar time in Athens, Greece (longitude 24° east, LSM 30° east) at 14:00 on 15 November.

Activity 12

Solar time

Calculate the solar time at sunrise on 21 December in Athens, Greece (latitude 38° north), as well as the azimuth of the sun at sunrise on that day.

Activity 13

Solar position

Find the time that the sun is due west on the 21 June, as well as the solar height at that time.

Activity 14

Solar radiation

Find the direct normal solar radiation intensity, as well as the diffuse solar radiation intensity on a horizontal surface, on 21 June in Athens (altitude 0) when the sun is due west on a day with low visibility.

Activity 15

Solar radiation

Find the direct solar radiation intensity on a south-west (45° west of south) facing vertical wall in Athens, Greece, when the sun is facing west.

Activity 16

Solar radiation

Find the total solar intensity on the plane of the previous example, assuming diffuse solar radiation intensity of 150W/m^2 and ground reflectivity of 0.25.

Activity 17

Shading coefficient

Find the shading coefficient:

(a) for a reflective one-pane window ($\rho(0) = 0.4$, $\alpha(0) = 0.1$);
(b) for an absorptive one-pane window ($\rho(0) = 0.1$, $\alpha(0) = 0.4$).

Activity 18

Shading coefficient

Find the shading coefficient of a double-glass window whose external pane is the absorptive glass described above and whose internal pane is double-strength glass. Assume that the internal and external properties of both windows are the same and that the thermal resistance of the gap $1/h_g$ is equal to $0.16m^2.K/W$.

Activity 19

Sol-air temperature

Find the sol-air temperature of a horizontal roof when the solar intensity on a horizontal surface is $800W/m^2$, the air temperature is 32° Celsius and the mean radiant temperature of the sky is 20° Celsius lower than the air temperature:

(a) when the roof is white ($\alpha = 0.25$);
(b) when the roof is dark ($\alpha = 0.7$).

The radiative heat transfer coefficient can be considered equal to $5W/m^2.K$ and the convective one $20W/m^2.K$.

Answers

Activity 1

$$r = r_1 + r_2 = \frac{L_1}{\lambda_1} + \frac{L_2}{\lambda_2} = \frac{0.1}{1.4} + \frac{0.05}{0.03} = 1.74\text{m}^2.\text{K/W}$$

$$R = \frac{1}{h_o} + r + \frac{1}{h_i} = 0.04 + 1.74 + 0.12 = 1.90\text{m}^2.\text{K/W}$$

$$U = 1/R = 1/1.90 = 0.52\text{W/m}^2\ K$$

$$Q = -U(T_o - T_i) = -0.52\ (20 - 5) = -7.8\text{W/m}^2$$

$$T_{is} = T_i + \frac{Q}{h_i} = 20 + (-7.8)(0.12) = 19.1°\ \text{Celsius}$$

$$T_{os} = T_o - \frac{Q}{h_i} = 5 - (-7.8)(0.04) = 5.3°\ \text{Celsius}$$

Note that:

- Most thermal resistance is due to the insulation.
- Without the insulation, the U-value would be 4.32W/m^2.K, the flux 64.8W/m^2, the internal surface temperature 12.3° Celsius, and the external one 7.6° Celsius – illustrating that the insulation is necessary not only to reduce the heat flux, but also to reduce the difference between the air and the surface temperature.

Activity 2

(a) Insulation external:

$$T_c = (\frac{1}{h_o} + \frac{L_1}{2\lambda_1})\ \rho_1 c_{p1} L_1 + (\frac{1}{h_o} + \frac{L_1}{\lambda_1} + \frac{L_2}{2\lambda_2})\ \rho_2 c_{p2} L_2 =$$

$$= (0.04 + \frac{0.05}{2 \times 0.03}) \times 25 \times 1000 \times 0.05 \times (0.04 + \frac{0.05}{0.03} + \frac{0.15}{2 \times 1.4})\ 2100 \times 653 \times 0.15 =$$

= 363,114 seconds = 100.9 hours

$$T_i\ (12h) = 20 + 10[1 - \exp(-12/100.9)] = 21.1°\ \text{C}$$

(b) Insulation internal:

$$T_c = (\frac{1}{h_o} + \frac{L_1}{2\lambda_1})\ \rho_1 c_{p1} L_1 + (\frac{1}{h_o} + \frac{L_1}{\lambda_1} + \frac{L_2}{2\lambda_2})\ \rho_2 c_{p2} L_2 =$$

$$= (0.04 + \frac{0.15}{2 \times 1.4}) \times 2100 \times 653 \times 0.15 + (0.04 + \frac{0.15}{1.4} + \frac{0.05}{2 \times 0.03})\ 25 \times 1000 \times 0.05\ = 12{,}245s = 3.4h$$

$$T_i\ (12h) = 20 + 10[1 - \exp(-12/3.4)] = 29.7°\ \text{C}$$

(c) Insulation in the middle:

$$T_c = (\frac{1}{h_o} + \frac{L_1}{2\lambda_1})\ \rho_1 c_{p1} L_1 + (\frac{1}{h_o} + \frac{L_1}{\lambda_1} + \frac{L_2}{2\lambda_2})\ \rho_2 c_{p2} L_2 =$$

$$(\frac{1}{h_o} + \frac{L_1}{\lambda_1} + \frac{L_2}{\lambda_2} + \frac{L_3}{2\lambda_3})\ \rho_3 c_{p3} L_3 =$$

$$(0.04 + \frac{0.075}{2 \times 1.4}) \times 2100 \times 653 \times 0.075 + (0.04 + \frac{0.075}{1.4} + \frac{0.05}{2 \times 0.03})\ 25 \times 1000 \times 0.05 + \ (0.04 + \frac{0.075}{1.4} + \frac{0.05}{0.03} + \frac{0.075}{2 \times 1.4})\ 2100 \times 653 \times 0.075$$

$$= 188{,}284s = 52.3h$$

$$T_i\ (T_i(12h) = 20 + 10[1 - \exp(-12/50.4)] = 22.1°\ \text{C}$$

Note that after 12 hours, when the insulation is on the internal part, the internal temperature is almost equal to the external one. In the two other cases, it has hardly moved from the initial temperature of 20° Celsius.

Activity 3

(a) Floor hotter than the air: $L^3 \Delta T = 54\text{m}^3.\text{K}$ (i.e. the flow is turbulent).

$$h_c = 1.52\ (2)^{1/3} = 1.92\text{W/m}^2.\text{K}$$

(b) Floor colder than the air:

$$h_c = 0.59\ (2/3)^{1/4} = 0.53\text{W/m}^2.\text{K}$$

(c) Vertical wall: flow is turbulent according to the criterion,

$$h_c = 1.31\ (2)^{1/3} = 1.66\text{W/m}^2.\text{K}$$

Activity 4

(a) $U \times L = 2 \times 0.5 = 1\text{m}^2/\text{s}$ (i.e. the flow is laminar). The heat transfer coefficient is $2\ (0.5/2)^{1/2} = 1\text{W/m}^2.\text{K}$. Note that in that case, the contribution from free convection might be appreciable and should be calculated from the equations of the previous section.

(b) $U \times L = 5 \times 10 = 50\text{m}^2/\text{s}$ (i.e. the flow is turbulent)

$$U = 6.2\ (5^4/10)^{1/5} = 14.2\text{W/m}^2.\text{K}$$

Activity 5

The air flow Q, being proportional to ΔP^n, n can be found from the data above:

$$n = \ln(Q_{50}/Q_{10})/\ln(50/10) = 0.68$$

Extrapolating to $\Delta P = 4\text{Pa}$, one obtains that:

$Q_4 = 1\ (4/10)^{0.68} = 0.53$ ACH or $0.53 \times 500/3600 = 0.074\text{m}^3/\text{s}$.

From Equation 35, the effective leakage area can be calculated:

$$\text{ELA} = (0.074/(8/1.2))^{1/2} = 0.105\text{m}^2$$

Activity 6

(a) From Equation 36 the stack ventilation is:

$$Q_{st} = 0.25\ ELA\sqrt{gH_L|\Delta T|/T} = 0.25 \times 0.105 \times \sqrt{(4 \times 10/298 \times 9.8)} = 0.03\ \text{m}^3/\text{s}$$

The wind effect ventilation is:

$$0.3 \times 0.105 \times 0.5 \times (6/10)^{0.5} = 0.012\text{m}^3/\text{s}.$$

$$Q = \sqrt{Q_{st}^2 + Q_w^2 + Q_{vent}^2} = 0.032\ \text{m}^3/\text{s}$$

From Equation 38:

The total ventilation is $0.032\text{m}^3/\text{s}$ or 0.23 ACH.

(b) For that case, the wind effect ventilation is $0.12\text{m}^3/\text{s}$. Again, from Equation 38, the total ventilation is $0.125\text{m}^3/\text{s}$, or 0.9 ACH.

(c) From Equation 38, with the additional 2 ACH, one finds that the total ventilation and infiltration rate is 2.2 ACH ($0.305\text{m}^3/\text{s}$).

Activity 7

Floor to ceiling: from Figure 8, setting $X/D = 3.33$ and $Y/D = 1.67$, one finds after interpolation that the shape factor $F_{\text{floor}\rightarrow\text{ceiling}} \approx 0.45$.

From Figure 7.9, for the shape factor between the floor and the 5m × 3m wall, setting $Y/X = 2$, $Z/X = 0.6$, one obtains that the shape factor $F_{\text{floor}\rightarrow 5\times 3\text{wall}} \approx 0.09$.

From Figure 7.9, for the shape factor between the floor and the 10m × 3m wall, setting $Y/X = 0.5$, $Z/X = 0.3$, one obtains that the shape factor $F_{\text{floor}\rightarrow 10\times 3\text{wall}} \approx 0.185$.

Note that:

$$F_{\text{floor}\rightarrow\text{ceiling}} + 2\ (F_{\text{floor}\rightarrow 5\times 3\text{wall}} + F_{\text{floor}\rightarrow 10\times 3\text{wall}}) = 1,$$

in accordance with Equation 47 ($\sum_{i=1}^{n} F_{ji} J_i = 1$).

Activity 8

Using Equation 48:

$$q_1 = \frac{\sigma\,(T_2^4 - T_1^4)}{1/\varepsilon_1 + 1/\varepsilon_2 - 1} = 10 \times 10 \times 5.68 \times 10^{-8} \times$$

$$[(273.15 + 40)^4 - (273.15 + 20)^4]/(1/0.9 + 1/0.22 - 1) =$$

2727W

Activity 9

The area of the floor is 100 square metres, whereas the area of the four equilateral triangles is $4 \times (10 \times 10 \times 0.866/2) = 173\text{m}^2$. Thus, applying Equation 48 one obtains:

$$Q = \frac{10\times 10\times 5.6\times 10\times^{-8}\times[(273.15 + 40)^4 - (273.15 + 20)^4]}{(1 - 0.22)/0.22 + (1 - 0.9)/0.9(100/173) + 1}$$

= 2748W

Note that without the aluminium foil on the floor, the flux would have been:

$$Q = \frac{10\times 10\times 5.6\times 10^{-8}\times[(273.15 + 40)^4 - (273.15 + 20)^4]}{(1 - 0.9)/0.9 + (1 - 0.9)/0.9(100/173) + 1}$$

= 10,780W

(i.e. four times larger).

Activity 10

Using Equation 51:

$$h_r = 4 \times 5.68 \times 10^{-8} \times [273.15 + (20 + 40)/2)]^3$$

$$/(1/0.22 + 1/0.9 - 1) = 1.36\text{W/m}^2.\text{K}$$

$Q = 1.36 \times (40 - 20) \times 10 \times 10 = 2720\text{W/m}^2$, which is very close to the result of the exact method used in the previous section.

Note that without the aluminium foil the radiative heat transfer coefficient would have been:

$$h_r = 4 \times 5.68 \times 10^{-8} \times [273.15 + (20 + 40)/2)]^3$$

$$/(1/0.9 + 1/0.9 - 1) = 5.18\text{W/m}^2.\text{K}$$

Note, however, that one is usually more interested in the *total* heat transfer coefficient by both *radiation and convection*, which is affected less by the use of aluminium foil since the *convective* heat transfer coefficient is unaffected by it.

Activity 11

The equation of time is +15min (Equation 55):

$$\text{AST} = \text{LST} + \text{ET} - 4(\text{LSM} - \text{LON}) = 14{:}00 + 0{:}15$$

$$- 4(24 - 30) = 14{:}39.$$

The declination, δ, is defined as the angle between the rays of the sun and the plane of the equator and is given by:

$$\delta = 23.45° \sin(2\pi \frac{n + 284}{365}) \tag{56}$$

It is 0° on the spring and autumn equinox days (21 March and 21 September), minimum (–23.45°) on the winter solstice (21 December) and maximum (23.45°) on the summer solstice (21 of June).

Activity 12

Setting $\beta = 0$ and $\delta = -23.45°$ from Equation 57:

$$H = \text{acos}(-\tan(-23.45)\tan(38)) = -70°$$

(the minus sign because one refers to the morning)

$$\text{AST} = 12 + H/360 = 7{:}20$$

$$\varphi = \text{acos}(-\sin\delta/\cos L) = -60°$$

(again the minus sign means east of south)

Activity 13

From Equation 58, setting $\cos\varphi = 0$, one finds that:

$$\sin\beta = \sin\delta/\sin L = 0.6465,\ \beta = 40°$$

From Equation 57:

$$H = \text{acos}([0.6465 - \sin(23.45)\sin(38)]/\cos(23.45)\cos(38)) = +56°$$

$$\text{AST} = 12 + H/360 = 15{:}45$$

Activity 14

From the values presented in the section entitled 'Solar radiation on clear days', for mid-latitude summer, r_o = 0.96, r_1 = 0.99 and r_k = 1.02.

From these values and with $A = 0$, $a_o = 0.026$, $a_1 = 0.81$ and $k = 0.755$, from Equation 59, the solar constant is 1308W/m². From Equation 60, with $\sin\beta = 0.6465$ (as calculated in the previous section), the direct normal solar radiation intensity is 440W/m² and from Equation 61 the diffuse radiation on the horizontal plane is 277W/m².

Note that for the high visibility, the ratio between direct normal and diffuse is completely different: the direct normal radiation intensity would be 750W/m² and the diffuse 86W/m².

Activity 15

From Equation 65 and using the solar height calculated in the previous section:

$$(\beta = 56°\ \text{C})\ \cos\theta = \cos(56)\cos(90 - 45) = 0.395$$

Thus, the direct solar intensity on that wall is: $440 \times 0.395 = 174\text{W/m}^2$.

Activity 16

Using Equation 64, the total horizontal radiation is:

$440 \times \sin 56 + 277 = 717\text{W/m}^2$.

From Equation 62: $I_{tot} = 174 + 277 \times (1/2) + 0.25 \times 717 \times (1/2) = 402\text{W/m}^2$.

Activity 17

By applying Equation 68 with $U/h_o = 1/4$, one obtains for the reflective window $SC = 0.61$ and for the absorptive window $SC = 0.70$. The difference is relatively small, since in both cases $\tau(0) = 0.5$.

Activity 18

From Equation 69:

$$SC = [\frac{0.5 \times 0.86}{1 - 0.4 \times 0.08} + (0.4 + \frac{0.5 \times 0.08 \times 0.4}{1 - 0.4 \times 0.08})\frac{1}{8} +$$

$$\frac{0.5 \times 0.06}{1 - 0.4 \times 0.08}\frac{5}{8}] / 0.86 = 0.6$$

Activity 19

For the first case, Equation 74 gives $32 + 0.25 \times 800/25 + 5 \times (-20)/25 = 36^o$ Celsius.

For the second case, the same equation gives 50.4^o Celsius.

8

Applied Lighting Technologies for Urban Buildings

Sašo Medved and Ciril Arkar

Scope of the chapter

People tend to receive information about their surroundings through visual means. The way in which the form and the surface of an object appear in our surroundings depends upon the way in which these objects are illuminated and how our eyes perceive the light that is reflected from their surface. Visual comfort defines the quality of this process. Besides ensuring better working conditions, light also affects the natural level of people's health. The scope of this chapter is to introduce the importance of quality natural and artificial lighting in ensuring visual comfort with respect to urban buildings, and the interaction between lighting and energy use in these buildings.

Learning objectives

During the study of this chapter, readers will be acquainted with:

- the physical basics of our perception of light and the physical and physiological values that are used in lighting techniques;
- sources of natural light and numerical models for evaluating these sources;
- the basics of artificial lighting and lighting efficiency of different types of lamps;
- the requirements for lighting comfort and natural lighting in the living area, together with architectonic and technological measures for lighting comfort improvement;
- the importance of efficient lighting with regard to sustainable energy use in buildings.

Key words

Key words include:

- human sight;
- photometric quantities;
- light sources;
- daylighting;
- artificial lighting;
- visual comfort;
- lighting requests for buildings in the urban environment;
- light pollution;
- energy use in buildings.

Introduction

We can define two basic types of light: daylight, which is emitted by the sun and reaches the Earth either directly from the surface of the sun, or indirectly due to its diffusion in the atmosphere; and artificial light, which is produced by luminaries. Contemporary illumination devices transform electrical energy into light; therefore, their use affects a building's energy consumption. Because daylight does not provide permanent and even illumination in a building, it is always combined with artificial lighting. Nevertheless, the provision of sufficient suitable natural illumination and exposure to sunlight is one of the most important considerations when planning a building. Added to this, we must ensure that artificial lighting is not too expensive in terms of the energy it consumes. The urban environment is different from the rural environment in a number of respects. Typically, cities have a wide variety of building types – residential, business and public. Because of this, the demands in terms of light comfort vary as well. Buildings are built in a wide variety of shapes in which lighting conditions can differ. The enormous consumption of energy and high traffic density in cities causes a substantial amount of pollution in the atmosphere. A consequence of this is that the direct and indirect light flux from the sun's radiation is reduced, and with it the quality of daylight.

This chapter begins by describing the fundamentals of human light perception, followed by a section on photometric quantities used in planning a lighting environment. Next, the demands that have to be fulfilled in the course of planning are presented. The chapter ends by emphasizing the correlation between lighting and energy use in buildings. For further assistance, some computer tools have been added to the CD accompanying this book.

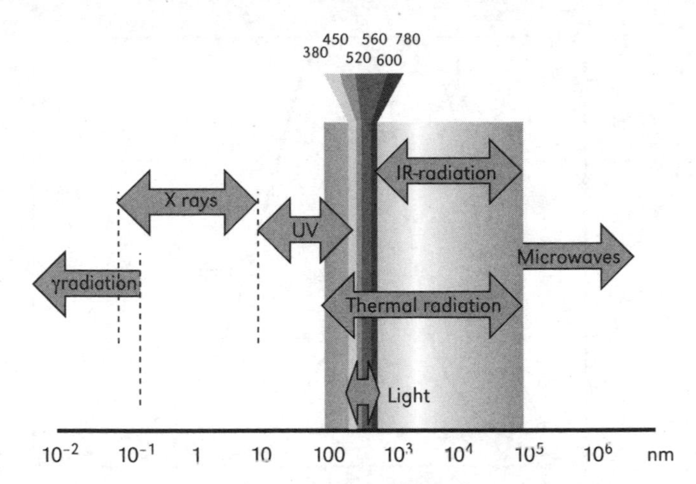

Figure 8.1 *Different parts of the electromagnetic wave spectrum*

Light

Ionized bodies or bodies heated above their local temperature transmit energy to their surroundings. We can consider this energy in terms of electromagnetic waves with different wavelengths, or as energy packages without mass, known as photons. In either case, no substance is required for the transfer of energy, so this energy transfer can also take place in a vacuum. The way in which people perceive electromagnetic waves depends very much upon the wavelength of the radiation and the senses used for the perception. For example, we see radiation with our eyes but also feel it with our skin. Electromagnetic waves with wavelengths of between 380 and 770 nanometres (nm) can be seen by the eyes, and this type of radiation is called light.

Human sight and its characteristics

The human eye can be compared to a camera. The lens, the iris and the film of the camera, are, in the human eye, replaced by the lens, the iris and the retina. By contracting the lens (accommodation), the eye adapts to be able to see objects at different distances, and by expanding or narrowing of the pupil (adaptation) the eye regulates the light flux falling on the retina. The human eye is, however, not equally sensitive to the different wavelengths of light. To get the impression of equivalent luminance from a surface, the eye must expand to allow a bigger light flux if the surface is coloured red or violet-blue than it would if the surface were coloured green or yellow. This characteristic of our eyes is called the spectral luminous efficiency $V(\lambda)$. Figure 8.3 shows the relative visibility factor for an eye adjusted to bright surroundings compared with one adjusted to living in the dark.

Measured from the centre of vision, the field of view is seen as approximately 180° wide in the horizontal direction. The eyebrows and cheekbones limit the field of peripheral vision to about 130°, and because of the retina's characteristics we see objects in this area differently. Sharp (foveal) vision occurs in a 2° cone around the centre of vision; but visibility is still quite high in the foveal surround, which is within a 30° cone around the centre of vision. The location and the brightness of the objects in this field of vision are very important.

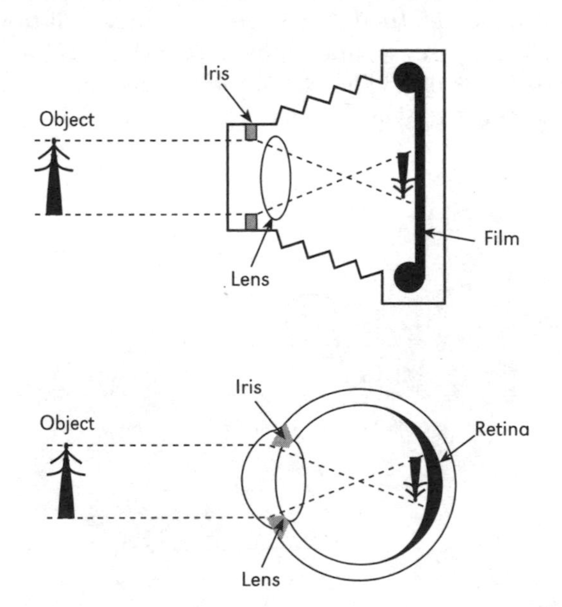

Figure 8.2 *The human eye functions like a camera*

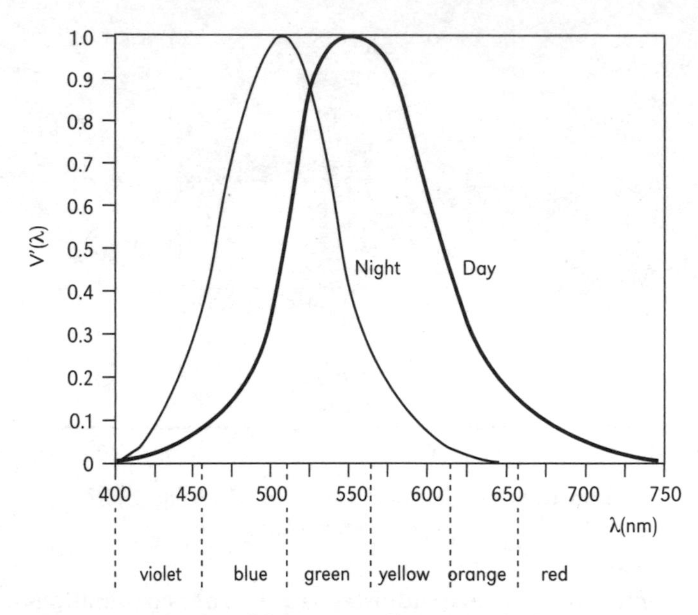

Figure 8.3 *Spectral luminous efficiency of a human eye*

The most important factors for our quality perception of our surroundings are:

- the sharpness of sight;
- the speed of sight perception, which is related to the time between the appearance of the object and its recognition;
- depth distinction (we can see a 0.4mm divergence between two objects at a distance of less than 1m; at a distance of 1000m, we can only discern if the two objects are separated by more than 275m; at a distance of 1300m the ability to distinguish depth does not exist any more);
- the geometry (spherical aberration) and colour (chromatic aberration) of the picture of the object we are observing (this is a consequence of the sun's rays refracting as they pass through the eye's lens).

Photometric quantities

In order to evaluate light sources and the visual environment, we use scotopical and photo-optical quantities. We can express the first of these with units of energy; however, with the second, we also consider the effect of light on the human eye or on our light perception. For both quantities the systems are related by the spectral luminous efficiency $V(\lambda)$.

Before examining the basic photometric quantities, we should determine the space solid angle. This angle is expressed in terms of the ratio of a sphere's surface area to the square of its diameter. The unit of solid angle is the steradian (sr). The steradian represents the solid angle of a sector from a sphere with a radius of 1m that occupies $1m^2$ of the sphere's surface. If the light source is a point, it emits light into the entire solid angle, which measures 4π or 12.5sr; but if the light source is a plane, it emits light into half of the solid angle, which is 2π or 6.25sr.

The spectral luminous efficiency value of the eye is at a maximum when illuminated with light with a 555nm wavelength (for normal vision). It has been experimentally determined that at this wavelength the radiation flux emitted by a surface is the same as a light flux of 683 lumens (lm). This value is the maximum value of the luminous efficiency of radiation Km for this monochromatic radiation. But because light sources normally emit light in the visible spectrum, we define the entire lumination flux using the following expression:

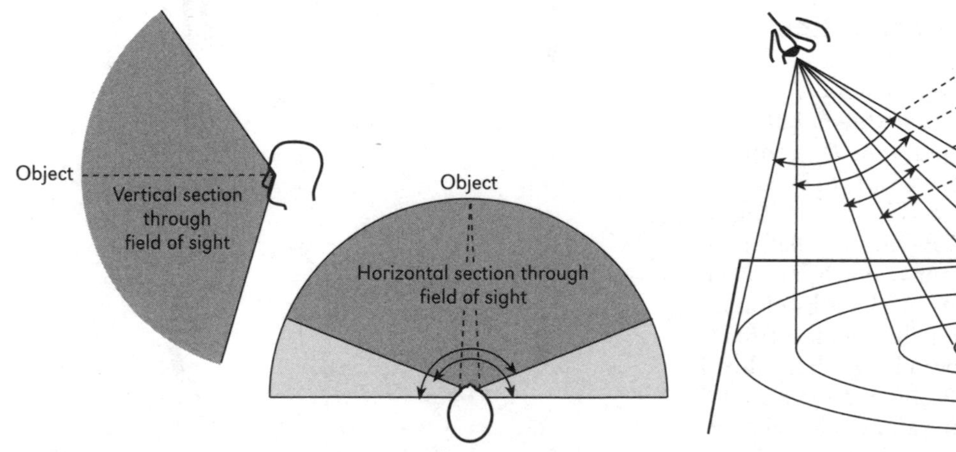

Source: Lechner, 1991

Figure 8.4 *Field of view and centre of vision*

$$\Phi = K_m \cdot \int_{\lambda=380}^{\lambda=760} \frac{d\Phi_\lambda}{d\lambda} \cdot V_\lambda \cdot d_\lambda \ (lm)$$

where K_m is the luminous efficiency of the radiation (lm/W) and $d\Phi_\lambda/d\lambda$ is the spectral distribution of the radiant flux (W/nm).

Light flux is emitted by an illuminating device into the entire solid angle. The light flux emitted by an illuminating device into a unit of solid angle is called the light intensity, which is measured in candelas (cd). Therefore, the light intensity of a certain source is equal to:

$$\mathrm{I} = \frac{d\Phi}{d\Omega} \left(\frac{\mathrm{lm}}{\mathrm{sr}} \equiv \mathrm{cd} \right)$$

where: $d\Phi$ is the infinitesimal light flux (lm); $d\Omega$ is the infinitesimal solid angle (sr).

If we observe the light source with a certain light intensity from different distances, we discover that its luminance reduces in proportion to the square of the distance between the eye and the source. *Luminance* is the only photometric quantity that our eye can perceive. The luminance of a source with an unchanged light intensity also reduces if the centre of vision does not coincide with the normal to the illuminant's surface. Luminance is then defined by the following expression:

$$L = \frac{\mathrm{I} \cdot \cos\theta}{r^2} \left(\frac{\mathrm{cd}}{\mathrm{m}^2} \right)$$

where θ is the angle between the normal to the illuminant and the direction of the beam (degrees); r is the distance between the light source and the point of observation (metres).

Let us suppose that two different large planes receive equal light flux. Because the surfaces are not equal, the density of the light flux will depend upon the surfaces of each plane. The value of the light-flux density is called the *illuminance* and is measured in lux (lx). Determining the quality of the illuminated environments, the illuminance of the surface – for instance, a work plane – is one of the basic parameters of visual comfort. Illuminance also depends upon the angle between the light-flux beam and the normal to the surface:

$$E = \frac{d\Phi \cdot \cos\theta}{dA} \left(\frac{\mathrm{lm}}{\mathrm{m}^2} \equiv \mathrm{lx} \right)$$

where dA is the infinitesimal surface receiving an infinitesimal light flux (square metres).

Therefore, a black sheet and a white sheet of paper on our work plane will both be equally illuminated; however, their luminance will not be equal. The illumination of the surfaces can also be expressed as a relative value. This method is normally used when referring to space illumination by daylight. As a result, we call it the *daylight factor* (*DF*). It is determined by the expression:

$$DF = \frac{E_{p,i}}{E_{h,e}} \cdot 100 \quad (\%)$$

where $E_{p,i}$ is the illuminance of a selected point in space (lx); $E_{h,e}$ is the illuminance of a horizontal unshaded outdoor surface (lx).

The sun's radiation entering a space is not just important for visible comfort, it also affects people's health and their general feeling of well-being. Therefore, the time a certain space is exposed to solar radiation in the course of a year is also important. This time period of direct solar radiation is called the sunlight duration. It can only be determined on the basis of geometrical relations, which are linked to the position of the sun, any obstacles in the surroundings, and the solid angle in which the window on a façade can receive direct radiation from the sun. We can, therefore, assume that the meteorological parameters of the environment do not affect the sunlight duration.

Sources of light

Daylighting

In the world that surrounds us, the most powerful source of light is the sun. As the Earth rotates every 24 hours, dividing the day into periods of light and dark, we also name this source daylight. The position of the sun at any given time is determined by the altitude of the sun and its azimuth. The altitude of the sun is determined from the angle between the sun's rays and the horizon. The maximum height of the sun is at noon, and at sunrise and sunset the height is equal to zero. The azimuth of the sun is the angle between a sunray projection on the horizontal plane and the southerly direction. By general agreement, the azimuth of the sun is negative before noon, positive in the afternoon, and equal to zero when the sun is shining from the south. Besides the time of the day, the declination of the Earth's rotation axis and latitude of the place (L) from which the sun is observed

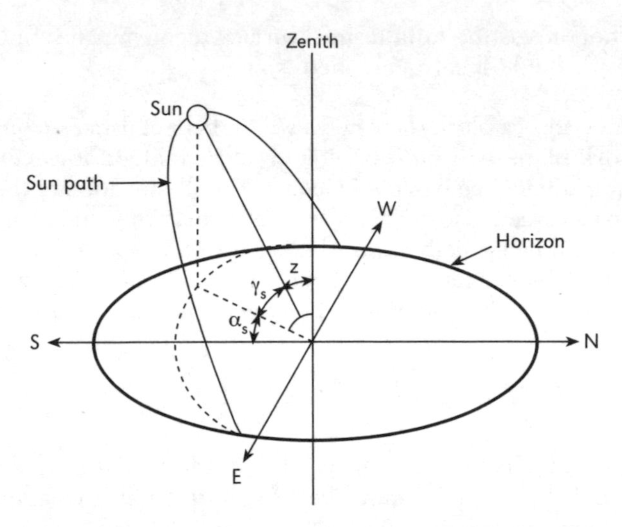

Figure 8.5 *The position of the sun in the sky can be represented by the altitude angle α_s and its azimuth angle γ_s; z is the sun zenith angle*

also have an effect on the position of the sun in the sky. The year-long positions of the sun in the sky can also be illustrated using a cylindrical coordinate system. This type of illustration is known as a sun-path diagram. Figure 8.6 shows the sun-path diagram for Athens (Greece, latitude $L = 38$ degrees) and Nordcap (Norway, latitude $L = 72$ degrees).

Sunrays, which are also referred to as direct solar radiation, are partly absorbed and reflected when they pass through the atmosphere because of the absorption and scattering caused by molecules of water vapour, ozone, carbon dioxide (CO_2), oxygen (O_2) and aerosols. This is how diffuse radiation originates. In nature, the magnitude of the radiation and the relationship between direct and diffuse radiation constantly change, and therefore certain points in the sky have periodically variable light intensities, which represent a major obstacle to daylight planning. In order to deal with this, three typical sky conditions that are characteristic for a particular place and time of year have been determined – overcast, clear and real sky. A certain type of sky is chosen as representative, depending upon the engineering problem we are solving. For instance, by dimensioning daily illuminations of an interior, we are able to predict that the sky is overcast. We determine the light intensity of the daylight sources as being different for the three different sky conditions.

Overcast sky

The characteristic of an overcast sky is that the luminance of the zenith is three times bigger than the luminance of the horizon. Points in the sky with the same zenith angle have the same luminance, which, therefore, does not depend upon their azimuth. For a particular point with a angle ε, the luminance L_p is equal to:

$$L_P = L_Z \cdot \left(\frac{1 + 2 \cdot \cos\varepsilon}{3} \right) \left(\frac{\text{cd}}{\text{m}^2} \right)$$

where: ε is the zenith angle (degrees) of the selected point; L_z is the luminance of the zenith for an overcast sky determined by the expression:

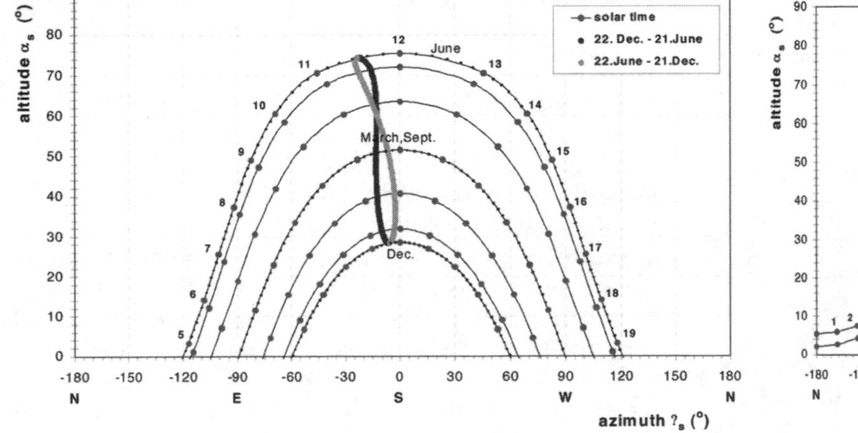

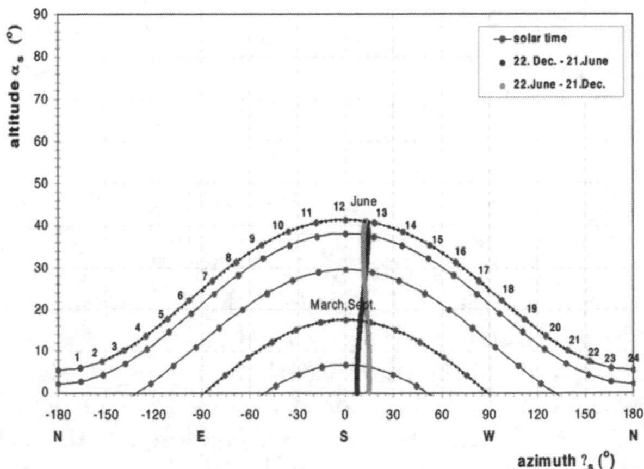

Source: Arkar, 2001

Figure 8.6 *Sun-path diagram for Athens* (left) *and for Nordcap* (right); *the looped curve represents 12:00 local time according to the equation of time and difference between standard and local meridian*

Figure 8.7 *Clear* (left), *cloudy* (right) *and overcast* (bottom) *skies*

$$L_Z = \frac{9}{7 \cdot \pi} (300 + 21000 \cdot \sin\gamma_S) \quad \left(\frac{\text{cd}}{\text{m}^2}\right)$$

The luminance of a horizontal, unobstructed plane, illuminated by an overcast sky, is determined by the following expression:

$$E_{h,e} = \frac{7 \cdot \pi}{9} \cdot L_Z$$

With the Daylight_sky.xls spreadsheet code, one can determine the luminance for a chosen city at any time of the year. For a selected example, the luminance of an overcast sky is shown in Figure 8.9. The standardized model of an overcast sky is made for a relatively clean atmosphere, so it does not distinguish between urban and rural environments.

Clear sky

The luminance of the points in a clear sky cannot be determined as easily as in the case of an overcast sky. It depends upon many geometrical and meteorological parameters that change with time and position. In general, we can state that the luminance of a point P in a clear sky depends upon:

$$L_p = f(L_z, \alpha_s \gamma_s, \varepsilon, \alpha_p, \gamma_p, \eta, \tau)$$

where γ_p is the azimuth of the observed point in the sky P (degrees); η is the angle between the sun and the observed point P (degrees); ε zenith angle (degrees) of the selected point P in the sky; τ is the transmittance of the atmosphere (1).

The illuminance of a horizontal plane from a clear sky also depends upon the condition of the environment, particularly upon the absorption and dissemination of light on gas molecules and particulate material in the atmosphere. The condition of the atmosphere can be evaluated with

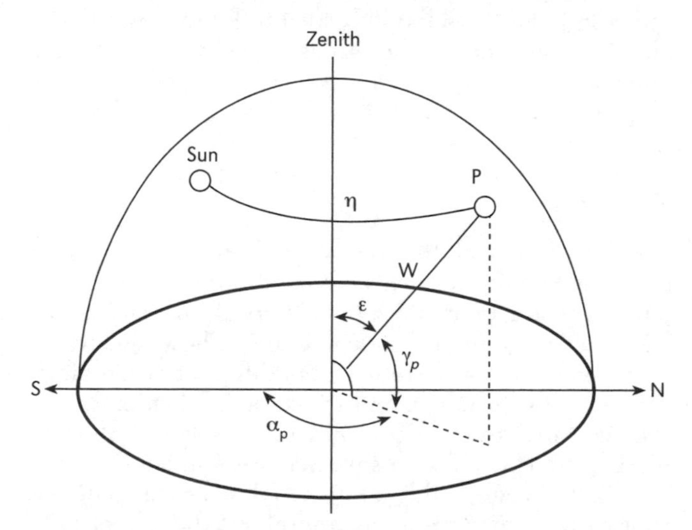

Figure 8.8 *Geometrical parameters of point P for a clear-sky luminance determination*

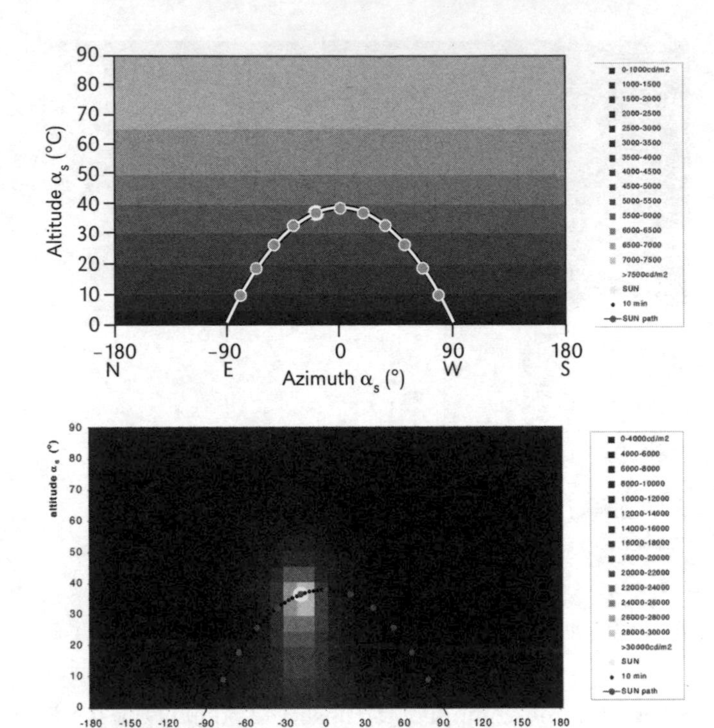

Source: Arkar, 2001

Figure 8.9 *Luminance of the sky determined with a model of overcast* (top) *and clear sky* (bottom) *for 21 March at 11:00 (sun time) for an urban area*

different physical measurements. In the spreadsheet Daylight_sky.xls, we can analyse the illuminance of horizontal planes in rural, urban and industrial areas, on the basis of a typical monthly total turbidity factor that is characteristic for these areas. An example of the luminance of the sky in an urban area is shown in Figure 8.9. The model of luminance is described in detail in Arkar (2001)

Cloudy sky

The conditions of the sky, described by the models presented above, are extreme conditions in nature and quite rare situations. With the cloudy sky model, we can also evaluate intermediate conditions. These models are based on a statistical evaluation of different meteorological data – for instance, an estimation of cloudiness, the relationship between global and diffuse solar radiation, or the probability of direct solar radiation in hourly intervals. In urban areas, the duration of direct sun radiation is shorter than in rural areas; therefore, the model for a cloudy sky also distinguishes between these two situations.

Luminous efficacy of solar radiation

In addition to the model for determining different skies' light intensity, we can also determine the light flux from the database of the solar radiation data. These data are available for most cities. Daylight data from the solar radiation data are determined with the luminous efficiency of the solar radiation K_s. This constant is defined by the relationship between the light and the radiation flux of the solar radiation, or the relationship between the illuminance and the density of the solar radiation on the observed plane:

$$K_S \left(\frac{\mathrm{lm}}{\mathrm{W}} \equiv \frac{\mathrm{lx} \cdot \mathrm{m}^2}{\mathrm{W}} \right)$$

The average values of the luminous efficiency of solar radiation for different meteorological conditions from a variety of authors are reported in Littlefair (1990). A summary is presented in Table 8.1.

Table 8.1 *The light effect of sun radiation for various sky conditions; γ_s is presented in Figure 8.5*

	K_s lumen per Watt (lm/W)
Cloudy sky	107
Clear sky, global sun radiation	$91.2 + 0.702.\cdot\alpha_s - 0.,00063.\cdot\alpha_s^2$
Clear sky, direct sun radiation	$51.8 + 1.646.\alpha_s - 0.,01513.\cdot\alpha_s^2$
Clear sky, diffuse radiation	144

Source: Littlefair, 1990

Artificial lighting

The discovery of fire at an early stage of mankind's development made it possible for people to illuminate and warm their homes. They were no longer dependent upon the natural light of the sun when performing tasks. Their living and working rhythms no longer relied upon the light and dark times of the day. These additional, or as we call them, the artificial light sources, changed throughout history. At first, individuals burned biomass and animal fat; later, they were mostly burning fossil fuels, oil derivates and gases. But only the invention of electrical energy and electric light bulbs at the beginning of the 20th century made quality illumination possible. From this point on it was possible to have light without polluting the internal spaces with emissions from burning fossil fuels. But for artificial lighting, we need energy, and electric light bulbs produce heat, which affects the thermal comfort of living spaces. So, in order to save energy when lighting buildings, it is very important to create harmony between natural and artificial lighting.

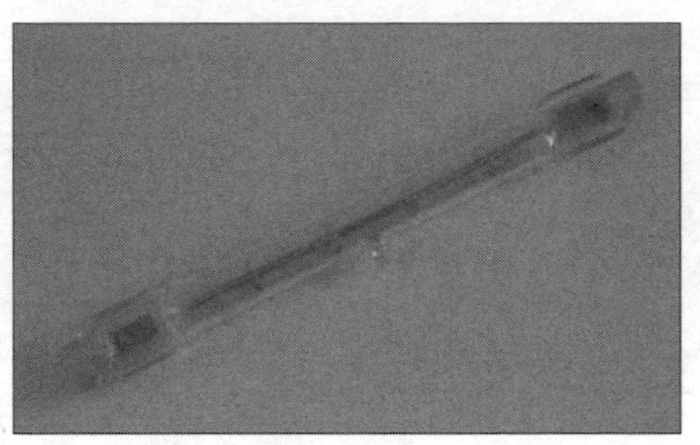

Figure 8.10 *A standard incandescent lamp* (left), *a halogen lamp with double-sided connection* (middle) *and a low-voltage halogen lamp with parabolic reflector* (right)

When referring to artificial lighting, we think of a system composed of electric light bulbs, lamps and control systems. Control systems are presented in Chapter 5, while the basic characteristics of light bulbs and lamps are covered in subsequent sections of this chapter.

Artificial light sources: Lamps

Lamps are devices that convert electrical energy into heat and light. Light sources can be divided into various types according to the physical principle by which the light is produced and their efficiency. The efficiency of a light source is defined as the relationship between the light that is emitted and the electricity that is consumed or, alternatively, the ratio of the emitted light to the heat produced. Theoretically, we could turn all of the electrical energy into light by emitting radiation with a wavelength of yellow-green light (monochromatic radiation with a wavelength of 523nm) that would be equal to 680lm/W. But the human eye is not equally sensitive to all the wavelengths that make up 'white' light. Therefore, the theoretical efficiency of a light source emitting only white light is just 200lm/W. However, as we will see, the efficiency of modern light sources is even lower than this.

For the past 100 years we have mainly used incandescent lamps. In an incandescent lamp, light is emitted by the electrical heating of a tungsten filament. This is a metal with a high melting point of 3680° Celsius. By heating the tungsten with an electrical current, the filament functions as an optical black body, reaching temperatures of up to 2700° Celsius and emitting a lot of heat, as well as light of all wavelengths. As a result, the lighting efficiency of these lamps is only about 15lm/W, and because of tungsten evaporation the life of these lamps is restricted to about 1000 hours. The evaporation of tungsten from the filament can be decreased by filling the bulb with an inert gas such as argon, or by adding halogens, especially bromine or iodine. This is how tungsten-halogen lamps are produced. Because the temperature of the filament is higher, about 2800–3000° Celsius, the lighting efficiency of these bulbs is between 16 and 18lm/W, and their lifetime is increased up to 2000 hours. The glass bulb is very hot (200–300° Celsius) and must be made from temperature-resistant quartz glass. These lamps are smaller in comparison to classical lamps, especially if adapted for use with low-voltage electricity (6, 12 or 24 volts). The filaments in these lamps are smaller and are more of a point light. With the use of mirrors, the light can be directed very accurately. Therefore, this type of light can be used in vehicles, projectors and lamps used to show details. The lighting efficiency of low-voltage halogen lamps is up to 25lm/W.

Despite a low lighting efficiency and a relatively short lifetime, incandescent lamps are still in common use because of their low price, because it is easy to control the emitted light flux simply by changing the voltage and because they are produced in many different shapes. They also have very good colour rendering, with prevailing wavelengths of orange and red.

A much better lighting efficiency can be achieved with fluorescent lamps. They are made from a glass tube that is coated with a fluorescent substance such as phosphorus. The inside of the lamp is filled with inert gas and low-pressure mercury vapour, and on each end of the glass tube there are built-in electrodes. When the electricity flows through the electrodes, it ionizes the mercury atoms that emit ultraviolet radiation with wavelengths of between 184nm and 254nm. The fluorescent coating turns this invisible radiation into light. With different fluorescent coatings we can affect the colour tone of the light emitted by the fluorescent lamps from warm light, with prevailing wavelengths of orange and red, to cool light, with the prevailing wavelength of blue light. Fluorescent lamps also require an external device called a ballast; this ballast can be made of a copper coil or a solid-state electronic device. The lighting efficiency of fluorescent lamps is 40 to 90lm/W, and they have a

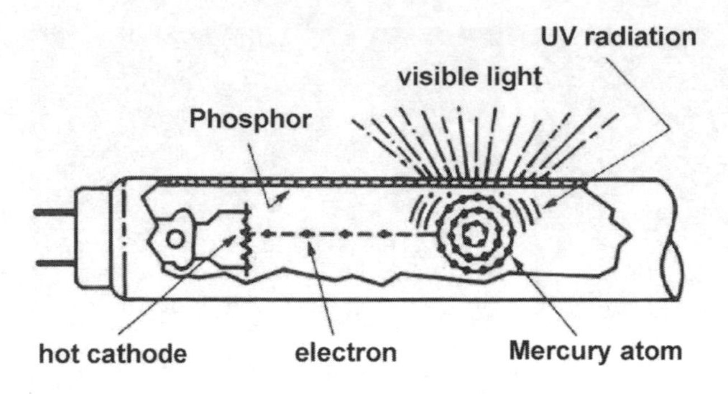

Source: Lechner, 1991

Figure 8.11 *Features of a fluorescent lamp*

lifetime of between 12,000 and 15,000 hours with the use of a modern ballast. Their long life and high efficiency normally offset the higher price.

The long tubes of fluorescent lamps mean they cannot easily replace classical incandescent lamps. Therefore, compact fluorescent lamps that join all of the necessary elements in their casing have been developed. Their lighting efficiency is comparable with classical fluorescent lamps; but their life is a little shorter because of their complicated structure – between 6000 and 10,000 hours.

With special types of bulbs, the light is emitted from a small arc tube, which is inside a protective outer bulb that is filled with an inert gas such as xenon, some mercury, and a substance that turns invisible mercury radiation into light. This substance is either a mixture containing iodine, for bulbs with metal halides, or it is sodium, under pressure, in high-pressure sodium lamps. This part is installed in a glass casing and has a form similar to an incandescent lamp. For their operation, they have similar requirements to fluorescent lights: they need a ballast. It is common for these lamps to require 5 to 10 minutes to

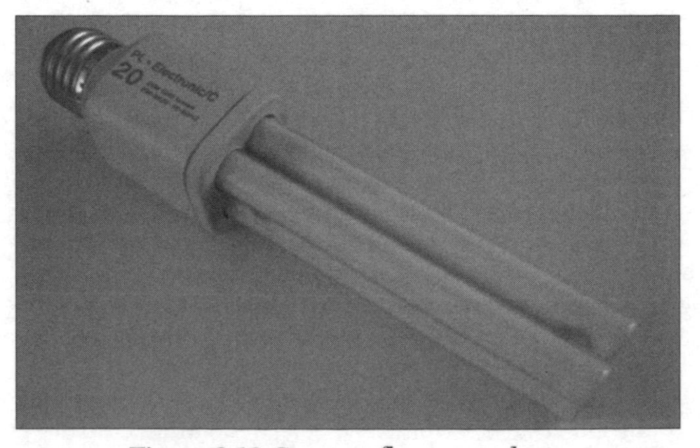

Figure 8.12 *Compact fluorescent lamp*

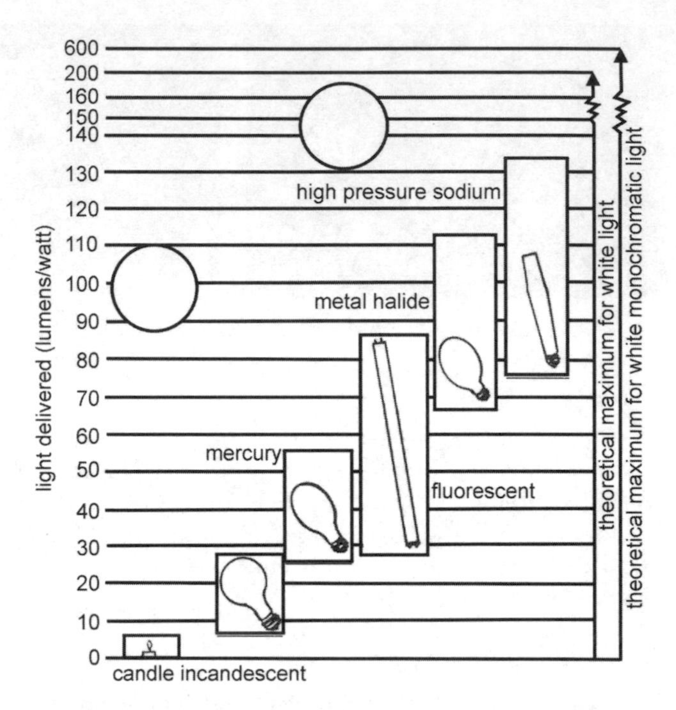

Source: Lechner, 1991

Figure 8.13 *Lighting efficiencies and lifetimes of different lamps*

reach their maximum light output, and they will not restart immediately after a voltage interruption. They require about 5 minutes to cool down before the arc can be restarted. Metal-halide lamps are distinguished by their high lighting efficiency of 80–125lm/W, their long life and their even light spectrum. As a result, they are used in places where high-quality illumination is required – for example, schools, shops and offices. High-pressure sodium lamps are also known for their high lighting efficiency (70–140lm/W) and long life (20,000 to 24,000 hours); but because of the light, mostly with wavelengths of yellow and red, they tend to be used in spaces such as warehouses and for outdoor illumination where the quality of the light is not so important. Lighting efficiencies and lifetimes of different lamps are shown in Figure 8.13.

Lighting fixtures/luminaries

Luminaries have three major functions: they protect the lamp and support it with a socket; they supply power to the lamp; and they control the direction of the light emitted by the lamp. In general, we divide luminaries into six categories that are defined by the way in which they distribute light – from direct (sending the light down on the work plane), general (distributing the light more or

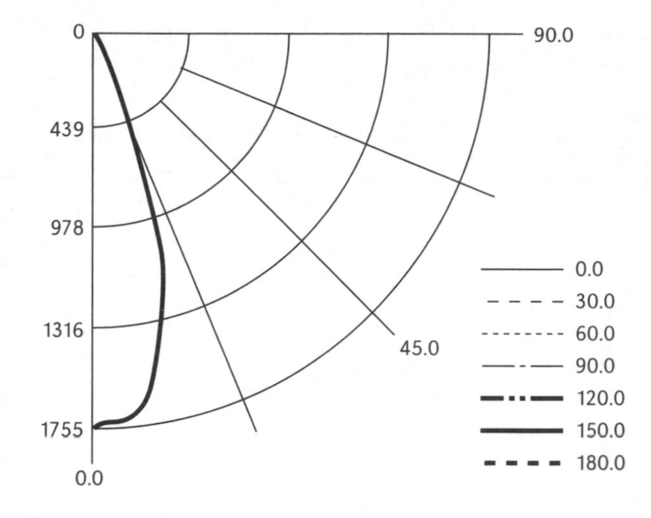

Source: Cerne and Medved, 2001

Figure 8.14 *Lamps with direct light-flux distribution and its photometric diagram*

less equally in all directions), to indirect (directing almost all of the light up to the ceiling). The producers of luminaries provide photometric data sheets for their products. On these sheets the lighting intensity at a certain angle is presented with the distance from the centre of the light source. With the help of the photometric diagram and the distance between the luminaries and the surface in the same angle direction, we can determine the distribution of the illuminance in space, or, for instance, in a work plane.

For the process of indoor- and outdoor-lighting design, the characteristics of the luminaries described in this chapter are not sufficient. Lighting designers must be aware of the requirements for visual comfort and the criteria that help designers to evaluate the state of the visual comfort in living and working spaces.

Visual comfort requirements

The demands of light comfort in indoor and outdoor areas result from the psychological and physiological needs of people. Psychological needs are satisfied if the lighting provides us with the following:

- a guide to our motion and orientation in space and a visual connection to our surroundings;
- connections to our bio-rhythms;
- an ability to make objects recognizable in space;
- an ability to direct attention and help us to classify the importance of the information received by sight;
- an assurance of a feeling of individuality with more or less illuminated parts of large spaces;
- a diversity of internal spheres and an absence of fear in dark spaces – where danger is normally expected.

Biological needs can be satisfied with good planning and checking of lighting areas. For this we use physiological requirements (the most important ones are presented later in this chapter). These are physical quantities adjusted to the characteristics of the human eye. We must be aware that this is a group of very different, but equally important, demands, which must be fulfilled individually.

Beside thermal comfort, sound protection and good indoor air quality, visual comfort comprises the fourth major field of a pleasant, healthy and productive indoor environment. Visual comfort must be fulfilled through several conditions, described below.

Illumination level

For the perception of objects in space and the performance of various tasks, a certain level of lighting is required. To recognize a human face, a face lighting of 10, or – calculated on a horizontal plane – a lighting of 20lx, is required; but for the recognition of details on a face, a tenfold larger illuminance of 200lx must be provided. The first value represents the minimum level for recognition, the second one a minimum level of illuminance for working areas. Generally, an optimum level of working-area illuminance for agreeable working conditions depends upon the task we are performing. These levels (presented in Table 8.2) are for artificially illuminated areas, and must be ensured for each of the working areas.

Table 8.2 *Recommended level of illuminance for working spaces with artificial illumination according to the needs of the work*

Activity space	Building type	Illuminance (lux)
Art, craft, teaching	Schools, colleges, hospitals, factories, offices	300–500
Laboratories	Hospitals, factories, offices	500–700
Staff rooms, common rooms	Hospitals, factories, offices	150–300
Offices	Hospitals, factories, offices, post offices, banks, education buildings	500
Computer rooms	Offices, banks, education buildings	500–750
Drawing rooms	Education buildings, offices, factories	500–1000

Source: J&J, 1993

Daylight factor (DF) criteria

We have already defined the daylight factor (DF) as a measure of daylight quality. Similar to the illumination level of artificial lighting, the appropriate daylight factor also depends upon the requirements of the work. During natural illumination, because the level of illumination changes considerably with the depth of the space, we evaluate the adequacy of the daylight with the average daylight factor DF_{avg}. The recommended values of DF_{avg} are stated in Table 8.3; however, the recommended daylight factors vary from country to country. For instance, in the UK the minimum DF is 2 per cent in kitchens, 1.5 per cent in living rooms and 1 per cent in bedrooms, while in Greece the recommended DF is 1.5 per cent in kitchens, 3.5 per cent in living rooms and 1 per cent in bedrooms.

The DF at a chosen point in a room should be calculated by considering nominal illumination of an outdoor unshaded horizontal plane, and the model of an overcast sky. The nominal illumination indicates the minimum illumination in a selected daily time interval. Figure 8.15 presents the nominal illumination of an outdoor, unshaded horizontal plane for different site latitudes and daily time intervals. For example, in mid-Europe, the illumination of the outdoor unshaded horizontal plane will be at least 54001x for 90 per cent of the time interval between 09:00 and 17:00 hours (as this time interval indicates daily working hours).

Table 8.3 *Recommended minimum and average daylight factor and evenness of lighting in different spheres*

Ambient	DF_{av} (%)	DF_{min} (%)
Offices	5	2
Classrooms	5	2
Living rooms	1.5	0.5
Bedrooms	1.0	0.3
Kitchens	2.0	0.6

Source: J&J, 1993

There are several parameters that influence the daylight factor at a chosen point in a room: the sky condition, nearby buildings and their surface reflectivities, the reflectivities of inner surfaces of the room and, of course, the transmittance of the window's glazing. In Figure 8.16 the daylight-factor components in the room are shown. The daylight factor is determined by adding the following components:

$$D \propto (D_s + D_{sky} + D_a + D_b)\, \tau_{g,vis} + D_i$$

where $D_a \propto \rho_a$ and $D_b \propto \rho_b$

where D_s is the sun component of daylight factor (%); D_{sky} is the sky component of daylight factor (%); D_a is the surrounding ground-reflected component of the daylight factor (%); D_b is the surrounding building-reflected component of the daylight factor (%); D_i is the internally reflected component of the daylight factor (%); ρ_a is the ground reflectance (albedo) (1); ρ_b is the surrounding-building reflectance (1); τ_{vis} is the light transmittance of the window glazing (1).

Table 8.4 *Typical reflectivity ρ_a in ρ_b*

Material	ρ_a, ρ_b
Aluminium	70–85
Asphalt	10
Concrete	0–50
Stone	5–50
Wood	5–40
Paints, white	70–90
light tones	50–75
dark tones	20–40
black	5
Brick, red	25–45
Snow, fresh	60–75
Glass, ordinary	7
reflex	20–40
Grass	10–30

Source: Medved, 1993

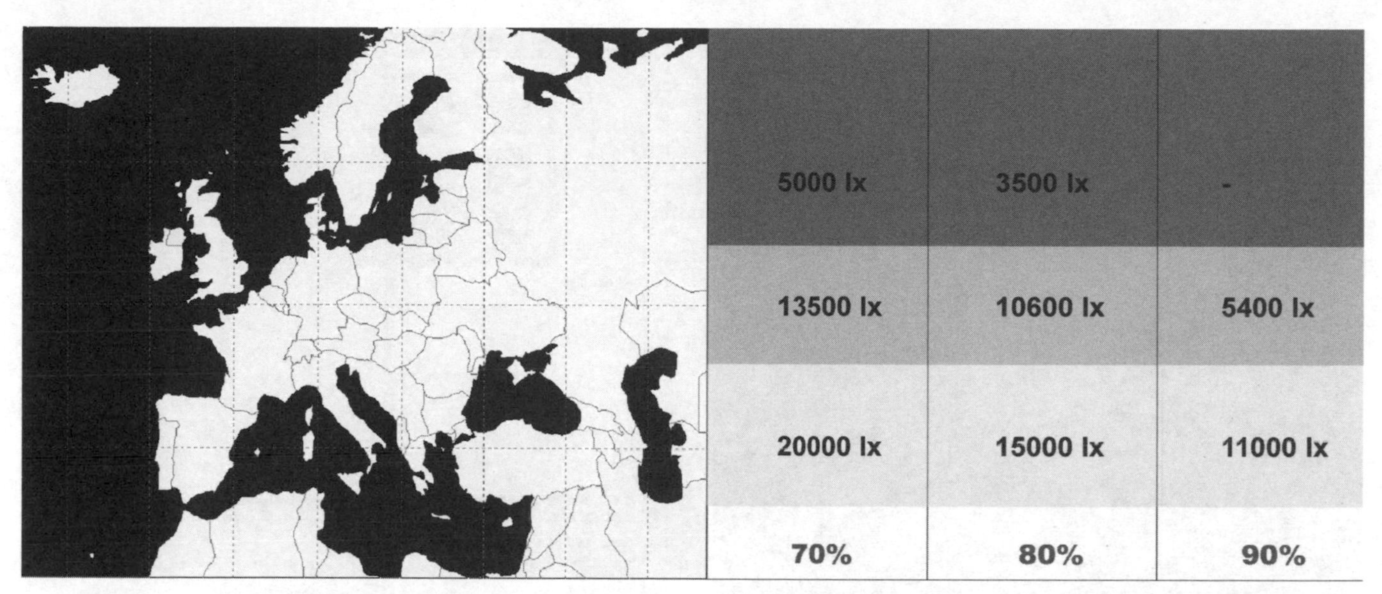

Figure 8.15 *Minimum illuminance of the outdoor horizontal unshaded plane under overcast sky during 70, 80 or 90 per cent of daily working hours (09:00 to 17:00) in different EU regions*

Table 8.5 *Thermal and optical properties of the different glasses*

Type of glass	U (W/m²K)	τ_{vis} (1)	g (1)
Double glazing	3.0	0.82	0.76
Triple glazing	1.8	0.74	0.70
Double, low-e glazing	1.9	0.77	0.60
Double, low-e, Ar	1.1	0.74	0.60
Solar protection glazing	1.1	0.65	0.33

Note: U = thermal transmittance of the glazing; g = total energy transmittance of the glazing. These indicate the total heat flux into the living space indicated by solar radiation.
Source: Medved, 1993

The listed transmissions of the types of glazing in Table 8.5. are valid for clean glazing. Because of the higher levels of air pollution in cities, the fraction of light transmitted through glass decreases with time faster in cities than in other areas. Measurements show that the transmittance of vertical windows decreases by 5 per cent over a period of six months, for windows sloping at 60° the reduction is 25 per cent, and for windows at an angle of 45° the amount of transmitted light drops to 45 per cent during the same time period. We should also mention that glasses with a catalysator layer, which have a solar-assisted auto-cleaning feature, are already in production. Besides improving the daylight-transmittance properties, these glasses also reduce costs and are environmentally friendly because they reduce the amount of detergents needed for cleaning.

Tall buildings in cities form canyons, which tend to lower the sky component of the daylight. However, we also have to take into account the fact that the materials used in cities – such as asphalt, concrete, glass or metal – can increase the daylight component that is reflected from the surroundings and nearby buildings. The amount of daylight that a building with a north-facing façade in a street canyon receives is strongly dependent upon the reflectivity of the opposite, south-facing façade, especially on sunny days.

A wide range of nomograms are stated in Littlefair (1990) for determining the daylight factors for different geometrically shaped canyons and atriums. Because the daylight factor depends upon a large number of meteorological, geometrical and optical parameters, each case is

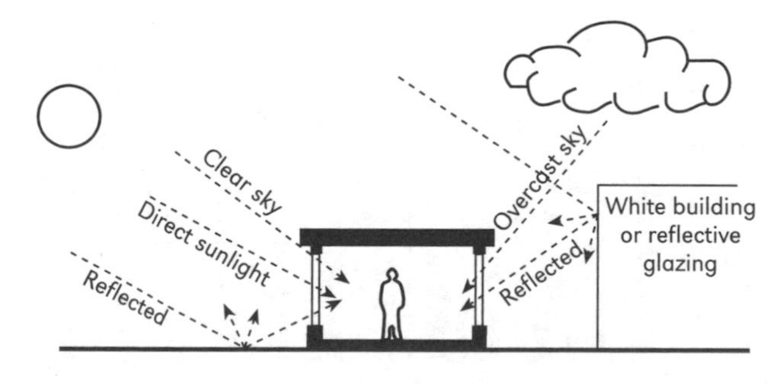

Source: Lechner, 1991

Figure 8.16 *Components of the daylight factor (DF) in an urban environment*

Figure 8.17 *A typical urban street canyon*

unique. However, it can be analysed with various computer tools. For the case of an overcast sky, the daylight factors can be analysed with the DFCAD calculation tool (DF.xls) on the accompanying CD (Arkar, 2001).

Figure 8.18 indicates the difference between the illuminance of a north-facing façade in a street canyon and an unobstructed north-facing façade in plane surroundings on sunny days. In the case of highly reflective façades

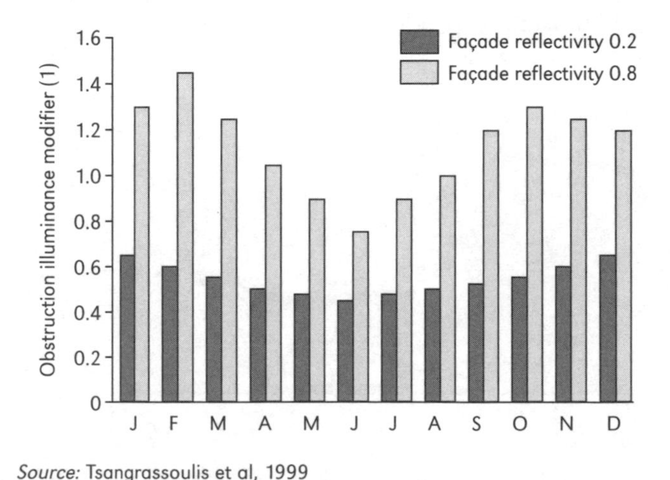

Source: Tsangrassoulis et al, 1999

Figure 8.18 *The obstruction illuminance multiplier*

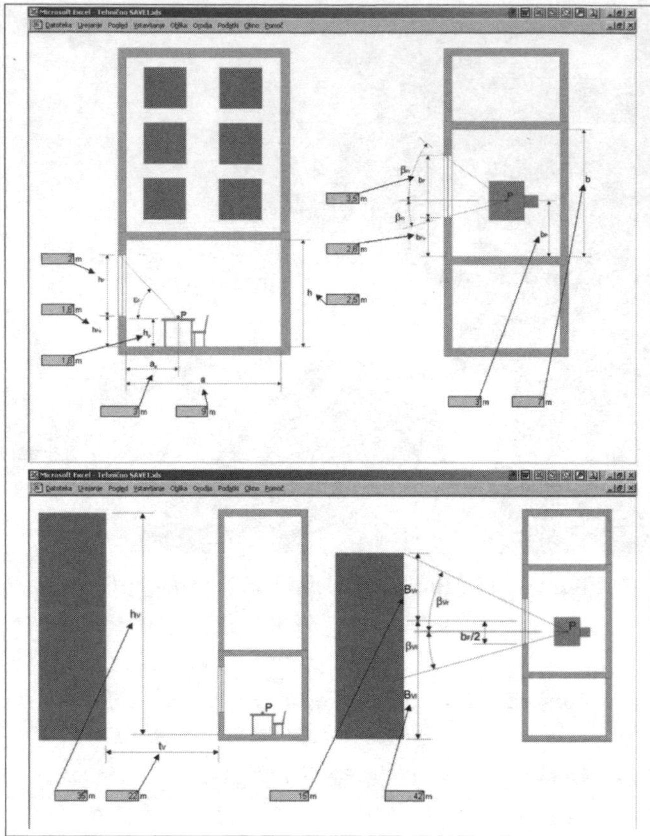

Source: Arkar, 2001

Figure 8.19 *Input windows of the DF.xls spreadsheet, on the left for entering geometrical data relating to the surrounding room and the observed point in the room; and on the right for the canyon data input*

the illuminance of the north-facing façade during winter months is even greater due to the strong reflections from the opposite south-facing façades.

Uniform illumination criteria

The required illumination should be ensured for all working planes, regardless of their position inside the living space. Uniform illumination must be evaluated using the relationship between the minimum illuminance of a working plane E_{min} and the average illuminance of all working planes in the room E_{av}. With artificial lighting, it is relatively easy to achieve uniform illumination. However, it is much more difficult to ensure this with daylight because the condition of the sky changes constantly and because the room illumination decreases rapidly with the distance from the window. In this case the uniformity of the illumination can be improved by the following:

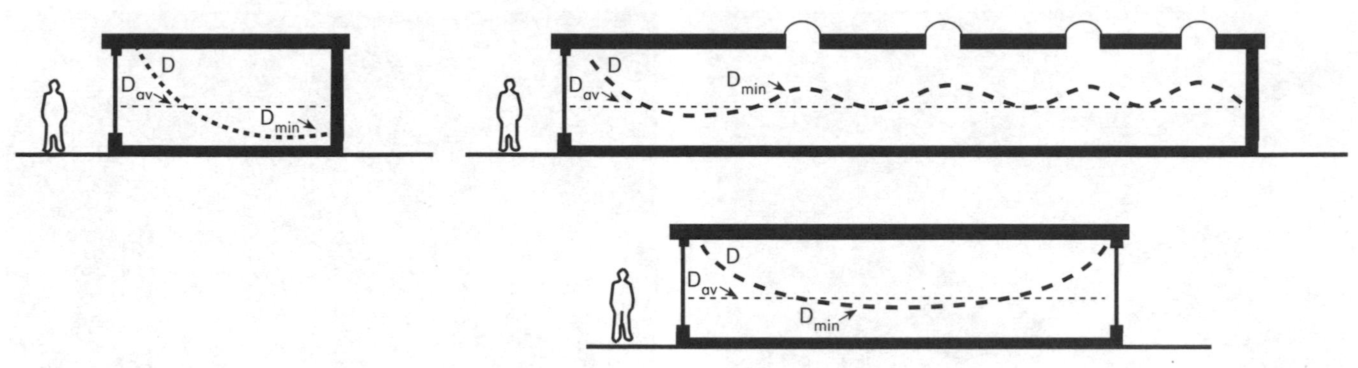

Source: Lechner, 1991

Figure 8.20 *The uniformity of the daylight illumination inside a living space can be significantly improved by well-considered arrangement of the windows*

- Locating the windows to parallel walls and installing a skylight if the depth of the interior is more than about three times the height of the window.
- Diverting the sun's rays to the ceiling with reflective jalousies, prismatic elements and holographic films. These elements use two basic optical principles: reflection and refraction. The optical characteristics of these elements must be adapted to the orientation of windows and the longitude (L) of the site where the building will be constructed.
- Attaching highly reflective surfaces to the inner or outer sides of the window, such as windowsills and light shelves.
- Incorporating light pipes, light ducts and fibre optics – elements that track the sun and beam the sun's rays into the room. This method of improving the quality of daylight is especially efficient in cities because light pipes positioned on the roof, which is normally less shadowed than the windows in the sidewalls, can conduct the light.
- Considering the colour of the walls and the interior. The colours should be chosen in such a way that the light reflexivity on the ceiling is between 80 and 85 per cent, on the side walls between 50 and 60 per cent, and the reflexivity of the ground between 15 and 30 per cent.

Glare

Glare occurs when we perceive a source of strong light in our *central field of vision*. Glare is one of the main reasons for light discomfort. It can be caused by the brightness of light sources (daylight and artificial) or by a strong light flux reflected from surfaces in the space and the surroundings – especially from windows, glass, façades and metal

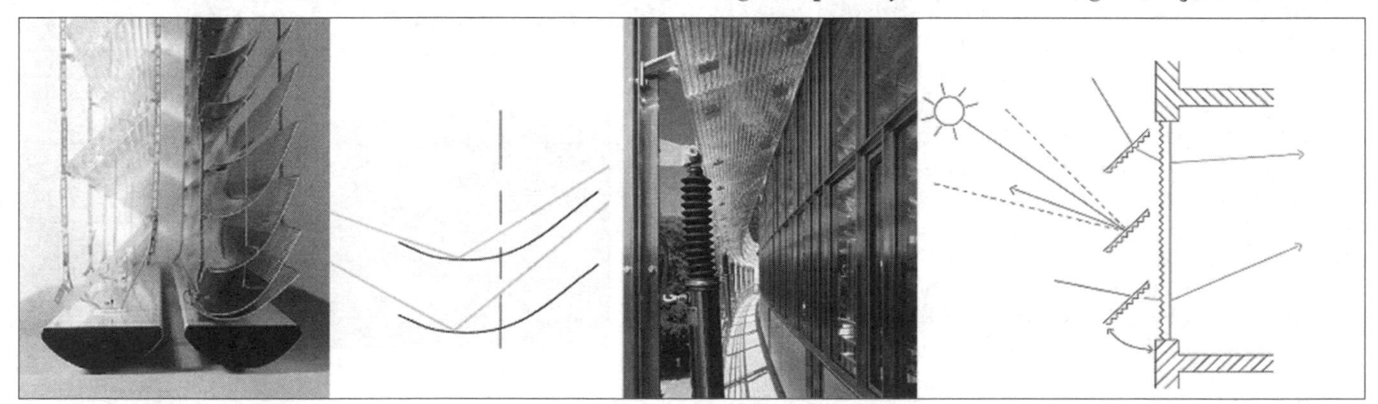

Note: By using holographic films, these three-dimensional bodies can be replaced with a two-dimensional plane that has the same optical characteristics.

Source: Muller, 1991

Figure 8.21 *The operation of the reflective jalousies, shades and prismatic elements, which divert daylight deep into the room space, use two basic optical principles: reflection (reflective shades) and refraction (prismatic elements)*

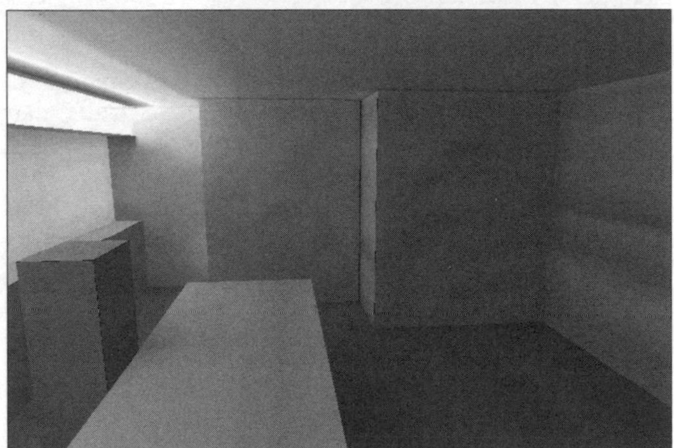

Figure 8.22 *In the office shown, the uniformity of the daylight illumination increased from 1:6 to 1:3.5 after the light shelf was installed; computer simulations and a photo of the window with and without light shelf are shown*

linings in the urban environment. In the first case, we are referring to direct glare, and in the second case, to reflected glare. By observing the glare caused by the sun and the sky, and the glare caused by lamps, it has been demonstrated that people are more sensitive to the glare caused by lamps. Consequently, glare is evaluated with the glare index (GI) for artificial lighting and the DGI for daylight glare. In Table 8.6, the criteria for the evaluation of the GI formed by the Illuminating Engineering Society (IES) are stated.

Glare as a consequence of daylight can be decreased with the architectonic design of buildings and spaces, or by the implementation of different devices, as follows:

- Decrease the windows' surface area and/or rearrange their positions.
- Decrease the transmittance of the glass with more absorption of the sun's radiation or with a higher level of reflectance on the outer surface of the glass.

Table 8.6 *Description of the glare perception and the allowed glare index (GI) and daylight glare index (DGI) according to the different interiors and spaces*

Glare	GI	DGI	Interior or space
Imperceptible	10	16	Art galleries
Just acceptable	16	20	Offices, museums, classrooms
	19	22	Laboratories, offices, hospitals
Uncomfortable	22	24	Factories – drawing offices
	25	26	Factories – rough work
Intolerable	28	28	

Source: J&J, 1993

Source: Schmitz-Gunter, 1998

Figure 8.23 *With light pipes it is possible to bring lead daylight to spaces that are 10m or more below the ground; above is a system with parabolic tracking mirrors* (bottom left) *a combination of skylights and optical light pipes for lighting a corridor 4m below roof level* (top right and left)

However, care is required here because this can also affect daylight illumination. In the case of the absorptive glasses, the temperature of the glass is higher, and this affects the operating temperature and can cause local thermal discomfort. The transmitivity of glass can be changed automatically depending upon the conditions in nature – for example, the density of the light flux (photochromatic glazing); or the temperature (thermochromatic glazing); or by the user themself; in the case of windows with a liquid-crystal layer; metal oxides that change colour when bonded into an electric circle (electrochromic glazing); or by using different substances (gasochromic glazing). We often refer to these types of windows as 'smart' windows.

- Attach shades to the inner or outer sides of the glass, which can be movable or fixed. Externally fixed shades are very efficient for shading windows on south-facing façades. The most appropriate for the daylighting of spaces are louvers, which pass the light from the brightest part of the sky at the zenith and shade the spaces from direct solar radiation, especially in the summer when the sun is high and there is no fear of glare. Glare can be avoided efficiently with the use of movable shades – for example, jalousies, shutters, roller blinds or blades – that can be attached on the inner or outer side of the window. However, external shading devices are more efficient because they keep the sun's radiation outside the space, while internal shades allow individual adjustment of the illumination.

Figure 8.24 *North-facing windows in a shopping centre ensure quality daylight without causing glare and overheating problems*

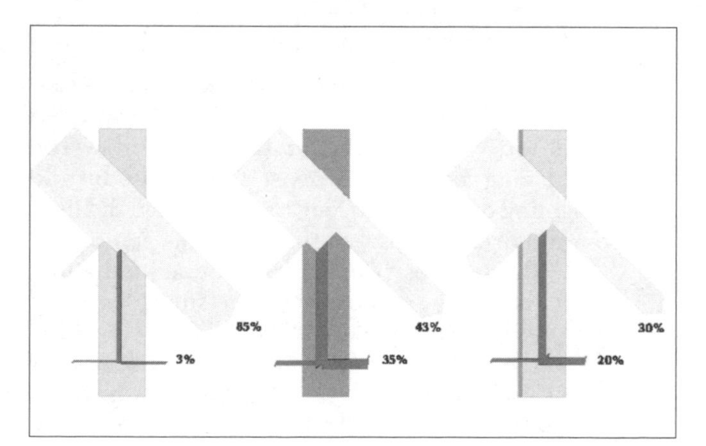

Figure 8.25 *The transmittance of solar radiation through common absorbtive and reflective glass; in the case of absorbtive glass the transmittivity of solar radiation decreases, but the heat flux into the interior is increased by radiation and convection heat transfer*

Note: In this case the glare caused by the sun was decreased by the screen printed on the upper part of the glazing, and the glare caused by reflection from the ground by the screen printed on the lower part of the glazing. But, simultaneously, the inner surface temperature of the glass was up to 15° Celsius higher on a sunny day compared with clear glass. This caused thermal discomfort because of the higher radiant temperature.

Figure 8.26 *With printed screens the transmittivity of glass is decreased only on the parts where glare is expected* (right)

With artificial lighting attention must be paid to areas where the lamp emits light flux in the direction in which we can perceive direct glare when our sight is directed horizontally, in the direction which can cause veiling glare because of the reflection of light from the working surface. The glare from lamps can be prevented by proper light positioning, with a proper choice of the photometric characteristics of the lamp and with ribs, lenses or diffuse mirrors, which restrict direct sight of the lamp or the lighting source.

Requests with reference to daylighting and the duration of sun exposure for urban areas

One of the specific problems related to ensuring natural lighting of buildings in the urban environment is a consequence of the high construction density and the tall buildings. These surrounding buildings cover part of the sky, shade the windows from the sun and therefore reduce the daylight factor and duration of the sunshine. As we

Table 8.7 *Maximum obstruction angle measured 2m above the ground*

City latitude (degrees)	City	Obstruction angle (degrees)
35	Kreta	40
40	Madrid	35
45	Ljubljana	30
50	Stuttgart	25
55	Edinburgh	22
60	Stockholm	20

Source: Littlefair, 2001

Figure 8.27 *Fixed shades in the form of louvers on a south-facing façade (8 October; south façade; 13:00; L = 46 degrees; γ_s = 7 degrees; α_s = 35 degrees) such devices efficiently shade interiors and still allow quality daylighting; but at lower altitudes of the sun (α_s < 30 degrees) they do not prevent glare*

know, the daylight factor is determined by the overcast-sky model; therefore, it is proportional to the angle of the sky that is visible from the selected point. The luminance of an overcast sky depends only upon the latitude of the observation site. Therefore, the maximum allowed obstruction angles measured from ground level and the top of the surrounding buildings vary according to site latitude. The maximum recommended values are presented in Table 8.7.

Sunlight also has an effect on people's health because sunrays give us the feeling of warmth and a healthy environment. The criteria that state the necessary amount of sunlight are, therefore, more subjective, and demands vary considerably. The duration of sunlight is the time when a certain point – for instance, the centre of the window – can be sunlit despite its position on the building or nearby obstacles. The veiling of the sun because of clouds is not considered in this case.

Note: Internal movable shades enable an individual regulation of natural lighting and glare, but they heat up because of the absorbed solar radiation and after that they radiate heat into the space. Even highly reflective roller blinds (bottom) at a low outside temperature (5° Celsius) heat up to 45º Celsius and can cause local thermal discomfort.

Figure 8.28 *Movable shades efficiently decrease the visual discomfort caused by glare*

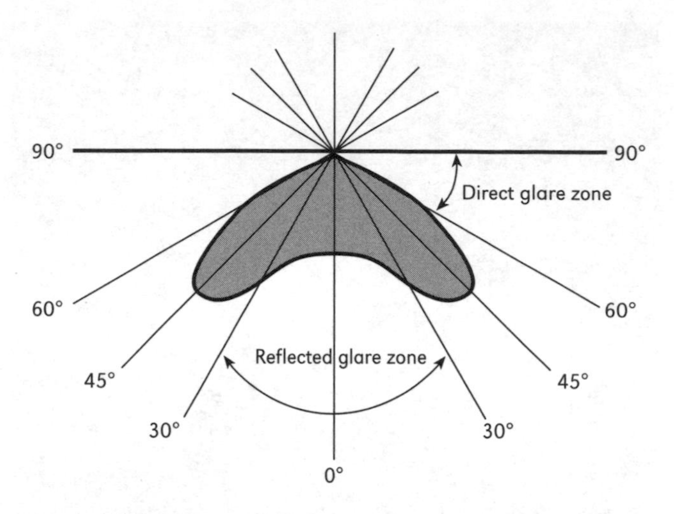

Source: Lechner, 1991

Figure 8.29 *Photometric diagram of a lamp with areas of direct- and reflected-glare zones*

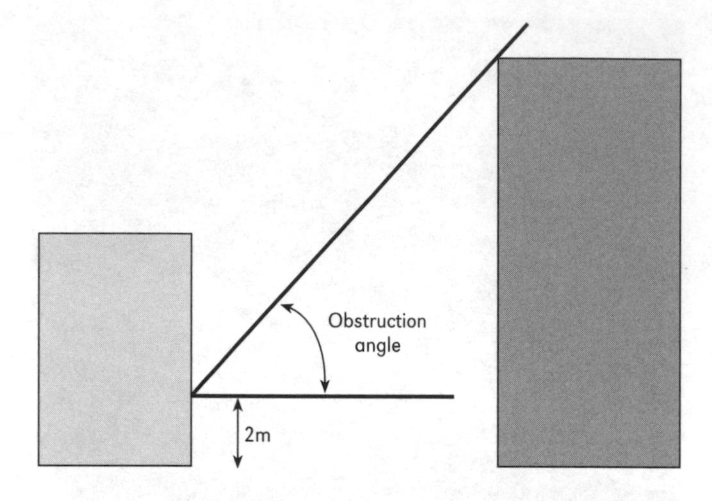

Figure 8.30 *Presentation of the obstruction angle*

Tables 8.8 and 8.9 summarize the criteria for selected countries and site latitudes.

The effect of nearby obstacles can be analysed with different computer programmes (for a review see J&J, 1993). Here we often face problems that relate to the exact listings of complex building geometry and vegetation. We can analyse the surroundings more accurately with a visual treatment of the landscape. Figure 8.31 shows the surroundings of an observed point, photographed with a camera, and a computer review of the surroundings in a cylindrical diagram (Medved et al, 1992).

Table 8.8 *Minimum sunlight duration stated in the regulations of different countries*

Country	Duration of sunlight
UK	> 25% annual probable sunlight hours in six months
Germany	> 4 hours of sunlight on 20 March
The Netherlands	> 2 hours of sunlight on 19 February
Slovenia	> 1 hour of sunlight on 21 December; > 3 hours of sunlight on 21 March and > 5 hours of sunlight on 21 June

Table 8.9 *Minimum sunlight duration depending on the latitude (L) of the site*

City latitude (degrees)	Sunlight hours
40°–50°	>2 hours sunlight on 19 February
+ 50°	25% annual probable sunlight hours spread over six months of the year

Graphically, we can relatively easily and quickly evaluate the sunlight duration with a sun-path diagram onto which we superimpose the surrounding obstacles. For buildings that use the sun for heating, it is very important that the windows, sun spaces and sun walls face south, and that they are not shadowed during the winter. Therefore, in the area between south-east and south-west (in the area of the azimuth between –45 degrees and +45 degrees) in front of the building, the altitude angle of the obstacle should not be higher than the minimal noon altitude of the sun. Because of the surroundings, buildings in an urban area are rarely obstructed to such a small extent, and therefore it is more suitable to use the criteria stated in Table 8.10.

Light pollution

People enjoy looking at the night sky to see the stars and the planets. With the increase in electric lighting in urban areas, the view of the night sky has gradually decreased. Astronomers call the wasted light that goes up into the sky *light pollution*. About 99 per cent of the European population lives in areas where the night sky is above the

Table 8.10 *Highest obstruction angles in the area between south-east–south–south-west in front of a solar-heated building*

Latitude (degrees)	Critical angle (degrees)
< 45	66.5 (latitude)
45–50	20
> 50	70 (latitude)

Source: Littlefair, 2001

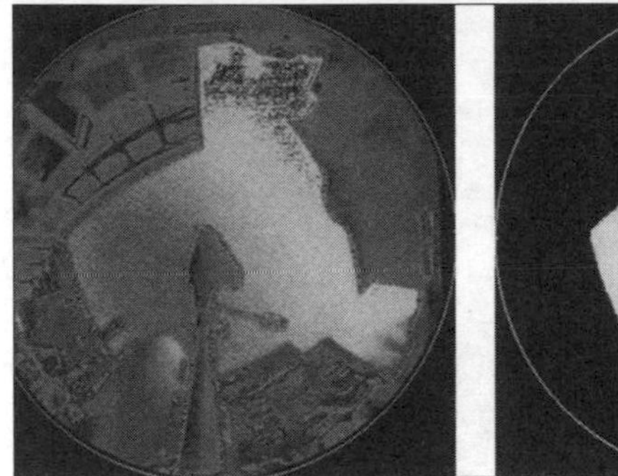

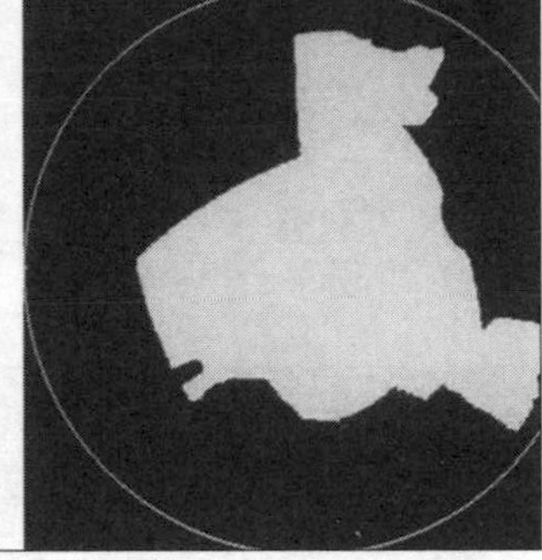

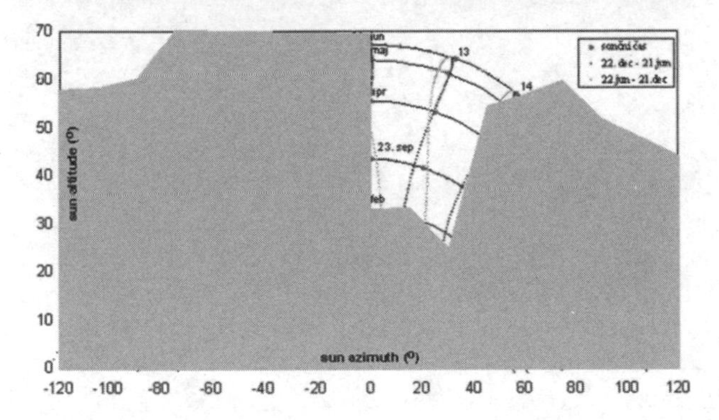

Source: Medved et al, 1992

Figure 8.31 *A photo of the internal yard* (top), *digitized and computer-analysed picture in a spherical mirror* (middle left) *and a sun-path diagram with surroundings* (bottom)

threshold set for polluted status, and more than half of the population have lost naked-eye visibility of the Milky Way because of the night-sky brightness. Maps of the artificial night-sky brightness at zenith at sea level are now available to compare levels of light pollution in the atmosphere, to recognize polluted areas.

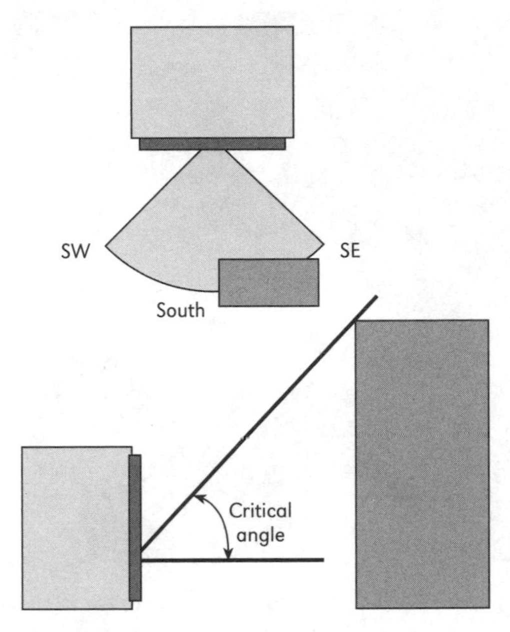

Figure 8.32 *Area in front of the solar heated building where the obstruction angle should be as minimal as possible*

The use of efficient outdoor lighting can greatly reduce this problem since 'overlighting' and 'uplighting' are avoided. The efficiency of outdoor lighting is determined by the control of light disbursement, lamp efficiency and the appropriateness of the level of illumination. The following list describes the characteristics of efficient light disbursement and control:

- Fixtures for parking areas, walkways and street lighting should direct all of their light to the ground.
- Fixtures should illuminate signs and billboards from above rather than from below.
- Architectural lighting should illuminate only the intended target. Spill lighting is to be avoided.
- Lights should automatically be turned off or dimmed according to the natural illumination.

The most efficient lamps (bulbs) for outdoor lighting are the low-pressure sodium type. The most efficient lamps commonly used for outdoor lighting are high-pressure sodium as they give good colour rendition, and metal-halide lamps, which are favourites for use in public and commercial parking areas where colour rendering is paramount.

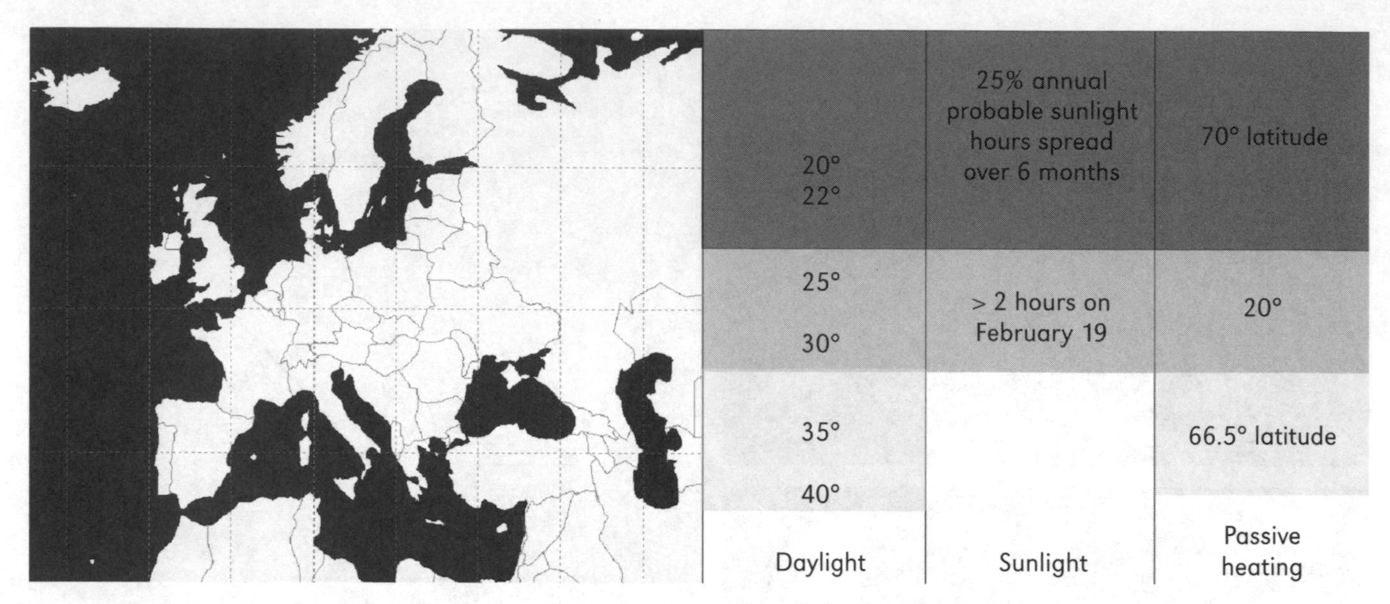

Figure 8.33 *A summary of the criteria of maximum recommended obstruction angles regarding daylighting, sunlight duration and passive solar heating*

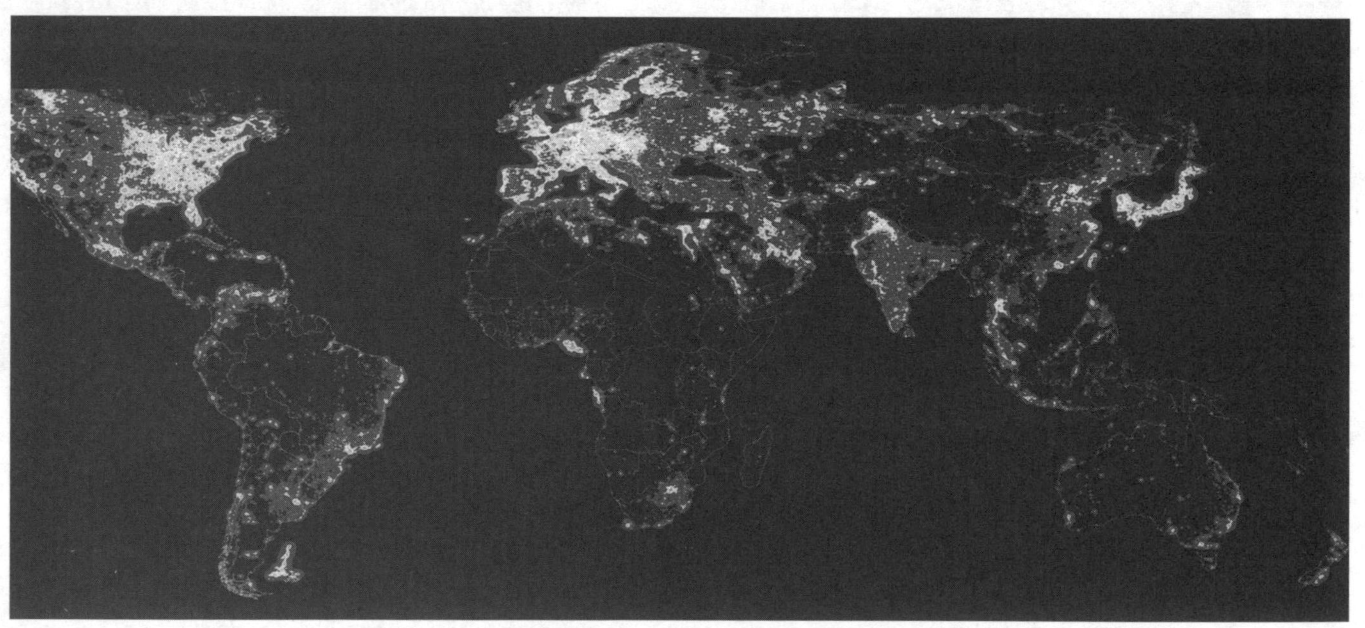

Source: Cinzano et al, 2001

Figure 8.34 *World map of night-sky brightness; bright pixels represent the most critical areas*

Lighting and the use of energy in buildings

The illumination of buildings and surroundings is connected to the use of energy. In residential buildings, the use of energy for illumination does not exceed 5 to 10 per cent of all the energy used; but in commercial and public buildings the energy consumption for artificial lighting could be as high as 50 per cent (Schmitz-Gunter, 1998)

As a result, indoor and outdoor illumination have an important effect on the rational use of energy, especially because lamps turn high-exergy electricity into light. The electricity is also produced with low efficiency and therefore with a high impact on the environment. Measures that help to decrease the consumption of energy for illuminating buildings include the following:

- Choose bulbs with a high light efficiency. Table 8.12 shows a comparison between installed electric power and the annual electricity consumption per square metre of office-floor area using different lamp technologies.
- Studies show that it will be possible to reduce electricity use by an additional 15 to 20 per cent with improved luminary light-flux distribution using better shapes and materials of reflectors.
- Regulate the light flux emitted by lamps according to the daytime (time-based control) presence of people (presence-based control) and the required level of lighting with automatic control systems (see also Chapter 5).
- Combine daylighting and artificial lighting; however, introducing daylight will have an effect on the higher use of energy for heating during the winter because windows have higher thermal losses than opaque building constructions. There will also be an effect during summer because of the high thermal load from solar radiation. Therefore, glass should have a low thermal transmissivity (U), a high light transmissivity (τ_{vis}) and a low total energy transmissivity (g) (see Table 8.5).

Table 8.11 *Electricity use for artificial lighting in different buildings as a percentage of the end-use energy consumption*

Type of building	Share of end-use for lighting (per cent)
Residental buildings	10%
Schools	10–15%
Factories	15%
Hospitals	20–30%
Commercial buildings	50%

Source: Schmitz-Gunther, 1998

Table 8.12 *Installed electrical power of lamps and the use of electrical energy for artificial illumination of the office building*

Lamp type	Luminary electric power (W/m²)	Yearly specific end-use of electricity (kWh/m² year)
Standard incandescent lamp	25	9.6
Halogen lamp	20	7.6
Fluorescent lamp	6	2.3

Note: For illumination of the work plane 500 lux (lx), daylighting is calculated to be sufficient in 80 per cent of working hours.

The process of planning a building should be an integral process in which, on the one hand, we consider the mutual influence of the thermal and the optical characteristics of the glazings; and the other the planning should involve different experts. As an example of successful integral or holistic planning, two cases are presented: a typical office in a commercial building with different double façades and the case of a shopping centre with a different skylight area.

In Figure 8.35 a three-storey commercial building with a south-oriented double façade is presented. The building is located in a site where DD = 2500K per day per year. During the process of design, the energy consumption for heating and lighting of the typical office was analysed. The area of a typical office is 20 square metres. The low-e double glazing was between the office and the single-glazed façade. High reflective and white painted fixed louvers installed inside the double façade were analysed. Other parameters (e.g. temperature, inner heat sources, interior, ventilation and occupancy patterns) were typical for commercial buildings. The specific yearly energy consumption for heating a typical office was analysed for north- and south-orientated offices without a double façade and for south-oriented offices with a double façade. Relative electricity consumption for artificial lighting was determined for the same offices according to the time period when the daylight factor (DF) is lower than 4 per cent.

The results are presented in Table 8.13. The double façade reduced energy consumption for heating but increased the use of artificial lighting. Overall performance shows that the double façade can be treated as an energy saving element.

Table 8.13 *Energy required for heating and relative electricity consumption for artificial illumination of a typical office in the building depicted in Figure 8.35*

	(kWht/m²a)	(¢kWhe/m²a)
North-oriented office without double façade	49.0	+ 0
South-oriented office without double façade	26.0	+ 0
South-oriented office with double façade and white louvers	11.5	+ 7.5
South-oriented office with double façade and reflective louvers	11.0	+ 6.5

Figure 8.35 *Commercial office with south-oriented double-glazed façade*

The second case illustrates the energy consumption of the shopping centre shown in Figure 8.37. The energy use for heating, cooling and lighting were analysed according to the size and quality of the skylights. The area of the shopping centre is 1530 square metres, and the indoor temperature is 20° Celsius in winter and 24° Celsius during the summer. The results are presented in Table 8.14 for roof skylight thermal transmittance U = 2.8W/m^{2}K, light transmittance τ_{vis} = 0.80 and energy transmissivity g = 0.75. The required illumination inside the shopping centre is 900lx. No shading device on the skylights was used.

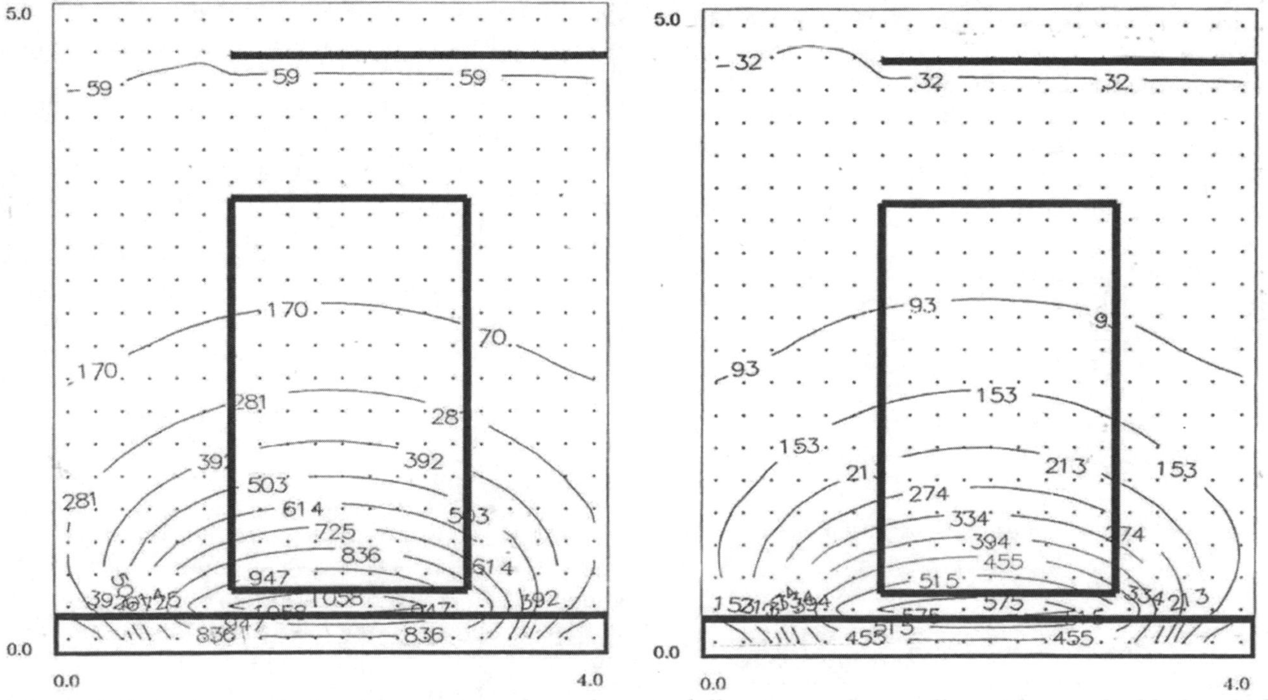

Figure 8.36 *Iso-luxes calculated for overcast sky and nominal illumination for an office without a double façade* (left), *and for an office with a double façade and reflective louvres* (right)

Figure 8.37 *Shopping centre with central shopping area with flat roof and skylights: view from the south* (left), *construction of skylights* (right)

Table 8.14 *Energy use for heating, cooling and lighting of the shopping centre shown in Figure 8.37*

Area of the skylights (% of floor area)	(DF_{avg}) (%)	Heating (kWh_t/m^2a)	Cooling (kWh_t/m^2a)	Lighting (kWh_e/m^2a)	Σ (kWh/m^2a)
0.0	0.0	50.3	4.3	70.8	125.4
1.6	1.3	50.8	8.0	49.3	108.1
2.7	1.8	51.3	10.8	41.3	103.4
4.9	2.8	51.1	16.5	33.3	100.9

Synopsis

Light comfort is one of the four elements of living comfort (besides thermal, sound and air-quality comfort) and is equally important to the other three. It is important, despite the basic rules, for each building to be treated as an individual case. Only in this way can we successfully solve the main task of building construction: to supply the best residential area with the smallest possible energy consumption and influence on the environment in terms of energy use. This is particularly true for buildings in cities, where conditions in the building often have strong links to the more or less shady surroundings and the urban climate.

References

Arkar, C. (2001) Daylight Computer Code, University of Ljubljana, Faculty of Mechanical Engineering, Ljubljana

Baker, N., Fanchiotti, A. and Steemers, K. (1993) *Daylighting in Architecture: A European Reference Book*, CEC, DG XII, James & James, London

Cerne, B. and Medved, S. (2001) *Influence of the Quality of the Building Envelope on Energy Use in Shopping Centres*, Thesis, University of Ljubljana, Faculty of Mechanical Engineering, Ljubljana

Cinzano, P. F., Falchi, C. D. and Elvidge, C. D. (2001) 'The first world atlas of the artificial night sky brightness', *Monthly Notices of the Royal Astronomical Society*, vol 328, pp689–707

'Daylighting: Areas for technical development', www.europa.eu.int/comm/energy_transport/atlas

Deutsches Institut für Normung e.V. (1985) *Tageslicht in Inneraumen*, Berechnung, DIN 5034, Teil 2

Deutsches Institut für Normung e.V. (1985) *Tageslicht in Inneraumen*, Grundlagen, DIN 5034, Teil 2

Energy Research Group (1994) *A Termie Programme Action: Daylighting in Buildings*, Energy Research Group, School of Architecture, University College Dublin, EC, DG DGXVII, Dublin

Hsieh, J. S. (1986) *Solar Energy Engineering*, Prentice-Hall, Englewood Cliffs, New Jersey

Lechner, N. (1991) *Heating, Cooling, Lighting: Design Methods for Architects*, John Wiley and Sons, New York

Littlefail, P. (1990) 'Measurements of the luminous efficacy of daylight', *Light Research and Technology*, vol 13, pp192–198

Littlefair, P. (1990) 'Innovative daylighting: Review of system and evolution method', *Lighting Research and Technology*, vol 22, no 1, pp1–15

Littlefair, P. (2001) 'Daylight, sunlight and solar gain in the urban environment', *Solar Energy*, vol 70, no 3, pp177–185

Medved, S. (1993) *Solar Engineering*, University of Ljubljana, Faculty of Mechanical Engineering, Ljubljana

Medved, S., Arkar, C., Cerne, B. and Vidrih, B. (2001) *Computer and Experimental Analyses of Thermal and Lighting Characterised in Selected Offices of the Commercial Buildings HIT*, University of Ljubljana, Faculty of Mechanical Enginnering, Report, Ljubljana

Medved, S., Griãar, P and Novak. P. (1992) 'Video and computer aided optimization of the solar radiation in urban environment', *Renewable Energy, Technology and the Environment*, Pergamont Press, vol 4, pp2244–2248

Mid-career Education, Technology Module 4: Daylighting and artificial lighting, Altener Project, Esbensen Consulting Engineers, Copenhagen

Mid-career Education, Technology Module 8: Calculation and design tools, Altener Project, CIENE, University of Athens, Athens

Moore, F. (1991) *Concepts and Practice of Architectural Daylighting*, Van NostrandReinhold, New York

Muller, H. F. O. (1991) *Intelligente beleuchtung von raumen*, Institut fur Licht und Bautechnik an der Fachhochschule Koln, Essen

Santamouris, M., and Asimakopoulos, D. (1996) *Design Source Book on Passive Solar Architecture*, EC, DC XVII, Altener, Athens

Schmitz-Gunter, T. (1998) *Lebensraume: Der Grosse Ratgeber für Okologisches Bauen und Wohnen*, Konemann Verlagsgesellschaft mbH, Köln

Tsangrassoulis, A., Santamouris, M., Geros, V., Wilson, M. and Asimakopoulos, D. (1999) 'A method to investigate the potential of south-oriented vertical surfaces for reflecting daylight onto oppositely facing vertical surfaces under sunny conditions', *Solar Energy*, vol 66, no 6, pp439–446

Recommended reading

1. Lechner, N. (1991) *Heating, Cooling, Lighting: Design Methods for Architects*, John Wiley and Sons, New York
   This book deals with the heating, cooling and lighting of buildings, not as discrete and isolated problems, but in the holistic sense of being integral parts of the larger task of environmental manipulation. The three chapters on light, electic lighting and daylighting present information required by the designer to create a quality lighting environment. Architects will find the most relevant information and practical tools available to them when designing heating, cooling and lighting systems. The design tools are mainly concepts, guidelines, handy rules of thumb, examples and physical modelling.
2. Baker, N., Fanchiotti, A. and Steemers, K. (1993) *Daylighting in Architecture: A European Reference Book*, CEC, DG XII, James & James, London
   This book is aimed at the architect and engineer who wants to acquire an in-depth understanding of the principles of daylighting. The concept of the book is also a good indication of the current scientific and design support work in European countries. The book is divided into several chapters, comprising light and human requirements according to the physiological and psychological background to the visual process; photometry of materials, with precise photometric descriptions of material surfaces and their optical properties; light transfer models with a review of techniques; and simplified tools and computer codes. A major role of this reference book is to provide design guidance.
3. Moore, F. (1991) *Concepts and Practice of Architectural Daylighting*, Van NostrandReinhold, New York
   The primary goal of this book is to familiarize the practising architect with the concepts and the analytical procedures needed to confidently design daylit buildings. Mathematical theory has been replaced wherever possible with analogies and graphic explanations. Graphic and computer design tools have been emphasized in lieu of tables and formulae. Intended as a practical introduction for the serious professional, this comprehensive book brings together a full range of design tools, as well as practical hints.

Activities

Activity 1

Describe the physical phenomena of light and human vision.

Activity 2

Compare physical and photo-optical quantities in lighting technology.

Activity 3

Describe the basic differences between numerical models of different natural sky approximations.

Activity 4

What are the most important features of electric lamps for lighting designers?

Activity 5

What are the most important requirements in order to ensure good indoor visual comfort?

Answers

Activity 1

The way in which people recognize electromagnetic waves depends very much upon the wavelength of the radiation and the senses used for the perception. For example, we see radiation with our eyes but feel it with our skin. Electromagnetic waves with wavelengths of between 380 and 770nm can be seen by the eyes, and this type of radiation is called light. The human eye can be compared to a camera. The lens, the iris and the film of the camera, are, in the human eye, replaced by the lens, the iris and the retina. By contracting the lens (accommodation), the eye adapts to be able to see objects at different distances, and by expanding or narrowing of the pupil (adaptation) the eye regulates the light flux falling on the retina. The human eye is, however, not equally sensitive to the different wavelengths of light. To get the impression of equivalent luminance from a surface, the eye must expand to allow a bigger light flux if the surface is coloured red or violet-blue than it would if the surface were coloured green or yellow. This characteristic of our eyes is called the spectral luminous efficiency $V(\lambda)$. Measured from the centre of vision, the field of view is seen as approximately 180 degrees wide in the horizontal direction. The eyebrows and cheekbones limit the field of peripheral vision to about 130°, and because of the retina's characteristics we see objects in this area differently. The sharp (foveal) vision occurs in a 2° cone around the centre of vision; but it is true that awareness is still quite high in the foveal surround, which is within a 30° cone around the centre of vision. The location and the brightness of the objects in this field of vision are very important.

The most important factors for our quality perception of our surroundings are:

- the sharpness of sight;
- the speed of sight perception, which is related to the time between the appearance of the object and its recognition;
- depth distinction (we can see a 0.4mm divergence between two objects at a distance of less than 1m; at a distance of 1000m, we can only discern if the two objects are separated by more than 275m; at a distance of 1300m the ability to distinguish depth does not exist any more);
- the geometry (spherical aberration) and colour (chromatic aberration) of the picture of the object we are observing (this is a consequence of the sun's rays refracting as they pass through the eye's lens).

Activity 2

In order to evaluate light sources and the lighting environment, we use scotopical and photo-optical quantities. We can express the first of these with units of energy; however, with the second, we also consider the effect of light on the human eye or on our light perception. For both quantities the systems are related by the spectral luminous efficiency $V(\lambda)$.

The spectral luminous efficiency value of the eye is a maximum when illuminated with light with a 555nm wavelength (for vision adapted to normal vision). It has been experimentally determined that at this wavelength the radiation flux emitted by a surface is the same as a light flux of 683 lumens (lm). This value is the maximum value of the luminous efficiency of radiation K_m for this monochromatic radiation. However, because light sources normally emit light in the visible spectrum, we define the entire lumination flux using integration over the whole visual spectrum.

Activity 3

In nature, the magnitude of the radiation and the relation between direct and diffuse radiation constantly change, and therefore certain points in the sky have periodically variable light intensities, which represent a major obstacle to daylight planning. In order to deal with this, three typical sky conditions that are characteristic of a particular place and time of year have been determined – overcast, clear and cloudy sky. The characteristic of an overcast sky is that the luminance of the zenith is three times bigger than the luminance of the horizon. Points in the sky with the same zenith angle have the same luminance, which, therefore, does not depend upon their azimuth. The luminance of the points in the sky cannot be determined as easily as in the case of an overcast sky. It depends upon many geometrical and meteorological parameters that change with time and position. The illuminance of a horizontal plane from a clear sky also depends upon the condition of the environment, particularly upon the absorption and dissemination of light on gas molecules and particulate material in the atmosphere. The variable conditions of the sky present extreme conditions in nature and are often quite rare situations. With the cloudy-sky model we can also evaluate intermediate conditions. These models are based on a statistical evaluation of different meteorological data – for instance, an estimation of cloudiness, the relationship between global and diffuse solar radiation, or the probability of direct solar radiation in hourly intervals.

Activity 4

Lamps are devices that convert electricity into heat and light. Light sources can be divided into various types according to the physical principle by which the light is produced and their efficiency. The efficiency of a light source is defined as the relationship between the light that is emitted and the electricity that is consumed or, alternatively, the ratio of the emitted light to the heat produced. Theoretically, we could turn all of the electrical energy into light by emitting radiation with a wavelength of yellow-green light (monochromatic radiation with a wavelength of 523nm) that would be equal to 680lm/W. But the human eye is not equally sensitive to all the wavelengths that make up 'white' light. Therefore, the theoretical efficiency of a light source emitting only white light is just 200lm/W. However, the efficiency of modern light sources is even lower than this.

Incandescent lamps have a lighting efficiency of between 15 and 25lm/W and their lifetime is increased to 2000 hours. A much better lighting efficiency can be achieved with fluorescent lamps (between 40 to 90lm/W) and they have a lifetime of between 12,000 and 15,000 hours with the use of a modern ballast. Metal-halide lamps are distinguished by their high lighting efficiency of 80 to 125lm/W, their long life and their even light spectrum; as a result, they are used in places where high-quality illumination is required – for example, schools, shops and offices. High-pressure sodium lamps are also known for their high lighting efficiency (70 to 140lm/W) and long life (20,000 to 24,000 hours); but because the light mostly has wavelengths of yellow and red, they tend to be used in spaces such as warehouses and for outdoor illumination where the quality of the light is not so important.

Luminaries have three major functions: they protect the lamp and support it with a socket, they supply power to the lamp and they control the light emitted by the lamp. In general, we divide luminaries into six categories that are defined by the way in which they distribute light – from direct (sending the light down on the work plane), to general (distributing the light more or less equally in all directions), to indirect (directing almost all of the light up to the ceiling). The producers of luminaries provide photometric data sheets for their products. On these sheets the lighting intensity at a certain angle is presented with the distance from the centre. With the help of the photometric diagram and the distance between the luminaries and the surface in the same angle direction, we can determine the distribution of the illuminance in space, or, for instance, in a work plane.

Activity 5

For the perception of objects in space and the performance of various tasks, a certain level of lighting is required. In order to recognize a human face, face lighting of 10lx, or – calculated on a horizontal plane – a lighting of 20lx, is required; but for the recognition of details on a face, a tenfold bigger illuminance of 200lx must be provided. The first value represents the minimum level for recognition, the second one a minimum *level of illuminance* for working areas. Generally, an optimum level of working-area illuminance for agreeable working conditions depends upon the task we are performing. Similar to the illumination level of artificial lighting, the *appropriate daylight factor* also depends upon the requirements of the work. Because with natural illumination the level of illumination changes considerably with the depth of the space, we evaluate the adequacy of the daylight with the average daylight factor (DF_{avg}). The required illumination should be ensured for all working planes, regardless of their position. *Uniform illumination* must be evaluated using the relationship between the minimum illuminance of a working plane (E_{min}) and the average illuminance of all working planes in the room (E_{av}).

With artificial lighting it is relatively easy to achieve uniform illumination; however, it is much more difficult to ensure this with daylight because the condition of the sky changes constantly and because the room illumination decreases rapidly with the distance from the window. One of the main reasons for light discomfort is *glare*. It can be caused by the brightness of light sources (daylight and artificial), or by a strong light flux reflected from surfaces in the space and the surroundings – especially from windows, glass, façades and metal linings in the urban environment. In the first case, we are referring to direct glare, and in the second case, to reflected glare. By observing the glare caused by the sun and the sky, and the glare caused by lamps, it has been demonstrated that people are more sensitive to the glare caused by lamps. Consequently, glare is evaluated with the glare index (GI) for artificial lighting and the daylight glare index (DGI) for daylight glare. Sunlight has an effect on people's health because sunrays give them the feeling of warmth and a healthy environment. The criteria that state the *necessary amount of sunlight* are, therefore, more subjective, and demands vary considerably. The duration of sunlight is the time during which a certain point – for instance, the centre of the window – can be sunlit despite its position on the building or nearby obstacles.

9
Case Studies

Koen Steemers (editor)

Scope of the chapter

This chapter describes the design and performance of nine case study buildings in urban environments. Climates represent the range of conditions found in Europe and projects are located in Greece, Slovenia, Germany and the UK. The building types primarily include office and residential use, which together make up the largest proportion of the built environment.

Learning objectives

The primary purpose of this chapter is to provide real examples of successful energy-efficient urban buildings. It is not intended to be comprehensive in terms of representing all building types or microclimates, but to demonstrate the application of key design principles in real building projects.

Key words

Key words include:

- case studies;
- offices;
- residential and educational buildings.

Introduction

The case studies presented demonstrate a combination of integrated design strategy with the appropriate application of technology in order to achieve low-energy buildings. They also usefully demonstrate – though not exhaustively – a rich range of concepts that have been developed in response to the specific contexts and briefs, and not applied in a deterministic manner. The range of architectural solutions highlights that low-energy urban design carries with it no stylistic or architectural constraint – in fact, it can be argued that the solution space is increased.

Nine building case studies are described. They have been chosen to provide a representative range of uses, including:

- office;
- residential;
- educational.

The case studies are located in a range of European climates:

- southern;
- continental;
- coastal.

The projects also demonstrate a number of integrated low-energy strategies:

- natural ventilation (cross-ventilation, atrium, double skin, stack);
- daylighting strategies (including switching systems for artificial lights);
- integration of passive design with mechanical systems.

The case studies are as follows:

1 'Meletikiki' office building near Athens, Greece;
2 'Avax' office building in Athens, Greece;
3 'Ampelokipi' residential building in Athens, Greece;
4 'Bezigrajski dvor': energy-efficient settlement in Ljubljana, Slovenia;

5 Commercial building with a double façade in Nova Gorica, Slovenia;
6 EURO centre commercial building with atrium in Ljubljana, Slovenia;
7 Potsdamer Platz: office and residential development in Berlin, Germany;
8 School of Engineering, De Montfort University in Leicester, UK;
9 Inland Revenue Office Headquarters with stack ventilation in Nottingham, UK.

Each has been chosen to represent a distinct set of conditions (e.g. combination of climatic and urban context, function or mix of functions, energy strategy, etc.), and has been selected on the basis of the appropriate level of information available. More importantly, each case study offers a specific insight and environmental approach that is successful, related typically to lighting, ventilation or mechanical systems concepts for urban buildings.

The format of the case study descriptions follows a general outline, which enables some degree of easy comparison between the contexts and solutions offered by each project presented:

- *Title*: name, type (office, retail, etc.) and location (city, neighbourhood) of building.
- *Introduction*: brief statement of the key challenge or innovation (e.g. natural ventilation in a noisy urban environment, daylighting strategy with high obstructions, refurbishment of a redundant urban building, etc.).
- *Location and climate*: climate (latitude, longitude, monthly temperatures, solar radiation, etc.), microclimate (local data where available to compare with synoptic data), physical context, development/planning constraints, etc.
- *Building description*: building form, space planning, façade design, systems, details, etc.
- *Key features*: special aspects that are innovative or interesting with particular reference to the urban context and issues.
- *Performance*: data related to the thermal, lighting and energy performance (actual or targeted) and comments on occupant satisfaction.

Case study 1: Meletikiki office building

Table 9.1 *Meletikiki office building*

Name:	Meletikiki Ltd: A. N. Tombazis and Associates, Architects
Type:	Office building at Polydrosso
Location:	Northern part of Athens: residential area

Introduction

The building is located in Polydrosso, Halandri, a residential area in the northern part of Athens, and is the office building of the architectural group A. N. Tombazis and Associates – Meletikiki Ltd.

The building presents some important design characteristics from an energy point of view. It is well insulated, with increased insulation provided to all exterior walls and roofs.

For reasons of bioclimatic design, the building is narrow and long. This shape is efficient for natural daylighting purposes. Furthermore, all windows are shaded by individual exterior vertical motorized shading devices. The natural light is reflected from fabric panels below the skylights of the roof. Ceiling-recessed downlights and task lighting compose the artificial lighting system. The control of the artificial lighting system is manual on/off. The building is also equipped with seven split and two single package air-to-air heat pumps for cooling and heating purposes.

The design of the building and its structural characteristics permit the application of various strategies oriented towards reducing the energy consumption of the building. The application of night ventilation techniques is an important measure in order to reduce the cooling load of the building. Another measure that is important for reducing the energy consumption for cooling is the installation of ceiling fans. By using the ceiling fans, it is possible to increase the set-point temperature during the summer season and, therefore, to reduce the operation time of the cooling system.

This particular building also permits the use of daylight because the workstations are located close to the building openings. Additionally, in order to increase the penetration of daylight into the upper part of the building, skylights have been installed.

Location and climate

The building is located in Polydrosso, Halandri, a residential area in the northern part of Athens, with relatively low traffic (see Figure 9.1).

The climatic characteristics of Athens are presented in Figures 9.2 to 9.4. The climatic data are for the Athens test reference year.

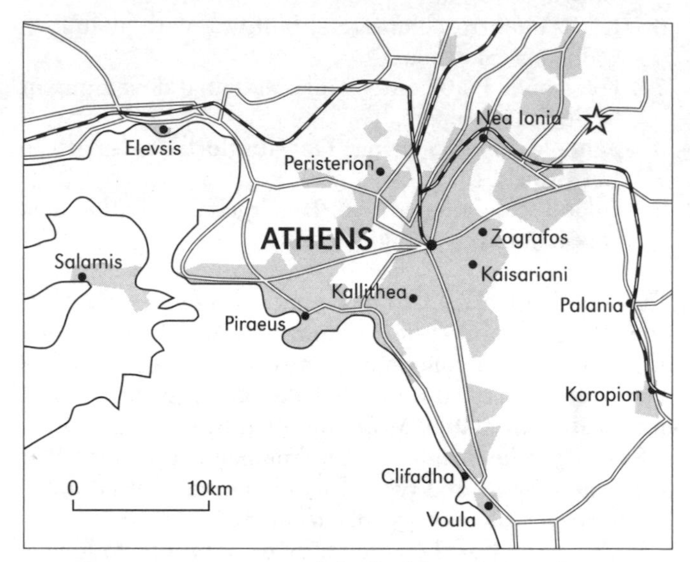

Figure 9.1 *Location of the building (star indicates the location region)*

Figure 9.2 illustrates the monthly average global and diffuse solar radiation. According to this figure, the global solar radiation varies between 200W/m^2 during the winter period and about 500W/m^2 during the summer season (maximum during August).

Figure 9.3 presents the monthly average ambient temperature and the humidity for the city of Athens. The minimum ambient temperature is close to 8° Celsius during February, while the maximum is about 28° Celsius during August. With regard to the outdoor temperature,

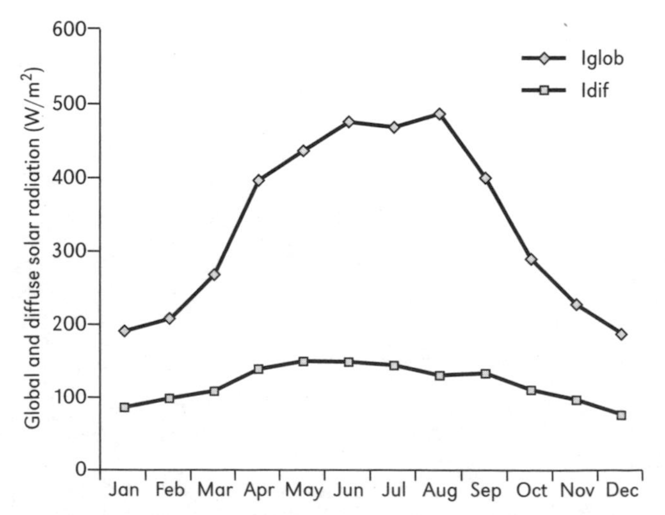

Figure 9.2 *Monthly average global and diffuse solar radiation on a horizontal plane in Athens*

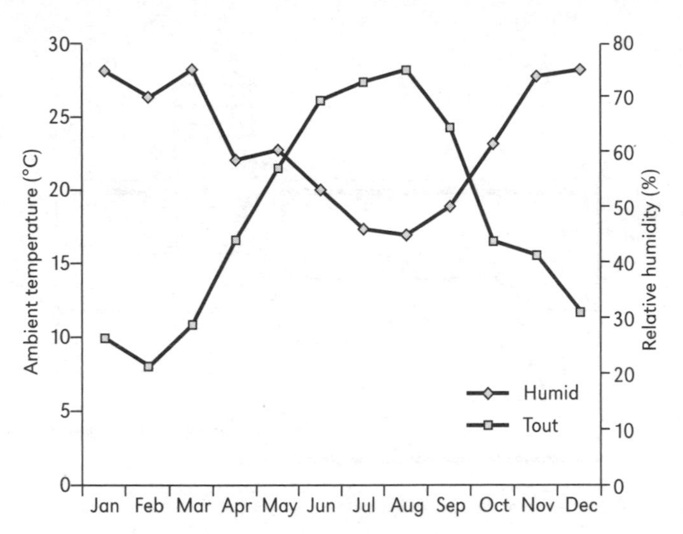

Figure 9.3 *Monthly average external temperature and external relative humidity in Athens*

Source: A. N. Tombazis and Associates

Figure 9.5 *Meletikiki building*

the hourly maximum value is close to 40° Celsius because the summer season is quite warm in Athens. On the other hand, during the winter the minimum hourly value of the temperature is close to 0° Celsius. The seasonal variation of the relative humidity is fairly negligable. During the summer period the relative humidity varies between 45 to 60 per cent, while during the winter the variation is between 60 to 75 per cent.

Finally, Figure 9.4 illustrates the monthly average wind speed and direction. These measurements concern the centre of Athens. During the entire year the wind velocity varies between 3 and 4.5 metres per second. For the same period the wind direction is from south-east to south-west.

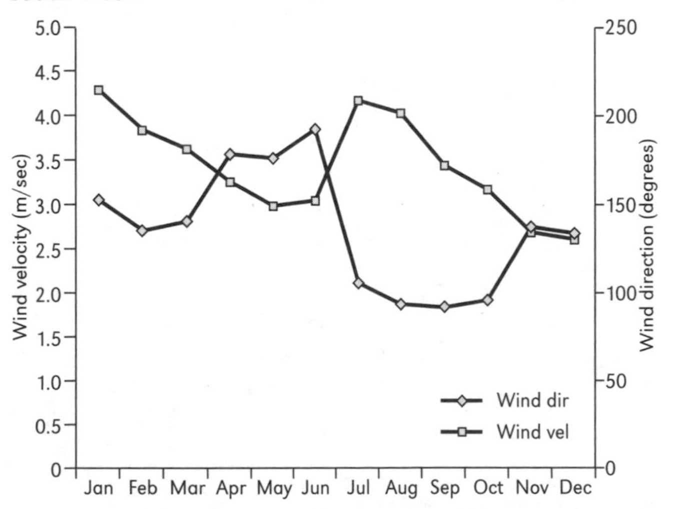

Figure 9.4 *Monthly average wind velocity and wind direction in Athens*

Building description

The building was constructed in 1995. It is the office of A. N. Tombazis and Associates (Figure 9.5). The same architect, Alexandros Tombazis, designed the adjacent buildings, which are used as the offices of a construction company. It has three floors and a basement. It is a unique space in terms of interior design, space manipulation and daylight design.

The long axis of the building faces north and south. The east façade is the main entrance. The west and south façades of the building overlook an open space. The east façade is partly adjacent to a rectangular building of 11.45m in height, 22.7m in length and 10m in width (see building 1 in Figure 9.6). The remaining façade overlooks the paved semi-urban space. Furthermore, at the opposite side of the office, there is a second building (see building 2 in Figure 9.6).

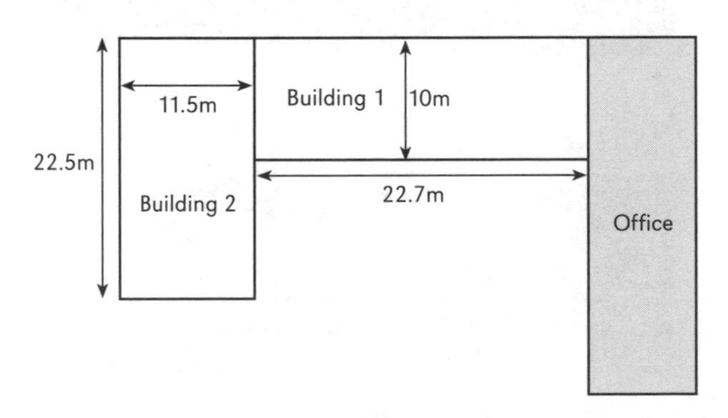

Figure 9.6 *Location of the building and neighbouring buildings*

Source: A. N. Tombazis and Associates

Figure 9.7 *The building complex where the office is located*

The entrance to the building is through an open space that has a small pond (Figure 9.7). The route is through Monemvasias Street to the patio and through the wooden bridge on the level of 1.6m. The entrance door is through a corridor 1.75m × 4.75m, which has service areas (lifts and toilet) on the opposite side. The building has a garage on the level of –2.73m and one basement on the level of –1.63m. The level above the entrance is at 1.6m (Figures 9.8 and 9.9). The building has three floors, which have two levels each, and are jointed by a staircase. Level 1.6

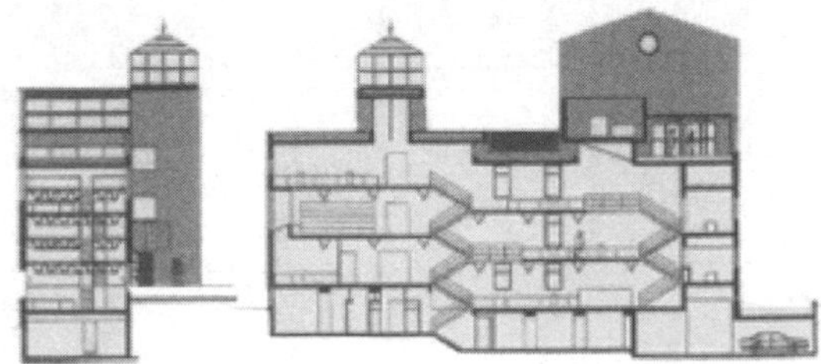

Source: A. N. Tombazis and Associates

Figure 9.8 *External views of the building*

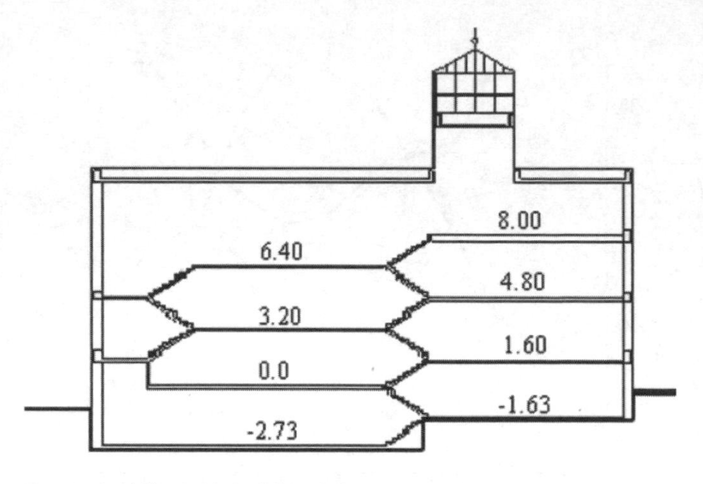

Source: A. N. Tombazis and Associates

Figure 9.9 *Section of the building showing the different levels*

has access to level 3.2, on both sides of the platform. From level 3.2, there is access to level 4.8, and from level 4.8 to 6.4. The staircase from level 6.4 leads to level 8. The building has skylights on the top floor (levels 6.4 and 8). The space is used for drawing offices, computer-aided design, a meeting room and a library. The total floor area of the building (except the floor area of the garage, which is an unairconditioned zone) is close to 500 square metres.

The space is open plan (see Figure 9.11). Each floor has an area close to 100 square metres and the height of each floor is 3m. The building has a rectangular shape of 7m in width, 29.6m in length and 10.6m in height. Its section is very interesting, being divided into semi-levels, which create a unique quality of space. These different

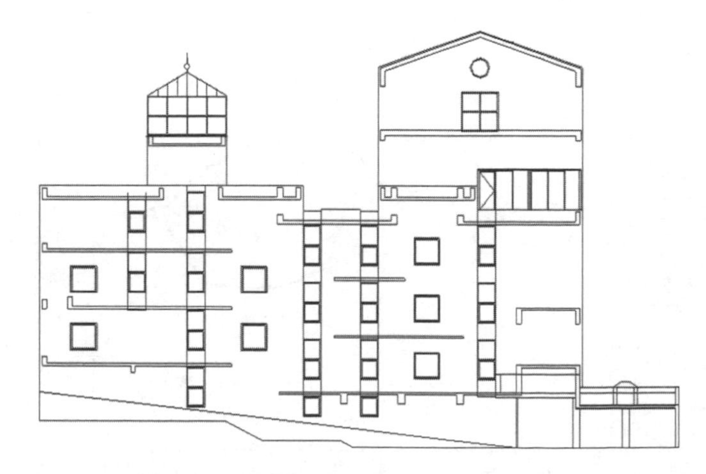

Source: A. N. Tombazis and Associates

Figure 9.10 *A section of the building*

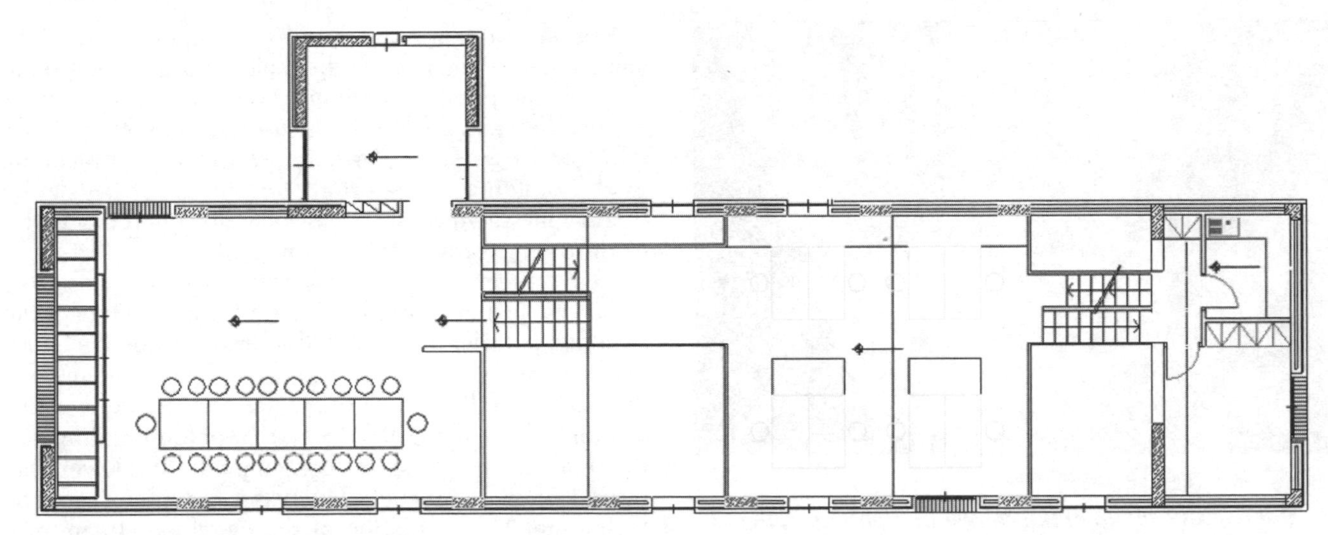

Source: A. N. Tombazis and Associates

Figure 9.11 *Plan of the building, showing the levels 3.20m and 4.80m (library, meeting room, drawing boards)*

Source: A. N. Tombazis and Associates

Figure 9.12 *Photo of the interior space of the building*

levels express individuality of function. Staircases of a height of approximately 1.60m divide each level from each other. There are no interior walls in this building (Figure 9.12).

Key features

The shape of the building and the location of the workstations, near the windows, allow an efficient use of daylight. The building is entirely day lit from both sides by windows and from the roof by way of clerestories (see Figure 9.13). Clerestory lighting is also provided above the administrative area on the ground floor. Apertures purposely have been kept small in order to provide sufficient, but not excessive, daylighting because there are many days of clear sky conditions; this avoids overheating. Furthermore, the clerestories are equipped with reflective blinds in order to distribute the daylight 'deeper' into the building.

An important element of the building is the shading devices, which comprise exterior blinds for the windows and interior blinds for the skylights. The material is white plastic cloth, and its height can be adjusted through a rotation mechanism. On the top floor of the building, skylights are constructed in the middle of the ceiling. They are covered with panels of cloth, which hang from the ceiling.

Another important feature is the presence of ceiling fans that aim to reduce the cooling load of the building. These fans operate during the summer season and they are controlled by the building energy management systems (BEMS) of the building. This strategy permits an increase

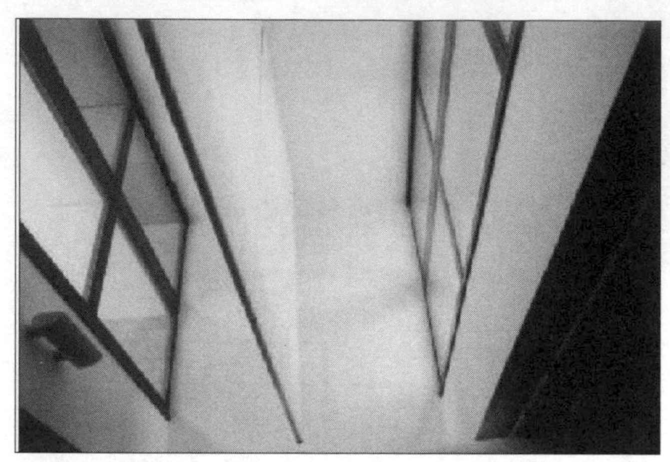

Source: A. N. Tombazis and Associates

Figure 9.13 *Photo of the clerestory lighting on the roof of the building and the reflective curtains*

in the Heating, Ventilation and Air-conditioning indoor set-point temperature for cooling. The comfort zone is more extended than the normal summer comfort zone and rises up to 29° Celsius, as illustrated in Figure 9.14.

As noted, the ceiling fans allow the system settings to be raised cooling by 1° to 3° Celsius. This results in a saving of about 8 per cent of the cooling load. Another advantage of the ceiling fans is that they can improve natural ventilation during pleasant summer days by circulating the air that passes through the opened windows.

The application of night ventilation techniques during the cooling period is also important. During summer nights, ventilation provides a cooling effect by using the outdoor air to carry away the heat from the building. The efficiency of the night ventilation is strongly related to three parameters: the relative difference between the indoor and outdoor temperature; the supply of fresh air; and the part of the mass of the building that is exposed to the low temperature airflow. The lower the outdoor temperature and the higher the fresh air supply, the more effective the system is. Furthermore, the interior planning of the building is important for the exposure of the thermal mass, which determines the flows and the 'paths' of air passing through the building.

The current building uses two extractor fans (2 × 25,000 cubic metres per hour) installed on the roof of the building. In order to facilitate the airflow inside the building, a number of windows (four or five) remain open during the application of the technique. These two fans are controlled by the BEMS system of the building and they operate only if the outdoor temperature is lower than the indoor one. This control increases the efficiency of the technique. The insulation of the envelope (10cm) also improves the effect of night ventilation.

Performance

Figure 9.15 illustrates the energy use in the building for cooling, heating, lighting, equipment, etc. As the building is designed to reduce the energy consumption for cooling, this parameter represents only 5 per cent of the total energy consumption. Additionally, the energy consumption of the artificial lighting system is also low (4 per cent), proof of effective natural lighting design.

With regard to the application of night ventilation, the performance of this technique has been evaluated by using the transient system (TRNSYS) simulation tool (Klein, 1990). In order to develop a model for the simulation of the building, the indoor temperature has been measured during the application of the technique. Furthermore, the measurements were compared with the simulation results (see Figure 9.16).

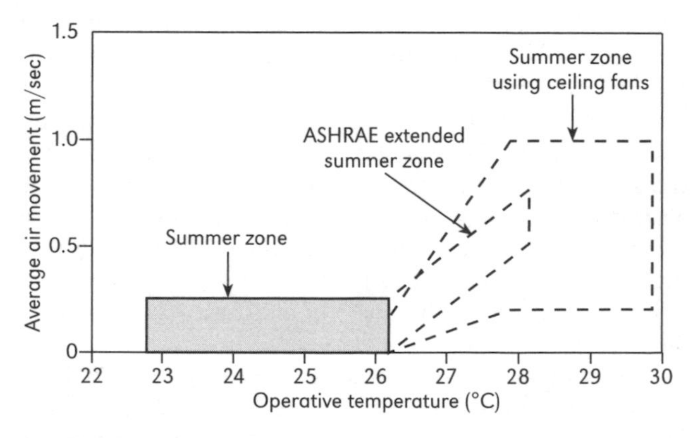

Source: Rohles, 1983

Figure 9.14 *Comfort zone when ceiling fans are operated*

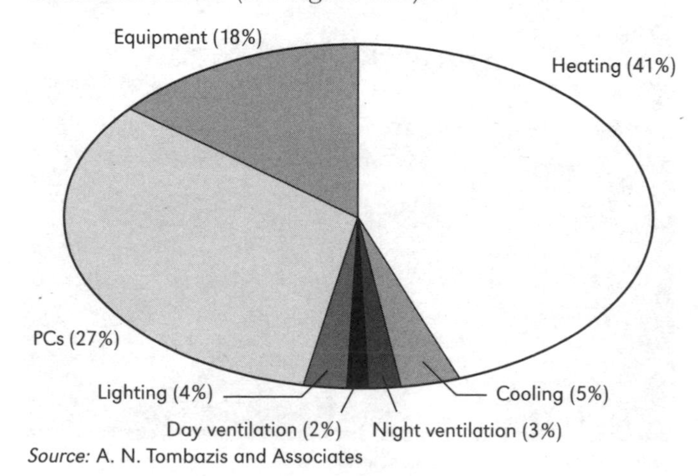

Source: A. N. Tombazis and Associates

Figure 9.15 *Energy use*

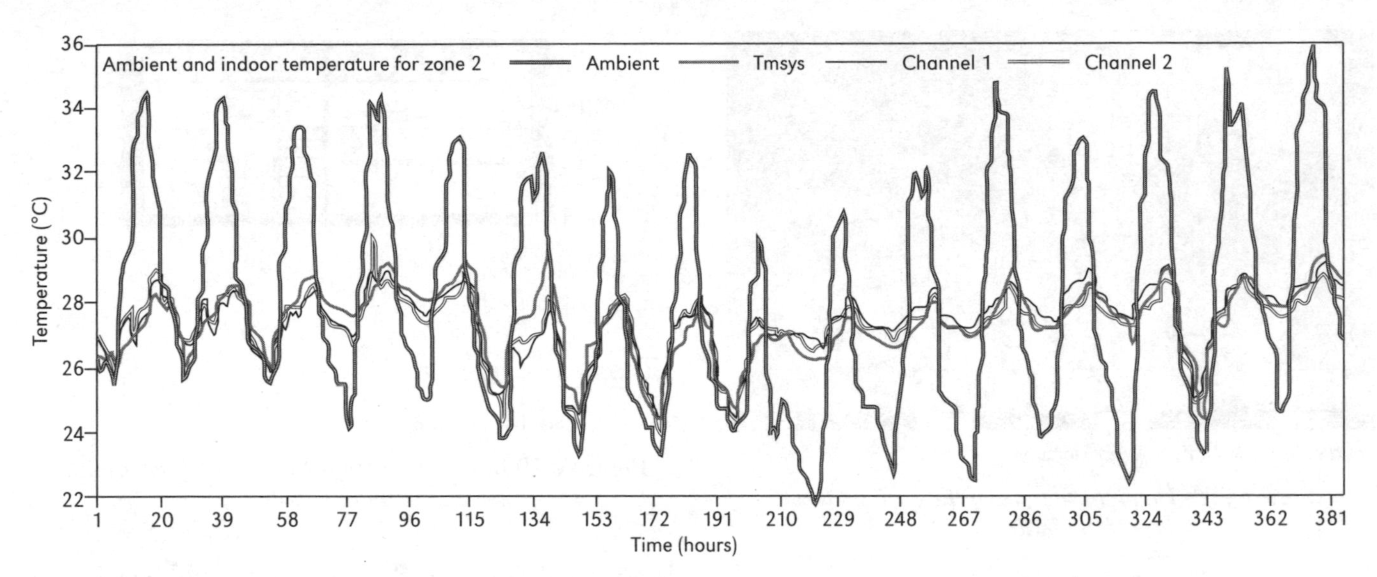

Source: Geros, 1999

Figure 9.16 *Measured (channel 1 and 2), simulated (TRNSYS) and outdoor air temperature (ambient) for level 0.0 of the building*

Figure 9.16 also clearly shows the application period of the night ventilation (when the indoor temperature is almost equal to the ambient one during the night). By using the developed model, the coefficient of performance of the night ventilation was calculated (see Figure 9.17). The current results refer to a set-point temperature equal to 27° Celsius. This analysis takes into account four different airflow rates during the application of the technique (5, 10, 20 and 30 air changes per hour) and shows that when the airflow is increased the coefficient of performance (COP) is reduced as it requires a bigger size of fans; therefore, the energy consumption for the ventilation is increased.

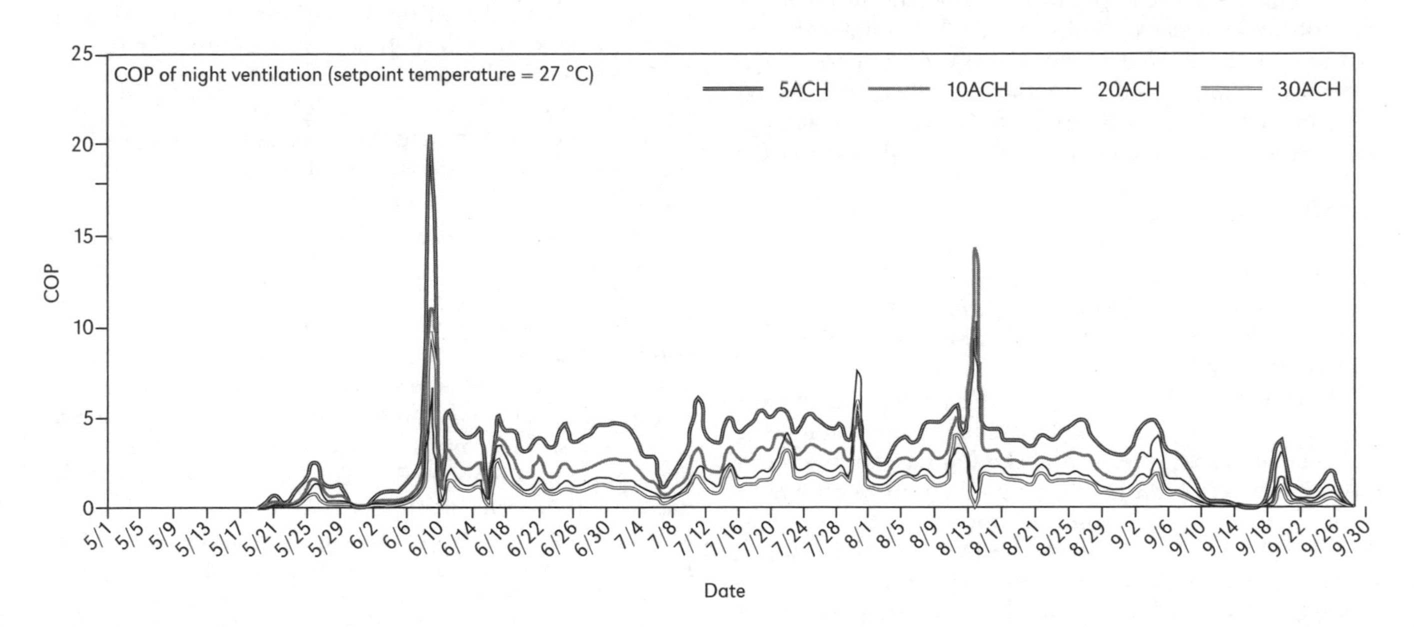

Source: Geros, 1999

Figure 9.17 *Coefficient of performance (COP) of the night ventilation*

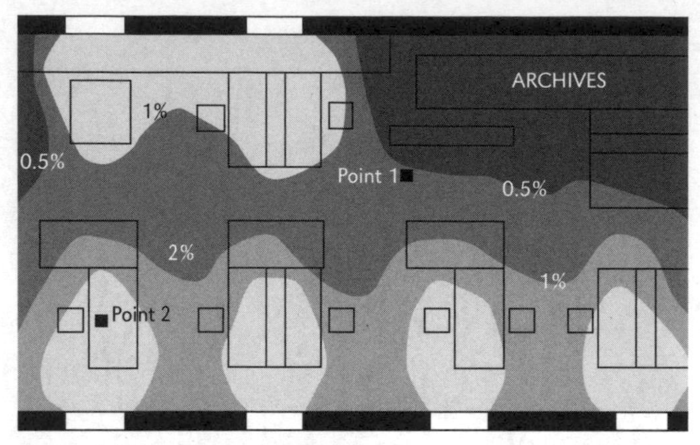

Source: A. N. Tombazis and Associates

Figure 9.18 *Daylight factors of the –1.63m level of the building*

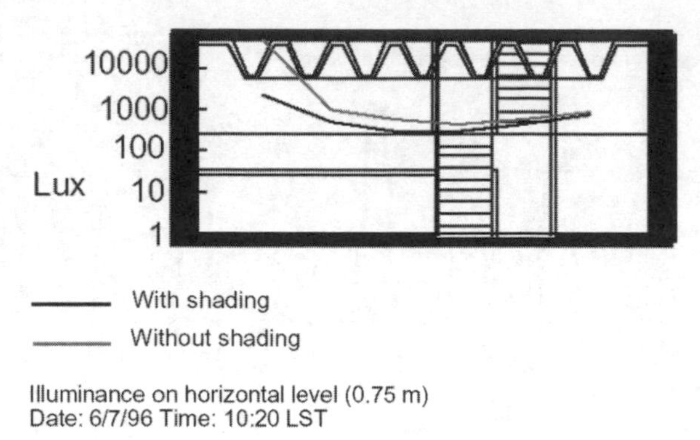

Source: A. N. Tombazis and Associates

Figure 9.19 *Illuminance on the horizontal level with and without shading*

Due to the shallowness of the building and the provision of windows on both sides, the daylight environment seems to be relatively homogenous. Because it is an open space and the interior surfaces have high reflectance, good daylight penetration is ensured. Areas with daylight factors of less than 0.5 per cent (see Figure 9.18) are used as archives and for photocopier services. Generally, there is symmetry in the distribution of the daylight factors between different levels. This is because all levels have the same geometrical configuration. Evidence from measurements of the upper levels shows the strong influence of the clerestory window on daylighting levels. On western windows the daylight factors are 22 per cent. On the north windows, due to the presence of external obstructions, they are 16 per cent. The eastern windows, which face the common courtyard, have a daylight factor of 12 per cent.

All apertures facing east and west are provided with manually controlled vertical exterior electric motored fabric blinds positioned approximately 15cm from the wall, providing both shading and glare control. The large clerestory over the main stairwell is provided with a light shelf on the south side.

With the blinds closed the levels of illumination are reduced to 95 per cent of their original values when there is direct sun incidence. In the shaded parts of the room the reduction of light levels is inversely proportional to the distance from the window. Sun enters the building after 15:00 local standard time (LST) during the summer solstice and after 14:30 LST during the winter solstice. Using the shading devices can eliminate the presence of sun patches on working surfaces. Eastern windows encounter a different problem. Because they face the common courtyard the sun is blocked during most of the day. Sun patches on working surfaces would be particularly annoying during the summer solstice in the early morning (until 9:00 LST).

Conclusion

This office building is an interesting example of a well-designed building from an architectural and energy point of view. The shape and the orientation, as well as the interior (open space) planning of the building, allow the local climate to be exploited for energy conservation and daylighting purposes. The installation of a BEMS gives the possibility of controlling various systems (such as ceiling fans, night ventilation fans, etc.) installed in the building. The studies and measurements conducted in the building show that from the daylight and solar protection point of view, the building is well balanced. The positioning of the workstations near the windows and the manually controlled shading devices give the possibility of controlling or utilizing daylight.

The building is also well insulated through the use of ceiling fans and the application of night ventilation techniques to reduce the energy requirements for cooling. The results of the studies performed for night ventilation show that the ventilation system used for this purpose is oversized. The two extract fans that operate during the summer nights give airflow rates of close to 25 air changes per hour (ACH). The coefficient of performance (COP) for this air supply is quite low regarding the lower airflow rates (which require lower fan power) that have a higher COP (see Figure 9.17). On the other hand, the COP for 25 ACH is generally higher than 2 and is comparable with the COP of the HVAC system. This experience shows that it is important to perform detailed studies before applying similar cooling techniques.

Case study 2: Avax office building

Table 9.2 *Avax office building*

Name:	Avax S. A. (construction company)
Type:	Office building on Lycabettus Hill
Location:	Centre of Athens

Introduction

This building is located in the centre of Athens on Lycabettus Hill and is the headquarters of a major Greek construction company.

The target of this construction was to develop an environmentally sound building that accommodates the company's increasing needs, promotes its professional identity and provides excellent comfort conditions for its users.

A prime design concern was flexibility. As a result, the building is responsive to natural conditions, and is capable of controlling itself and of keeping its energy consumption low, while preserving the environment from unnecessary pollution.

Regarding its aesthetic identity, the building is intended to be representative of a (young) dynamic construction company and to express a modern, technologically oriented architectural vocabulary. These objectives had as much influence on the entire design procedure as did the materials and methods of construction.

The main design concepts for this building concern the use of daylight, while the artificial lighting system is considered a backup system. The building is equipped with a movable type of side-fins shading device in order to control the solar heat gains. During summer nights the building is ventilated (night ventilation) as a pre-cooling strategy by using a mechanical ventilation system. Additionally, ceiling fans are installed in the office spaces of the building, which are manually controlled. The operation of these devices extends the comfort zone of the building. Finally, another important element of the building is the use of a central cooling system (air/water heat pumps), combined with a cold storage system (ice banks).

Location and climate

The building is located in downtown Athens, on the slopes of Lycabettus Hill (see Figure 9.20). The area is densely built-up, with an open view towards the east, while neighbouring buildings shade the western elevation. The climatic characteristics are as for Case Study 1.

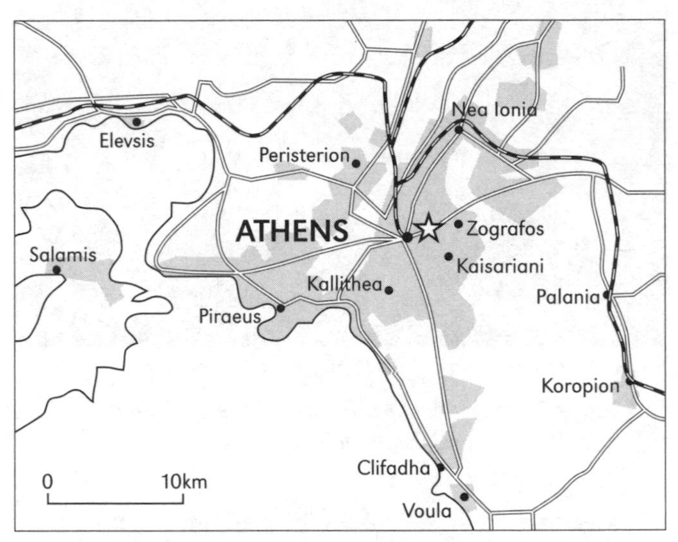

Figure 9.20 *Location of the Avax building (star indicates the location region)*

Building description

The Avax building was constructed in 1998 (see Table 9.3). The architect of the building was Alexandros N. Tombazis, assisted by: N Fletoridis (Meletikiki: Alexandros N. Tombazis and Associates Architccts Ltd). The site is in downtown Athens on the slopes of Lycabettus Hill and is approximately 500 square metres in area, situated in a densely built-up area with an open view towards the east. It is exposed to the sun throughout the morning. While the eastern elevation is open, neighbouring buildings shade the western elevation.

The wedge-shaped site, the orientation and the variable but predominantly warm/dry climate of Athens dictated the major architectural decisions.

The building is oriented east (see Figures 9.21 and 9.22) along a north–south axis, with its back and narrow sides touching adjoining buildings. It makes full use of the total surface permitted by planning regulations to be built upon, with free space on ground level in front and at the back, as well as an interior garden.

Table 9.3 *Avax project time scale*

Design:	February 1992–October 1992
Construction:	February 1993–May 1998
Occupation:	June 1998
Monitoring:	September 1998–September 1999

Source: A. N. Tombazis and Associates

Figure 9.21 *The eastern façade of the building when the glass panels are closed*

Source: A. N. Tombazis and Associates

Figure 9.22 *The eastern façade of the building when the glass panels are open*

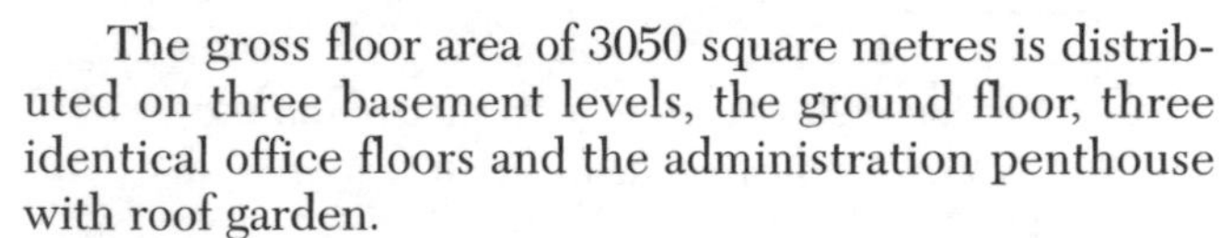

The gross floor area of 3050 square metres is distributed on three basement levels, the ground floor, three identical office floors and the administration penthouse with roof garden.

All main areas (offices, meeting rooms) are situated along the glazed front façade and are separated from secondary spaces (toilet, kitchenettes and vertical circulation shafts) by a circulation area. Each level is, thus, divided into a front/activity zone and a service/back-up area. The degree of openness of the layout varies from floor to floor and can be modified in the future. Floor-to-ceiling glass dividing panels and free-standing furniture have been used in order to give the interior a high degree of openness and transparency.

The building was basically conceived as an east-facing linear climate sensor. The front east-facing elevation is designed to act as a double skin – a diaphragm to protect the inner glazing – with selective control of thermal gains and daylight. The front is dominated by five 16m high concrete columns, which exemplify the structural system of the building, and by the custom-made solar fins or glass panels, shown open in Figure 9.22 and closed in Figure 9.23, designed to perform as shading devices for the middle floors.

Source: A. N. Tombazis and Associates

Figure 9.23 *Photos from the interior space of the building*

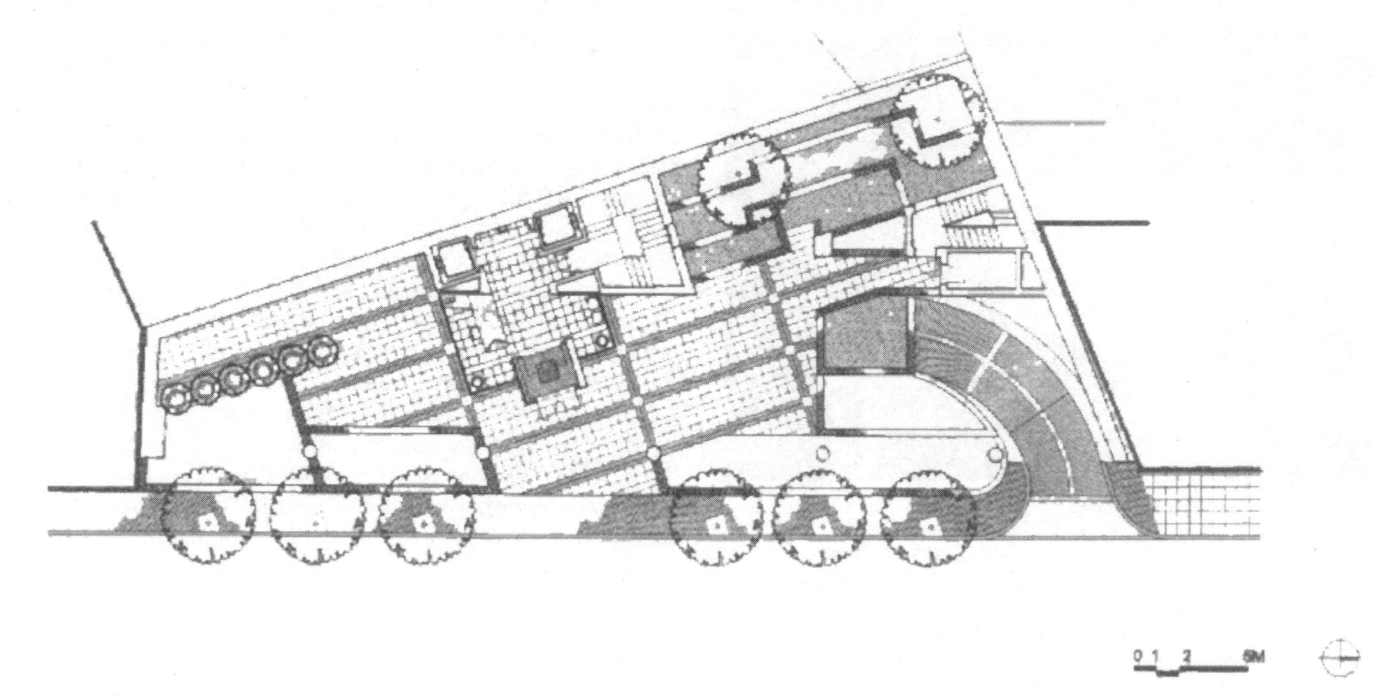

Source: A. N. Tombazis and Associates

Figure 9.24 *A plan of the Avax building*

Some interior views of the building are presented in Figure 9.23. The first one shows clearly the effect of the glazing panels when they are closed.

A plan and section of the building are also presented in Figures 9.24 and 9.25.

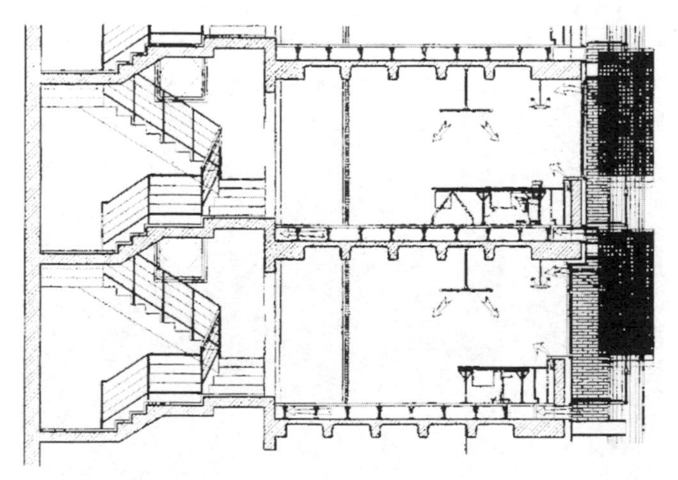

Source: A. N. Tombazis and Associates

Figure 9.25 *A section of the Avax building*

Key features

The building is designed to exploit the natural lighting potential of the area. It is designed by coupling two important elements: the layout of the office spaces, which is narrow (the workstations are located close to the windows; see Figure 9.26), and the 'intelligent' permeability of the eastern façade (see Figure 9.27). The dimensions of the standard offices (7m wide, 3m deep) increase the levels of daylight in these spaces. It is important to underline that the western façade requires no special shading because of its proximity to the adjacent buildings.

The solar fins are the dominant daylighting features, installed on the eastern façade (see Figure 9.27). They comprise vertical glass silk screen-printed panels, and they have sufficient performance to shade the building façade and to protect the indoor spaces from solar radiation. Another important feature is that they are controlled automatically in response to temperature and solar radiation. In order to ensure the proper operation of the devices, manual override is provided via infra-red remote controls. A series of fixed, white horizontal metal grills is also attached to these devices in order to provide shading. The top and mezzanine floors are shaded by conventional, external Venetian blinds that are also controlled by the central system and have manual override.

Source: A. N. Tombazis and Associates

Figure 9.26 *Location of the workstations*

Windows and skylights (one for each office bay) are manually operable. The window elements have a total height of 1.7m. They are divided into an upper part that concerns the daylight (one-third of window height) and a lower vision pane (two-thirds of window height). The glazing ratio is only 10 per cent on the western elevation of the building, while on the eastern elevation it is 45 per cent. All windows are clear, double-glazed units with a *U*-value of 2.8W/m^{2}K.

The high performance of the daylighting system permits users to consider the artificial lighting system as a backup. In order to minimize energy consumption, various design concepts have been adapted. The walls and ceiling are painted in light colours in order to maximize the diffusion of the light. The workstations are equipped with task lighting devices under the control of users, while the general indirect lighting level is low (200–250 lux, on average). The most commonly used luminaries are high-efficiency fluorescent lights. Special automation systems control the level of artificial lighting, according to the natural lighting level. In order to ensure the performance of the lighting systems, the users, via infra-red remote controls, can manually override the automatic operation of these systems. The installation of presence detectors ensures the switching on/off of the general lighting system depending upon the presence of occupants. Furthermore, a lighting control system is installed in the parking area of the building, which switches on the lighting system only if cars are moving.

The HVAC system is provided as a back-up. The raised false floor permits the installation of local air-conditioning units that cover the local load demand, fresh air, as well as inlet and extraction fans for natural ventilation. The raised floor acts as an air plenum.

The energy design of the building aims to reduce/avoid the operation of the air-conditioning system by using various passive-cooling techniques. In order to be more specific, the external shading fins are used during warm days to control the solar gains, while the internal gains due to the operation of the artificial lighting system are minimized; therefore, the cooling load of the building is reduced. Furthermore, night ventilation techniques are applied as a 'pre-cooling' strategy by operating a mechanical ventilation system at a rate of 30 air changes per hour (ACH), from 21:30 to 7:00, when the outdoor temperature is lower than the indoor one. Another hybrid cooling strategy applied is the operation of ceiling fans during the summer season (see Figure 9.26), which are manually controlled. This operation extends the comfort zone from

Source: A. N. Tombazis and Associates

Figure 9.27 *The movable solar fins for outside and inside the building*

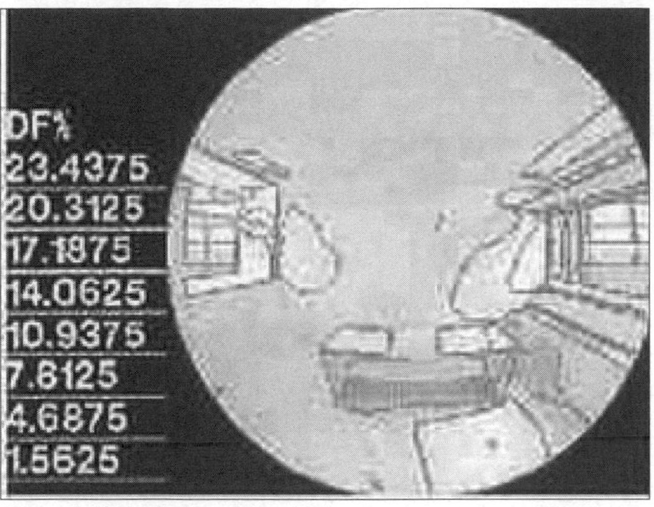

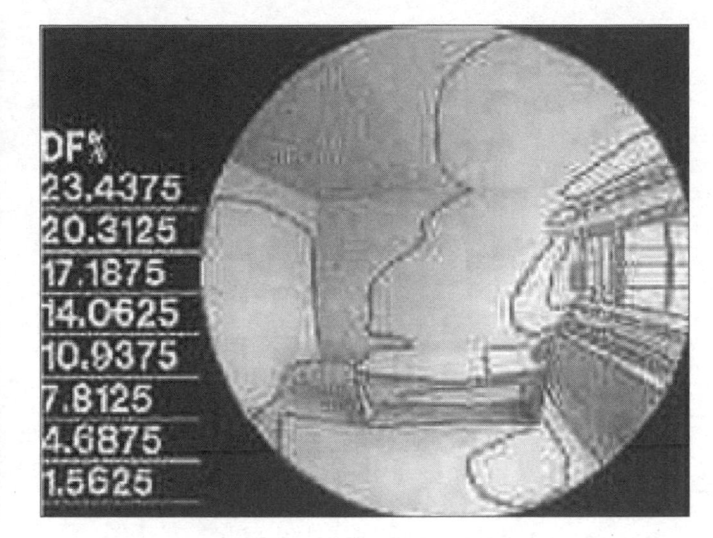

Source: A. N. Tombazis and Associates

Figure 9.28 *Daylight factor distribution in a room with opaque and translucent partitioning walls*

25° to 29° Celsius. Finally, the central cooling system (air/water heat pump) is combined with a cold storage system (ice banks).

The installation is controlled and programmed by a central building energy management system (BEMS). The operation of the BEMS can reduce the cooling load by the appropriate control of the solar shading and the night ventilation. On the other hand, occupants are also expected to cooperate in minimizing the energy consumption (by opening windows and using the ceiling fans or air-conditioning units when required).

Performance

As stated earlier, the dominant daylighting feature is the solar fins. These shading devices provide a shading coefficient of 70 per cent and are automatically rotated in response to temperature and solar radiation. In order to evaluate the performance of the solar fins, extensive simulations have been performed by using the radiance software tool. The calculated daylight factor for a representative space of the office is presented in Figure 9.28. According to this figure, the daylight factors vary from 2 to 23 per cent, with a value close to 10 per cent on the height of the workstations.

The actual performance of the building was monitored continuously after the opening of the building. The measured parameters comprise the following:

- ambient air temperature and relative humidity;
- indoor air temperature and relative humidity;
- actual energy consumption.

According to the results of the monitoring period, the indoor temperature per floor (in various zones) is quite homogenous. The temperature difference between the various spaces varies from 0.5° to 1.0° Celsius, due to the different operation schemes. The mean indoor temperature varies from 21.5° Celsius during the winter, to 28.5° Celsius during the summer period. These levels are acceptable for the thermal comfort point of view, and take into account the operation of the heating and cooling system.

For this experimental period, the measured indoor relative humidity is presented in Figure 9.30. The measurements show that the indoor relative humidity varies from 23 to 61 per cent.

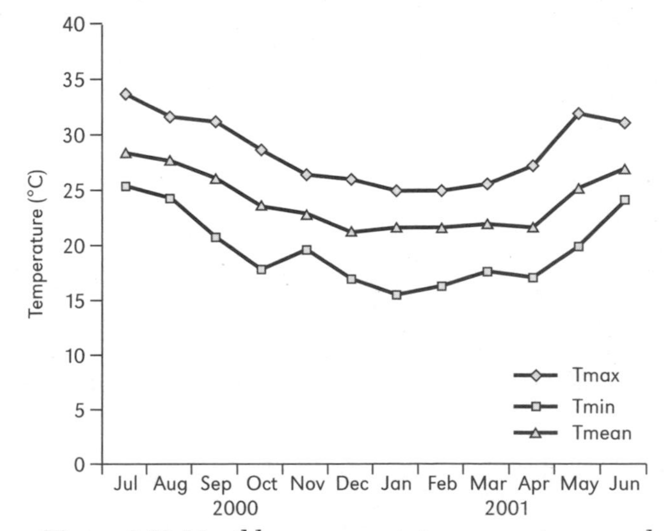

Figure 9.29 *Monthly average minimum, maximum and mean indoor air temperature*

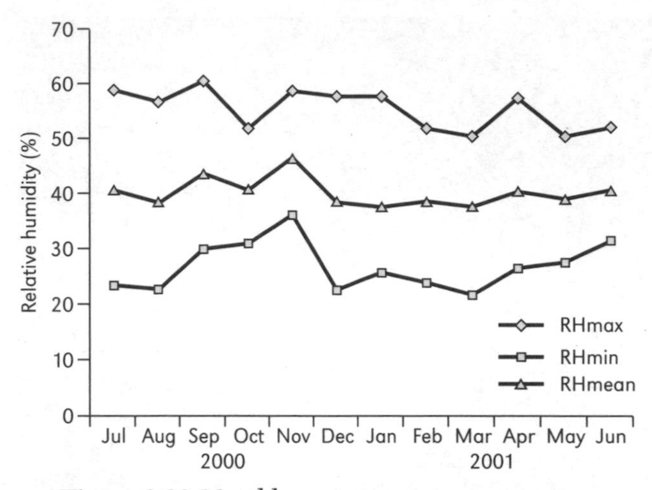

Figure 9.30 *Monthly average minimum, maximum and mean relative humidity*

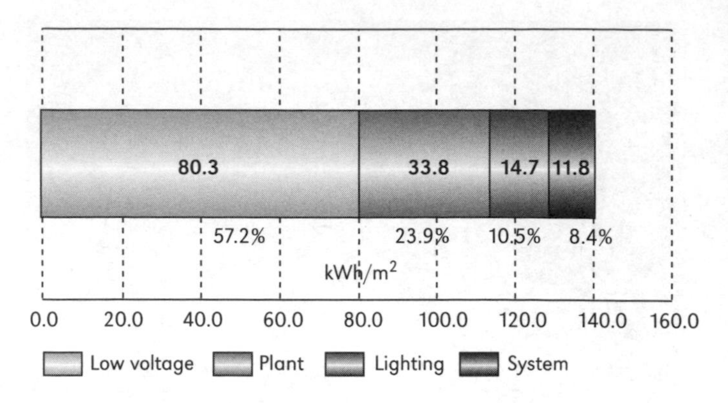

Figure 9.31 *Annual energy consumption per use*

The annual energy consumption per use is illustrated in Figure 9.31. It is important to underline that cooling and heating consumption are not separate, because the same heat pump unit is used for both heating and cooling. Therefore, the BEMS system records the total energy consumption of this unit. The following figure shows that the annual low-voltage consumption is close to 80kWh/m^2 (consumption of equipment, elevators and any energy consuming device except the cooling, heating and lighting equipment). The HVAC system contains a central heat pump that consumes about 34KWh/m^2/year for the heating and the cooling operation. The energy consumption of the artificial lighting system is close to 15KWh/m^2, while the fan-coil units and the distribution system consume about 12KWh/m^2.

The high energy consumption of the low-voltage appliances is due to the operation of the ventilation system for the night ventilation (it is a separated system from the HVAC), the ice banks, the ceiling fans, the operation of the solar fins, etc. The annual energy consumption is close to 140KWh/m^2.

Conclusion

The Avax office building is an example of a building that combines various passive and hybrid techniques. During the design process, the following environmental issues were taken into account wherever possible, in addition to those already discussed:

- embodied energy of materials and components;
- reduction of CO_2 emissions;
- avoidance of ozone layer-depleting substances;
- use of natural materials;
- use of wood from managed forests.

The operation of the solar fins shows that the control of the solar gains and the exploitation of natural lighting can be effective. But the right design and sizing is very important. For example, the increased airflow rates during the application of the night ventilation require the installation of high-capacity fans. This can result in higher energy consumption, which is due to the operation of the fans rather than the reduction of the cooling load that occurs the next day after the application of night ventilation. Furthermore, another possible problem for this type of installation is the high noise levels of the fans, especially if the building is located in a residential area. For this particular building, neighbours from the surrounding residential buildings frequently complained about the high noise levels of the fans during the summer nights.

The manual operation of the ceiling fans is a possible strategy; but, generally, it is more effective to connect these fans with the BEMS.

Use of the occupancy sensors is a favourable solution for reducing the artificial lighting system consumption; but the designer should be careful not to install this type of sensor in spaces where people are frequently present. These sensors are efficient for spaces such as corridors, garages, etc.

The installation of a BEMS system that controls ventilation, shading, lighting and the back-up air-conditioning system can reduce energy consumption and permits the monitoring of the building, improving the operation and efficiency of various systems.

Case study 3: Ampelokipi residential building

Table 9.4 *Ampelokipi residential building*

Name:	Ampelokipi residential building
Type:	Residential building
Location:	Ampelokipi, Athens city centre

Introduction

This building is located in Ampelokipi, a residential area district in the centre of Athens, near the Hilton Hotel. It is used for housing.

A vertical approach resulting from the small available lot and the need to create the necessary solar-energy collecting surfaces is characteristic of the design.

The building presents important design features from the thermal point of view. The design characteristics include a greenhouse in the south façade of the building; trombe walls at certain available points in the external envelope of the building; solar collectors; an auxiliary heating system powered by natural gas and a biomass option with a traditional fireplace, using wood.

The target of this construction was to develop high thermal-energy savings during winter. Solar-air heating methods and energy-collecting elements were adapted to complement a townhouse design. Optimization of both architectural design and solar-collecting efficiency within given economic restrictions had to be achieved, which led to the elimination of active higher technology systems in favour of passive solar-energy collecting methods.

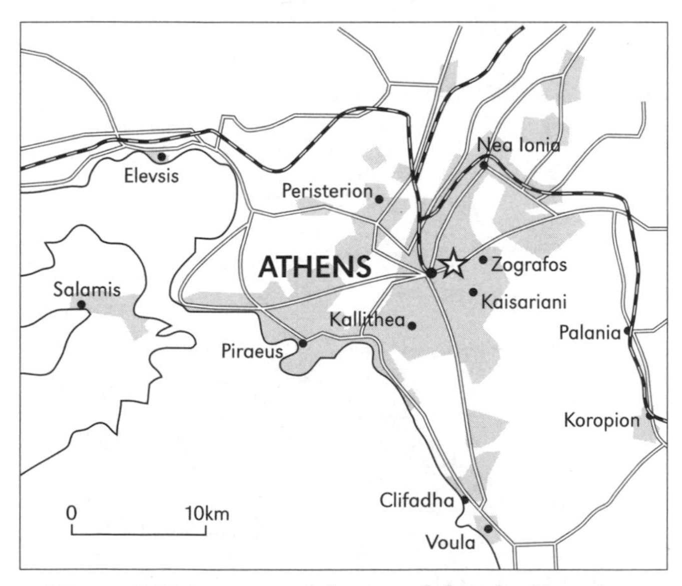

Figure 9.32 *Location of the Ampelokipi building (star indicates the location region)*

In order to keep a low initial cost and to free the whole project from restrictions and technological problems created by the use of heat tanks and forced air flows, simple day-to-day systems were used and most of the required elements were designed as part of the conventional construction of the building.

Location and climate

The building is located in Ampelokipi, Athens and is a residential district within a built-up urban environment and substantial traffic in nearby streets (see Figure 9.32). The climatic characteristics are as for Case Study 1.

Building description

The building was constructed in 1986; the architect was Katerina Spiropoulou. The building is situated in the Ampelokipi region, in the centre of Athens. It is a semi-detached house of 210 square metres in total area and 900 cubic metres in volume. The urban and building regulations, the tall surrounding buildings that form a densely built urban environment and the small plot have strongly influenced the vertical form and construction of the building (Figure 9.33). This has also influenced the organization of the residence's internal space.

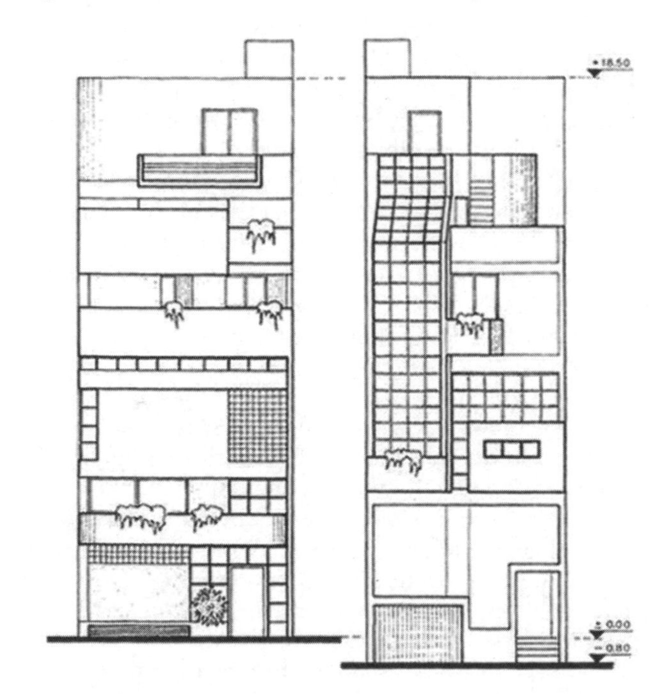

Source: Katerina Spiropoulou

Figure 9.33 *Sections of the Ampelokipi building*

Source: Katerina Spiropoulou

Figure 9.34 *The southern façade of the building*

The building, a four-storey house for a family of four, was first occupied in 1986. Its architecture is modern, compared with the neighbouring urban buildings, and is characterized by the applied passive solar systems, which are placed on the building's façades. The building's axis has been rotated in a north–south direction, so the main façades are not parallel with the road's axis, forming an 18 degree angle. The two main opposite façades are on Likias and Avlidos streets, while the other two sides adjoin the neighbouring semi-detached buildings.

The limited building plot, only 99 square metres in area, the needs of local families and the tall neighbouring buildings made it necessary to expand the building in a vertical direction. The ground floor is 5.5m in height and is used as a store. This area also houses the central heating installations and other auxiliary rooms. The main residence is therefore raised 5.5m above this and consists of three successive floors and a small mezzanine between the first and the second floor, 13 square metres in area. The living room, the dining room, the kitchen and a store are on the first floor (see Figure 9.36a). The children's bedrooms, a guestroom and a bathroom are placed on the second floor (see Figure 9.36b). The parents' bedroom, the necessary auxiliary spaces and an office are on the

Source: Katerina Spiropoulou

Figure 9.35 *Photo of the interior space of the building*

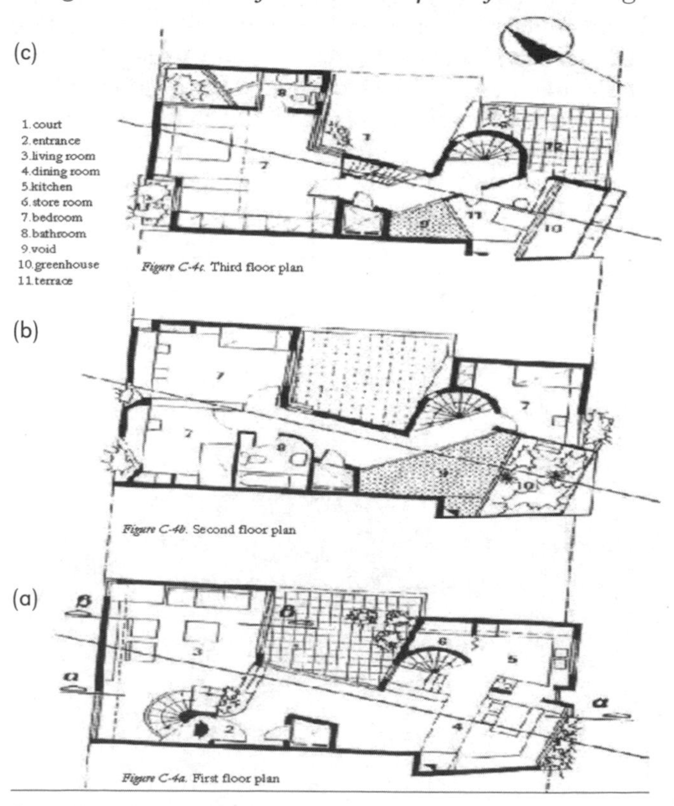

Source: Katerina Spiropoulou

Figure 9.36 *Floor plans of the building*

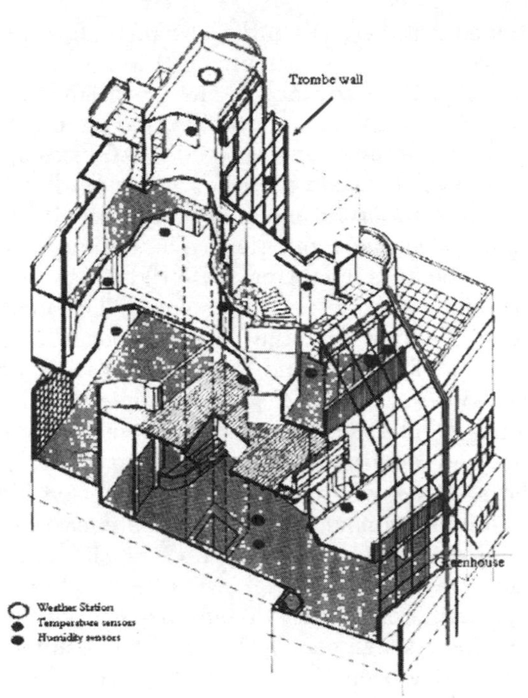

Source: Katerina Spiropoulou

Figure 9.37 *Axonometric plan of the Ampelokipi building; weather station and sensor's location in the building*

presents the weather station's position, and the temperature and humidity sensor's position in the building.

Key features

The building is designed to exploit solar energy for thermal energy saving and couples three important elements: the greenhouse/sunspace, the trombe walls at the envelope's alcoves and the direct solar gain through the southern openings.

One of the most important elements of the building is the sunspace. Through the south-oriented greenhouse indirect solar gains are obtained. The greenhouse is an intermediate space, bridging the external environment and the interior spaces. Being the reference point of the house, the sunspace can be seen from almost every inside or outside angle. It is planted in order to act as a garden and to give a sense of nature in this densely built urban area as a substitute for the missing panoramic view. The greenhouse, covering a glazed area of 19 square metres, extends 2.5 floors in height (9m) and consists of a metal-framed construction filled with single reinforced glass panels for security reasons. The surface between the greenhouse and the interior spaces is of single-glazed

third floor (see Figure 9.36c). The fourth floor, the smallest one, includes two study rooms. A sunspace stretches from a height of 9m (mezzanine floor) up to the bedroom floors. The floor plans of the building are depicted in Figure 9.36.

In the building there is a monitoring system that comprises 40 sensors connected to a data logger. The climatic parameters measured are the ambient air temperature; the total solar radiation; the diffuse solar radiation; the ambient air relative humidity; and the wind velocity and direction. The following quantities are measured in the internal spaces: air temperature on all of the building's floors and in the sunspace; air relative humidity; electricity consumption; gas consumption; and the space heating system's volumetric flow rate. Figure 9.37

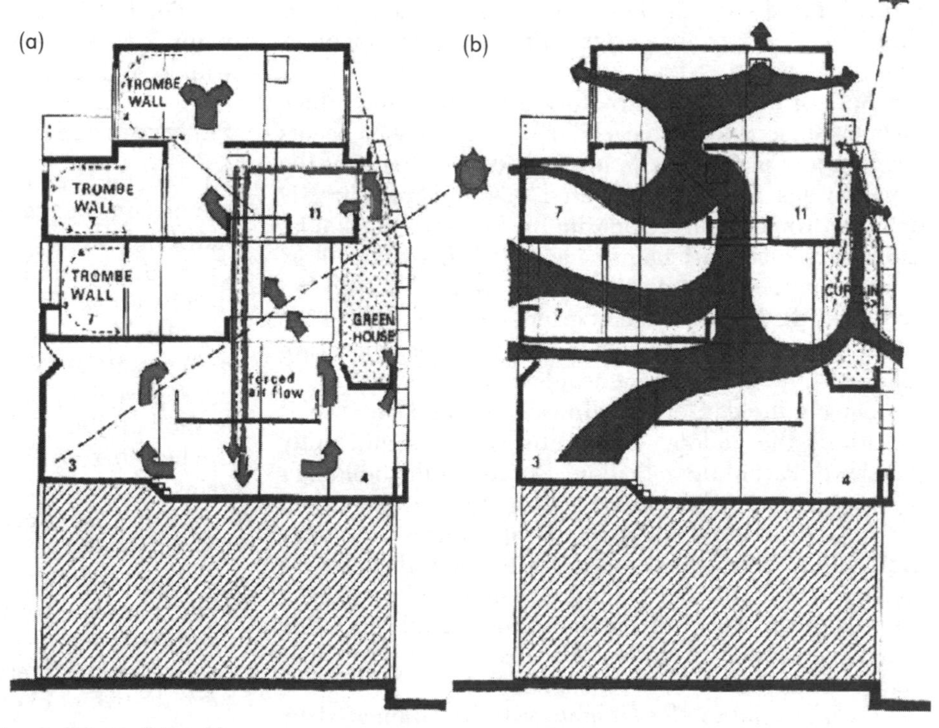

Source: Katerina Spiropoulou

Figure 9.38 *Interior airflow in winter and summer*

glass, providing the living room and the staircase with natural lighting – which is precious for the elongated building shape and the relatively confined façades. The air, as it warms up inside the greenhouse, moves to the upper floors where the bedrooms are located. In order to have hot air for the living room, an air-duct system is used with a small fan for forced air circulation (see Figure 9.38a).

Scattered openings in the glazed envelope of the greenhouse and dampers at the top prevent overheating, especially during the summer period (see Figure 9.38b). The shading of the greenhouse could be considered insufficient and an internal aluminium-coated curtain is the only solar protection measure.

Another key figure is related to the use of trombe walls, which aim to reduce the energy consumption for heating. Two trombe walls have been constructed at certain available points in the external envelope. The first one is east-oriented, covering an 8m^2 area (this covers part of the kitchen and the guest rooms' external walls), while the other is south-oriented, covering a 30m^2 area (this covers the bedrooms and the bathroom). A special cost analysis has been performed for these trombe walls because a significant part of them is shaded due to the building itself. The trombe walls were made of reinforced concrete and solid brick. Their external envelope is metal-framed, filled with single reinforced glass panes and is interrupted at certain points where external openings exist.

Both of them are multi-storied trombe walls, which give hot air separately to each floor. Air circulation occurs via dampers placed in the upper and lower parts of the trombe walls. The south-oriented windows open directly onto the trombe walls. Opposite the windows are similar external openings, which provide added ventilation of the system and the house during summer (Figure 9.38b), and help with the maintenance of external components of the walls. Between the two opposite windows, there are rollover shutters to prevent heat loss or overheating.

During the day, due to diffused and incidental solar radiation, the indoor temperatures are significantly increased. Part of the radiation is stored in the building's thermal mass, in the trombe walls. At night, the stored heat is returned, keeping indoor temperatures reasonably high, while the external roller shutters and the curtains are kept closed in order to minimize thermal losses to the external environment. In summer, cross-ventilation is the only natural cooling strategy because shading is inadequate, as has been previously mentioned.

Overheating of the trombe walls is eliminated by using the openings in the external glazed envelope, while keeping the dampers closed during the day. It should be mentioned that there is significant temperature stratification due to the open structure of the internal spaces, which helps the buoyancy of the warm air mass. The building therefore acts as a chimney and the upper floors are warmer. Measurements showed a 3° Celsius temperature difference between the lower and upper floors.

Another important element is the direct solar gain through the southern openings. The dimensions of the southern openings (total area of 19m^2) increase the levels of the solar radiation on these spaces. The mezzanine between the first and the second floor allows deep penetration of sunlight and free air circulation, unifying the interior spaces. The southern openings face 18° eastward. All glazed areas are metal-framed. The glazing is single and reinforced for security reasons, except in the northern façade, where double glazing is used. External roller shutters without core insulation are used for night-time insulation. Curtains offer extra shading in the interior.

Furthermore, in order to minimize energy consumption, various design concepts have been adapted. Solar collectors have been used to provide warm water for the user's needs. A traditional fireplace in the dining room is used sporadically. Its chimney is used for creating additional hot airflow in the rooms through which it passes.

Finally, there is also an auxiliary heating system. A natural gas boiler powers this auxiliary heating system with local thermostats for fuel saving.

Performance

During the heating period, indoor temperatures are significantly increased due to diffused and incidental solar radiation. Part of this radiation is stored in the building's thermal mass and particularly in the walls, floors and trombe walls. The percentage of opaque and transparent building elements in the building's envelope is presented in Figure 9.39.

Figure 9.40 shows the thermal energy consumption in the building. Solar systems contribute 42 per cent of the building's thermal energy demand, while incidental gains contribute 28 per cent, and the other solar gains contribute 2 per cent of thermal demand. The auxiliary heating demand is only 28 per cent.

Since the target of the construction was to develop high thermal energy saving during winter, the energy consumption from the auxiliary heating is low (28 per cent), proof of the right thermal design of the building. Seventy-two per cent of the building's heating requirements were provided by solar and internal thermal gains. The annual requirements for the same but conventionally

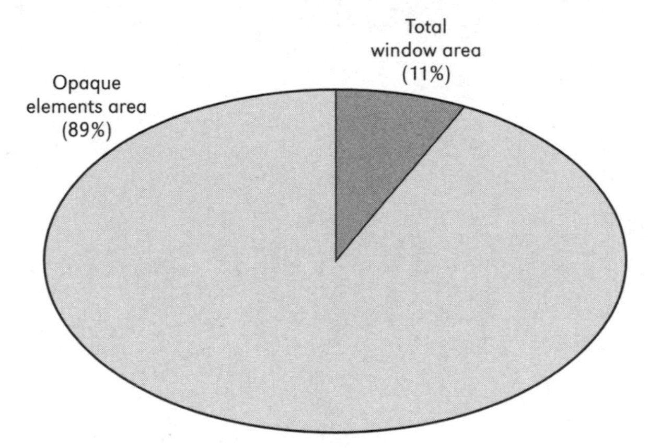

Source: Katerina Spiropoulou

Figure 9.39 *Percentage of opaque and transparent building elements in the building's envelope*

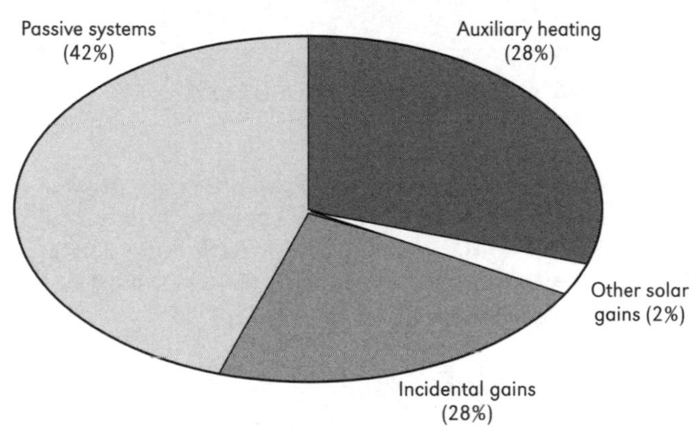

Source: Katerina Spiropoulou

Figure 9.40 *Thermal energy consumption*

designed house and the same thermal comfort levels are approximately 12,362.2KWh (59KWh/m^2/year), compared with 5380.2KWh (26KWh/m^2/year) of the actual annual auxiliary heating load of the building with passive solar systems.

Figure 9.41 presents the heating load in the heating period and the total energy gains from the passive systems in the building. The mean system performance is close to 74.8 per cent.

The thermal comfort conditions appeared to be satisfactory. The designers, who were also the inhabitants of the house, ascertained that vertical and horizontal cross-ventilation is very effective for natural cooling during the summer.

Indoor temperatures were within the limits of thermal comfort. Only during the summer vacations did the temperature reach 32° Celsius at the top of the fourth

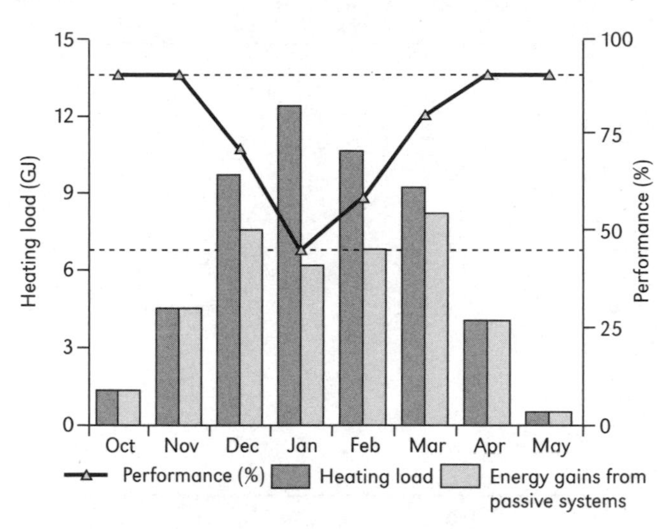

Source: Katerina Spiropoulou

Figure 9.41 *Monthly heating load and passive system's performance in the building*

floor during August and July, when the residents of the house were absent and natural ventilation did not take place. The comfort zone's limits have been assessed as from 19° up to 28° Celsius.

Conclusion

The *Ampelokipi* residential building is an example of a well-designed residential building with regard to thermal energy savings. The passive heating systems and the building's orientation are the most important design elements. The building's adaptation to the environment and to the local climate is also noticeable, as is its contribution to reducing CO_2 emissions.

The densely built environment, the narrow streets, the tall buildings and the narrow façades raised problems for passive solar design. These problems have been overcome by expanding the building in a vertical direction. The building's carefully insulated envelope and the small openings in the north façade ensure satisfactory internal thermal comfort conditions and low energy consumption in the winter. Solar protection during summer is rather inadequate, especially in the greenhouse, due to the lack of sun-shading devices. Implementing cross-ventilation through north and south external openings reduces summer overheating.

The floor spaces that overlook each over, connected vertically by an interior opening, give continuity to the space and provide the whole with good airflow and ventilation. This is very valuable during the hot summer months. Furthermore, solar radiation penetrates the

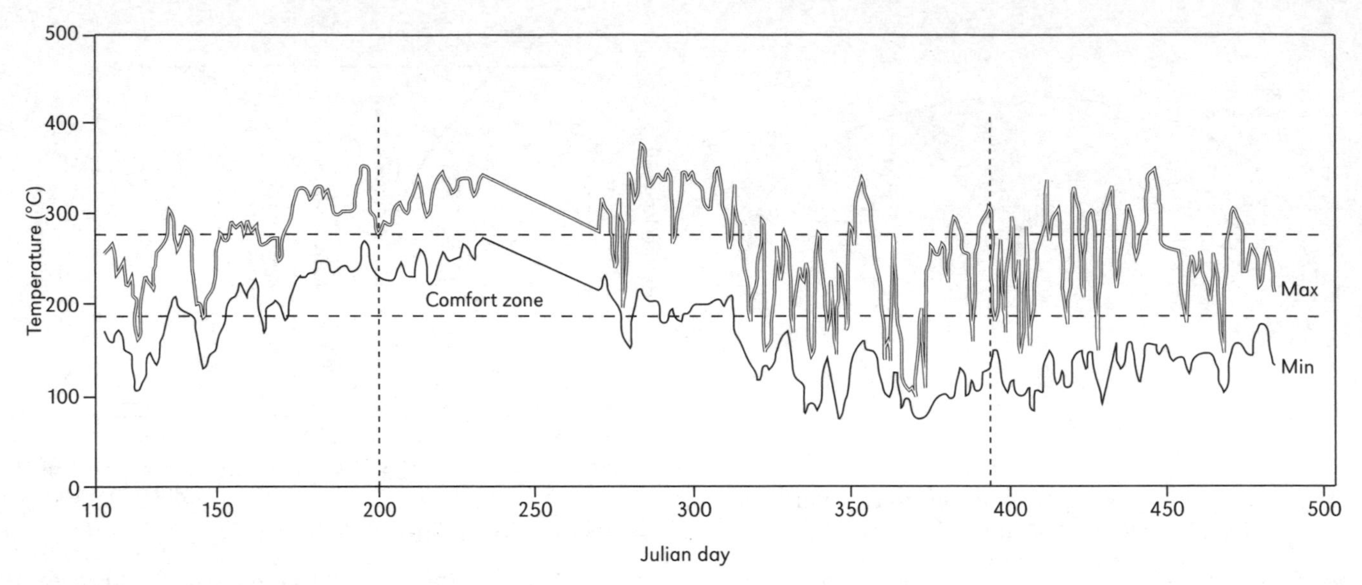

Source: Katerina Spiropoulou

Figure 9.42 *Daily maximum and minimum temperature values in different places in the building*

interior spaces so that there is an efficient use of daylight.

From the users' point of view, the frames of the external openings should be more weather stripped in order to reduce air infiltration losses during winter. Double glazing could possibly improve thermal comfort and reduce the noise level.

In summer, all external windows are open, as well as the dampers at the top and in between the trombe walls, in order to facilitate ventilation. Reflecting curtains between walls and glass panels are placed on most of the glass-covered areas.

Case study 4: Bezigrajski dvor: An energy-efficient settlement in Ljubljana

Table 9.5 *Bezigrajski building*

Name:	Bezigrajski dvor
Type:	Residential settlement
Location:	Centre of Ljubjana

Introduction

The Bezigrajski dvor settlement is located in the area of a former military barracks almost in the centre of the town.

At the design stage, there were no space limitations that would affect the design of the settlement. The planners wished to design a neighbourhood to a low density. They assured residents of privacy and also of good illumination and sunlight, with appropriate distancing between buildings.

Location and climate

The city of Ljubljana is located in the centre of Slovenia: latitude 46.5 degrees and longitude 14.5 degrees. The settlement location is shown in Figure 9.43. The city climate is continental, with 1660 sunshine hours per year and degree days of 3300 Kelvin (K) days/year. Average monthly temperatures and solar irradiation are presented in Figure 9.44. There are, on average, 160 rainy days with 1220mm of precipitation.

Settlement description

The settlement was constructed in 1998 by the architect Andrej Mlakar in Ljubljana. There are two taller commercial buildings (ground floor and six upper floors) along the traffic road at the south and north of the settlement that define the visual landscape. These two higher edifices also represent a sound barrier to prevent sound penetration into the settlement. Three groups of residential buildings inside the settlement are lower (ground floor and four upper floors) and are located around three squares. The minimum distance between buildings is equal to the height of the building, which makes enough space for sunlight, green areas, playgrounds and walking paths (Figures 9.45 and 9.46).

A two-storey car park is built below the entire neighbourhood. Approximately 80–120cm of soil has been placed on the roof of the car park in order to plant trees and shrubs. In this way, the heat island has been reduced.

The building interior is designed in such a way that each floor has four apartments. Each apartment has a lodge sunspace in the building corner for passive heating. Through these lodges (balconies), the apartments are linked to the outdoor landscape. Sunspaces are not heated. During the summer, apartments are shaded with external movable shading devices, which stretch over the entire glazed area (Figures 9.47 and 9.48).

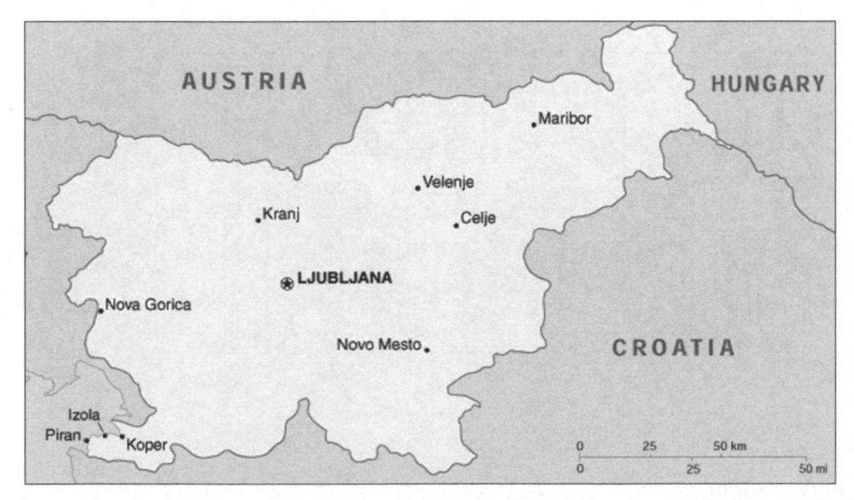

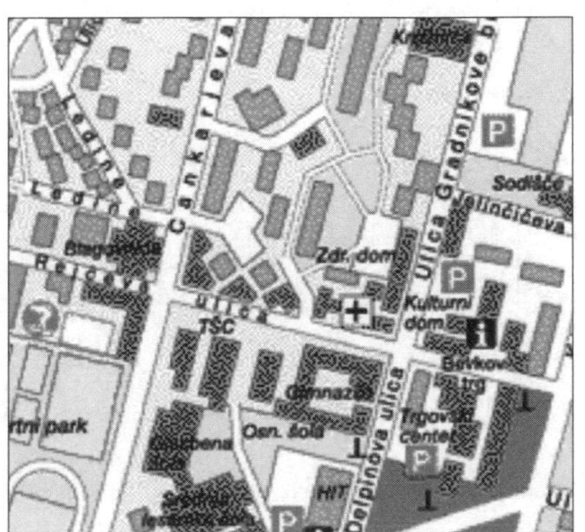

Figure 9.43 *Location of the settlement (map of Slovenia* (left) *and map of Ljubljana* (right))

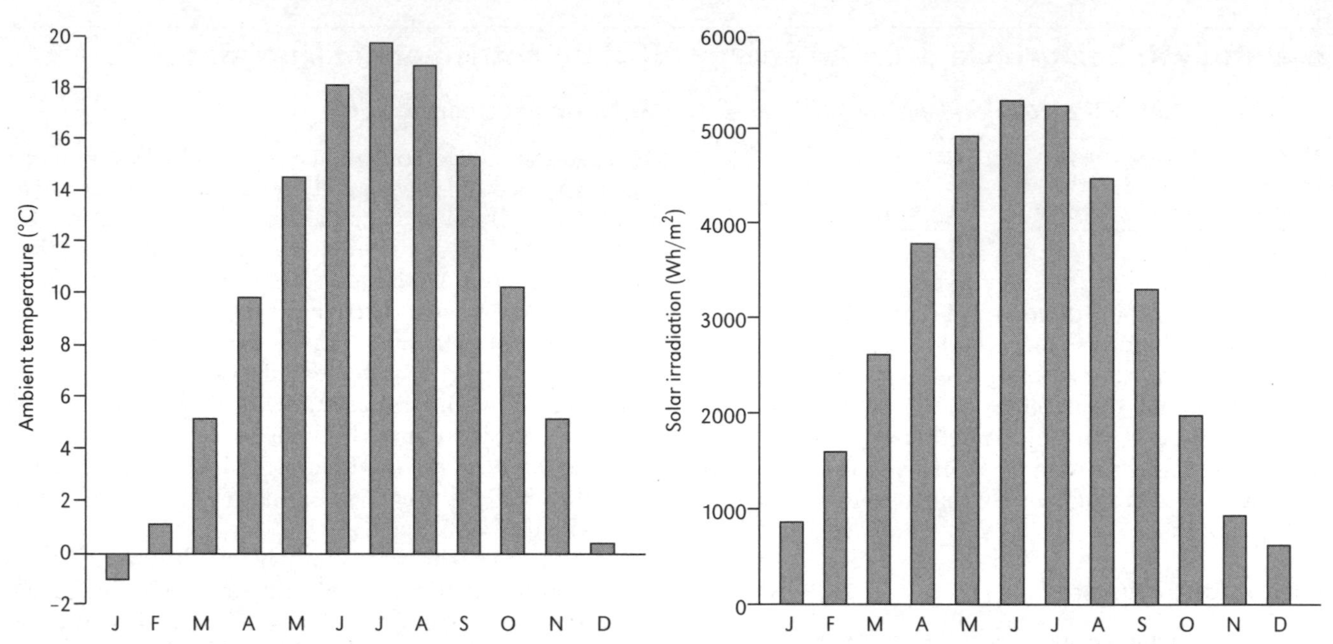

Figure 9.44 *Monthly average air temperature and daily solar irradiation in Ljubljana*

Heat supply

The settlement is connected to a distant heating system, 172km long. The system largely concentrates on supplying hot water, while steam is also supplied, to a lesser degree, to industry users (Figure 9.49). The distant heating system operates throughout the year and has two central heating plants. Both heating plants simultaneously produce heat and electricity. Medium-weight oil and coal are used as fuel.

Although traffic has increased, measurements show that concentrations of pollutants in the air have decreased during the last decade. Figure 9.50 presents the decrease in sulphur dioxide (SO_2) concentration in the city centre during the last decade.

Emission decrease in the city centre is primarily a result of the following:

- expansion of the distant heating network (heating power has increased by 29 per cent; the heating

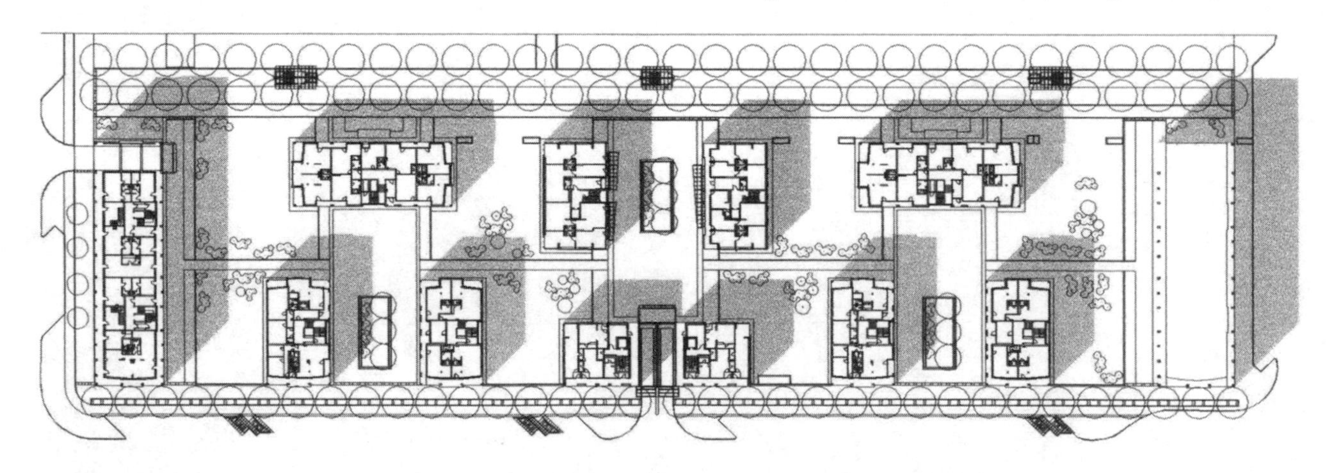

Source: Andrej Mlakar u.d.i.a., Bureau Krog d.o.o., Ljubljana

Figure 9.45 *Floor plan of the Bezigrajski dvor settlement; shades of the building are plotted at a sun altitude of 45 degrees and an azimuth angle of –45° (10 April at 10:00)*

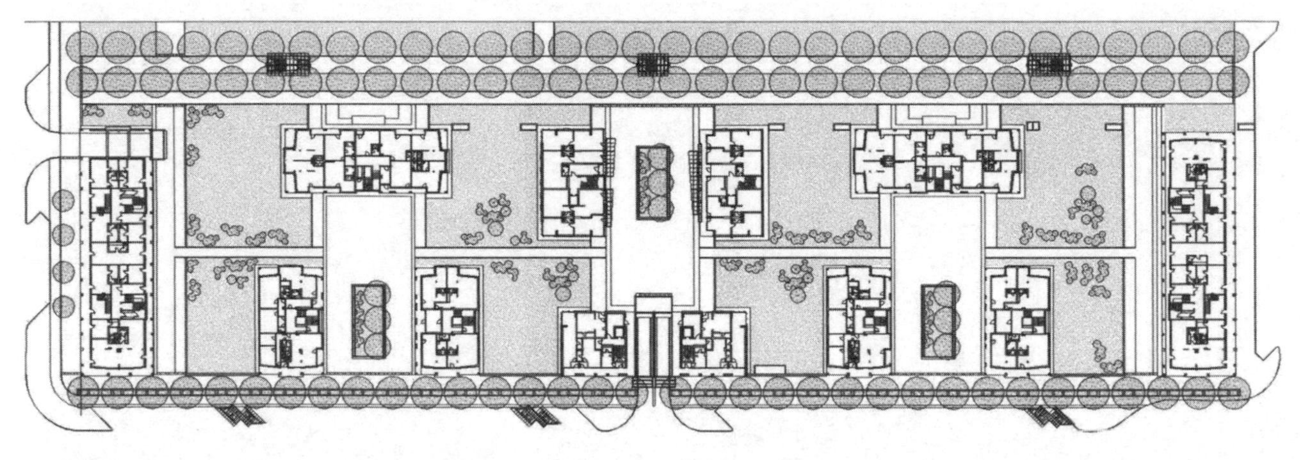

Source: Andrej Mlakar u.d.i.a., Bureau Krog d.o.o., Ljubljana

Figure 9.46 *Green areas reduce the heat island in the settlement*

power of suspended separated boilers was 110 megawatts);

- construction of a gas network for natural gas distribution (during the last ten years, the gas network in the city was enlarged by 150km);
- fuel quality improvement, which is used in both heating plants (on average, sulphur in medium weight oil has decreased from 1.5 to 0.7 per cent and in coal from 2.5 to 0.3 per cent).

Performance

Energy efficiency for the settlement has been evaluated using LT Method 4.0.

The LT method is a manual design tool for calculating energy performance in buildings. It is based on a computer model developed both at Cambridge Architectural Research and the Martin Centre, University of Cambridge. It is a simple manual procedure that requires only a pencil and a calculator. The LT method allows the designer to predict the energy consumption for a proposed building during the early stages of the design. LT Method 4.0 is sensitive to parameters such as building form and size; *U*-values of a building's construction; area, type and distribution of glazing; air infiltration; internal heat gains; the presence of sunspaces or an atrium and a heating system; and provides an output for annual primary energy use and CO_2 production under the end uses of heating, lighting and cooling. Results of different design options are combined in a single worksheet, giving

Figure 9.47 *Picture of the Bezigrajski dvor settlement from the north-east*

Figure 9.48 *View of the south and east façade with sunspace, and the east façade with external shading devices and windows with night insulation*

Source: Institute of Public Health, Republic of Slovenia and Municipality of Ljubljana, 1998

Figure 9.49 *Hot water (black) and steam (grey) network of distant heating system with both heating plants and the Bezigrajski dvor settlement (black dot)*

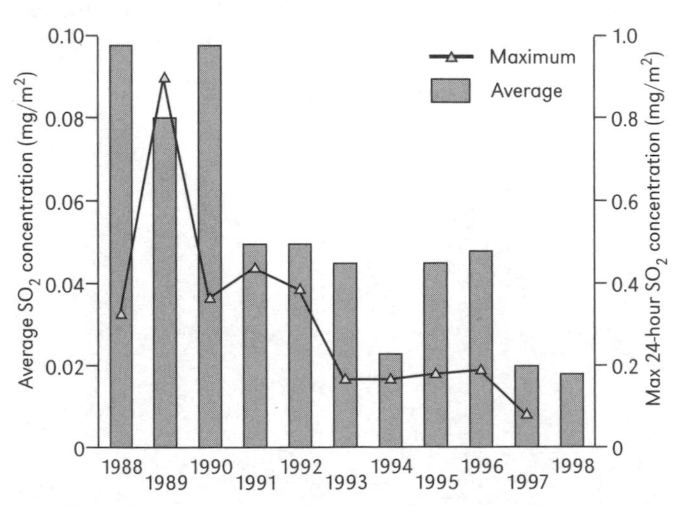

Source: Institute of Public Health, Republic of Slovenia and Municipality of Ljubljana, 1998

Figure 9.50 *Average and 24-hour maximum SO_2 concentrations in the city centre of Ljubljana*

a picture of the relative importance of various energy components. The appropriate worksheet has to be selected depending upon climatic conditions and building envelope surface-to-volume ratio. Heating energy is influenced primarily by winter temperatures, represented by heating degree (days) and, secondarily, by the availability of solar radiation. These two climatic variables are used to characterize three climate zones. Each climatic zone has three different worksheets for a different range of building envelope surface-to-volume ratios.

Figure 9.52 shows the results for the selected building presented in Figure 9.48. The opaque walls of the building are thermally well insulated (U = 0.42W/m^2K), while windows and sunspaces are double glazed (U = 2.8W/m^2K). Other important parameters are stated on the worksheet.

Figure 9.51 *Heat sub-station of the Bezigrajski dvor settlement*

Conclusion

From the results it can be seen that the building's heat consumption without internal and solar gains is approximately 110KWh/m^2 each season. Internal gains from people, electrical lighting and equipment, as well as solar gains from south-oriented windows and sunspaces, reduce energy need for heating by approximately 40 per cent. Taking into account the efficiency of the district heating system according to the LT method, the energy value or annual energy use of the building is 57KWh per square metre of floor area.

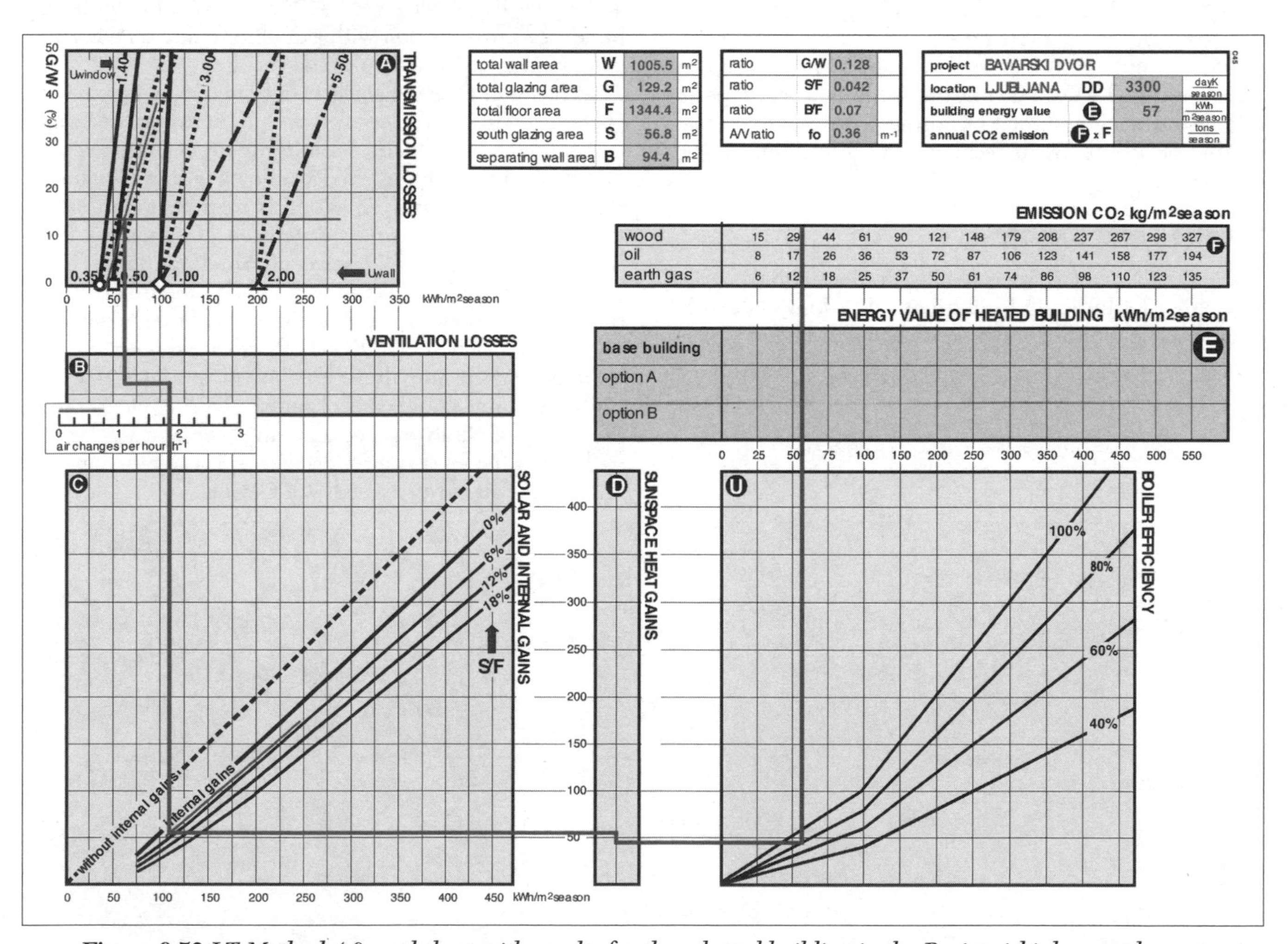

Figure 9.52 *LT Method 4.0 worksheet with results for the selected building in the Bezigrajski dvor settlement*

Case study 5: Commercial building with a double façade

Table 9.6 *HIT Center building*

Name:	HIT Center
Type:	Commercial building
Location:	Centre of Nova Gorica

Introduction

This commercial building is located almost in the city centre of Nova Gorica. It is a multi-storey commercial building that is occupied by different companies. The building façade, which faces the street, is south oriented. A double glass façade was built, firstly, because the south façade faces the street and is representative of the building, giving it substantial character. The glass façade also gives the building a suitable acoustic and privacy barrier from the street. The glass façade is long lasting and does not need special maintenance.

The double glass façade also enables efficient adaptation to optimal working conditions for different types of offices in the commercial building. A properly designed double façade helps to reduce overall building energy consumption.

Location and climate

The city of Nova Gorica is located in the western part of Slovenia: latitude 45.6 degrees and longitude 13.4 degrees. The building location is shown in Figure 9.53.

Building description

The building was designed in 1997 by a group of architects from the Stolp bureau from Nova Gorica, led by Aleš Šuligoj, and was finished in 1999. The building was integrated within the existing commercial and residential environment. Its main façade is south oriented and faces the street; the opposite north façade faces the parking area and other commercial and residential buildings.

The building has two basement levels with parking places. The east part of the ground floor is occupied by a commercial bank; there are also some small shops and a restaurant. Three office floors are used for administration and business activities and include different offices, a conference hall and a television centre. In the attic there are primarily service rooms. Figure 9.54 presents the building's cross-section with a double façade on the south side. Figure 9.55 is a plan of the building.

The double façade consists of a single glass pane with 60cm of air space. The separation wall between the façade and the interior is made from lightweight prefabricated elements and double-glazed windows with a *U*-value of 1.3W/m^{2}K. The outer envelope is made from an 8mm-thick, high-efficient solar protection glass pane with a light transmittance of 50 per cent. Shading of individual offices is provided by Venetian blinds on the inner side of the outer envelope glass, which are controlled manually.

The air cavity is closed on both sides and is open at the bottom. On the top there are louver shutters that are controlled by a central control system. During the summer, the louvers are fully opened all of the time; during the rest of the year, the louvers are closed except when the temperature in the air cavity exceeds 40° Celsius.

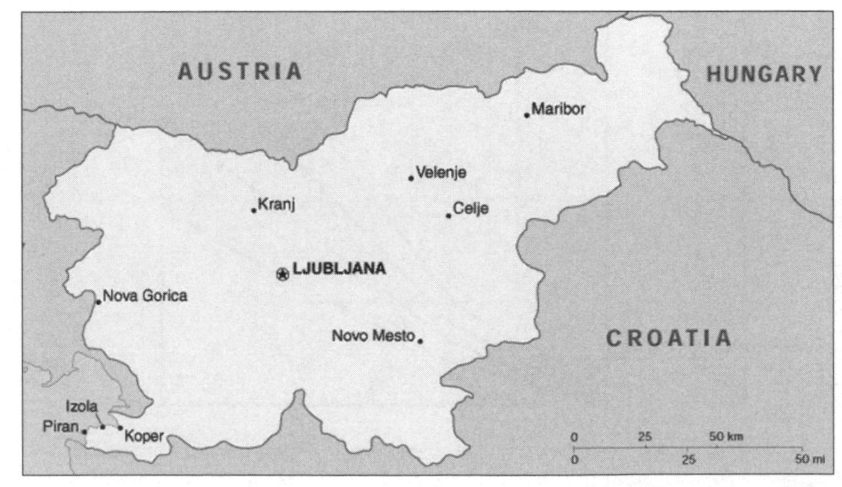

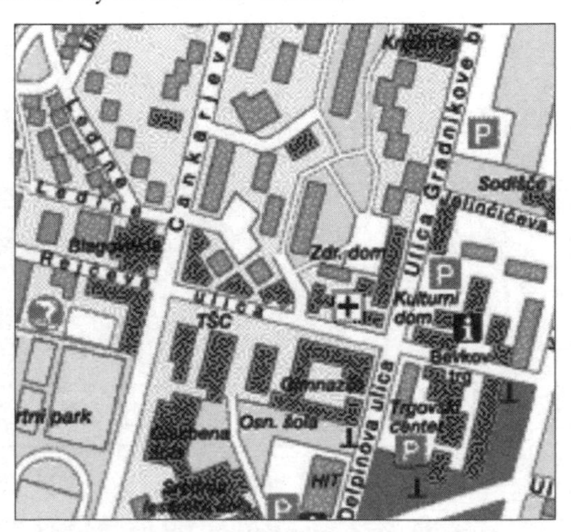

Figure 9.53 *Location of the commercial building (map of Slovenia* (left) *and Nova Gorica* (right))

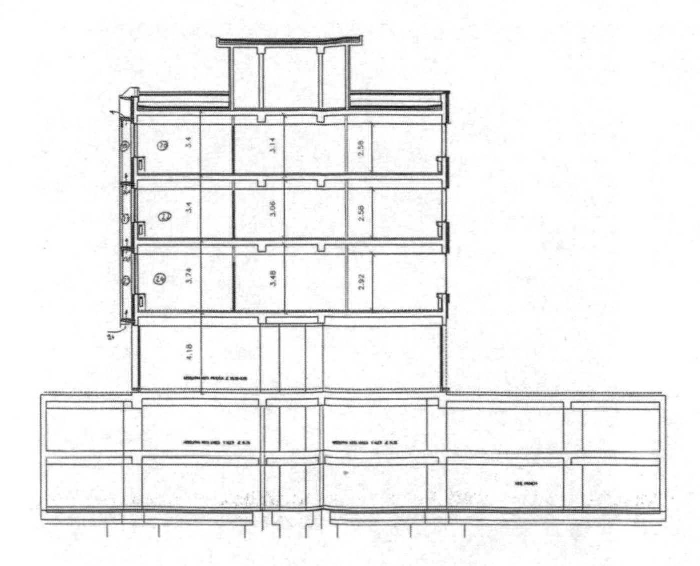

Source: Aleš Šuligoj, Bureau Stolp d.o.o., Nova Gorica

Figure 9.54 *Building cross-section; the three-storey double façade is on the left*

Key features

The building envelope is a building component that has the largest influence on thermal, lighting and acoustic comfort. As a rule, the modern architecture of commercial buildings uses large glazed elements. However, the performance of such (façade) elements in the context of comfort is not always optimal. Large glazed areas enable suitable natural lighting and passive solar heating during cold days; but they could be the reason for glare and overheating, and thermal discomfort due to temperature asymmetry. In spite of intensive development of new technologies, 'smart windows or façades' are not yet available on the market for a reasonable price. The optimal solution regarding energy use for heating, cooling and lighting could be achieved by incorporating different elements and concepts, such as selective transparent glazing, shading devices and controlled ventilation.

During the design stage for this building, several different options were analysed. One of them queried what type of glass should be used to prevent building overheating and to provide efficient daylighting. Another concerned arrangement and size of openings for controlled natural ventilation, including optimization of distance between the inner and outer façade. Different configurations of bottom and top openings, as well as openings in the outer glass façade, have been analysed using computational fluid dynamics (CFD) tools. Using CFD simulations, the airflow rate in the natural ventilated gap of the double façade was approximated. This approximation was later included in the transient heat transfer simulation programme TRNSYS for analyses of building heating and cooling demands. In natural lighting studies, different types and position of blinds were analysed using the daylighting simulation tool LUMEN.

Performance

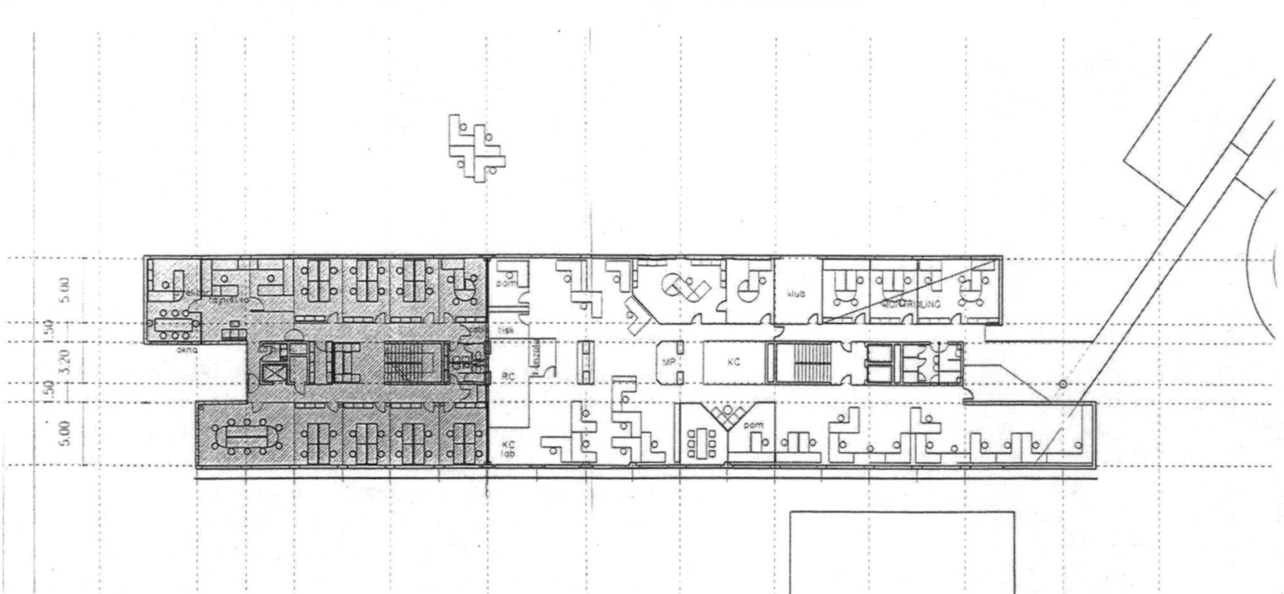

Source: Aleš Šuligoj, Bureau Stolp d.o.o., Nova Gorica

Figure 9.55 *First-floor plan of the building*

Figure 9.56 *Commercial building's south façade*

According to the results of numerical simulations, a solar protective glass with a total energy transmittance of 0.27 and visual transmittance of 0.5 were selected. Manually driven blinds are mounted on the inner side of the façade glass in front of each window. The air cavity is opened at the bottom and closed by an automatically regulated mechanical shutter at the top.

Figure 9.57 *Computational fluid dynamics (CFD) analyses of temperature in a double-glazed façade*

In Figure 9.58, ambient temperature and air temperatures in a double façade on each floor are presented, together with heat gains for each office attached to the double façade for the hottest week in the year according to a test reference year.

HVAC systems are controlled by a building management system that also monitors temperatures in the double façade. Figure 9.59 presents temperatures at different heights inside the double façade for the hottest summer day of 2002. The temperature gradient during the day confirms the efficiency of the natural ventilation inside the gap.

Climatic data monitored by a building management system (BMS) were used to compare measured and simulated temperatures in the double façade. Results are presented in Figure 9.60 for one week in June 2001.

The impact of a double-glazed façade on the energy demand for heating was tested numerically by looking at the energy consumption of north- and south-oriented offices without a double façade, compared with that of a south-oriented office with a double glazed façade. Results show that the heating season for south-oriented offices is approximately two months shorter than for the north-oriented offices. Figure 9.61 shows the energy

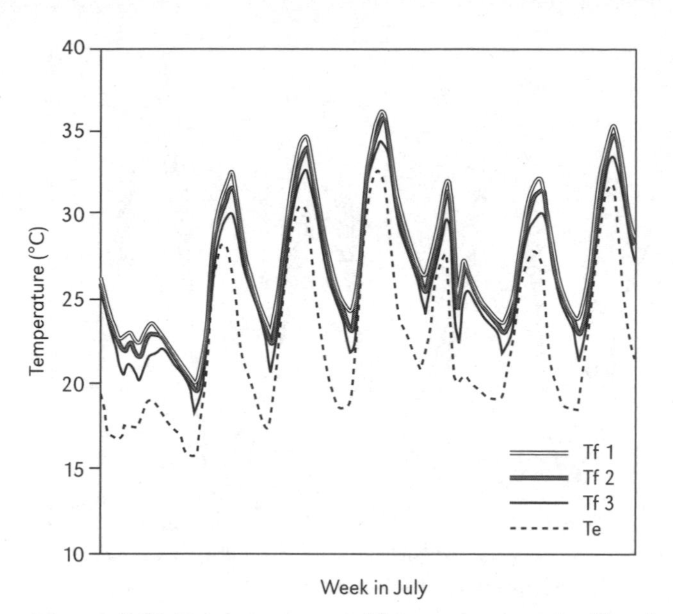

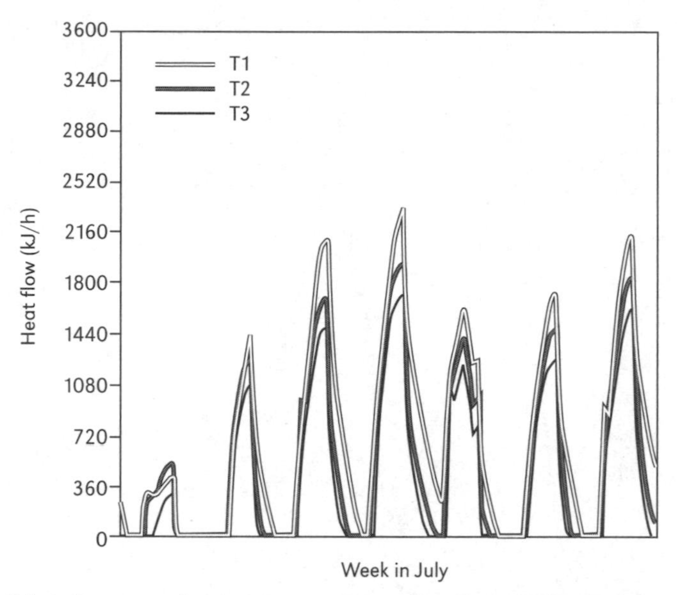

Figure 9.58 *Temperatures and heat gains in a double-glazed façade. Te: ambient temperature; Tf1–Tf3: temperatures in a double-façade cavity on different heights (floors); T1–T3: heat gains to a typical office on different floors*

requirements for heating each floor (except the ground floor) of the building. Much less energy for heating is required for south-oriented offices, especially when a double façade is attached.

Data from Figure 9.61 can be corrected according to the fact that the average temperature of a heated space was 23° Celsius, which gives overall specific energy use of 46.8kWh/m²/year. Natural gas consumption is monitored on a monthly basis. The data for the year 2001 are presented in Figure 9.62. Calculated specific end energy use considering boiler efficiency was 50.5kWh/m²/year.

Solar protection glass and Venetian blinds in a double façade reduce natural lighting. Different studies of natural lighting have been made using a lighting simulation program. The results for an office with a double glazed

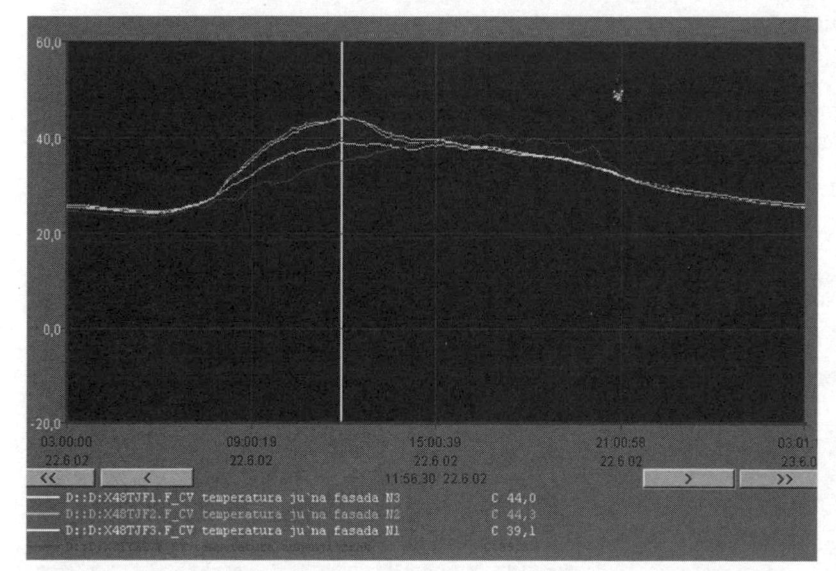

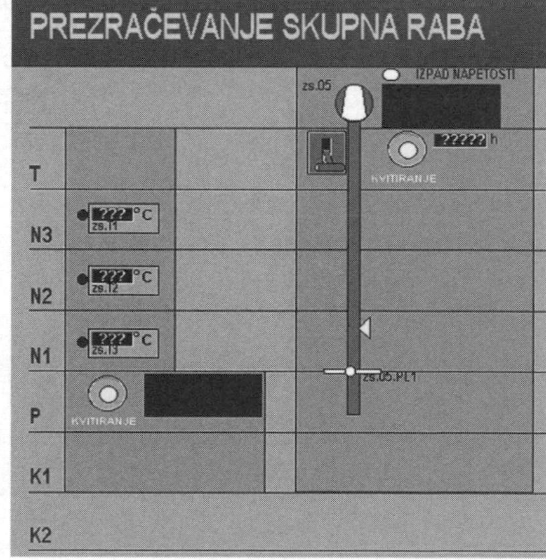

Figure 9.59 *Measured temperatures in a double-glazed façade on 22 June 2002; image shows air temperatures in the façade cavity of each floor and ambient air temperature. Figure on the right is a screen of a building management system for temperatures in a double façade*

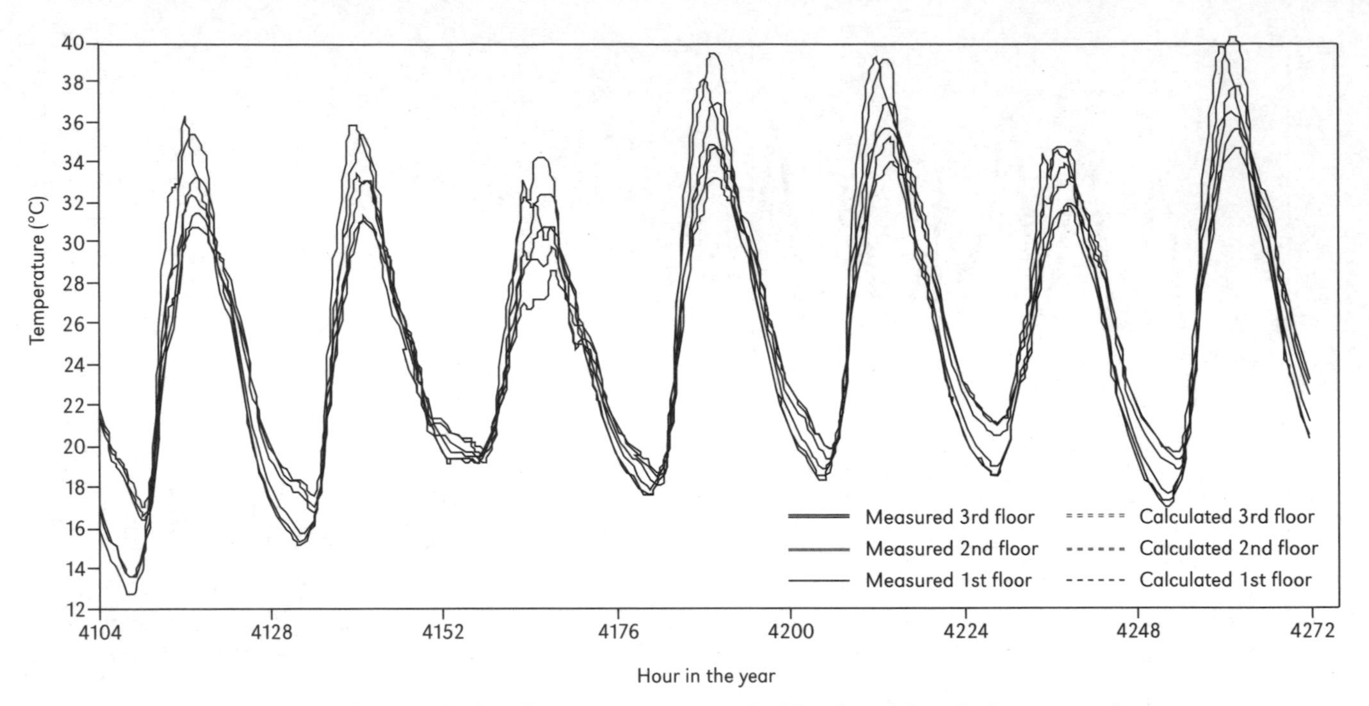

Figure 9.60 *Measured and calculated temperatures in a double-glazed façade for one week in June 2001*

façade with reflective (left) and white (middle) blinds, and for an office without a double glazed façade are presented in Figure 9.63, both for a cloudy day and for natural daylight with horizontal luminance of 5000lx.

Results show that in the case of a daylight factor of 4 per cent on a yearly basis, south-oriented offices need 400 to 500 hours of artificial lighting more than similar offices without a double façade. This equals 4 to 5kWh/m^2 additional energy use for lighting compared with single-façade offices; but this is substantially less than energy savings for heating. Because of less uniform illumination, two-zone artificial lighting has been suggested and applied in the south-oriented offices.

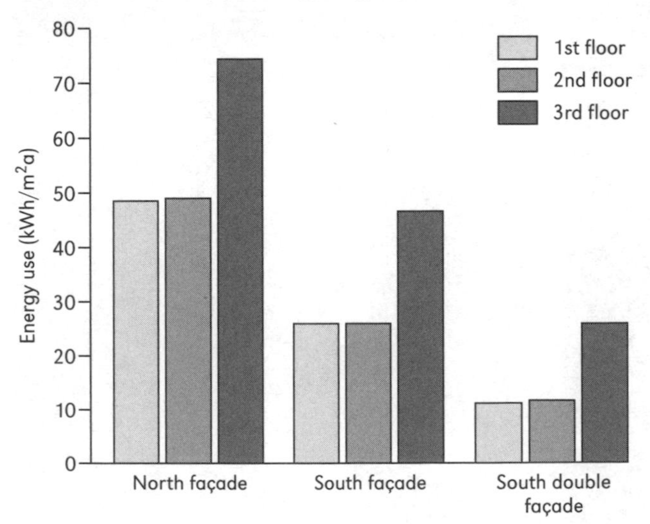

Figure 9.61 *Energy use for heating each square metre of office floor area calculated on the basis of data from test reference year and a room temperature of 20° Celsius*

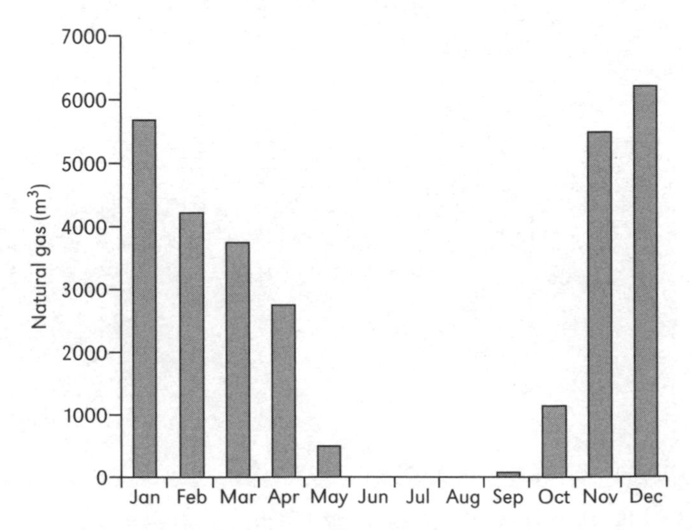

Figure 9.62 *Earth gas consumption for office building heating*

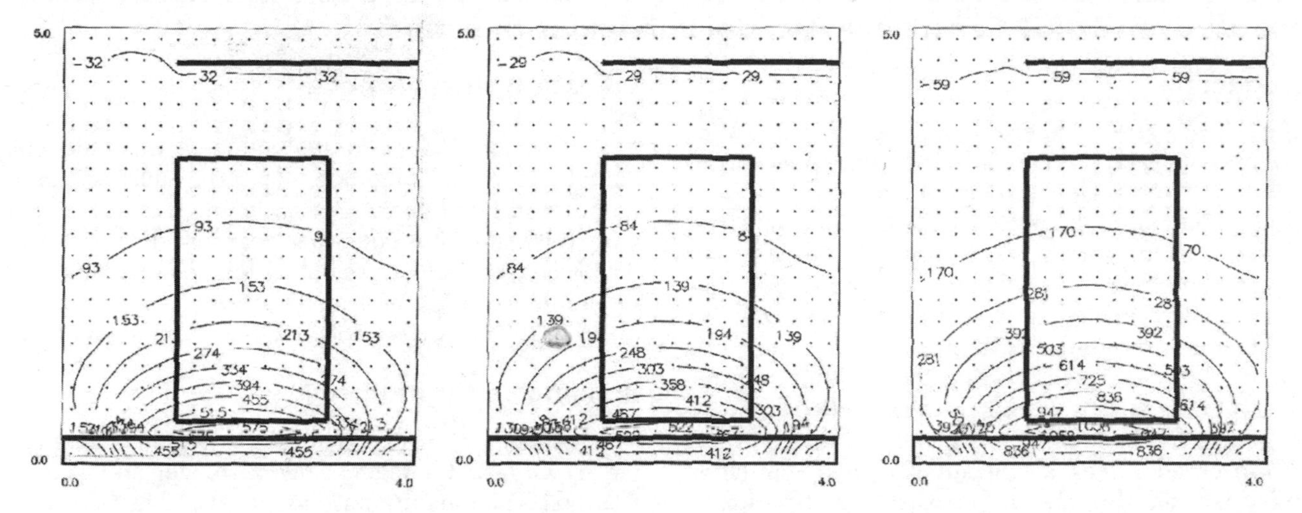

Figure 9.63 *Natural lighting in a typical office in the HIT Center building;* (left) *office with double façade and reflective blinds;* (middle) *office with double façade and white blinds;* (right) *office without double façade*

Conclusion

This project has demonstrated that double-glazed façades substantially reduce energy needs for heating and thus could be classified as an element that saves energy in spite of slightly greater electricity use for lighting. Shading with outer glass and Venetian blinds, and cooling of south-oriented double façades with natural ventilation, are also effective.

Positive experiences have led to the decision to construct a double façade on a second commercial building owned by the same company as a solution for a temporary overheating problem during the summer.

Table 9.7 *Euro Centre building*

Name:	EURO Centre
Type:	Commercial building
Location:	Centre of Ljubljana

Case study 6: EURO Centre commercial building with atrium

Introduction

The EURO Centre commercial building is located in the centre of Ljubljana. It is a multi-storey commercial building with an 'eye-shaped' atrium in the centre of the building. The building's north façade faces a main street with heavy traffic. On the south side, the building ends near neighbouring residential buildings, forming a narrow canyon corridor in an east–west direction. The building was integrated within the existing commercial and partly residential environment. During the building's design stage, several innovative solutions were applied to reduce energy use for heating, cooling and lighting. For the south-oriented offices, a specially designed ventilation system is used to prevent overheating. Shading is provided with perforated, reflective-shading roller blinds inside the office. Space between the shading device and the window is connected to the exhaust duct in order to prevent hot air from entering the office. For the atrium, different natural lighting and ventilation strategies have been analysed. The result was a specially designed mechanical ventilation system. The building has been occupied since mid 2002, and different system operation parameters are monitored by a central building management system (BMS).

Location and climate

The city of Ljubljana is located in the centre of Slovenia: latitude, 46.5°, and longitude, 14.5°. The building location is shown in Figure 9.64. The city climate is continental, with 1660 sunshine hours per year and degree days of 3300Kday/year. Average monthly temperatures and solar radiation are presented in Figure 9.65. There are, on average, 160 rainy days, with 1220mm of precipitation.

Building description

The building was designed in 1999 by a team of architects led by Gorazd Groleger, Samo Groleger and Davorin Gazvoda. The building was integrated within the existing environment (mostly commercial buildings) and was finished in mid 2002.

The building has three basement levels that serve as a car park. The ground floor is occupied by small shops, a restaurant and a control centre. Six office floors are used for commercial purposes. Figure 9.66 presents the building's cross-section without basement levels and Figure 9.67 presents a plan of the building's first floor.

On the north and west side is a glass façade with a coated raster. The U-value is 1.8W/m^{2}K. Internal shading is provided for those offices.

South-oriented offices have windows with a U-value of 1.4W/m^{2}K and an aperture area of 1.8m^2. Shading is provided with a perforated, reflective-shading roller blind inside the office. Space between the shading device and the window is mechanically ventilated for cooling. Fresh air for ventilation enters the offices from the north side of the office, providing displacement ventilation.

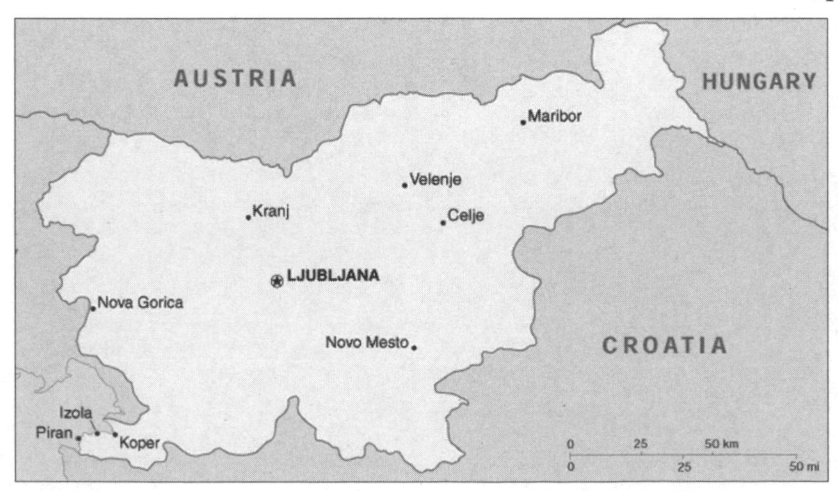

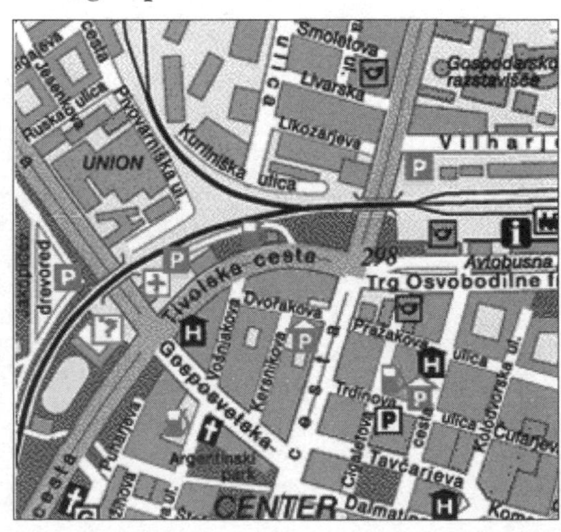

Figure 9.64 *Location of the EURO Centre commercial building (map of Slovenia* (left) *and map of Ljubljana* (right))

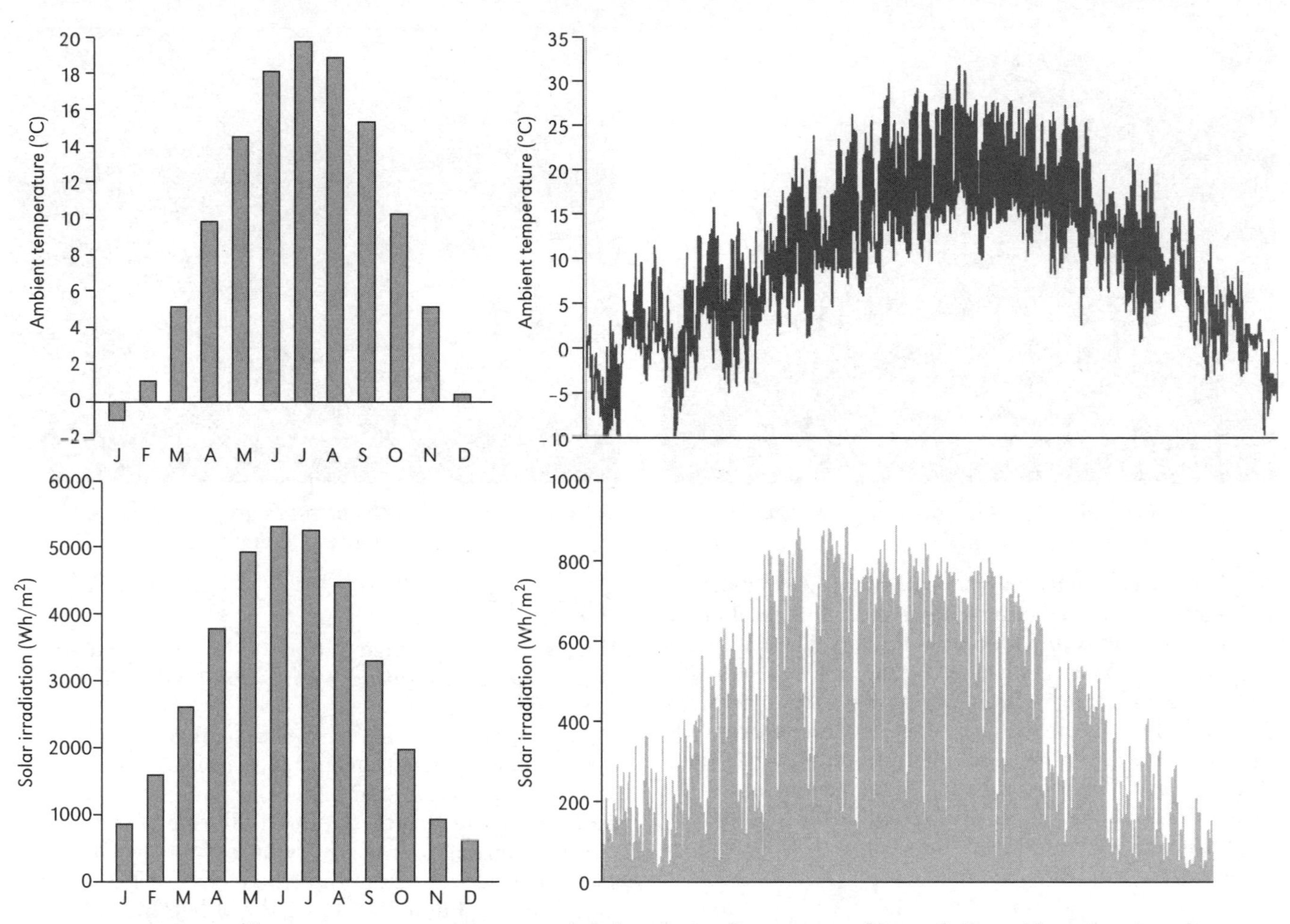

Figure 9.65 *Monthly average air temperature and daily solar radiation in Ljubljana* (left) *and hourly values from test reference year for Ljubljana* (right)

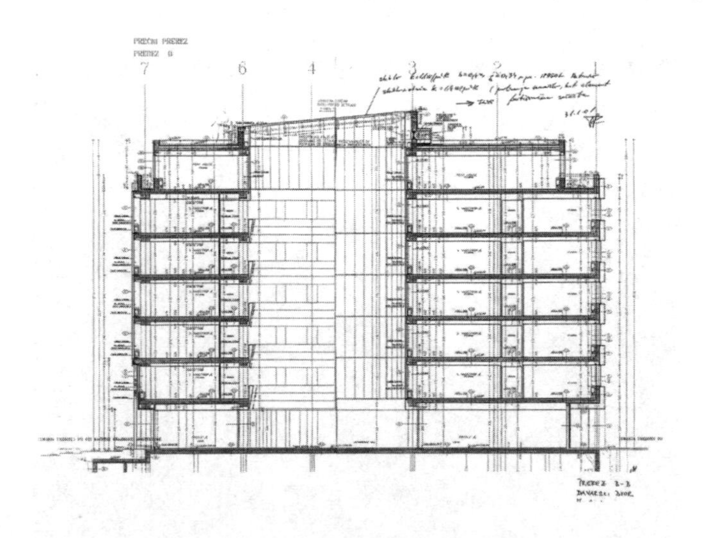

Source: Gorazd Groleger, Samo Groleger and Davorin Gazvoda

Figure 9.66 *EURO building's cross-section*

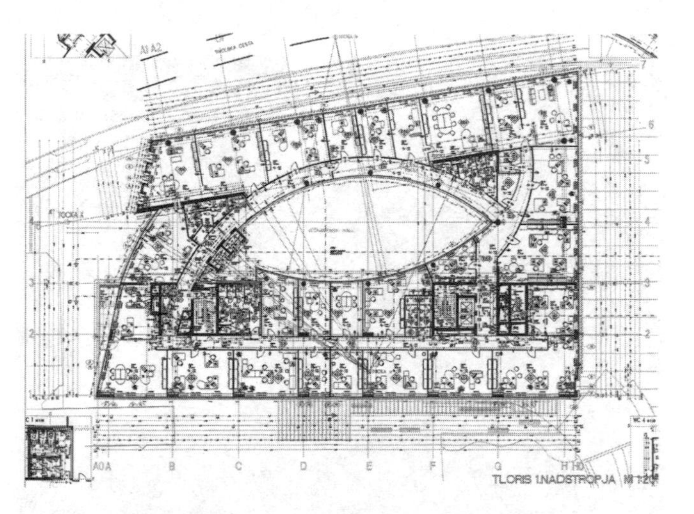

Source: Gorazd Groleger, Samo Groleger and Davorin Gazvoda

Figure 9.67 *EURO building's first-floor plan*

Figure 9.68 *EURO building's north façade*

Figure 9.70 *Part of the atrium's artificial lighting that simulates daylight*

The building atrium is 'eye-shaped', with 300m^2 of floor area, and is a communication and social focal point. Natural lighting is provided from the top. The lighting of the basement floor is enhanced by the reflective surfaces of glass separation walls. Cooling the atrium was not planned during the design stage; so an appropriate mechanical ventilation and shading strategy has been designed to prevent overheating in the atrium and in the upper floors of the building. The upper three floors are mechanically separated from the atrium with glazing, as shown in Figure 9.69.

Key features

During the design stage of the building, different strategies concerning lighting, ventilation and cooling were analysed.

Daylighting of the atrium base floor is enhanced through high visual-transmittance roof glazing without any shading devices, and high reflective-glass separation walls inside the atrium. On the north side of the atrium, the glass separation wall covers all six floors, while on the south side only the upper three floors have a glass separation wall (see Figure 9.70). Artificial lighting simulates natural conditions with focused lights and mounted mirrors.

Figure 9.69 *EURO building's 'eye-shaped' atrium*

Figure 9.71 *Openings on the rim of the atrium before bi-directional fans were installed*

Atrium cooling during the summer time is provided by ventilation instead of external shading because of architectural demands. Computational fluid dynamics (CFD) and transient heat transfer analyses have been used to ensure that temperatures and air velocities in the atrium will not cause thermal discomfort. Bi-directional operating fans on the rim on the top of the highest floor provide necessary cross-ventilation (Figure 9.71). The same fans are used in fire protection. On the basis of numerical simulations presented in Figure 9.72, the glass separation wall has been recommended and applied on the fourth to sixth floors to prevent overheating in adjacent offices.

Following the architect's demands, the south façade is without any outside shading device to provide a smooth surface. As a result, an inner shading device is used with the specially designed ventilation system. The exhaust ventilation air flows between the windows and the reflective blinds to remove the heat from the blinds, providing lower air and radiant temperature. The system is planned according to CFD simulations, presented in Figure 9.73.

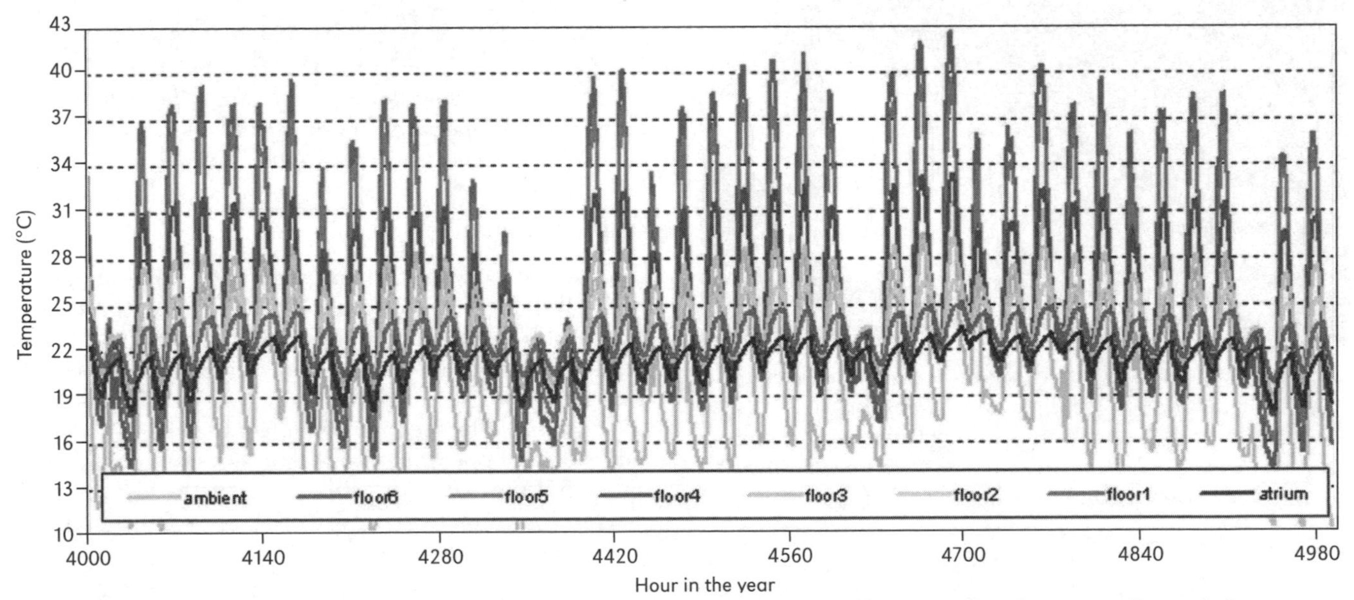

Figure 9.72 *Predicted temperatures in the atrium in summer conditions: cross-ventilation on the sixth floor (0.14 cubic metres per second) and displacement ventilation in the atrium (4 cubic metres per second)*

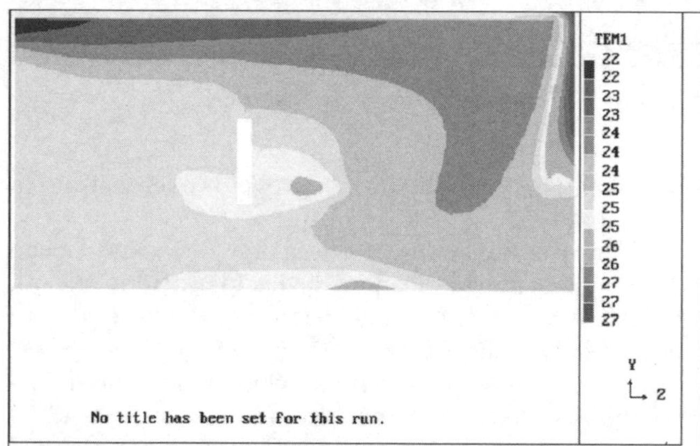

Figure 9.73 *South office ventilation system; concept of ventilation* (top), *computational fluid dynamics (CFD) analysis of indoor temperatures on a summer day at 12 noon* (middle); *and part of the system at the construction phase are depicted* (bottom)

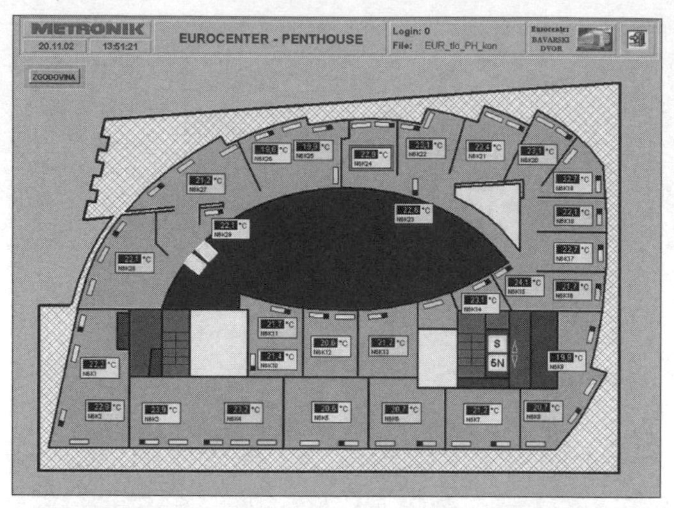

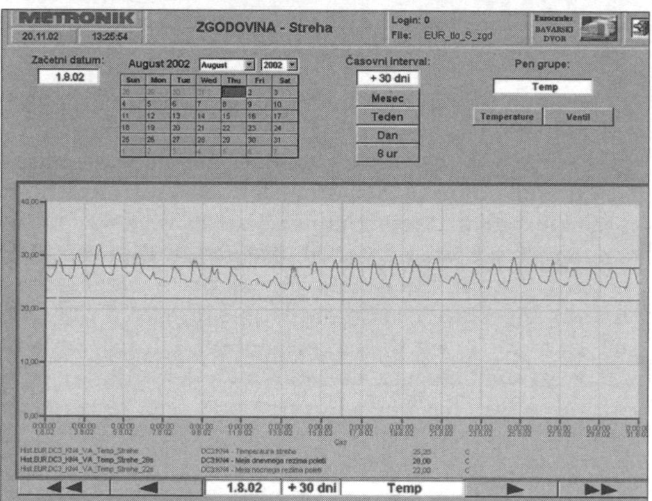

Figure 9.74 *Temperatures in offices on the sixth floor* (top) *and temperatures in the atrium at the top level during selected months* (bottom)

Performance

The building was finished and occupied in the summer of 2002. Building service systems are monitored and regulated by a building management system (BMS). Via the BMS, each occupant can adjust microclimatic conditions in the office. Current parameters (see Figure 9.74) can be adjusted by the intervention of the energy manager from a central operating room. Another important advantage of BMS is that it is possible to show the history of the measured parameters during past use of the building services systems (see Figure 9.74).

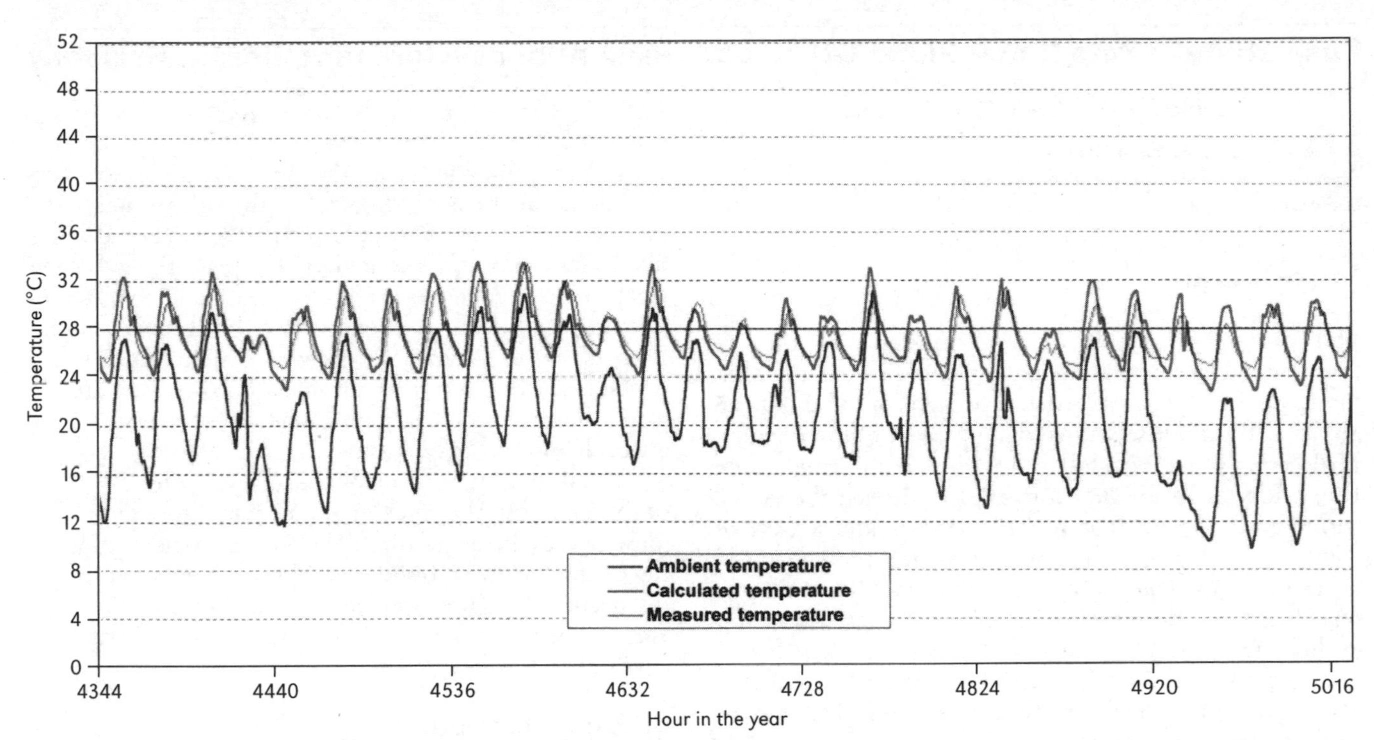

Figure 9.75 *Comparison of measured and calculated temperatures in the atrium on the sixth floor for July 2002*

Real climatic data were used to confirm numerically determined indoor environmental conditions. Figure 9.75 shows good agreement between measured and numerically determined temperatures in the atrium in July 2002, the first month of the building's operation.

Conclusion

The design of EURO Centre commercial building is a good example of integrated building design. Specialists from different fields cooperated with architects from the start of the project. Different technologies for energy conservation were applied, such as energy-efficient lighting and passive solar heating. Their effect will be evaluated during the building's monitoring process in the following years.

Case study 7: Potsdamer Platz: Office and residential development, Berlin, Germany

Table 9.8 *Potsdamer Platz building*

Name:	Potsdamer Platz
Type:	Office and residential building
Location:	Berlin

Introduction

This project is part of a larger master plan for the Potsdamer Platz development, set in the heart of Berlin, the capital of Germany, with a population of about 3.5 million. It is a densely developed project of three courtyard forms some 32m tall and with narrow streets only 13m wide on three sides. To the south-east there is a wider green space that provides views and access to daylight, air and morning sun, to which the courtyard forms open up. The three buildings (two offices and one residential) are designed to be naturally ventilated, even in this urban context. The architect was Richard Rogers Architects, in conjunction with RP+K Sozietat Engineers and Cambridge Architecetural Research Ltd (CAR) Consultants.

Location and climate

The climate of north-east Germany is 'continental' and thus has warm summers and cold winters, with strong seasonal variations. The buildings respond to minimizing heat loss in winter, as well as minimizing the need for air conditioning in the summer. In January and February, mean temperatures are below 0° Celsius, rising to 20° Celsius in the summer. Rainfall is moderate, with an annual total of 500–750mm falling during all months. Humidity is always high, with fog in the autumn, and winter can be overcast for long periods. Snow lies for long spells, and when the wind blows from the north very cold weather follows.

In microclimatic terms, there is no evidence for very significant heat island effects; but the microclimate can be characterized as overshadowed and noisy. Some vehicular traffic is present, particularly to the north, with some local traffic (e.g. deliveries) on the south and east streets. The development is well served by public transport in the form of an underground railway, which limits traffic problems. However, the presence of a shopping arcade, cafés, theatre and other public amenities means that this area is lively and busy.

Local planning constraints, as well as those prescribed by the master plan for the whole area (designed by Piano and Kohlbecker), put severe limits on the geometry of the buildings. Requirements that building occurs up to the site frontage and is up to 32m high aim to ensure that there is urban spatial integrity and continuity.

Building description

The project consists of three courtyard buildings, with two office blocks (with retail on the ground and first floor) to the north, and one residential block (with retail on the lower three floors) to the south. There were a number of key design strategies that developed from the simple master, and which were all informed by energy and environmental performance concerns.

The first design decision was to cut the closed courtyards open to the south-east, facing the park. This meant that, although some floor space was lost, it was now possible to have views towards the park from the courtyard. These openings (in the form of a stepped-back façade in

Table 9.9 *Climate data for Berlin*

Month	Average sunlight (hours)	Temperature (deg C)		Relative humidity (%)		Average precipitation (mm)
		ave min	ave max	am	pm	
January	2	-3	2	89	82	46
February	2	-3	3	89	78	40
March	5	0	8	88	67	33
April	6	4	13	84	60	42
May	8	8	19	80	57	49
June	8	12	22	80	58	65
July	8	14	24	84	61	73
August	7	13	23	88	61	69
September	6	10	20	92	65	48
October	4	6	13	93	73	49
November	2	2	7	92	83	46
December	1	-1	3	91	86	43

Source: Richard Rogers Partnership

Figure 9.76 *South-east façade of the Potsdamer Platz development showing the stepped opening in the elevation*

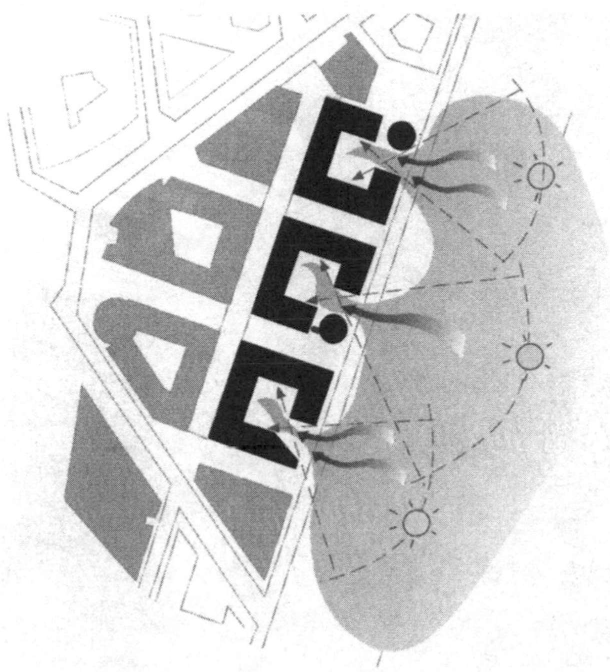

Source: Richard Rogers Partnership

Figure 9.77 *Environmental design strategies for the building form*

the south-east elevation) furthermore enabled increased winter morning sunlight, daylight and ventilation to reach the north-west sides of the blocks (Figures 9.76 and 9.77). Furthermore, offices to the rear of the courtyard could now have views out across the courtyard and to the park. Finally, the opening in the façade became the main entrance route to the blocks via the courtyards. Thus, a number of benefits were achieved with one strategic formal move.

The next design decision was to glaze the courtyards in order to create an atrium (Figure 9.78). The environmental advantages include a thermal buffer to heat loss in winter; preheated ventilation air in winter; thermal buoyancy-driven ventilation in summer; an acoustic buffer to the streets; and an intermediate environment for informal uses (e.g. café and garden). With the reduced heat losses through the façades facing the atrium, it is possible to increase the amount of glazing with no heat-loss penalty. As a result, daylight availability can be improved upon (at least to compensate for the reduction in daylight caused by the atrium glazing and structure).

As the streets and courts are fairly narrow, the initial design strategy for the façades was to determine appropriate glazing and shading proportions as a function of the level of obstruction. Thus, on the top floors – where obstructions are small – it is important to reduce solar gains through a combination of shading and reduced glazed areas. However, shading to the north is likely to be different from the south. Similarly, at the bottom of the building where overshadowing is greatest, daylight availability suggests increased glazing areas and little need for shading. The same considerations can be applied to the façades facing in towards the atrium. As a result, a 'kit-of-parts' was developed for the façades based on simple modules and allowing for combinations of opaque, translucent or transparent elements, moveable exterior or interior shading devices, and fixed façade components that can open (Figure 9.79).

The interior arrangements are as follows: a 14m deep plan that can be cellular or open plan, with windows that can open at low and high level. Thermal mass is provided above a perforated, suspended ceiling, which allows ventilation and radiative exchange with the space, while limiting the negative acoustic aspects of hard surfaces and allowing for a range of internal fit-out options and lighting systems. Perimeter radiators provide heat in the mornings and in the winter, and avoid cold downdrafts from the façade and openings. There is no provision for mechanical ventilation; but a centralized chilled water system is available for chilled ceiling systems where these are seen as necessary (i.e. in high internal gains spaces). High-level windows are opened automatically to allow for night-time ventilation (Figure 9.80).

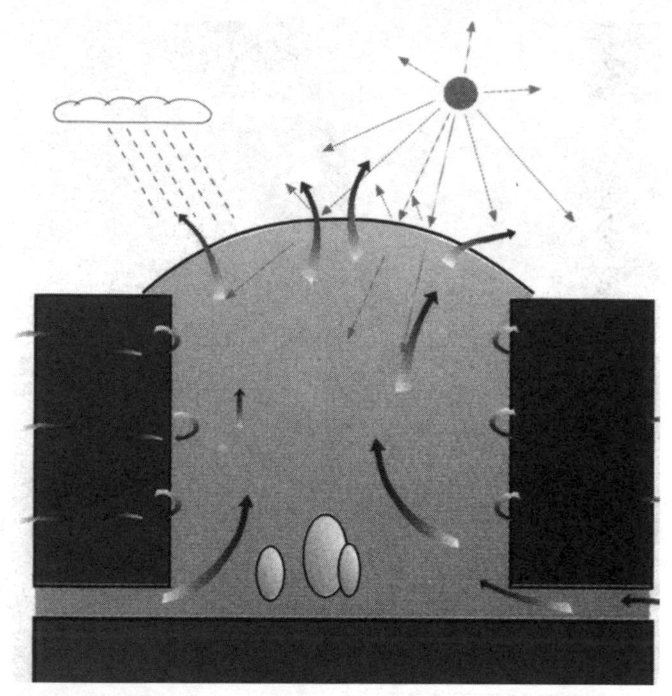

Source: Richard Rogers Partnership

Figure 9.78 *Environmental design strategies for the atrium*

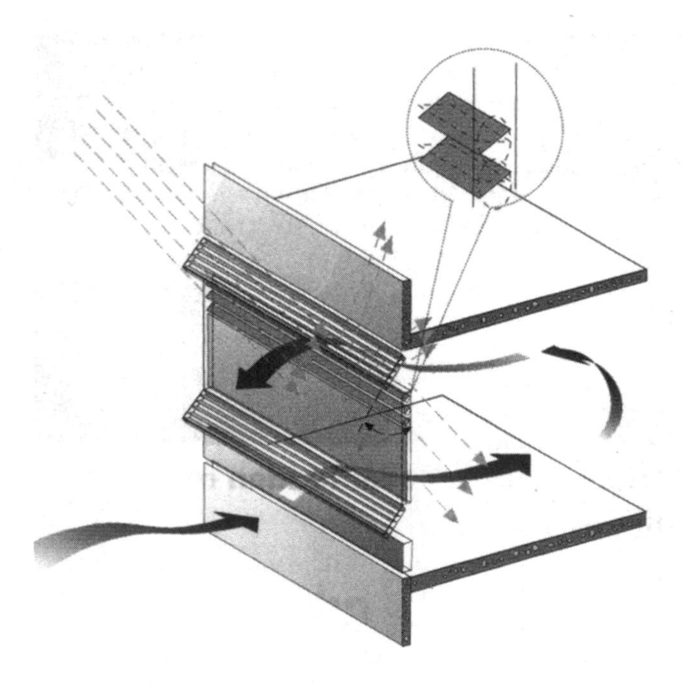

Source: Richard Rogers Partnership

Figure 9.79 *Environmental design strategies for the façade*

Source: Richard Rogers Partnership

Figure 9.80 *Environmental design strategies for the interior*

Key features

The outline description above indicates that the buildings integrate a range of low-energy strategies that are interdependent. Thus, there is no single feature that, on its own, determines the success of the solution. However, a primary consideration is clearly the ventilation strategy, particularly in the context of an urban site and with a speculative office brief. The key ventilation feature of the buildings is the use of a large (1.5m) plenum at the atrium floor level that penetrates right through the building (Figure 9.78). This ventilation zone enables fresh air to enter the atrium from three sides of the building in order to ensure that during summer there is sufficient fresh air to cool the atrium and the adjacent office spaces. The benefits of this plenum are numerous, including a noise reduction zone and a pre-cooling space. The distance from external louvers to the atrium is about 14m, which helps to significantly reduce noise transfer (combined with some acoustic baffling) from outside the atrium. The plenum is constructed of in-situ concrete, and when combined with night-time cooling can reduce the temperatures of the incoming air during the peak daytime period.

Performance

The buildings were recently completed and occupied, and have not been monitored. However, the predicted energy performance, assuming no use of chilled systems, is estimated to be 200kWh/m^2. This is approximately half that of a typical air-conditioned building. What is not clear is to what extent the tenants have opted for the chilled ceiling options and how this has affected the energy in use.

Overheating predictions show that the combination of natural ventilation, thermal mass, night-time cooling, daylight design and shading reduced the number of hours

Source: Richard Rogers Partnership

Figure 9.81 *Interior view of a typical atrium*

above 27° Celsius to 40 hours per year. This is well below the target figure of 5 per cent of the occupied period (i.e. 100 hours).

The costs of the building are approximately €2250 per square metre, slightly higher than for a typical office building. However, the architectural design qualities are high and these buildings are considered prestige projects.

Conclusion

This project has demonstrated that it is possible to design low-energy speculative buildings in an urban context. The increased costs associated with this superstructure – required to passively reduce the energy demand – are offset by reduced costs for services. The provision of optional cooling, though not a low-energy strategy, was seen as essential for commercial speculative offices of this quality. What is not yet clear is to what extent the cooling option has been adopted. One suspects that tenants unfamiliar with the passive systems and who are used to cooled office space will instruct their interior 'fit-out' designers to install cooling capacity. Thus, in providing the option for cooling, this is likely to result in a significant uptake of the option.

This project provides valuable generic insights for speculative as well as bespoke developments in similar urban contexts. The range of climate conditions also suggests that the concepts are applicable to a wide range of climates, from those predominantly in cold regions of Northern Europe down to mid-European regions such as France. The atrium strategy may prove to be inappropriate to southern climates where the thermal buffer benefits are not useful, and where summer conditions (particularly with warm nights) will require improved passive cooling strategies.

Case study 8: School of Engineering, De Montfort University, Leicester, UK

Table 9.10 *School of Engineering building, University of Leicester*

Name:	School of Engineering
Type:	Offices, workshops, auditoria
Location:	Leicester

Introduction

This university engineering department (designed by Short Ford Architects, in conjunction with Max Fordham Engineers) houses a multitude of functions, from auditoria and offices to computer rooms and engineering workshops. The combination of both noise-sensitive areas (such as the auditoria) and noise-generating spaces (the workshops) presents a range of environmental challenges beyond that of energy efficiency. The building is set in an urban environment and, yet, is totally naturally ventilated and largely day lit. It exploits passive strategies by opting for a convoluted plan form – as opposed to a deep plan – creating a number of microclimates, ranging from exterior courtyards to interior 'streets'.

The building accommodates 1000 occupants in 10,000m^2 over four floors. A significant level of internal gains from occupants and heavy machinery would conventionally lead to year-round cooling. The combination of shallow plans, thermal mass and stack ventilation enabled a passive, low-energy solution.

Location and climate

Leicester is a small city located in the centre of England (52.7 degrees north, 1 degree west). It has a moderate maritime climate but is protected from the ocean to the west. The annual temperature range is small, with a record low of –10° Celsius and a high of 34° Celsius. Rainfall varies little throughout the year, and is fairly low – typically less than 750mm annually. The UK's reputation for cloudiness is well merited, with a mean sunshine duration of 3.7 hours daily for Leicester.

Leicester is set in rolling, hilly farmlands and the city is primarily known historically for its textile industry. The site is located in Leicester city centre, is surrounded by two- to four-storey buildings and is adjacent to a street. With the potential of noisy traffic on the street, adjacent to the need for quiet auditoria, noise is an issue. Conversely, due to neighbouring buildings and the fact that the building accommodates noise-producing engineering facilities, reducing noise emission is also a key concern (Figure 9.82).

Source: Alan Short Associates

Figure 9.82 *View of the De Montfort engineering building in its context*

Building description

The building accommodates the following functions: lecture theatres (auditoria), classrooms, engineering laboratories, offices and a cafeteria. It is up to four storeys tall and is separated from the immediate context by streets and courtyards.

The plan is convoluted, adopting shallow proportions wherever possible in order to increase the availability of daylight and natural ventilation. The high internal gains associated with 1000 occupants and a large amount of heat-producing machinery would, particularly if lighting loads were added, create a continuous cooling demand in a conventional deep-plan engineering facility. Here, the typology of a shallow linear plan is twisted and turned to create courtyards and an internal concourse. Noise-sensitive areas where simple windows that open provide ventilation are orientated away from the street, whereas the noise-producing spaces and the auditoria exploit special soundproofing and ventilation strategies to enable them to face the street.

Façades are carefully articulated to reflect the functions and environmental criteria of the spaces within. A range of openings, from small view and ventilation windows to high-level roof lights, as well as ventilation louvers at the bottom and top of the building, provide the required environmental characteristics (Figures 9.83 and 9.84).

In the deeper plan areas or for the larger spaces (auditoria and engineering labs), the ventilation strategies

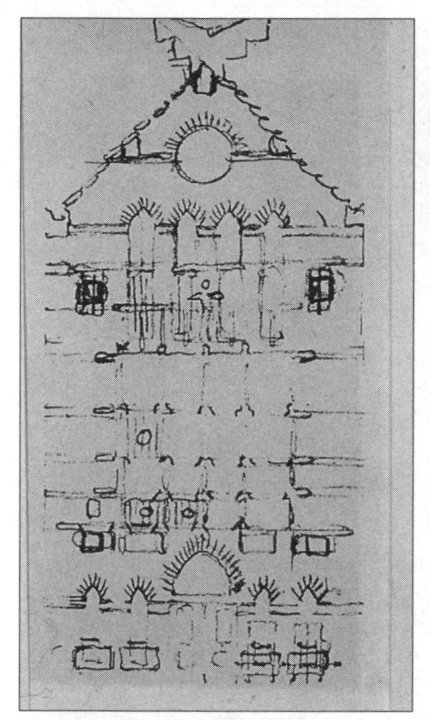

Source: Alan Short Associates

Figure 9.83 *Architect's sketch and final view of the façade*

Source: Alan Short Associates

Figure 9.84 *The roof profile* (left) *indicates the strong role that stack ventilation plays, while the section through the building* (right) *shows teaching spaces on the left with high glazing ratios facing in to a court, the tall concourse space in the centre with a stack, and an auditorium space on the right*

Source: Alan Short Associates

Figure 9.85 *Exterior, showing perforated brick peers for air inlets, and interior, showing high windows for good daylight penetration into the double-height mechanical spaces*

have a more significant implication for the building design. The use of ventilation stacks is evident from the overall view of the building, and these form a primary approach to ventilating key spaces naturally.

Key features

Generally, the main section of the building is ventilated from the perimeter – through windows or via acoustically lined louvered plenum spaces – to the centre and air is expelled via a series of stacks. The engineering laboratories are ventilated separately, but follow the same principles. Openings in the brick façades allow air to enter via an acoustically lined chamber (in this instance, to prevent noise from escaping from the interior to the exterior) into the space and it is evacuated via roof-mounted monitors. These spaces are single storey but double height (see Figure 9.85).

The heavy form of construction – primarily brick and concrete – provides ample thermal mass, which (when combined with night-time cooling controlled by the building management system) ensures that overheating is minimized.

The ability of the stack ventilation systems, combined with thermal mass and night-time cooling, to prevent overheating in an auditorium for 150 people is a particular achievement. One key conflict to overcome was the compromise between exposing sufficient thermal mass while having enough acoustic absorption to provide the appropriate reverberation characteristics. Careful calculations of both the thermal and acoustic performance were needed to reach a successful balance.

The daylighting strategies vary according to function and orientation. In the computer spaces, small view windows are used at low level, and light shelves with increased glazing above aim to provide good daylight distribution while limiting sunlight and glare. The mechanical laboratories (see Figure 9.85) have high-level glazed gables and roof lights. The deep overhangs prevent direct sunlight penetration.

Generally, artificial lighting is automated, except in the smaller office spaces, and a centralized control switches all lights off at 22:00. Light and movement sensors are provided and manual override is also available.

Performance

The building is still undergoing post-occupancy analysis. However, estimates show that energy savings will be in the order of 50 to 75 per cent of a conventional engineering building with mechanical cooling. Capital cost reductions of 35 to 24 per cent of the building costs have been achieved through reducing electrical and mechanical services.

Conclusion

This project demonstrates that a complex brief in an urban setting can be resolved to produce a low-energy, naturally ventilated building. Careful planning – particularly with respect to natural ventilation and noise – is essential, and a large surface area-to-volume ratio used here offers a series of conditions and microclimates that can be exploited. The use of stacks, even for ventilating auditoria and other spaces with high internal gains, has been shown to be effective. However, careful modelling and simulation of ventilation, acoustics and lighting were an essential part of the design process, and were necessary to balance and fine-tune the design strategies (not to mention fire safety).

The overall strategy is, in principle, applicable to other climates, although the effectiveness of night-time cooling and stacks will need to be assessed. Where diurnal fluctuations are smaller than in the UK, this strategy will not be able to cope with the level of internal gains of this building.

Case study 9: Inland Revenue Office Headquarters, Nottingham, UK

Table 9.11 *Inland Revenue headquarters, Nottingham*

Name:	Inland Revenue Office Headquarters
Type:	Offices
Location:	Nottingham

Introduction

The six Inland Revenue buildings are designed to accommodate the UK's tax office in 40,000 square metres and on three to four floors, set on a 100,000 square metre site in the centre of Nottingham. The challenge was to design a low-energy, naturally ventilated and day-lit building in this urban context. The strategy was to adopt shallow plans for cross-ventilation and daylight availability, supplemented by stack ventilation and exposed thermal mass. The large site area enabled the urban design approach to create a quieter and cleaner microclimate for many of the buildings' façades, limiting the issues of noise and pollution as constraints on the natural ventilation approach. The buildings were designed by Michael Hopkins Architects, in conjunction with Ove Arup Engineers.

Location and climate

Nottingham is in the same climate region as Case Study 8 in Leicester, so the synoptic climate data are the same. The local conditions differ in that the site is next to two major roads, one of which is raised and runs along the short length of the site: a railway line to the south and a canal to the north. The site is also large enough for a small urban enclave to be designed and thus to create its own microclimate of semi-pedestrianized streets and courtyards. The urban plan was conceived around maintaining views to the city centre and the castle.

Building description

A series of shallow plan (13.6m) court and L-shaped plans of three and four storeys are planned around new landscaped streets and squares. All wings and façades of the buildings are essentially the same, independent of orientations. The structure is heavyweight, with exposed pre-cast concrete ceilings to the lower levels (all but the top floor). The top-floor level is made of a lightweight form with a steel roof structure and roof lights for ventilation purposes. The upper floor is thus independently ventilated and does not rely upon the towers used for natural ventilation on the lower floors.

At the corners of each block are ventilation towers, which provide the stack effect of drawing air through the office space. These towers also accommodate the fire escape stairs.

Windows can be opened – in the form of large sliding glazed doors – to allow for natural ventilation, although alternative routes for fresh air supply have been

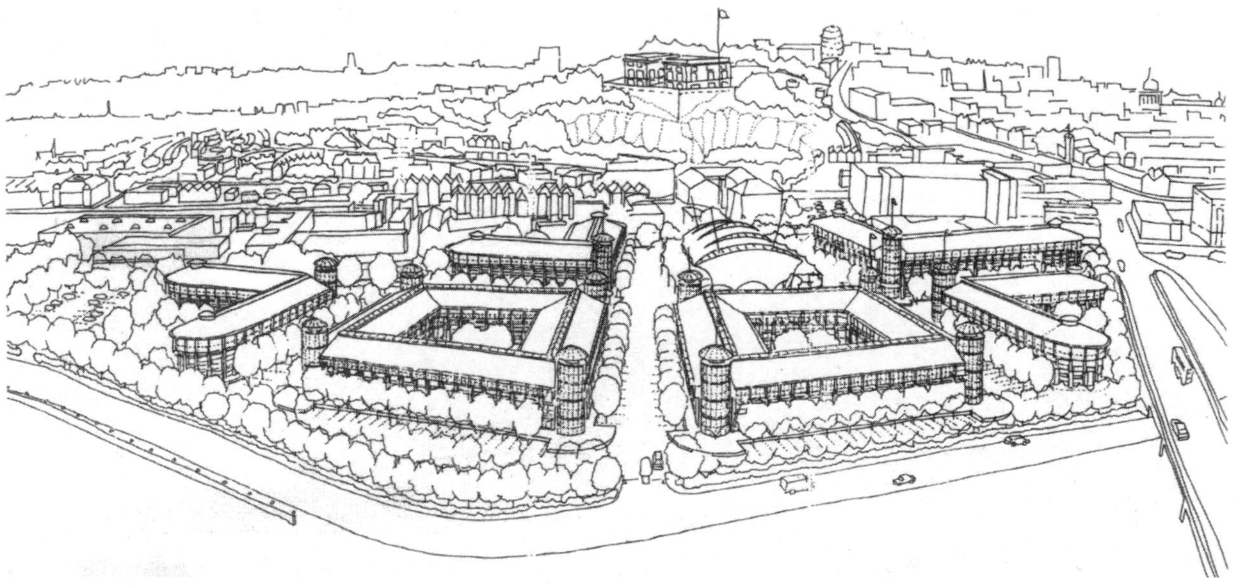

Source: DETR, 2000

Figure 9.86 *Overview of the site and views to the castle*

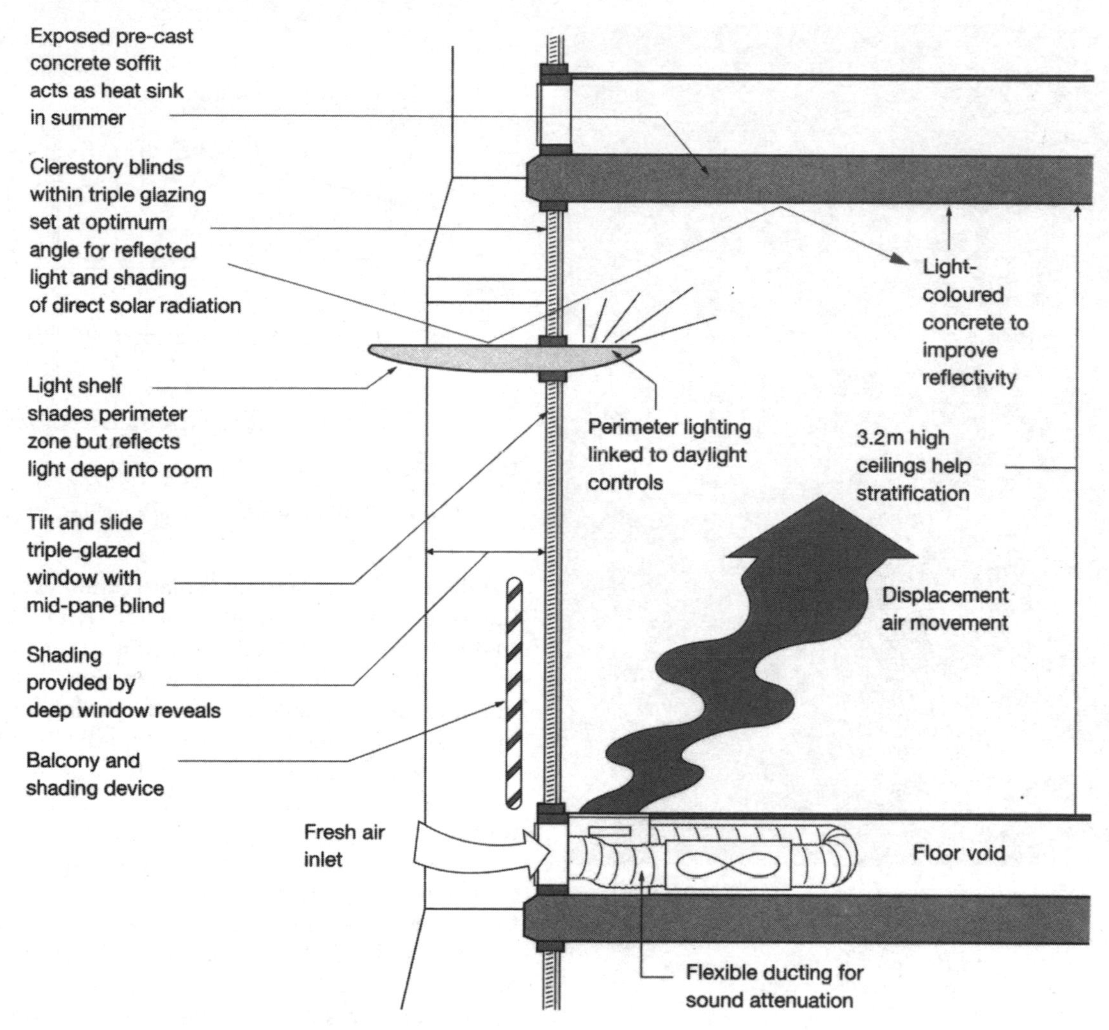

Source: DETR, 2000

Figure 9.87 *Perimeter environmental strategies*

integrated. All glazing is triple glazed, with an inner-sealed argon-filled unit and an outer cavity with adjustable blinds. Although there is a fairly large amount of glazing, the tall windows are shaded by a combination of fixed louvered balustrades and light shelves externally, as well as moveable inter-pane blinds to provide user control. High-level glazing above the external light shelves aims to improve daylight penetration deeper into the plan (Figure 9.87). Offices are generally open plan, though some cellular space is provided around the core spaces at each corner.

The environmental systems could be described as fan-assisted natural ventilation in that small air-supply fans are provided near the perimeter to provide minimal winter-time fresh air. Perimeter heating is also installed to counteract any heat loss in winter and to provide early morning preheating, when necessary.

Key features

The key energy features of this design are the integration of natural ventilation, thermal mass, shading and daylighting.

It is the ventilation strategy that is the most interesting and worth discussing in more detail. A combination of natural and mechanical ventilation is used, allowing different strategies for different times of the year. Natural ventilation is achieved by manually opening the glazing (either sliding doors on the lower floors or windows on the top floor) during warm weather. Alternatively, underfloor perimeter inlets can be opened to provide fresh air

Source: DETR, 2000

Figure 9.88 *The ventilation towers are a key architectural feature*

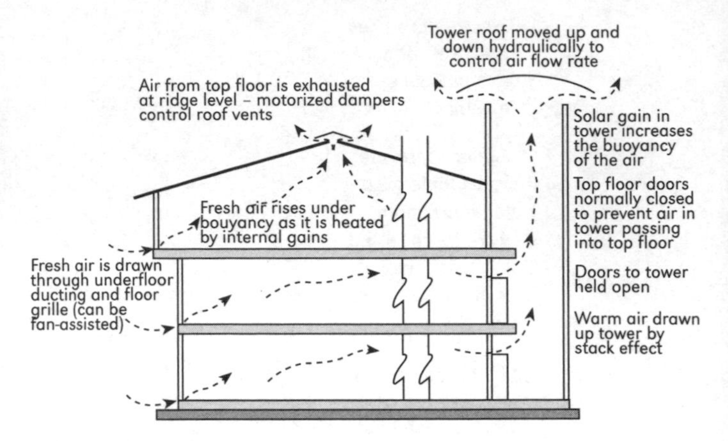

Source: DETR, 2000

Figure 9.89 *Overall natural ventilation strategy*

naturally, avoiding the potential problems associated with noise. The air entering via the perimeter inlets can also be mechanically forced in, using under-floor fans. This provides a degree of control to assist night-time ventilation in the summer and preheating in the winter (the under-floor systems include perimeter heating). Figure 9.87 shows the perimeter design principles.

The exhaust air can leave the building either under conventional cross-ventilation (i.e. the air enters one side and leaves via the opposite side of the plan) or by exploiting the stack effect created by the towers (Figure 9.88). However, cross-ventilation is only effective if windows or inlets are opened and there is sufficient pressure difference between the windward and leeward side (i.e. there is sufficient wind). A fall-back option, for when the wind pressures are low, is the use of the towers. Doors into the towers from the lower floors are permanently open (except in the case of a fire). These towers are constructed of glass blocks in order to exploit solar gains to assist in the thermal buoyancy (i.e. 'solar chimneys'). The tops of the towers can be gradually opened by raising the fabric roof upwards by 1m and providing a large area that can be opened to allow rising warm air to escape. The umbrella-like 'top hat' tower roof openings are controlled by the building management system (BMS). Only the top floor is ventilated independently from the towers via motorized roof vents (Figure 9.89). The same ventilation routes (i.e. from perimeter inlet, via office space, and out of towers or roof vents) are used for night-time cooling, controlled by the BMS. The exposed concrete ceilings in the offices are an important feature of the passive cooling strategy, and the shading and daylight strategies help to minimize thermal gains. The benefits of noise reduction achieved as a result of the perimeter system means that noise is not a significant constraint to natural ventilation.

Performance

One of the building's wings was monitored during 1996 and 1997. The summer temperature assessment showed that the top floor was always the warmest (24.3° Celsius average) compared with the ground floor (22.5° Celsius).

During the summer of 1996, a total of 25 working days had temperatures exceeding 27° Celsius on the top floor, compared with 13 hours on the second floor, 10 on the first and none on the ground floor. On average, the building exceeded 25° Celsius for 2.9 per cent of the occupied year and 28° Celsius for 0.3 per cent of the year – levels that are considered acceptable in the context of a criterion of no more than 25° Celsius for no more than 5 per cent of the year (Figure 9.90).

Air movement was lower than expected, particularly in the middle of the wings. Typical summer ventilation

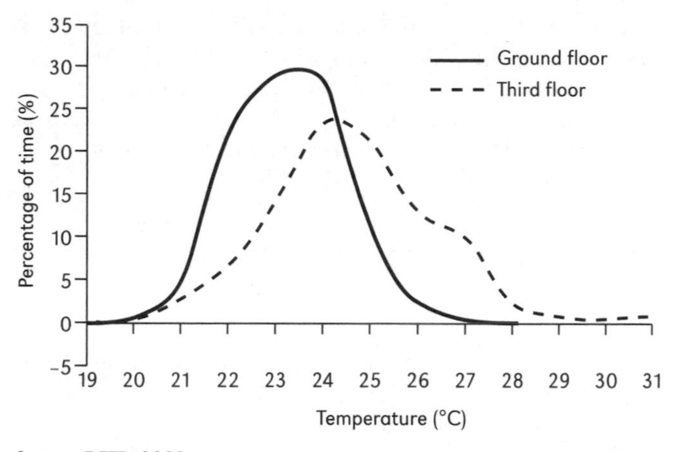

Source: DETR, 2000

Figure 9.90 *Percentage of internal temperatures recorded*

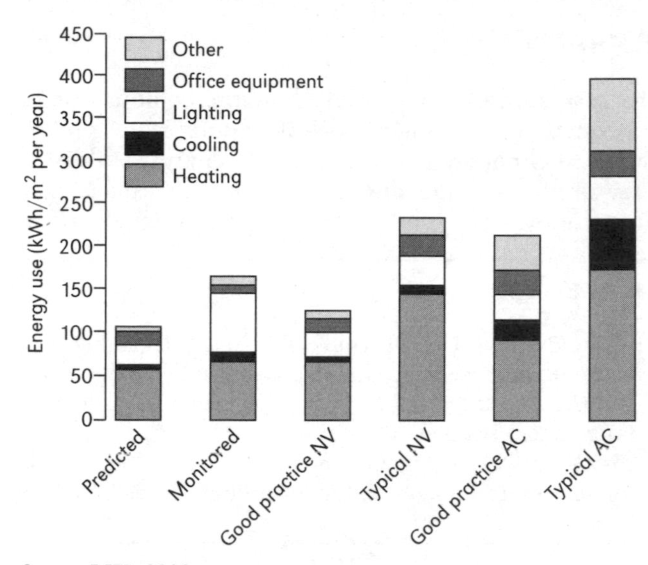

Source: DETR, 2000

Figure 9.91 *Comparisons of energy use*

rates of below 2 air changes per hour (ACH) were measured. Night cooling was found to be effective, achieving up to a 2° Celsius reduction in day-time temperatures.

The daylight performance is disappointing, largely due to the extensive shading strategies employed (the moveable mid-pane louvers are not retractable), and also due to the relatively dark concrete ceiling finish (original 'white concrete' was eliminated due to costs). However, the top floor with its roof light achieves average daylight factors of above 2 per cent, compared with below 1 per cent for the other floors.

The predicted annual total energy use figure was 94kWh/m^2 (excluding small power), consisting of 26 for lighting, 6 for pumps and fans and 57 for heating and hot water, and the rest for ancillary uses. The actual energy performance during the first year of operation was 157kWh/m^2 – about twice the predicted demand. Overall, the building uses almost 20 per cent less heating energy than an air-conditioned building, but between two and three times the lighting energy (Figure 9.91).

Occupant surveys reported reasonable general satisfaction, though high temperatures on the top floor and 'stuffiness', possibly due to limited ventilation rates, were mentioned. Occupants did not seem to be aware of the perimeter mechanical ventilation option – only one in five fans was in operation at any one time. Daylight was considered to be low, as reported in the measured surveys.

Conclusion

This project provides a number of valuable lessons. The integration of energy strategies has very significant theoretical potential, but complexities and compromises can occur. For example, excessive fixed and non-retractable shading resulted in disappointing daylight availability; the exposed thermally massive concrete ceiling has low reflectance, giving a darker visual environment and making light shelves less effective; occupant behaviour (of, for example, the ventilation controls) reduces the thermal and energy performance; and the stacks are too far apart to effectively ventilate the whole of the floor plan.

Many of the problems can be overcome in this building by remedial measures. They might include improved BMS control over the perimeter fans (rather than reliance on occupants); occupant and light sensors for the lighting controls; and zoned heating control as a function of floor-level and façade orientation. The decision to have the same façade design strategy on all façades has reduced the potential of fine-tuning the shading strategy. For example, little shading is required on the north side, which would benefit from increased daylight without a significant thermal gain penalty. Perhaps a more controllable opening window design – as opposed to large sliding doors – would have offered more manual control over the natural ventilation and limit draughts.

Although the design targets have not all been met, it is clear that this building provides an energy performance that is better than that of a standard solution, and that with remedial interventions, an overall improved performance is expected.

Synopsis

The case studies demonstrate how the combination of integrated design strategy with the appropriate application of technology can achieve low-energy buildings. They also usefully demonstrate – though not exhaustively – a rich range of concepts that have been developed in response to the specific contexts and briefs, that have not been applied in a deterministic manner. The range of architectural solutions highlights that low-energy urban design carries with it no stylistic or architectural constraint – in fact, it can be argued that the solution space is increased.

References

DETR (2002) 'The Inland Revenue Headquarters, Nottingham: Feedback for designers and clients', *New Practice Case Study 114*, Energy Efficiency Best Practice Programme, London

Geros, V. (1999) 'Ventilation nocturne: Contribution a la reponse thermique des batiments', PhD Thesis, INSA de Lyon, France

Klein, S. A. (1990) 'TRNSYS: A transient system simulation program', Solar Energy Laboratory, Report no 38-13, University of Wisconsin, Madison

Rohles, F. H. (1983) 'Ceiling fans as extensions of the summer comfort envelope', ASHRAE Transactions, vol 89, pp245

Recommended reading

A number of case study-based books exist that deal with energy efficient projects.

1 Jones, L. (1998) *Sustainable Architecture*, Laurence King, London
This book contains a rich variety of 'environmentally sensitive' case study projects, with excellent photographs, although not a significant amount of technical data. As a design source book, it is an attractive document.

2 Hawkes, D. and Forster, W. (2002) *Architecture, Engineering and Environment*, Laurence King, London
This book features case studies, and provides a useful critique of buildings that are of significant architectural merit. This book of case studies emphasizes the link between the architectural and engineering solutions to environmental design.

3 Fontoynont, M. (ed) (1999) *Daylight Performance of Buildings*, James & James, London
Finally, Marc Fontoynont has edited an excellent book on 'daylit buildings', which describes over 60 case studies from the perspective of their lighting performance. The buildings cover a wide range of types, periods and climates, and were monitored. Each is described in terms of its qualities, and the book gives details of the monitoring results, together with photographs. It complements the recently published *Daylight Design in Buildings* by Baker and Steemers (2002, James & James), which is a source book and design guide for daylighting.

10

Guidelines to Integrate Energy Conservation

Marc Blake and Spyros Amourgis

Scope of the chapter

This chapter presents a summary of key issues regarding environmental conditions and energy conservation of urban buildings. While each issue is dealt with in greater depth in other chapters, the merits of a comprehensive approach and utilization of all options available to the designer are discussed here as guidelines. As conditions vary from site to site and from one building function to another, it is up to the ingenuity of the designer to address all issues and factors and to resolve then through the building design.

Learning objectives

After studying this chapter, readers will understand how to address all of the factors affecting the environmental conditions of urban buildings, generally; and specifically through the building design process, readers will be able to search for those solutions that will conserve non-renewable energy resources.

Key words

Key words include:

- urban climate;
- heat and mass transfer in buildings;
- illumination of buildings;
- urban energy resources management;
- economic feasibility of buildings;
- intelligent controls and advanced building management systems (BMS);
- building design guidelines.

Introduction

Considerations for energy conservation include the energy 'embodied' by the building materials in their manufacturing process (see Chapter 4), the energy consumed during the construction process, and energy demands for the operation and maintenance of a building during its life cycle. Finally, energy is also consumed in the demolition of a building at the end of its useful life.

This chapter deals with the conservation of energy during the functional life of a building. The energy performance of a building depends, largely, upon the design decisions made at the planning stage and in the way in which the occupants interact with the building. The significance of making the right decisions in order to integrate the building concept with environmental and energy conservation issues is discussed in Chapter 2. Following these introductory comments, a number of guidelines are provided that highlight the various options available to integrate energy conservation measures during the initial stages of the building design.

There is no single method or means that could be applied generally to produce maximum energy savings in all buildings, as there are many variable factors that affect *the decision-making stage* of each building design problem. The best approach is to consider all of the possibilities and options to conserve energy in each building design situation and to study their synergy in producing cumulative results. Therefore, the issue is to integrate in the building design as many options as are feasible.

With regard to the issue of integration, it is important to distinguish between the *utilization and assimilation of a number of measures in formulating a building design concept* in order to decrease the consumption of energy during the life span of a building, and the 'piecemeal' approach of resolving the various issues separately before *applying elements in a building design* that improve one

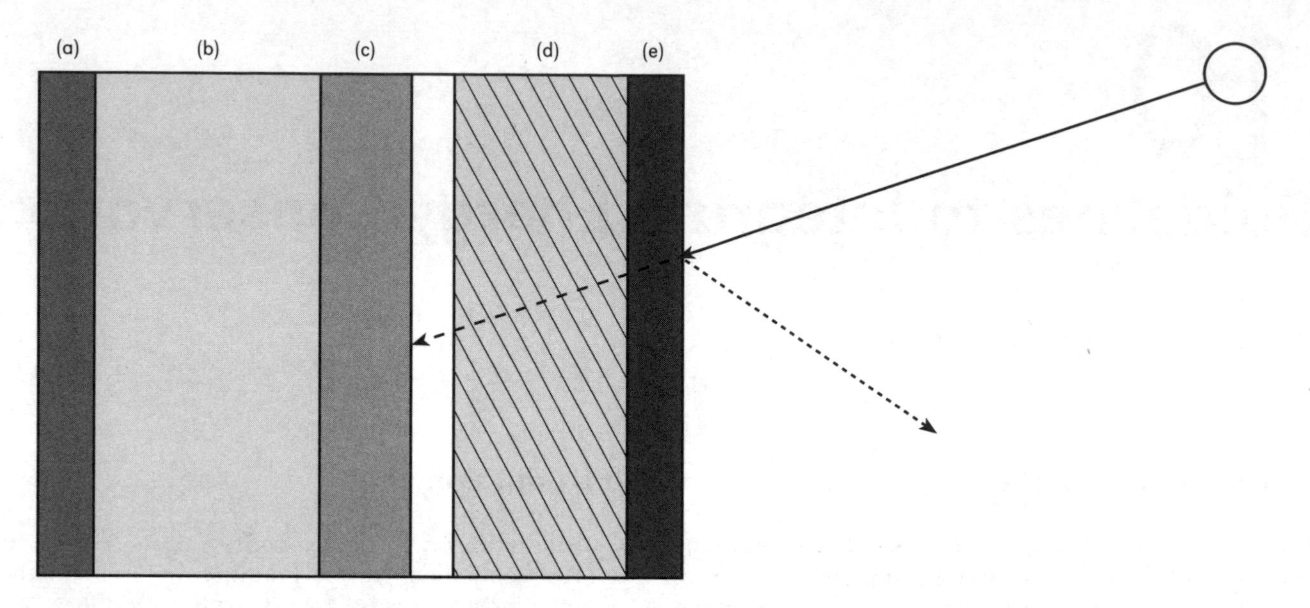

Note: The insulation (c) protects heat transfer through conduction to the mass of the interior wall (b) from heat gains by solar radiation to the external wall (d), while the high albedo white-washed plaster (e) partially mitigates radiation effects to the external wall (d) through reflectivity.

Figure 10.1 *Wall section: (a) shows interior plaster; (b) shows internal cavity wall skin (i.e. bricks); (c) shows thermal insulation material in the cavity; (d) shows external cavity wall skin (i.e. bricks); (e) shows white-washed plaster (high albedo)*

aspect only of the energy performance of a building. The first approach is more difficult and creates a real challenge for the building designer to be inventive. The second approach is easier since a building is designed conventionally and, at the end of the process, a number of environmentally corrective measures are added to the initial design. Such examples are universally common.

Environmental criteria and energy conservation are equally important to all other design criteria and factors that help to generate a building design. For this reason the guidelines in this text must not be seen as keys to solving only particular building functional problems, but, in their totality, as generating a building concept and function, as a whole.

The environmental design of a building is a rather complex task, as it must respond effectively to often conflicting environmental requirements. For example, the sun provides natural light to interior spaces while it creates glare. At the same time, it is desirable for the sun's rays to enter the interior of a building during the cold months, but undesirable for them to do so in the warm months. To address only one condition is not sufficient.

Finally, in dealing with urban buildings one faces a considerable number of limitations. These result from the great diversity of environmental conditions within the urban context, such as differing ground morphology, low to high building densities, fixed orientations of streets and private properties, the size and shape of building sites, as well as a variety of building heights and volumes. This diversity presents opportunities as well as constraints to the range of available options, from one urban site to another, for energy conservation and particularly for passive solar systems.

General issues

There are a number of issues that are important to understand and that make up the context to specific building design decisions. These are analysed in greater depth in various chapters of this book. In this chapter, the principles and findings of other disciplines that feed into the architectural design process are listed briefly. The designer's expertise is to integrate the findings of the specialist as guidelines in building design, and in complex situations to consult with the necessary experts in order to resolve the specific issues.

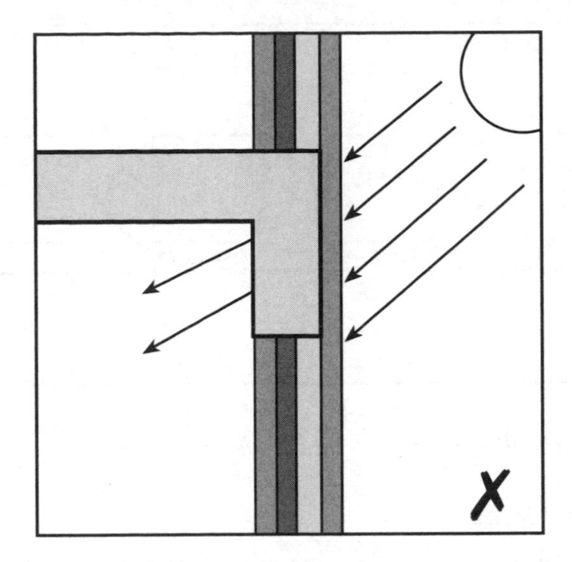

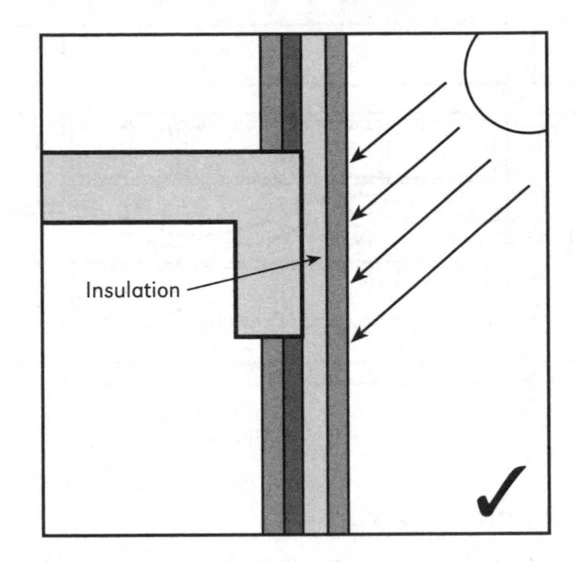

Note: (*left*) Incorrect as heat passes unabated into concrete beam and slab; (*right*) correct as heat is stopped by insulating material between the building skin and the concrete beam

Figure 10.2 *Two building section details which show how thermal bridges can occur if proper insulation is not provided at tectonic elements*

The urban climate

While land uses in rural areas are more even in texture and extent, in the urban environment they are so varied in their texture and density that microclimatic conditions are much more difficult to assess and predict. Therefore, it is important to emphasize a number of points. Some of these are useful regarding the benefits of a building and some relate to how a building design affects the microclimate of its surroundings.

In urban areas, as described in Chapter 6, the main differences between the urban environment and the surrounding rural areas are as follows:

- Air temperatures in cities are higher compared with the adjacent countryside (this difference is known as the 'urban heat island' effect).
- The speed of the wind in urban areas is usually less than the speed of the wind at the same height in adjacent rural areas.
- Urban pollution can reduce sunlight intensity in industrial areas.
- Direct sunlight and daylighting intensity varies from location to location as large buildings cast shadows over others (this condition also varies according to the orientation and siting of the buildings).

The airflow between buildings in the urban canopy is very difficult to assess. It is particularly difficult to assess the microclimatic prevailing conditions in areas with an uneven density of buildings unless one conducts extensive readings and tests. Other factors that make it difficult to assess and estimate air movements are the constant changes that take place in the urban environment, with demolitions of old buildings and the construction of new structures with a different volume and mass, as well as changes in plant materials.

Generally, improvement of the overall urban environment during the hot months can be achieved primarily by reducing solar heat gains as follows:

- using materials with a high albedo in buildings and where hard surfaces are necessary;
- planting trees, shrubs and ground cover;
- spacing buildings to allow for air movement.

Details regarding the albedo and emissivity of different materials are provided in Chapter 6.

Appropriate spacing of buildings and the use of deciduous trees may improve, during the cold months, solar heat gains and, of course, natural lighting conditions.

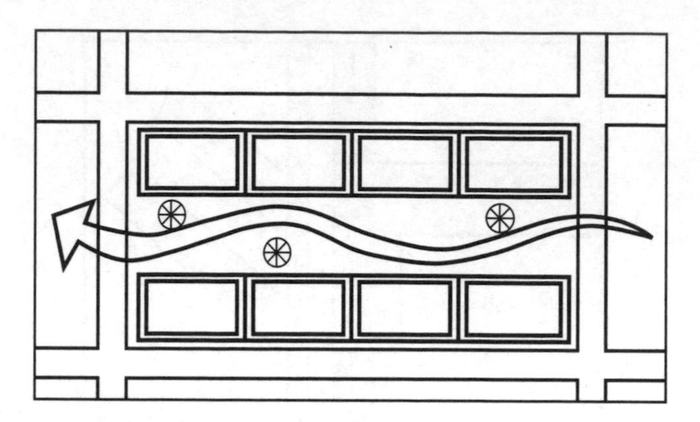

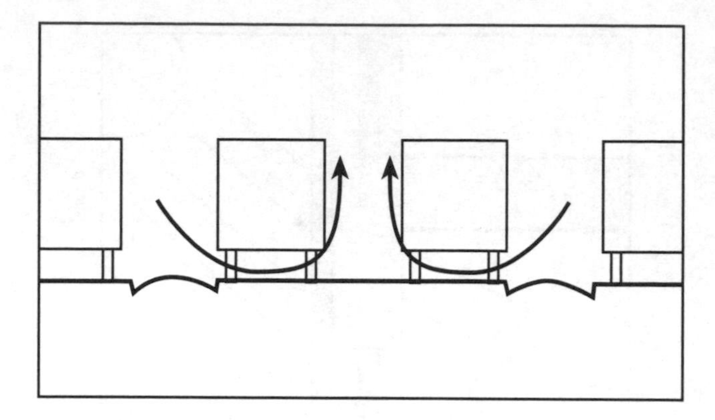

Figure 10.3 *Continuous building system layout of city block*

Heat and mass transfer

The principles to remember in designing a building are that heat is transferred (see Chapter 7) in three modes by:

- conduction;
- convection;
- radiation.

Of those modes, the first has to do with the heat transferred through the mass of a building: primarily the walls and roofs, and in some situations the floors. The second mode occurs in a building as a result of airflow and the third is emitted by matter and transported by electromagnetic waves, the sun being a major source of radiant heat. Figure 7.13 in Chapter 7 illustrates the heat transfer balance of a building.

Thermal insulation

The primary method of protecting a building from heat losses and gains is to insulate the building's various external elements. Patented thermal insulation materials provide protection primarily from conduction through the mass of the building envelope.

Heat gains

In hot weather the building envelope gains heat from solar radiation and thermal convection, when the external air temperatures are higher than the interior of the building. In order to avoid this, thermal insulation is used to protect from heat transfer. However, it is more efficient to consider also the external finish materials, colour and texture that provide protection from solar radiant heat. This can be achieved by increasing solar reflectivity by using high albedo materials (see Table 6.3 in Chapter 6).

Heat losses

In cold weather the interior mass of the wall (b) stores heat gained from convection and/or radiation from internal sources, and insulation (c) stops heat transfer to the exterior skin (d) by convection (see Figure 10.1).

However, if a building is occupied occasionally, it takes longer to heat up until the interior skin (b) absorbs enough heat to sustain internal temperatures at a constant level.

Thermal bridges

Thermal bridges occur when materials of uneven conductivity are placed next to each other or structural elements such as columns, beams or slabs interrupt thermal insulation layers. In these cases, provisions have to be made to reduce heat gains or losses where thermal bridges occur.

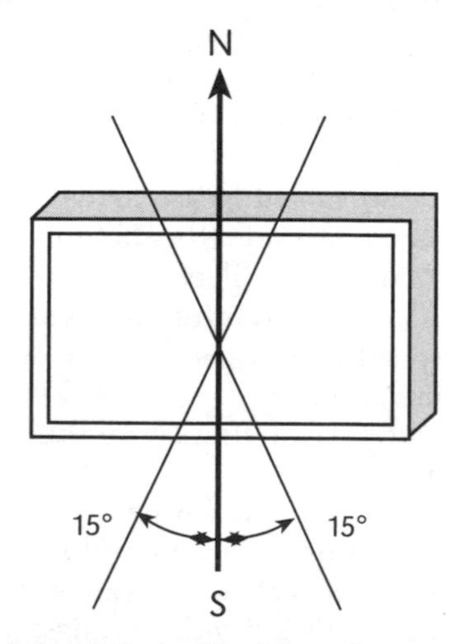

Figure 10.4 *Roof plan of building showing favourable orientation of the major façades to within 15° of the north–south coordinates*

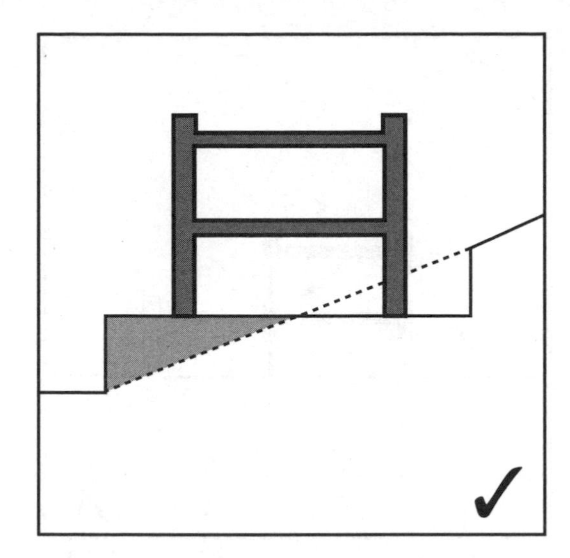

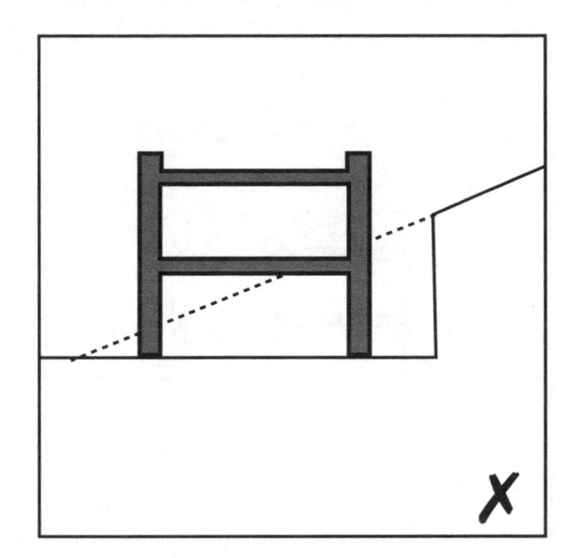

Figure 10.5 (left) *Correct building section showing how equal amounts of cut and fill are created during excavation;* (right) *incorrect building section showing how too much soil is exported from the site during excavation*

Illumination

Natural light is a free source during daytime. Artificial light is necessary during darkness. Over the last 40 or more years, artificial lighting sources have become a 'feature' during daytime, quite unnecessarily, in the interior of buildings!

Natural lighting

Illumination of the interior of buildings during daytime can be provided through natural lighting. In urban buildings orientation is given by the location of the site within the street pattern of the urban fabric. In disadvantageous situations, natural light can be improved by utilizing reflective light from surfaces that receive direct light. Visual comfort requirements are analysed in Chapter 8, as well as the components of daylight factor (see Figure 8.16) in the urban environment. Where possible, in the design of any urban building it is important to observe the obstruction angle between buildings, according to the latitude of the location (see Table 8.7 in Chapter 8). When designing new areas, care must be taken to space buildings at distances that allow exposure to the sun in winter months.

Artificial illumination

Artificial lights are used for after daylight hours. Considerable energy can be saved when using compact fluorescent lamps. Apparently, 20 per cent of all electricity used in the US is directly consumed for lighting. Actually, energy consumption is higher when the energy used to remove the heat generated from the lights is estimated (Von Weizsäcker et al, 1997, p36).

Generally, energy can be conserved by using timers, as well as sensors, for switches of temporary use spaces (such as toilets), photovoltaics for the exterior of buildings and, as mentioned before, compact fluorescent lamps.

Energy and resource management in the urban environment

Infrastructure

Where urbanization has taken place the opportunities to explore all available options are limited. In most of the urban centres except those that expand on new grounds, there is scope for corrective actions to improve infrastructure and to address issues such as recycling of waste; tapping into secondary sources (i.e. wasted energy from thermodynamic electricity plants); converting to less polluting energy sources, such as natural gas; improving on public transport (Von Weizsäcker et al, 1997, plate 11);[1] planting to improve the 'the heat island' effect; using photovoltaics for ancillary public lighting; water management; solar energy for domestic hot water; and other similar measures, mostly addressing general infrastructural improvements. Chapter 12 outlines a number of opportunities and examples that have been tried in several parts of Europe.

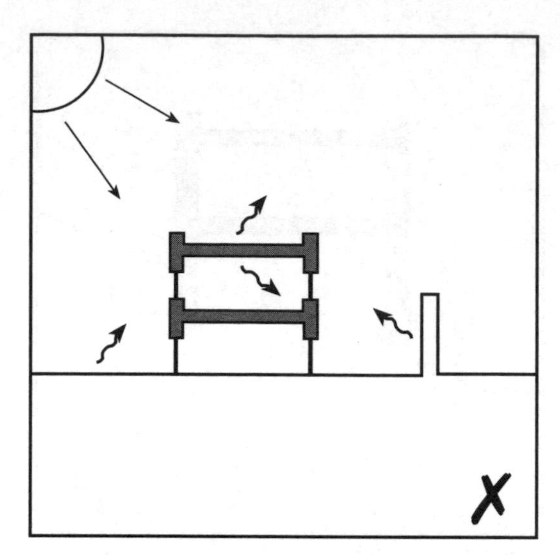

Figure 10.6 (left) *Correct building section showing how building surface and ground are placed to assist in the absorbtion of unfavourable heat gains;* (right) *incorrect building section showing how surfaces and ground absorb and radiate heat*

Building projects

Generally, most construction projects in existing cities are single buildings. Of those buildings in moderate and higher density areas, there is scope to address energy conservation issues through design as described in Chapters 2, 3 and 4. Each building project can contribute considerably to this effect through careful consideration of, primarily, energy-efficient interiors, as well as the management of resources, the remodelling of existing buildings and adapting them to contemporary needs.

Economic feasibility

Energy conservation can be measured in quantitative terms through cost and financial benefits. However, qualitative benefits are difficult to assess and some are not so apparent. The 'ozone hole' and the 'global greenhouse' effects were not clearly appreciated 50 years ago; the cost not to reduce but to halt the process now is, indeed, very high, if globally feasible. There is no doubt that to convince people to adopt different practices is to provide evidence of economic benefits for the medium term, if not

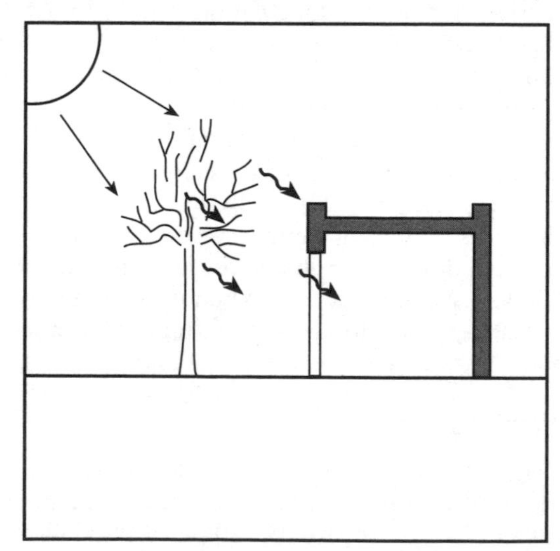

Figure 10.7 *Building sections showing how a deciduous tree* (left) *can shade a building in summer, and in winter* (right) *lose its leaves to allow the sun to pass through and warm the building*

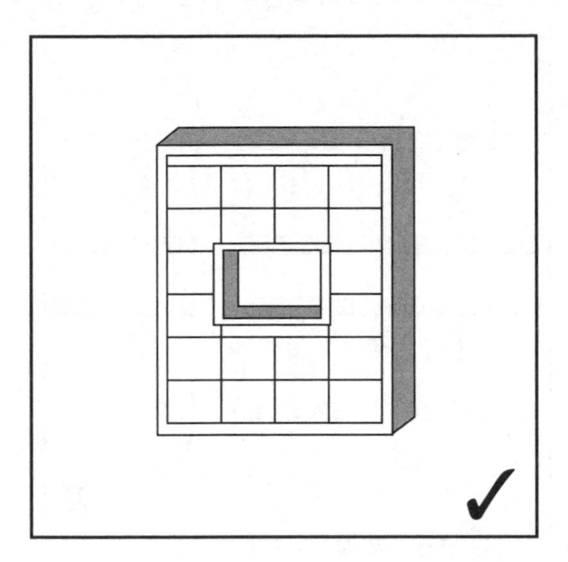

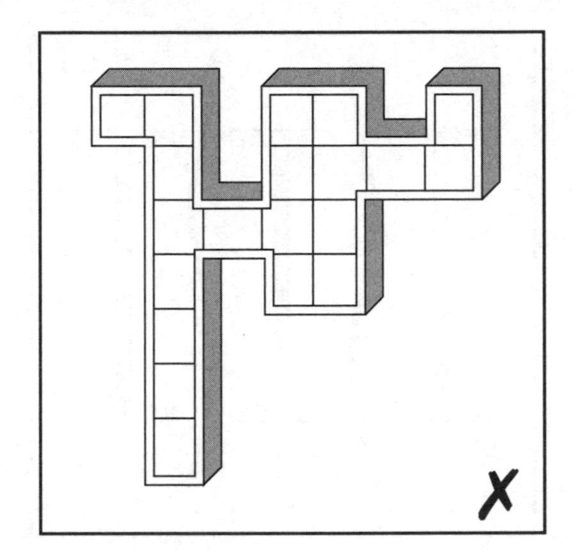

Figure 10.8 (left) *Correct building floor plan showing a low surface to volume ratio;* (right) *incorrect building floor plan showing a high surface to volume ratio*

for longer term periods. Therefore, economic analysis is also a very important tool for assessing a building project. Feasibility studies can help to avoid adopting excessive building standards as, indeed, can recycling existing buildings. Passive cooling and heating systems can improve indoor quality, as well, by eliminating the airborne diseases that sometimes are associated with mechanical systems. Furthermore, materials from renewable resources can be used to replace synthetic materials to create a healthier, non-allergenic interior environment.

Intelligent controls and advanced building management systems (BMS)

In urban buildings, simple building management systems have been used traditionally to control heating, air conditioning and lighting. Building management systems could be used to totally control a building environment. Regarding the latter, the aim is to achieve maximum efficiency and optimum environmental comfort conditions for humans (as described in Chapter 5).

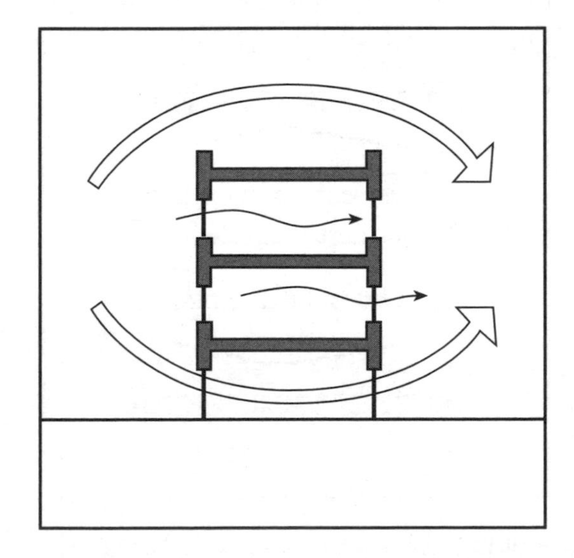

Figure 10.9 *Building section showing open pilotis allowing free airflow below the building*

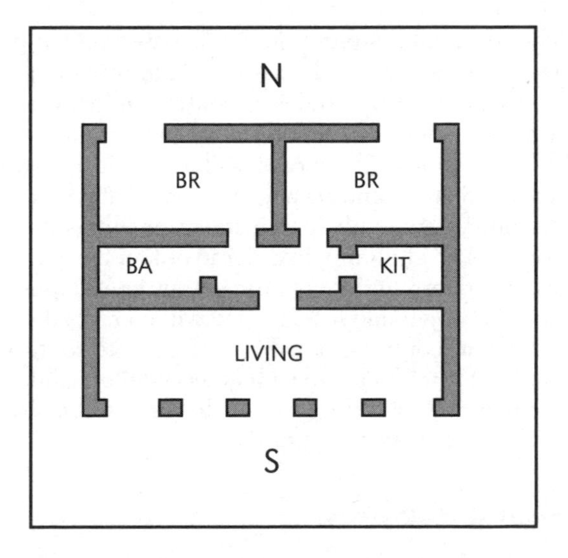

Figure 10.10 *Floorplan showing the ideal location of rooms in relation to north and south directions*

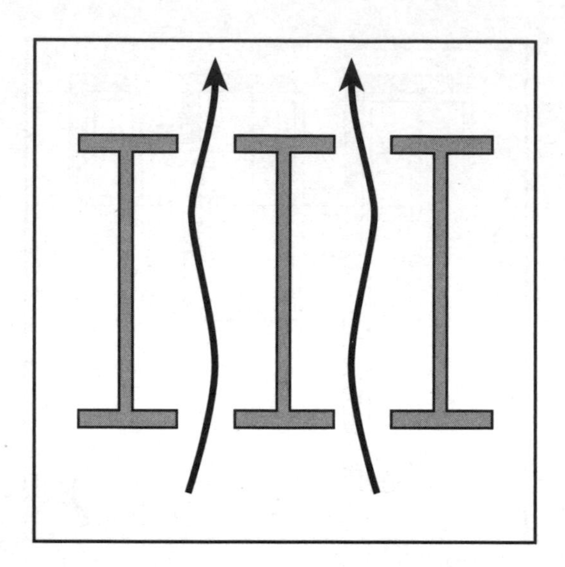

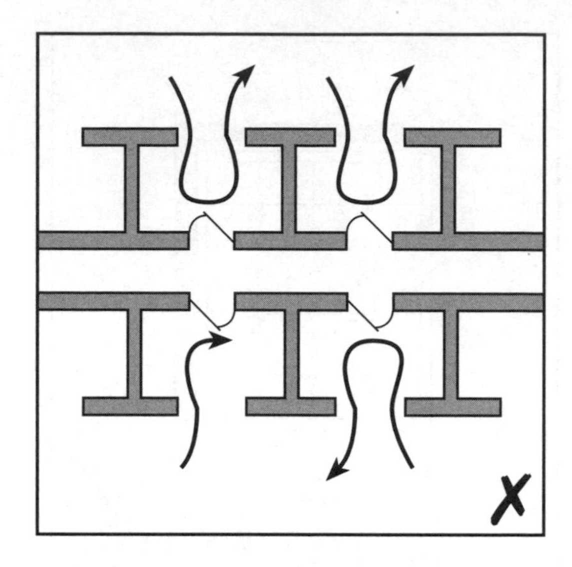

Figure 10.11 (left) *Correct building floor plan showing free flow of air currents through the space;* (right) *incorrect floor plan showing interrupted airflow due to the introduction of a corridor*

There are three basic categories of control systems:

- those that provide a dual-position regulating action (on/off);
- those that provide proportional control action;
- artificial intelligence programmes that respond simultaneously and to variable conditions.

The range of activities that can be controlled and monitored in a building is shown in Figures 5.2 to 5.4 in Chapter 5.

The term 'intelligent building' is based on the possibilities offered by advances in information and communication technologies to control and monitor the building environment, as well as a range of many other building functions. The most feasible use at the present times is to create a central control system that regulates an integrated system of all such installations that affect the environment of a building in order to obtain the maximum results of comfort, with minimum energy consumption.

Social changes and urban living will increase the need for intelligent controls; possibly with remote controls. It may, in the future, become widely feasible to monitor the environment of a building from a distance and to reduce even more energy consumption.

Design guidelines

There are two loosely defined categories of energy-saving approaches in a building (Chrisomallidou, 2001, p247). The first includes simple methods of conserving energy through the use of:

- thermal insulation of the external skin of the building;
- construction details that seal air leaks from the interior of the building;
- interventions in the mechanical systems of a building.

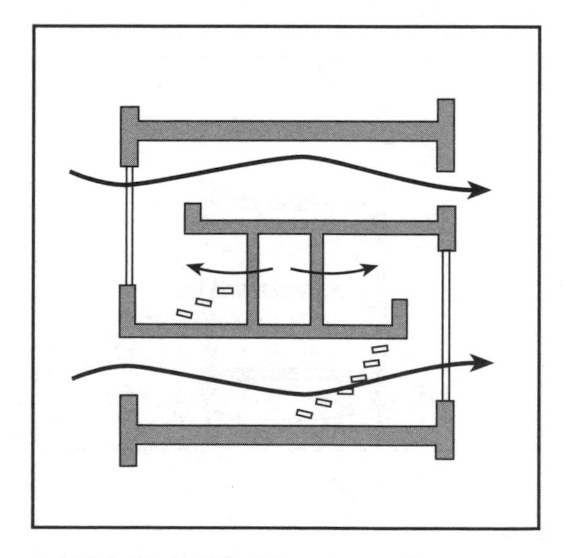

Figure 10.12 *Section of a building showing a compromise solution; two apartment units with a corridor between them, allowing for free flow of air in both units*

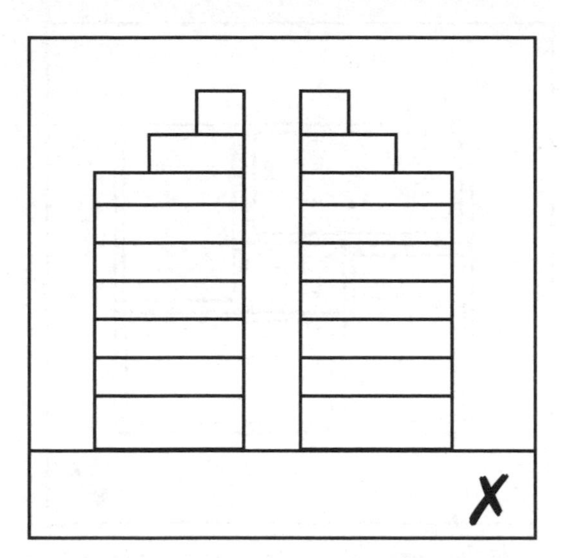

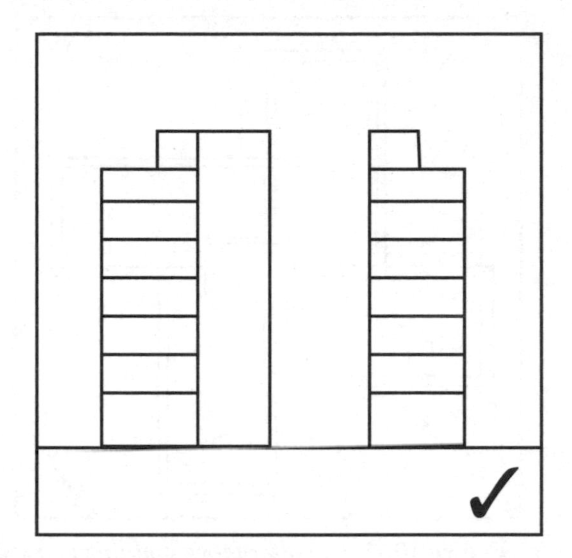

Figure 10.13 (left) *Incorrect building section showing narrow space between two building blocks;* (right) *correct building section showing wide space between two buildings, allowing better flow of air and use of natural daylight*

The second category includes approaches, such as:

- utilizing passive solar heating;
- natural cooling;
- natural ventilation.

Efficient thermal insulation of the external elements of a building may be seen as a prerequisite to applying any other additional techniques of the second category. Other than availability of materials and cost, there is no other basic limiting factor to the thermal insulation of a building. This is not true for the other approaches, where, as mentioned before, it is difficult to find in all sites or locations in the urban environment the essential environmental conditions in order to apply all available options.

The following guidelines highlight a number of key points. They must not be seen in isolation to each other by the designer, but should be combined and adjusted to each other, leading to optimum design solutions for a

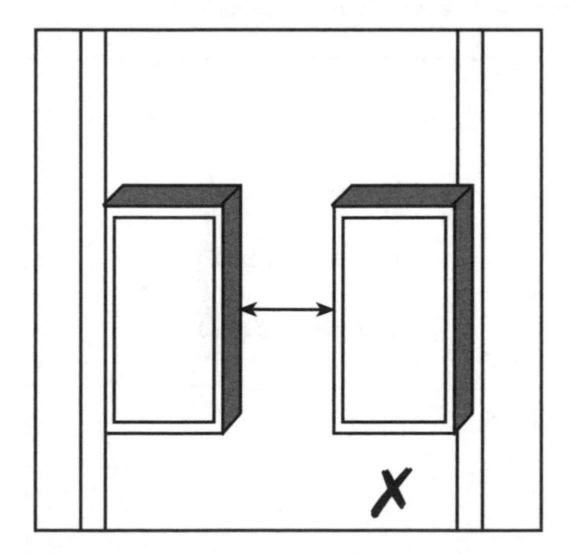

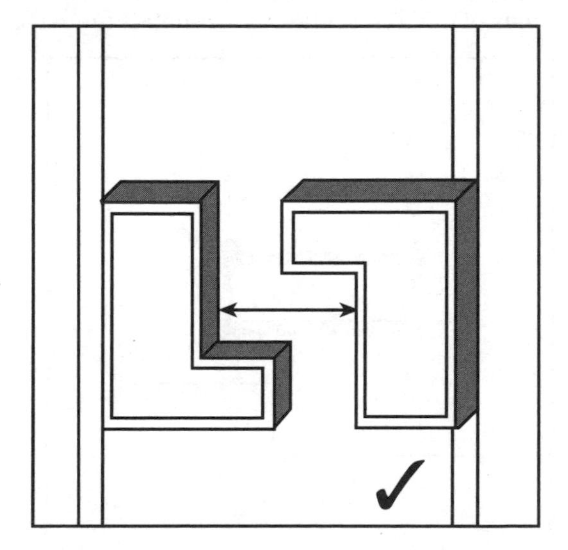

Figure 10.14 (left) *Incorrect building floor plan showing narrow gap;* (right) *correct building plan showing alternative arrangement with a wider gap between building blocks in order to allow better airflow and use of natural light*

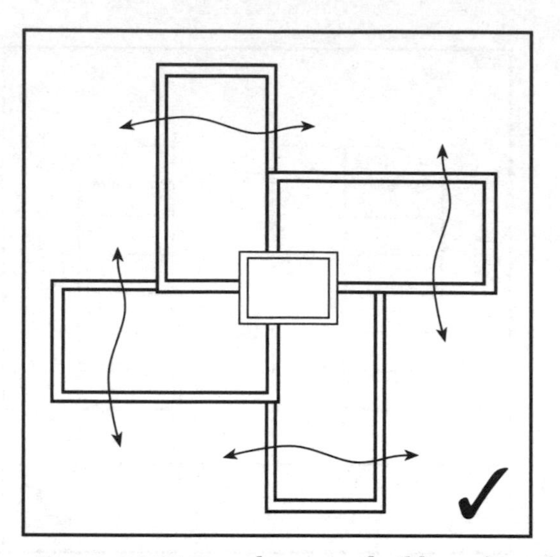

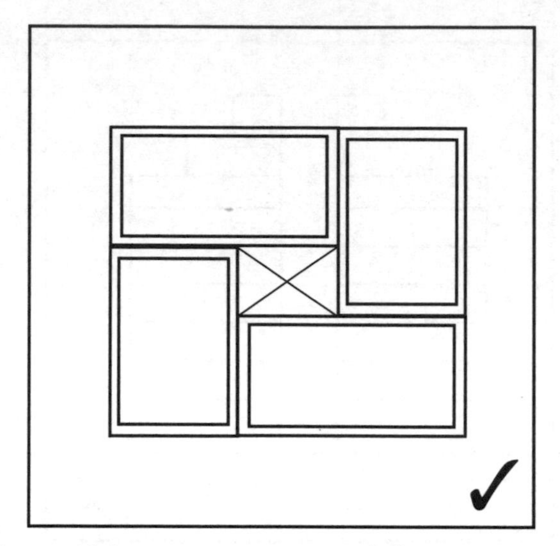

Figure 10.15 *Two alternative building plans showing the use of the same component in different ways in order to achieve free flow of air*

given site. North hemisphere references to orientation are used for all the examples mentioned in this chapter.

Site layout

Within the urban fabric there are very few large enough open spaces where one can design the site layout free of existing constraints. Most sites are tight spaces within built-up areas with fixed parameters for placing a building (building line, setbacks, rear yards, etc.).

An urban renewal project where clearing of old structures takes place is rare, and only new subdivisions and extensions of city limits offer opportunities for designing the site layout. In such cases, it is essential after the site analysis to work closely with the environment, while addressing the other factors of the programme, and to take advantage of:

- the topography, without disrupting the natural watershed;
- the local winds and wind movement between buildings;
- solar exposure and the distance between proposed buildings;
- the possibilities offered by planting.

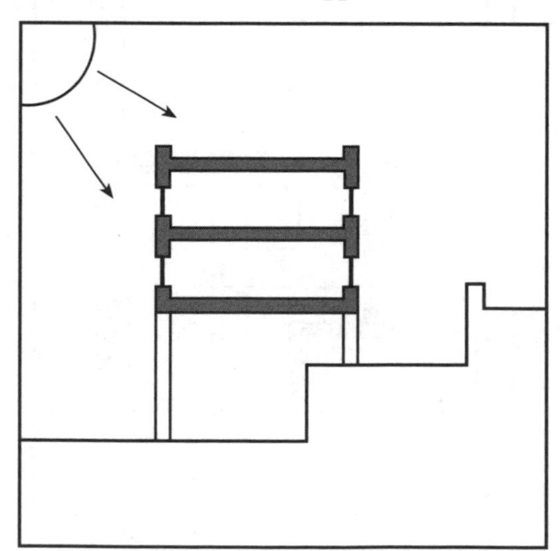

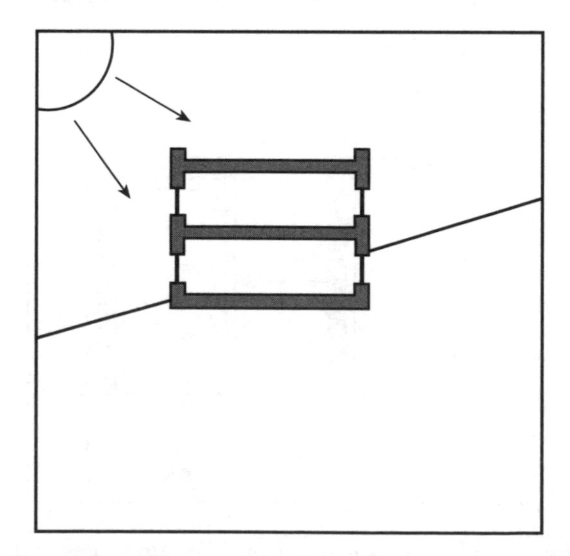

Figure 10.16 *Building sections showing various ways to place the building in relation to the ground*

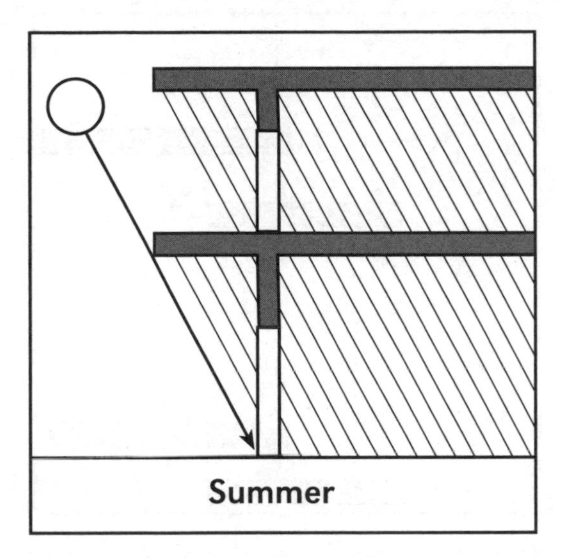

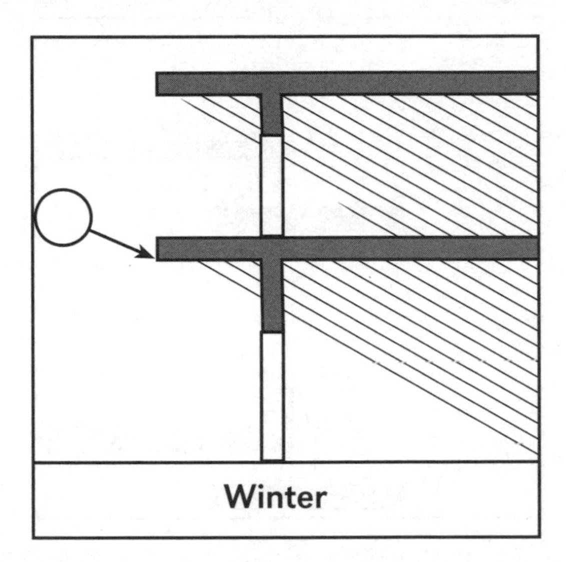

Figure 10.17 *Building sections showing how overhangs can assist in controlling sunlight*

Siting a building

The confines of building sites within the urban fabric often do not offer many options for placing the mass of the building upon a site. The following guidelines are applicable when certain conditions are available.

Orientation

Facing south or within 15 degrees of due south is advisable for optimal performance of heat-gaining spaces.

Excavations

Sloping sites offer opportunities for 'cut and fill' to save carting away and to reuse topsoil (when it is of usable quality).

Rear yards and building setbacks

Hard surfaces must be avoided. Precipitation must be allowed to percolate into the ground. Parking surfaces can be designed to provide a stable surface while allowing water absorption.

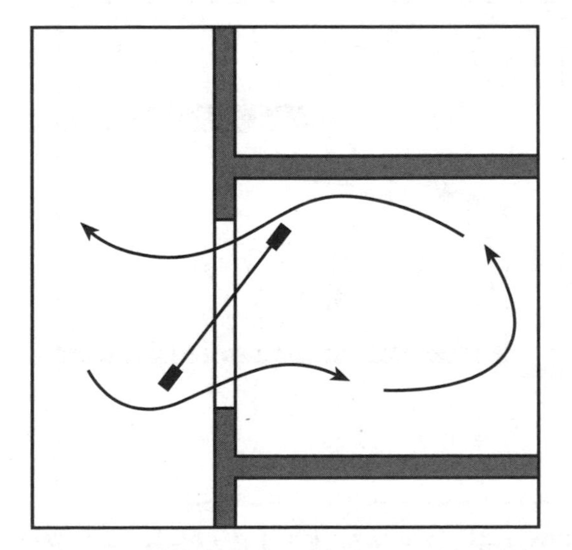

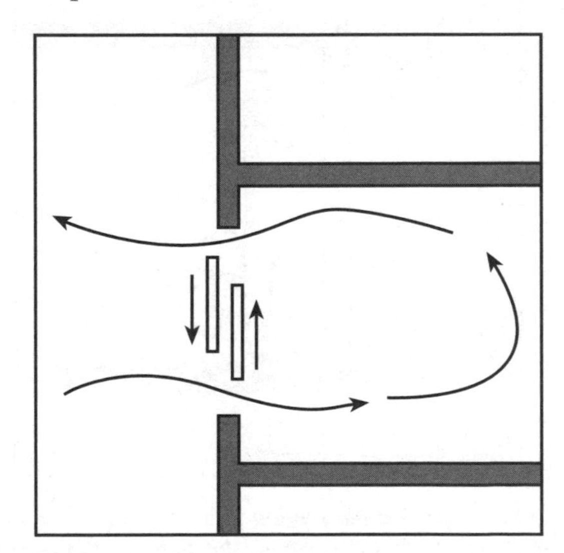

Figure 10.18 *Building sections showing various window types and the circulation of air,* (left) *pivoting and* (right) *double sash*

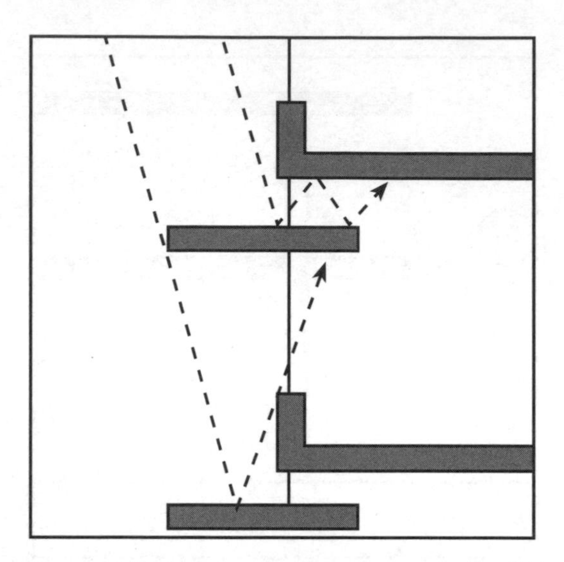

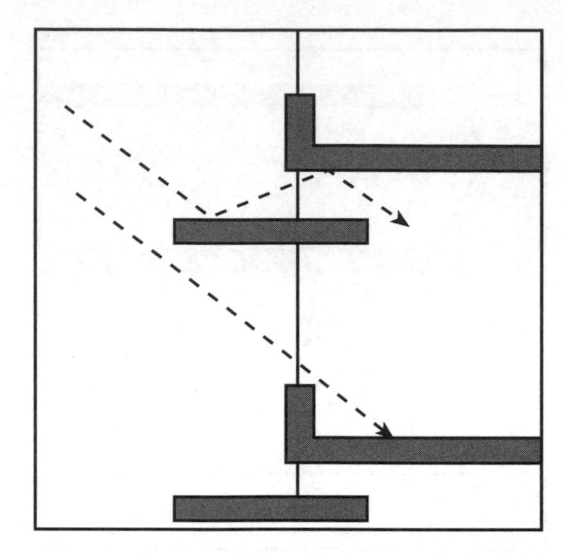

Figure 10.19 *Building section showing a fixed fin and how it assists in controlling sunlight coming into the building during summer and winter seasons*

Plants

Deciduous trees are very useful to shade the building and/or the ground in the summer without obstructing sunlight in the winter. Shading hard surfaces or the building may reduce air temperature by as much as approximately 7° Celsius during the warmest period.

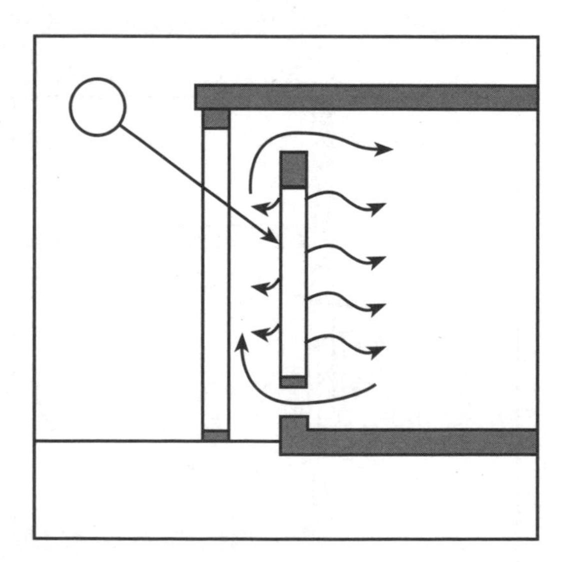

Figure 10.20 *Vertical section of a trombe wall showing how the sun's energy passes through a wall of glass, is stored in a mass and is slowly radiated to the adjacent room*

Building configuration

The shape of a building, the interior layout, the size and solar orientation are all factors that affect its energy consumption. To obtain maximum energy conservation, all four factors must be considered in an integrated manner.

Building shape

Considerable energy savings can occur by reducing the surface-to-volume ratio and getting the right solar orientation.

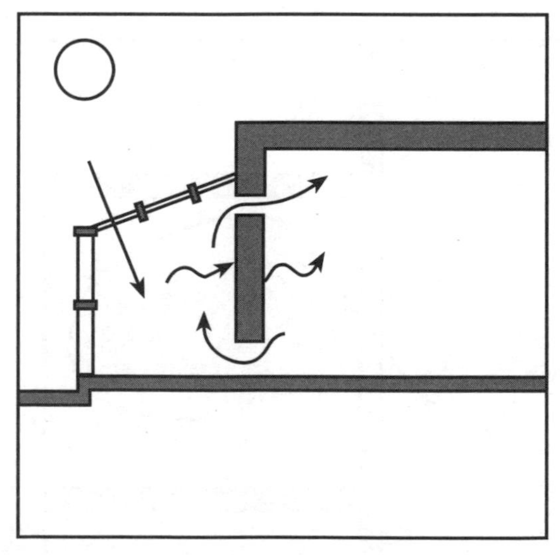

Figure 10.21 *Vertical section of an attached sun room showing how warm air is trapped in the volume of the glass structure so it can be transferred to the building*

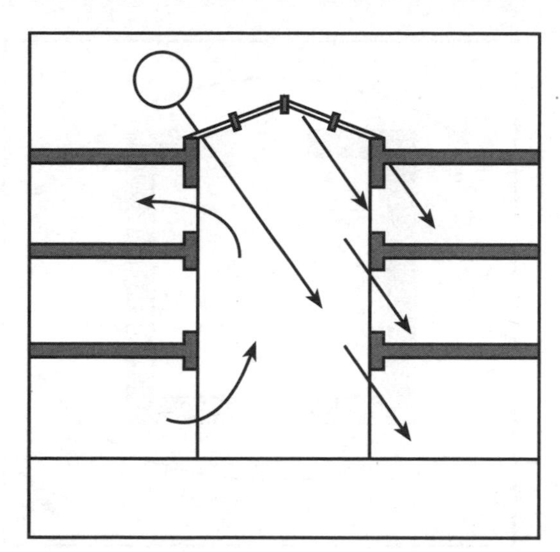

Figure 10.22 *Vertical section of an atrium showing how the sun's energy can be trapped in the space and transferred to the adjacent rooms*

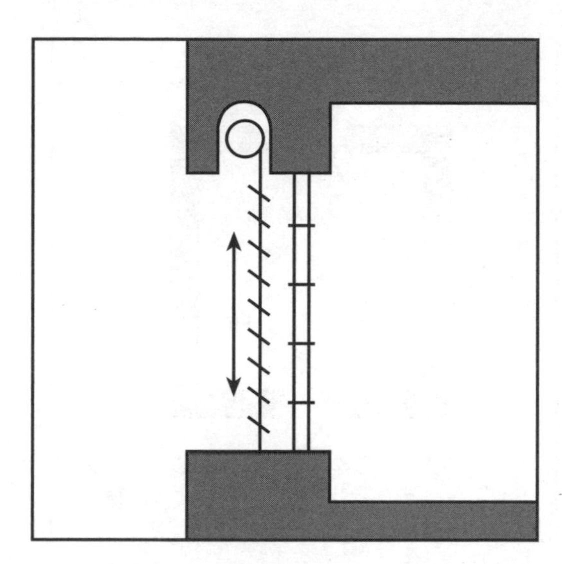

Figure 10.23 *Vertical window sections showing how an exterior roller shutter can be used to control sunlight before it enters the building*

Building on pilotis

Building on pilotis, when feasible, and leaving the ground floor open allows air movement between the street and the rear yards.

Interior layout

The layout in relation to the appropriate orientation can help to conserve energy for heating and cooling, as well as for lighting.

Relation to attached units

Avoid layouts of units without cross-ventilation in order to retain the free flow of air and to receive direct daylight in habitable rooms.

Tall buildings

Avoid compact building arrangements to allow for airflow and sun exposure. Where this is not possible, include southern light through the adjustment of the building section.

Relation of building to the ground

The further away the building is from the ground, the more it is subject to climatic conditions. When feasible, such as in suburban areas, keep the building in close contact with the ground; otherwise it is necessary to compensate with additional exterior building insulation in order to keep internal temperatures stabilized.

Designing wall openings

The position of openings, proportions and opening areas must be designed to meet interior space functional requirements (natural light and ventilation), views and aesthetic requirements of the exterior treatment of the elevations.

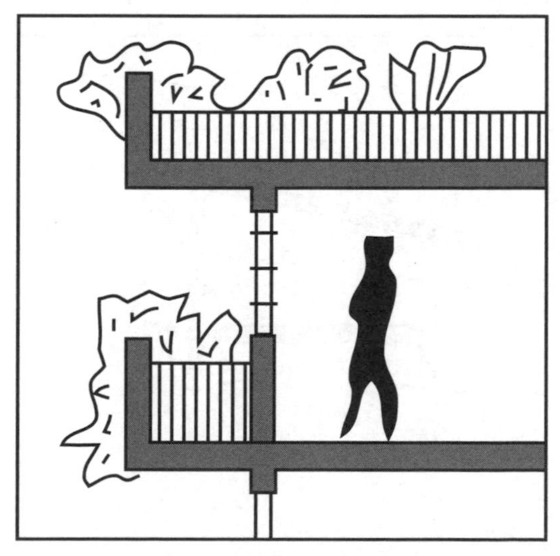

Figure 10.24 *Vertical building section showing how vegetation might be used to help absorb the sun's energy*

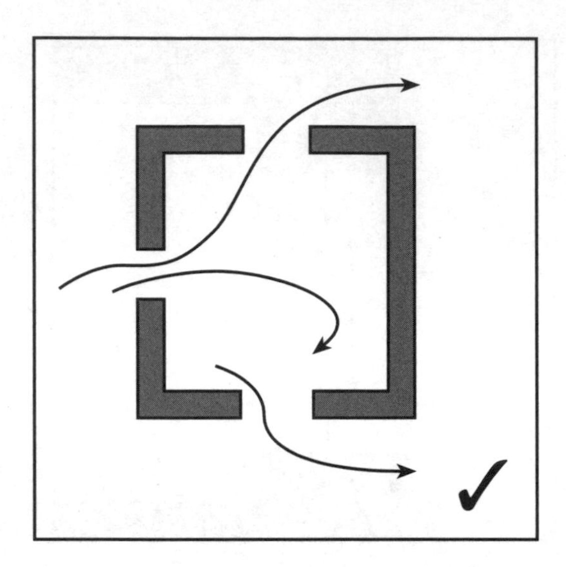

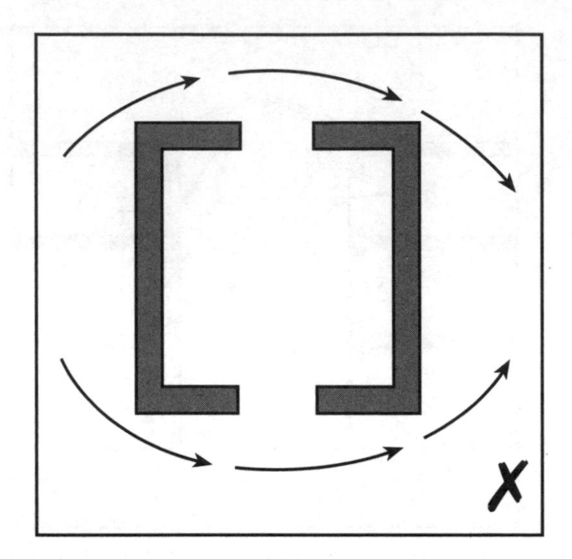

Figure 10.25 (left) *Correct placement of openings to assist in the free flow of air in a room;* (right) *incorrect placement of openings, hampering the free flow of air*

North openings

Generally, north openings provide steady natural light that is free of glare. Double glazing can reduce heat losses during the cold months. During the warm season, north openings are useful for natural ventilation.

East and west openings

In warm climates, east and west openings present problems of shading from the morning and afternoon sun, which is in a lower position. They need to be carefully considered (are they necessary for ventilation, natural light or views?) since protection from the sun is less easy than with southern orientations.

South openings

These are useful as the sun can enter during the cold months, while it is easier to block the summer sun. Glare is also an issue that needs to be resolved, particularly for working areas.

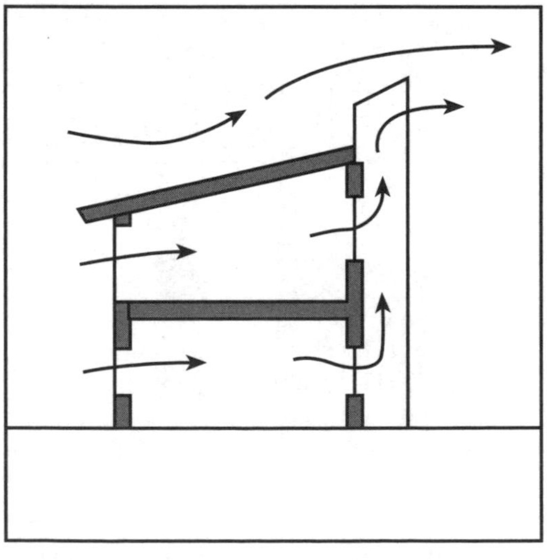

Figure 10.26 *Building section showing the use of a wind tower*

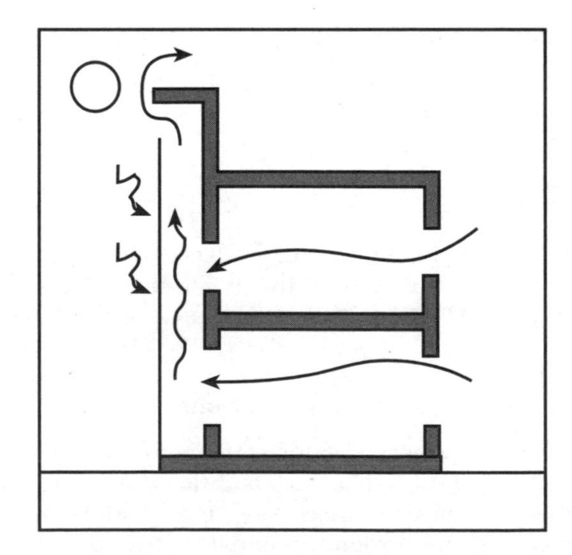

Figure 10.27 *Building section showing the use of a solar chimney*

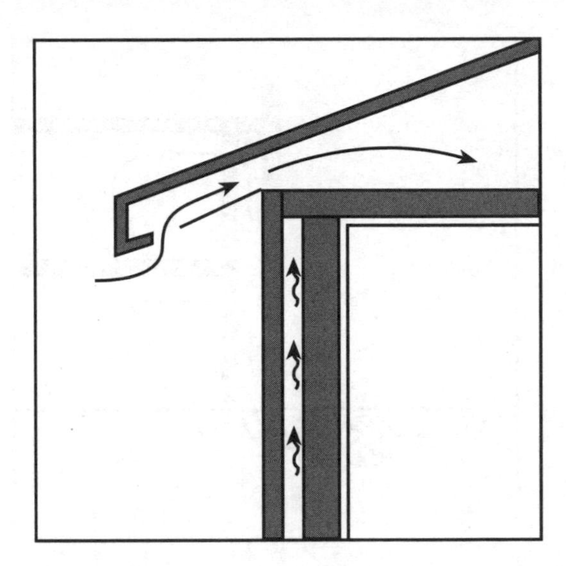

Figure 10.28 *Building sections showing various ways to insulate a building;* (left) *parasol-like roof to block sun and allow air movement;* (right) *cavities built in the wall and roof to allow for air movement*

Window systems

The choice of glazing systems can drastically alter the energy consumption of a building. The choice of fenestration can improve or mitigate natural ventilation. Window and door details may reduce or increase heat gains or losses through air 'draughts or hot–cold bridges'. Extensive glass surfaces must be double glazed in order to avoid unnecessary heat gains or losses. A number of contemporary hardware options allow windows to be flexible and operate as side-hung pivotal, depending upon the user's needs for ventilation.

Passive solar heating guidelines

Solar heat can be utilized through passive systems in a variety of ways to improve temperatures within the interior of a building.

Direct gain

Interior surfaces can be heated by means of properly designed south-facing windows.

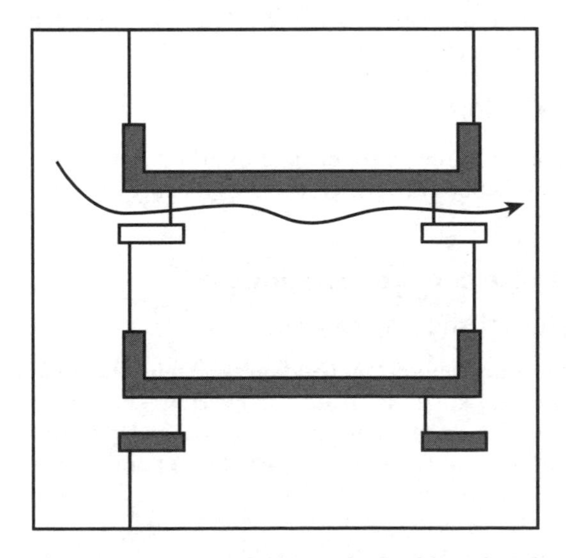

Figure 10.29 *Building sections showing* (left) *clerestory windows and* (right) *horizontal fins in the building façade*

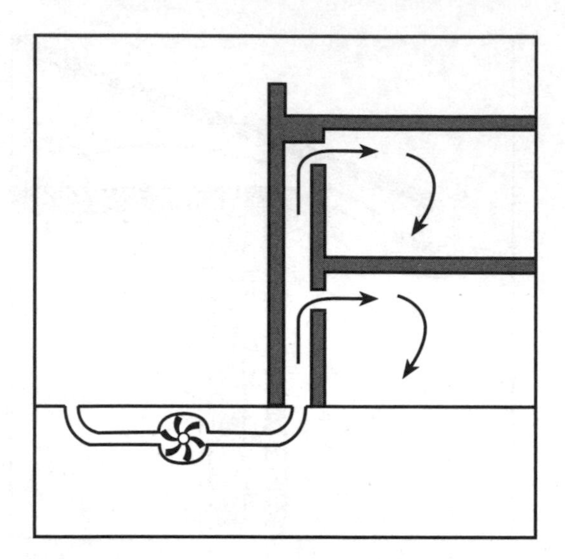

Figure 10.30 *Building section showing how cool air might be pumped into a building from underground pipes*

Figure 10.31 *Building section showing the planting of roof surfaces*

Trombe walls

Trombe walls can be used to protect interior spaces from direct solar gain, while simultaneously storing heat that can be used at a later time of day. The trombe wall operates as a mass capable of storing heat from the sun and then transferring it to the interior of the building, mostly by convection, within an enclosed area from the exterior system.

Attached sunroom

This works in a similar fashion to the trombe wall, although the warm air is collected in the sunroom which, in turn, heats the wall mass.

The atrium

Glazed multilevel rooms in a building can serve as collectors of solar energy, which is especially effective in cold climates.

Passive cooling techniques

External shading devices

There are several methods of reducing heat gain in the summer, such as fixed projections or breeze soleil (screens), or movable projections, such as awnings and operable shutters. There are also several new materials that provide similar effects.

Vegetation

As discussed earlier, by planting around the building and on roofs and walls, the heat load is reduced and internal temperatures are lower.

Air currents

The proper positioning and configuration of the building can enhance the flow or air currents, thus helping to keep the temperatures around and inside the building cooler.

External coatings

The choice of external cladding material and the colour of a building can affect the absorption of heat. Therefore, external coatings should be selected carefully.

Wind towers

Wind towers are simple structures that – through differential pressure – allow warm air to be passively extracted from a building, allowing for cooler air to be drawn in from lower parts of a building. They can be used in reverse by funnelling in cool winds from the towers' high opening and forcing the warm air out through the lower external openings.

Solar chimney

A metal shaft on the south side of a building is used to draw warm air out when external air currents are unavailable.

Figure 10.32 *Building section showing the use of water elements on roof surfaces*

Ventilated wall and roof construction

By providing a pocket of air between the inside and outside surfaces of a building, added insulation is provided.

Clerestories and roof vents

These simple devices can allow direct sun access or ventilation of internal spaces.

Ground pipes

Buried pipes can be used to pump latent energy from the ground to cool a building.

Roof configuration

Planted roofs

Planted roofs are an ideal method of insulating a roof, although the extra load on the structure and the cost of insulation against water intrusion often outweighs their benefits. This type of planting also aids in reducing the ambient temperature of urban centres.

Roof water elements

A roof spray or pond provides all of the benefits and drawbacks of planting, although the roof cannot be used as a habitable space.

Internal gains control

Through the use of energy-efficient lighting and sunlight, the heat gain inside a building can be considerably reduced.

Natural lighting

There are many means of using natural daylight to illuminate a building.

Artificial lighting

Low-efficiency artificial lights use more electricity than high-efficiency artificial lights since low-efficiency lights generate more heat in the interior of a building. Provision must be made to use high-efficiency artificial lights. Through the building design, it is important to make maximum use of natural light during the daytime.

Appliances and equipment

Most domestic appliances and equipment generate heat, which is undesirable during the hot months. The surplus heat from the appliances can be tapped during the cold months and directed to boost interior temperatures, while it can also be ventilated to the exterior during the summer months.

Natural air cooling

Air movement occurs naturally between areas of different pressure. When the difference in pressure is insufficient to cause the air to move naturally, then mechanical means may be used, preferably powered by renewable energy sources. As the air moves and ventilates the interior of a building, it drives away heat from the building and its occupants. The following options will assist air to move naturally and to ventilate a building:

- Position openings to allow cross-ventilation.
- Provide high-level openings to let rising hot air escape outside.
- Place openings between indoor–outdoor spaces of varying temperature levels, designed specifically as cool areas (i.e. shaded areas with water elements or ground cover plants) and generally design the interior of a building so that it is viewed as a sequence of spaces that facilitate air movement.

Synopsis

This chapter clearly shows the complexity of energy conservation in buildings as a result of using an integrated approach. A multitude of factors have to be considered during the design process in order to achieve the optimum cumulative results. The guidelines included in this chapter prove this point, rather than provide an exclusive list. The various factors that have to be considered range from general issues to very specific issues that deal with minute details of materials and constructions. The ultimate results, however, also depend upon the behaviour of the building's users.

An integrated design is a comprehensive approach:

> *... that considers the many disparate parts of a building project, and examines the interaction between design, construction and operations to optimize the energy and environmental performance of the project. The strength of this process is that all relevant issues are considered simultaneously in order to solve them* (Lopez Barnett and Browning, 1995, p100).

Note

1 The example of Curritiba, Brazil, is an excellent illustration of rational planning with environmental concerns in mind.

References

Chrisomallidou, N. (2001) 'Guidelines for integrating energy conservation techniques in urban buildings', in Santamouris, M. (ed) *Energy and Climate in the Urban Environment*, James & James, London, p247

Goulding, R. J., Lewis, J. O. and Steamers, C. T. (eds) (1993) *Energy in Architecture: The European Passive Solar Handbook*, B. T. Batsford Ltd, London

Lopez Barnett, D. and Browning, D. W. (1995) *A Primer on Sustainable Building*, Rocky Mountain Institute, Aspen, Colorado

Olgyay, V. (1963) *Design with Climate*, Princeton University Press, New Jersey

Roaf, S., Fuentes, M. and Thomas, S. (2001) *Ecohouse: A Design Guide*, Architectural Press, Oxford

Santamouris M. (ed) *Energy and Climate in the Urban Environment*, James & James, London

Von Weizsäcker, E., Lovins, A. B. and Hunter Lovins, L. (1997) *Factor Four: Doubling Wealth, Halving Resource Use*, Earthscan Publications, London

Watson, D. and Labs, K. (1992) *Climatic Buildings Design*, McGraw Hill Book Co., New York

Recommended reading

The following publications are recommended to the reader; each one provides an opportunity to expand on the issues and subjects dealt with in this chapter.

1 Von Weizsäcker, E., Lovins, A. B. and Hunter Lovins, L. (1997) *Factor Four: Doubling Wealth, Halving Resource Use*, Earthscan Publications, London
Design is a process of invention. To invent, one has to be in the right frame of mind. The information and analysis of various examples regarding energy conservation in the built environment contained in this book create a comprehensive frame of reference that inevitably captivates the focus of the designer. It is essential background reading before indulging in the design process with the environment in mind.

2 Santamouris M. (ed) *Energy and Climate in the Urban Environment*, James & James, London
This book comprises a detailed analysis of a whole range of issues regarding the built environment. A useful set of specific environmental references are included.

Activities

Activity 1

Show, through a rough sketch, at least two alternative solutions to how one might, during the summer months, allow south daylight into a building without permitting solar heat gain.

Activity 2

A developer has an option to build on two sites between two existing buildings in a city located in a 40-degree north latitude. Each is of similar size, price and location; but one is north–south facing and the other is east–west facing. Which of the two would you advise him to develop?

Activity 3

Describe at least two alternative ways of utilizing solar energy to heat spaces in cold climates.

Answers

Activity 1

By using dedicious trees to shade a building, direct sunlight is filtered, while ambient skylight is also permitted to pass through the building's openings.

In addition, by placing horizontal louvers on a building, direct sunlight is transformed into reflected light.

Activity 2

A north–south-facing site would be most advantageous because it is easy to control the sun's rays on the south-facing façade with horizontal louvers. The north façade is naturally protected from the sun. The east–west-facing site has problems due to the low level of the sun in the morning and the afternoon, thus requiring vertical louvers which can block the views from the building.

Activity 3

1. Trombe wall: the solar heat is stored in the mass of the wall. The cold air in the building is warmed by the mass and transferred mechanically throughout the day and night into the living spaces.
2. Direct gain through windows, clerestories or skylights provides another option.
3. The atrium: a glass-covered interior space stores radiant heat from the sun. If a multi-storey building is involved, the heat rising into the uppermost parts of the volume can be mechanically transferred to other parts of the building.

11

Indoor Air Quality

Vassilios Geros

Scope of the chapter

Indoor air quality (IAQ) is an important parameter in buildings, strongly related to the health of the occupants and therefore to their comfort. The aim of this chapter is to discuss the important role of IAQ and to present the ventilation systems and strategies that can be applied in urban buildings. Additionally, the chapter includes the requirement levels and the national standards of various pollutants and the procedures that can be used in order to evaluate the IAQ parameter.

Learning objectives

Upon completing a study of this chapter, readers will able to:

- understand various aspects related to indoor air quality;
- be familiar with the various indoor and outdoor sources and the relevant air pollutants;
- know the international standards for IAQ;
- understand the mechanisms for controlling indoor and outdoor pollutants;
- be familiar with various calculation models that concern IAQ.

Key words

Key words include:

- indoor air quality (IAQ);
- sick building syndrome (SBS);
- air pollutants;
- ventilation;
- IAQ calculation models.

Introduction

IAQ is an important parameter that characterizes the indoor environment and is strongly related to the health of a building's occupants. Additionally, the operation of the ventilation system of a building is also important, since the primary role of ventilation is to improve the building's indoor climate. Therefore, the performance of the ventilation system directly determines the IAQ of an indoor space. Furthermore, the air flows and paths inside the building have an important influence on the thermal comfort of the occupants, especially during the summer season. The impact of ventilation on the energy performance of the building is also crucial because when the outdoor/fresh air supply is increased, the cooling and heating energy requirements are also increased when the cooling/heating system is operating. On the other hand, there are energy conservation strategies, such as free cooling operation, that reduce the cooling load of the building by increasing the amount of outdoor air that is handled by the ventilation system. Therefore, it is important to appropriately design and operate the ventilation system by taking into account the specific characteristics of the urban microclimate (such as wind speed and direction, air temperature, pollutant concentration, etc.) in order to ensure the optimal performance of this system.

The content of this chapter includes building-related illnesses, presentation of the main indoor and outdoor pollutant sources, their main sources (including occupants, material emissions, etc.), the role of ventilation and the techniques/strategies that should be adopted in order to improve IAQ, the standards for IAQ and, finally, some simple calculation procedures.

Indoor air quality

In modern societies, people spend close to 90 per cent of their time indoors. It is therefore very important to ensure the quality of the indoor environment by taking into account parameters such as thermal and visual comfort, without ignoring factors that concern the quality of the air. The use of various ventilation techniques and strategies for ventilating indoor spaces, in order to increase the indoor air quality by replacing the indoor with outdoor air, has been in existence since humans brought fire into their caves and used an opening at the top of the caves to reduce harmful contaminants. Of course, the outdoor environment in big cities is also polluted; but it is a fact that frequently the air within buildings can be more seriously polluted than the outdoor air (sometimes even in the largest and most industrialized cities). Therefore, the risk to health may be greater when people are exposed to indoor air pollution. In 1984 the World Health Organization concluded that up to 30 per cent of new and remodelled buildings worldwide might be the subject of excessive complaints related to IAQ.

The use of increased ventilation rates, on the other hand is, in most cases, in opposition to energy efficiency as the fresh air energy load can be an important part of the whole building's energy consumption. Therefore, it is important to control the ventilation rates in a building by achieving, at the same time, acceptable levels of IAQ. Frequently, problems related to IAQ appear when a building is operated, managed or maintained in a way that is not in accordance with its original design or the appropriate operating procedures. Of course, IAQ problems can also be a result of poor building design or occupant activities.

Sick building syndrome and building-related illness

The quality of indoor air plays an important role in the health of a building's occupants. There are two general categories of 'diseases' that are connected with IAQ. The first one, called sick building syndrome (SBS), concerns situations in which the building's occupants experience acute health and/or discomfort effects that are apparently related to time spent in a building, and it is not possible to determine any specific illness or cause. The problems may be localized in one or more specific rooms or areas, or may extend to the whole building. But, in any case, the symptoms are not diagnosable and may correspond to the following 'indicators':

- The occupants of the building complain of lethargy, headaches, difficulty in concentrating, runny nose, dry throat, eye and skin irritation, dizziness and nausea, and sensitivity to odours.
- The cause of the symptoms is unknown.
- Symptoms frequently disappear soon after leaving the building.

When the term building-related illness (BRI) is used, then the symptoms concern diagnosable illness and can be attributed directly to contaminants of the indoor air. A BRI situation refers to the following aspects:

- Building occupants complain of symptoms such as cough, chest tightness, fever, chills and muscle aches.
- The symptoms can be clinically defined and have clearly identifiable causes.
- Complainants may require prolonged recovery times after leaving the building.

Of course, these complaints may result from other causes that are not related to IAQ. For example, symptoms may be linked with illnesses contracted outside the building, acute sensitivity (e.g. allergies), job-related stress or dissatisfaction or other psychosocial factors. However, it is possible that symptoms are caused or exacerbated by indoor air quality problems.

Every building has a number of potential sources of indoor air contaminants. Some of them are continuously emitted, such as building materials and furnishing, while others, such as cooking, smoking and the use of solvents, paints and cleaning products release contaminants intermittently. Grouped by origin, the most important sources of pollutants are:

- *Human and animal metabolism*: there is an association between oxygen consumption and carbon dioxide (CO_2) release that occurs due to human and animal metabolism. In addition to CO_2, some volatile organic compounds (VOCs) are also produced by this process. Generally, metabolic gases can lead to air quality and odour problem; but health hazards occur only at high concentrations. In this case, the ventilation requirements are usually low.
- *Occupant activities*: the air quality of an indoor space is strongly related to the use of the space and, of course, the occupant activities. For example, activities such as smoking, cooking and cleaning directly affect IAQ and can contribute to an increase in the concentration of various pollutants.

- *Building materials and equipment*: building materials and equipment are also important sources of pollutants. Carpets, furniture, paints, varnishes, etc., are emitters of pollutants and, depending upon their chemical composition, can have a significant effect on IAQ levels. The use of low-emitting materials can lead to a reduction in the ventilation requirement and, therefore, the energy consumption of the building.

Many factors can cause and/or contribute to sick building syndrome; the most common are listed below.

Ventilation system

The airflow rates of the ventilation system are strongly related to IAQ. When reduced ventilation rates are used, IAQ levels may be inadequate to maintain the health and comfort of a building's occupants. The effective or ineffective distribution of the air inside the building by the Heating, Ventilating and Air Conditioning (HVAC) system is another factor in SBS. 'Isolated' spaces in a building, where the air of the ventilation system cannot reach, may have serious problems from the IAQ point of view. According to the American Society of Heating, Refrigerating and Air-conditioning Engineers (ASHRAE), the minimum requirements are 27 to 30m^3 per hour of outdoor air per person (close to 36m^3 per hour per person in office buildings). In addition, if smoking is allowed (i.e. increased production of indoor contaminants), the minimum requirements are close to 110 cubic metres per hour per person. Generally, the rate of the outdoor air depends upon the activities that normally occur in that space (see ASHRAE Standard 62, 1989). And, of course, the maintenance of the duct and filtration systems are also important factors that can increase or reduce IAQ levels.

Indoor contaminants

Generally, the majority of the indoor air pollutants come from sources located inside the building. For example, VOCs, including formaldehyde, may be emitted by adhesives, carpeting, upholstery, manufactured wood products, copy machines and various other mediums. Also, tobacco smoke increases the levels of VOCs, other toxic compounds and respirable particulate matter. The effect of high concentrations of VOCs can be to cause chronic and acute health problems, and some are known carcinogens. When the concentration levels of VOCs are low to medium this may also produce acute reactions. In addition, unvented kerosene and gas space heaters, woodstoves, fireplaces and gas stoves can emit carbon monoxide, nitrogen dioxide and respirable particles, which are combustion products. Generally, the construction of the building as well as the use and the activities that take place in the indoor environment can seriously affect the IAQ levels.

Outdoor contaminants

The role of the ventilation system is to introduce outdoor air into the building in order to replace a part (in most cases) of the indoor air and increase the IAQ levels. The introduction of outdoor air can occur through natural means (natural ventilation: e.g. single-sided or cross-ventilation) or through mechanical means (mechanical ventilation: use of supply and/or exhaust fans and/or duckwork). In this case, the outdoor air that enters a building can be a source of indoor air pollution as it transfers outdoor contaminants. For example, if the ventilation system has a poorly located intake ventilator, windows or other openings, pollutants from motor vehicle exhausts, plumbing ventilators and building exhausts (e.g. bathrooms and kitchens) can enter the building.

Biological contaminants

This kind of contaminant includes bacteria, moulds, pollen and viruses. These contaminants may breed in stagnant water that has accumulated in ducts, humidifiers and drain pans, or where water has collected on ceiling tiles, carpeting or insulation. It is also possible for insects or bird droppings to be a source of biological contaminants. Legionnaire's disease and Pontiac fever are caused by one indoor bacterium named *Legionella*. Some of the physical symptoms related to biological contamination may include coughing, chest tightness, fever, chills, muscle aches and allergic responses such as mucous membrane irritation and upper respiratory congestion.

Research shows that ventilation (by introducing outdoor air indoors) as a measure of maintaining satisfactory IAQ levels represents only one aspect of a very complex problem. The previous mentioned elements may act in combination and may supplement other complaints, such as inadequate temperature, humidity or lighting. Nevertheless, it seems that ventilation has, in these problems, a more significant role when ventilation rates are very low (a fact that also decreases employee productivity). But it is also important that even after a building investigation, the specific causes of the complaints may remain unknown.

Indoor air quality design

Building investigation procedures

A building investigation procedure is necessary in order to identify and solve complaints related to indoor air quality issues by preventing them from recurring and by avoiding additional problems. A first and important phase of the investigation process is to discover if each of the complaints is related to indoor air quality or if the source of the problem is different. A second phase requires the identification of the source that causes the complaint(s); finally, the investigator should determine the most appropriate actions to correct the problems. Typically, the investigation procedure begins with a walk through the building or the area where IAQ problems are presented in order to gather information about four basic factors: the occupants, the HVAC system, the possible pollutant pathways and the possible contaminant sources.

The 'walkthrough' audit should collect information that concerns the history of the building and the reported complaints, as well as the HVAC zones and the complaint areas. This audit involves visual inspection of critical building areas and consultation with occupants. During the initial walkthrough, the investigator can develop some possible explanations for the complaints by using the gathered information in order to formulate a possible solution. In a second step, he/she can test the solution and examine whether the problem can be solved.

In the case where the collected information from the 'walkthrough' audit is insufficient to formulate a possible solution, or if a tested solution failed to solve the problem, the investigator should collect additional information in order to define a new or alternative solution. In this kind of approach the whole procedure of formulating solutions, testing and evaluating them continues until the final solution is achieved.

Another approach concerns air sampling for contaminants in order to identify the contaminant that causes the complaints; but sometimes it is quite difficult to determine exactly the source of the problem. This approach can include the measurement of parameters such as temperature, relative humidity, CO_2 and air movement that may determine the conditions of the building or the area. But, in this case, sampling for specific pollutant concentrations, without understanding how the building operates and the nature of the complaints, can be misleading. The experimental measurement of various parameters and the sampling strategy should be performed when the investigator has collected all of the necessary information and clearly understands the operation of the building.

Some of the actions that can be undertaken to resolve indoor air quality problems are:

- routine maintenance of HVAC systems;
- periodic cleaning or replacement of filters;
- replacement of water-stained ceiling tiles and carpeting;
- institution of smoking restrictions;
- venting contaminant source emissions to the outdoors;
- storage and use of paints, adhesives, solvents and pesticides in well-ventilated areas;
- use of these pollutant sources during periods of non-occupancy;
- allowing time for building materials in new or renovated areas to off-gas pollutants before occupancy.

Of course, the simplest solution (supposing that the maintenance of ventilation system is as it should be) is to increase the outdoor air ventilation rates, and this action can often be a cost-effective means of reducing indoor pollutant levels. HVAC systems should be designed, at least to meet ventilation standards, according to local building codes. Nevertheless, there are HVAC systems that are not operated or maintained properly, and therefore the design ventilation rates are not ensured. To achieve adequate IAQ levels, the HVAC system should be operated at least to meet the design standards; this is very often sufficient to ensure the quality of the indoor air. When there are important pollutant sources, the local ventilation system, if it exists, can be used to expel the exhaust-contaminated air into the outdoor environment.

The use of various filters and/or devices for air cleaning can be a suitable solution but presents certain limitations. For example, the use of a typical furnace filter (an inexpensive particle-control device) cannot effectively capture small particles. Additionally, mechanical filters cannot remove gaseous pollutants. On the other hand, high-performance air filters can capture small inhalable particles; but they are relatively expensive to install and operate. Moreover, adsorbent beds can be used to remove some specific gaseous pollutants; but these devices are also expensive and require frequent replacement of the absorption material.

Finally, another important issue is the education of the occupants, management and maintenance personnel, as well as the well-organized communication between the involved actors. The avoidance of IAQ problems can be achieved if these actors understand the causes and consequences of IAQ problems and work more effectively together.

Box 11.1 DETERMINING IAQ DESIGN TARGETS

- Determine the overall purpose, as well as the IAQ required levels.
  - Identify what the client accepts to do in order to achieve good IAQ (this can help to establish the IAQ goals).
  - Determine the wishes of the client concerning the IAQ levels (standard, above average, outstanding).
- Identify the indoor environmental quality concerns of the client.
  - ✓ Reduce occupants' sick building syndrome symptoms.
  - ✓ Reduce occupants' absenteeism.
  - ✓ Reduce occupants' odour-related complaints.
  - ✓ Reduce occupants' allergic or asthmatic complaints.
  - ✓ Reduce health risks related to Legionnaire's disease and Pontiac fever, colds and flu, and hypersensitivity pneumonitis.
  - Determine the participation of the client in the planning and design process.
- Identify the spaces that require special protection for good IAQ.
  - Determine any sensitive populations who occupy the building. In this case, the spaces that occupy and/or access the occupation periods (e.g. visitors or permanent occupants) and the characteristics and the requirements of this kind of population should be identified.
- Identify the spaces that are possible sources of contaminants.
  - Record the functions that are performed in these spaces (major and routine activities, potential pollutant emissions from materials, equipment and processes).
  - List 'suspicious' materials, from the IAQ point of view, or any other pollutant sources that may be brought from outside into the building.
  - List dangerous or toxic chemicals that may be brought into the building.
  - Identify spaces or areas of the building where outdoor pollutant sources can be brought in.

Box 11.2 IDENTIFYING SITE CHARACTERISTICS RELATED TO IAQ

- Determine the preferable locations for outdoor air intakes and air cleaning and filtration, and schedule the operation of the relevant systems according to short-term outdoor air pollutant concentration peaks (e.g. heavy traffic periods).
- Survey any industrial or commercial activities nearby that might be sources of pollutant emissions (manufacturing, waste handling, dry cleaning, food preparation, etc.).
- Survey any potential sources of contaminants related to soil or groundwater.
- Determine the prevailing wind conditions in the building location.
- Determine traffic patterns.
- Determine the local ambient air quality.

Box 11.3 DETERMINING THE OVERALL APPROACH RELATED TO ENVIRONMENTAL CONTROL

- Define the ventilation system (fully mechanical, hybrid system, fully natural).
- Identify the distribution system.
- Identify the return and exhaust parts of the ventilation system.
- Determine the link between the cooling/heating system and ventilation system.
- Describe the occupant-level controls that concern the regulation of the thermal conditions, as well as other environmental parameters (e.g. noise levels).

Process of a good IAQ design

This part of the chapter describes a sequence of phases and questions that can be followed or answered during a typical IAQ design project. In the following boxes, the main steps of such a design project are presented in bullet format (Spengler et al, 2000).

BOX 11.4 DESCRIBING THE BUILDING

- Identify the general building massing, layout and exterior openings.
- Describe the ventilation system flows, equipment locations, and the location of major pollutant sources, the exhausts from the ventilation system, cooling towers, outlets of combustion exhausts, etc.
- Identify how the air is supplied to occupants.
- Identify the outdoor pollutant sources (including sources such as drift from cooling towers, exhaust air from toilets, kitchens, biological/chemical hoods, etc.).
- Identify the location of the air intakes (HVAC system, doors, windows, etc.) relevant to outdoor pollutant sources.
- Identify the location of the mechanical ventilation equipment.
- Describe the control of the ventilation system in various levels (user level, zone level, building level, etc.).

BOX 11.5 DESCRIBING THE BUILDING MATERIALS

- Describe the dominant materials and their properties related to IAQ (e.g. use of low-emitting materials).
- Determine the necessary cleaning materials (toxic, non-toxic) for walls and floors.
- Determine the maintenance requirements of the floor-covering materials (wooden floors, carpets, etc.).

BOX 11.6 DETERMINING VENTILATION SYSTEM OPTIONS AND EVALUATION

- Conduct a box model analysis in order to investigate the various alternatives and describe a set of best alternative scenarios.
- Evaluate various options that concern the ventilation system, such as user-level control of the ventilation, options concerning energy efficiency and improved IAQ, and the use of natural or passive ventilation techniques, as well as heat recovery systems and displacement ventilation.

BOX 11.7 IDENTIFYING TARGET MATERIALS FOR IAQ EVALUATION

- Select the building materials that are important for IAQ issues.
- Identify the building materials that are not concrete, masonry, metal, stone tile or glass.
- Analyse the area and mass of major materials per unit of volume in each zone of the building in order to determine the materials with the most extended use, the largest interior total surface area and the greatest mass.
- Identify the target materials (including dry and wet products) and estimate the impact that emissions from these materials have on indoor concentrations (by multiplying the emission factor by the areas or masses).

BOX 11.8 IDENTIFYING THE CHEMICAL CONTENT AND EMISSIONS OF THE TARGET MATERIALS

- Acquire the product sheets and the volatile organic compounds (VOCs) emission tests for dry products and chemical contents lists for wet products.
- Determine if composites of wood products are extensively used and if low-emitting products have been used.
- Examine if wood products are 'isolated' from indoor air (e.g. sealed, laminated, etc.).
- Examine if the use of wet products can be reduced.

BOX 11.9 IDENTIFYING THE CLEANING AND MAINTENANCE REQUIREMENTS

- Examine the cleaning and maintenance chemical products that are used on major surface areas (walls, floors, ceilings) and identify the chemical composition of these products.

BOX 11.10 REVIEWING CHEMICAL DATA FOR THE PRESENCE OF STRONG ODORANTS, IRRITANTS, ACUTE TOXINS AND GENETIC TOXINS

- Examine the chemicals that are emitted at important rates by using specific databases that include information on the chemical properties (odour, irritation and toxicity) of various products.

BOX 11.11 CALCULATING THE CONCENTRATION OF DOMINANT EMISSIONS

- Use an indoor air model to calculate emissions of worst-case chemicals during a 24-hour period and over 30 days for each zone of the building and examine which of the pollution sources have the greatest impact on IAQ.

BOX 11.12 COMPARING THE CALCULATED CONCENTRATIONS WITH IAQ REQUIREMENTS

- Use the IAQ standards in order to determine the criteria concentrations.
- Compare the calculated concentration of each zone of the building against the criteria concentrations in order to evaluate the levels of various contaminants.

BOX 11.13 SELECTING PRODUCTS AND DETERMINING INSTALLATION REQUIREMENTS

- Determine the specifications concerning the acquisition, storage, transportation, handling and installation of the selected products, primarily in the case of buildings under construction.
- Determine which products should be replaced and/or 'isolated' from the indoor air, including the procedures described in the previous point, in the case of existing buildings.

BOX 11.14 VENTILATION SYSTEM REQUIREMENTS

- Determine the required outdoor air under full- and part-load operation conditions.
- Identify the locations of the outdoor air intake.
- Identify the filters required (type, efficiency, installation/operation/maintenance guidelines).
- Determine the air-cleaning specifications if the outdoor air quality levels are low.
- Identify the return air system requirements.
- Determine the air distribution system characteristics under heating/cooling full- and part-load operation in order to ensure the compatibility with the IAQ standards.
- Identify monitored variables, sensor positions, alerts, calculation procedures and operation requirements if IAQ issues are to be integrated within a building management system (BMS).

Indoor pollutants and pollutant sources

Indoor pollutants result both from outdoor and indoor sources. Furthermore, indoor pollutants may have a natural origin or may result from occupant activities. Of course, air pollutant sources may differ from one building to another, according to building design, the building construction, the location of the building, the occupants' activities, etc.

Indoor sources may comprise construction materials and furnishings that release contaminants more or less continuously. In most cases, the sources that are related to the occupants' activities release contaminants intermittently. Some of the occupants' activities that influence IAQ include smoking, cooking and using paint and cleaning products. The presence or not of a ventilation system, as well as its operation, strongly affects the concentration of the pollutants in the indoor environment.

In the urban environment, outdoor sources present a particular importance since they can affect the quality of the air that enters the building. Of course, the ventilation system can control, to a certain degree, the quality of the outdoor air that enters the building through the fresh air (e.g. by using filters); but in naturally ventilated buildings, air cleaning is ineffective. The major outdoor sources concern the following:

- Industrial emissions (local or distant) can be responsible for high concentrations of oxides of nitrogen and sulphur, ozone, lead, volatile organic compounds, smoke, particles and fibres. These polluting effects depend upon specific climatic conditions, especially in urban areas where the influence of effects such as heat island effect and/or airflow distribution around the buildings is very important.

- Traffic pollution is another important source in urban areas and concerns a major part of the outdoor pollution close to streets, tunnels and parking areas. Some of the major pollutants due to traffic are carbon monoxide and dioxide, carbon dust, lead and oxides of nitrogen.
- Soil-borne sources of pollutants in the vicinity of the building may include radon (a naturally occurring radioactive gas), methane (a product of organic decay) and moisture.
- Nearby sources may involve combustion emissions from neighbouring buildings or installations, extract ventilation systems close to air supply intakes, etc.

Generally, some sources of pollutants may be common in all types of buildings; but the design, air handling systems and occupants' activities influence the pollutant concentrations, and this is the reason that some pollutant sources can be more dominant than others. In the following paragraphs most important pollutants are presented, grouped by source of origin.

Synthetic organic compounds

The use of synthetic chemical compounds has increased during the last few decades, and various products exist as a result of organic chemistry.

Volatile organic compounds (VOCs)

The term volatile organic compounds refers to carbon-containing chemicals that participate in photochemical reactions in the ambient air. They are organic compounds that can be found in the vapour phase at ambient temperatures. VOCs have various sources related to building materials, such as paint, solvents, fuel storage, carpeting, adhesives, motor vehicles, tobacco smoke, toiletries and cosmetics, and cleaning supplies. Generally, they are produced when certain products are consumed. Studies show that indoor concentrations of most VOCs are two to ten times higher than outdoor ones. They are measured by using sampling and analysis methods and the sum of all measurements is referred as TVOCs (total volatile organic compounds).

The World Health Organization (WHO) classification system defines VOCs as organic compounds that are divided into four categories according to their boiling point as follows:

- very volatile (gaseous) compounds (boiling point: < 0° to 50–100° Celsius);
- volatile organic compounds (boiling point: 50–100° Celsius to 240–260° Celsius);
- semi-volatile organic compounds (boiling point: 240–260° Celsius to 380–400° Celsius);
- organic compounds associated with particulate matter or particulate organic matter (boiling point: > 380° Celsius).

There are three main categories of VOCs:

1. *Halogenated compounds*: these compounds are, in general, hydrocarbons with one or more halogens, usually chlorine. They can be found in products such as degreasers, cleaning agents and propellants in aerosol-type products. Long-term exposure to halogenated hydrocarbons may cause cardiac arrhythmias, affect kidney and liver functions, damage the nervous system and affect the reproductive system. Some of them are known or are suspected to be human carcinogens.
2. *Aromatic compounds*: these are, for example, substitute benzene rings. The basic aromatic VOCs are benzene (constituent in fuels, emitted as a combustion by-product) toluene and xylene (major components of adhesives, spray paints, art supplies, etc.) and styrene (constituent of tobacco smoke). Acute effects of exposure to aromatic VOCs include upper respiratory system irritation, eye irritation, headaches and fatigue. Long-term exposure may cause central nervous system damage and cancer.
3. *Aliphatic compounds*: these can be short, straight chain compounds containing carbon and hydrogen, hydroxyl or carbonyl functional groups (alcohols and aldehydes) or formaldehyde. Aliphatic hydrocarbons are components of gasoline and other fuels and they can be used as solvents. Formaldehyde is a colourless gas with a characteristic pungent odour. Indoor sources of formaldehyde can be adhesives used in building materials, insulation materials, tobacco smoke and combustion appliances. Indoor formaldehyde concentration is generally higher than outdoor, especially in new buildings with extensive use of particleboards. Symptoms of formaldehyde exposure include respiratory and eye irritation, headache, nausea and fatigue. Formaldehyde is also a known sensitizer (someone exposed to a high dose can become sensitive to lower doses).

Pesticides

Pesticides are, in general, chemicals that have different modes of action and are used to control, keep away or eliminate pests. They are sub-classified according to their action into many classes. There are close to 600 different pesticides and more than 45,000 formulations. Pesticides may include inorganic compounds (such as salts of arsenic, sulphur and chlorine), organo-chlorine compounds, organo-phosphorus compounds and pyrethroids. Generally, pesticides are used in agriculture; but a major indoor use involves pest control (e.g. in wood-frame buildings). Exposure to pesticides can be through inhalation (e.g. airborne pesticides during the application of the product), by dermal absorption and by ingestion.

Pesticides affect the nervous system, the liver and may cause cancer. Symptoms may include fatigue, loss of appetite, nausea and headaches.

Combustion products

Combustion products are produced during the burning process of any fuel, including tobacco. They are a mixture of gaseous and particulate materials. Water vapour and carbon dioxide (CO_2) are the two major products emitted by burning organic fuels in any state (solid, liquid or gaseous). Generally, the most important pollutants emitted during the combustion process include carbon oxides (CO and CO_2), nitrogen dioxide (NO_2) and polycyclic aromatic hydrocarbons.

Symptoms of exposure to combustion pollutants include respiratory and eye irritation, while long-term exposure can cause serious illnesses, such as heart and lung diseases.

Oxides of nitrogen

The main anthropogenic sources of NO_x production are the combustion of coal, oil, natural gas and vehicle fuel, and therefore NO_x can have both indoor and outdoor sources. For NO_2, the rate of production is related to the amount of oxygen, the flame temperature and the rate of cooling of the combustion products.

Oxides of carbon

CO and CO_2 are produced, indoors and outdoor, when carbonaceous material is burned in the presence of oxygen. Carbon monoxide, which is a colourless and odourless toxic gas, is produced by incomplete combustion of any carbonaceous fuel containing carbon. CO outdoor sources include vehicle and industrial emissions. Major indoor CO sources include unvented/malfunctioning gas cooking and heating appliances, wood stoves and emissions from garages attached to homes. CO_2 is also a colourless and odourless gas, which forms part of the atmosphere of the Earth. CO_2 is not generally considered a toxic gas; but in high concentrations it affects breathing. Indoor sources of CO_2 are associated with the presence of people (emission through breathing; see Table 11.1); therefore, the monitoring of CO_2 concentrations is important for ventilation studies.

Table 11.1 *CO_2 production rate for various activities*

Activity	Metabolic rate (W)	CO_2 production rate (l/s)
Sedentary work	100	0.004
Light work	150–300	0.006–0.012
Moderate work	300–500	0.012–0.020
Heavy work	500–650	0.020–0.026
Very heavy work	650–800	0.026–0.032

Combustion particulate matter

Generally, the particles of indoor combustion products are small (diameter less than 10μm) and are, therefore, breathable. In addition, various chemical compounds are absorbed within them and can cause damage to people exposed to them.

Environmental tobacco smoke (ETS)

This pollutant concerns the combination of smoke breathed out by active smokers (mainstream smoke) and the smoke that is released directly from the burning end of the cigarette (side-stream smoke). In general, the exposures of involuntary and active smoking differ quantitatively and qualitatively. Tobacco smoke consists of thousands of chemical components and in the indoor environment increases the levels of respirable particles, nicotine, polycyclic aromatic hydrocarbons, carbon monoxide, nitrogen dioxide and many other substances.

Heavy metals

Sources of heavy metals occur both indoors and outdoors. These pollutants include lead and mercury, which can be found in paint products and leaded gasoline. When the sources are outdoors, such as contaminated soil and dust, foot traffic can increase the levels of these contaminants in the indoor environment. Heavy metals affect the nervous system and can lead to foetotoxicity, teratogenicity and mutagenicity. Long-term exposure can increase blood pressure and can affect the kidneys and the nervous system.

Bio-aerosol contaminants

Bio-aerosols are airborne particulates composed of, or originating from, living organisms. They can include micro-organisms, fragments, toxins and waste products of living organisms. The sources of bio-aerosols are the outdoor and the indoor environment, as well as the occupants and the building. Outdoor sources include fungi, bacteria and plant pollen. Occupants can also be sources of bacteria and viruses. Concerning the building-related sources, HVAC systems can support the growth of fungi and bacteria, especially the components of these systems that have increased levels of humidity (e.g. cooling towers, humidifiers, cooling coils and air filters). Furthermore, elements such as wet carpets, various building materials and furnishings can be sources of microbiological growth. In general, outdoor bio-aerosols cannot be controlled and therefore one should prevent their entrance into the indoor environment. Chronic exposure can lead to permanent lung damage, hypersensitivity pneumonitis, 'humidifier fever' and asthma. Legionnaires' disease (caused by a bacterium called *Legionella pneumophilia* found in cooling towers, evaporative condensers and other warm water systems) and Pontiac fever (associated with exposure to airborne *Legionella*) are also diseases related to bio-aerosols.

Respirable particles

Respirable particles (dust) are less than 10μm in aerodynamic diameter. In the indoor environment, sources of respirable particles are unvented or malfunctioning combustion appliances, smoking, cooking, house dust, aerosol sprays and fibreglass insulation. Exposure to respirable particles is associated with respiratory illnesses, bronchoconstriction and aggravation of symptoms in asthma patients. Chronic exposure may lead to illnesses such as emphysema and chronic bronchitis.

Ozone

Ozone is mainly a secondary outdoor pollutant that is formed from a mixture of VOCs and NO_x exposed to sunlight. Ozone concentration has a direct dependence on sunlight intensity and the ratio of NO_2 to NO. Therefore, ozone concentrations have a strong peak during daylight hours. In general, high concentrations are found close to large cities. Of course, ozone affects the occupants of a building through the ventilation system and infiltration only when the outdoor concentration exceeds the health standards. But ozone is also a strong oxidant, reacting rapidly with gases and readily decomposing by reaction at surfaces. Inhaling ozone can increase the susceptibility of the lungs to infections and increase responsiveness to allergens and other air pollutants.

International standards of indoor air quality

There are many organizations that determine specific levels for the presence of contaminants that are related to indoor air quality. There are also various national regulations that determine certain standards, primarily concerning the concentration of pollutants in the indoor environment.

In general, national and international standards recognize almost the same physical or chemical impurities as pollutants that can cause a potential health risk in a building's occupants. Furthermore, a group of six pollutants called criteria pollutants are used as an indicator of indoor air quality, and comprise carbon monoxide, nitrogen dioxide, ozone, lead, particulate matter and sulphur dioxide. Of course, there are many more pollutants that are harmful to human health than these six criteria pollutants; but these concern the outdoor air used for ventilation indoors. Another approach is defined in Standard 62 of ASHRAE, which defines acceptable indoor air as 'air in which there are no known contaminants at harmful concentrations as determined by cognizant authorities and with which a substantial majority (80 per cent or more) of the people exposed do not express dissatisfaction'.

In Tables 11.2–11.8, reviews of international standards for indoor air quality are presented (the research was performed by the International Energy Agency). Three main indices are highlighted:

- the maximum allowable concentration (MAC) in the work space for an eight-hour period;

Table 11.2 *International standards for CO_2 concentration levels*

Country	MAC (parts per million)	Peak limit (parts per million)	AIC (parts per million)
Canada	5000		1000–3500
Germany	5000	2 x MAC	1000–1500
Finland	5000	5000	2500
Italy			1500
The Netherlands	5000	15,000	1000–1500
Norway	5000	MAC + 25%	
Sweden	5000	10,000	
Switzerland	5000		1000–1500
UK	5000	15,000	

Source: WHO, 1984; Santamouris, 2001

Table 11.3 *CO concentration levels for various areas*

Area	Concentration range (parts per million)
Natural base level	0.044–0.087
Rural areas	0.175–0.435
Industrial areas	0.87–1.75
Downtown levels	up to 40

Source: WHO, 1984; Santamouris, 2001

Table 11.4 *International standards for CO concentration levels*

Country	MAC (parts per million)	Peak limit (parts per million)	ME value (parts per million)	AIC (parts per million)
Canada	50	400		9
Germany	30	2 x MAC	8–43	1–18
Finland	30	75		8.7–26
Italy	30			
The Netherlands	25	120		8.7–35
Norway	35	+50%		
Sweden	35	100		12
Switzerland	30		7	
UK	50	400		

Source: WHO, 1984; Santamouris, 2001

- the maximum environmental (ME) value;
- the acceptable indoor concentration (AIC):, a threshold value of concentration below which the negative health effects are insignificant or at least tolerable.

Table 11.2 presents the carbon dioxide-related standards for various countries.

Table 11.3 presents the recommended higher concentration ranges for carbon monoxide according to the area.

Table 11.4 summarizes the carbon monoxide requirements for various countries.

Table 11.5 presents the values for nitrogen dioxide.

Finally, Table 11.6 gives the national standards for various countries for formaldehyde (HCHO).

A review of the standards of organizations such as the World Health Organization and the US Environmental Protection Agency (USEPA) is presented for five of the six criteria pollutants (carbon monoxide, nitrogen dioxide, ozone, lead and sulphur dioxide) in Table 11.7. Table 11.8 presents the US National Ambient Air Quality Standards (NAAQS) for all six pollutants.

Modelling indoor pollutants

Modelling indoor pollutants can be very effective for the study of indoor air quality. The main result of these models concerns the prediction of pollutant concentration. Therefore, it is easy to evaluate the occupants' exposure to various indoor pollutants, even before the construction of the building. In addition, by using this kind of model, it is easy to study various control strategies of various systems, such as the ventilation system, and evaluate their efficiency.

In general, there are two categories of models: steady state and unsteady state. The first category concerns the simplest approach to the problem and does not consider the time variation of parameters such as the concentration of indoor and outdoor pollutants, ventilation flows, etc. The unsteady state category of models is more complex and takes into account the time variation of time-dependent parameters.

Generally, steady-state models consider the air in the zone or the room that is studied to be well mixed. That means that the supply air is well mixed with the air in the zone. Certainly, this is not exactly the real situation since a certain amount of supply air bypasses the zone (with no mixture with the air in the zone) and is removed from the zone. A parameter that increases the efficiency of these mathematical models is the ventilation efficiency which

Table 11.5 *International standards for NO_2 concentration levels*

Country	MAC (parts per million)	Peak limit (parts per million)	ME value (parts per million)	AIC (parts per million)
Canada	3	5		0.3 (office) 0.052 (homes)
Germany	5	2 x MAC	0.05–0.1	
Finland	3	6		0.08 (daily average) 0.16 (hourly average)
The Netherlands	2			0.08–0.16
Sweden	2	5		0.15–0.2
Switzerland	3		0.015–0.04	
UK	3	5		

Source: WHO, 1984; Santamouris, 2001

Table 11.6 *International standards for formaldehyde (HCHO) concentration levels*

Country	MAC (parts per million)	Peak limit (parts per million)	AIC (parts per million)
Canada	1	2	0.1
Germany	1	2 x MAC	0.1
Finland	1	0.12 (new buildings)	0.24 (existing buildings)
The Netherlands	1	2	0.1
Norway	1	+100%	
Sweden	0.5	1	0.01–0.1
Switzerland	1		0.2
UK	2	2	

Source: WHO, 1984; Santamouris, 2001

determines the amount of supply air that actually accesses the occupied part of the zone.

Steady-state models

The simplest approach concerns the box model which considers a room that is well mixed and not pressurized with one inlet and one outlet. It is important that, in this case, no return air is considered (see Figure 11.1).

By taking into account a mass-balance equation and the fact that the same amount of air enters and exits the zone, the following equations describe the problem and calculate the indoor pollutant concentration in air (Ci):

$$Q \cdot C_o + S = Q \cdot C_i \Rightarrow C_i = C_o + \frac{S}{Q} \qquad (1)$$

where Q is the airflow rate of the outdoor and exhaust air; C_o is the pollutant concentration in entering air; S is the pollutant emission rate (mass emission rate); C_i is the indoor air pollutant concentration.

If there is a pollutant sink, Equation 1 is written as follows:

$$C_i = C_o + \frac{S - R}{Q} \qquad (2)$$

where R is the pollutant sink.

Equations 1 and 2 are valid only for systems without return air, which means that 100 per cent of outdoor air is supplied into the room. In the case where recirculation occurs and a filtration device is installed, the previous equations become more complex. When a filter is installed in the return air channel (see Figure 11.2), the next equations describe the problem and calculate the pollutant concentration in the room:

$$Q \cdot C_o + (1 - E_f) \cdot R \cdot Q_R \cdot C_i + S = Q_R \cdot \Rightarrow$$

$$C_i = \frac{Q \cdot C_o + S}{Q + E_f \cdot R \cdot Q_R} \qquad (3)$$

where E_f is the filter efficiency; R is the recirculation factor; Q_R is the volumetric airflow of the return air.

Table 11.7 *WHO guideline values for the criteria air pollutants*

Compound	Annual ambient air concentration (μg/m³)	Health endpoint	Observed effect level (μg/m³)	Guideline value (μg/m³)	Averaging time
Carbon monoxide (CO)	500–700	Critical level of carboxyhaemoglobin (COHb) < 2.5%		100,000 60,000 30,000 10,000	15 minutes 30 minutes 1 hour 8 hours
Lead (Pb)	0.01–2	Critical level of Pb in blood < 100–150μg Pb		0.5	1 year
Nitrogen dioxide (NO_2)	10–150	Slight changes in function in asthmatics	365–565	200	1 hour
Ozone (O_3)	10–100	Respiratory function responses		120	8 hours
Sulphur dioxide (SO_2)	5–400	Changes in lung function in asthmatics. Exacerbations of respiratory symptoms in sensitive individuals	1000 250 100	500 125 50	10 minutes 24 hours 1 year

Source: WHO, 1999; Sherman and Matson, 2003

Table 11.8 *USEPA national ambient air quality standards for the criteria air pollutants*

Pollutant	Standard Value	Averaging time
Carbon monoxide (CO)	9 ppm (10mg/m³) 35 ppm (40mg/m³)	8-hour average 1-hour average
Nitrogen dioxide (NO_2)	0.053 ppm (100μg/m³)	Annual arithmetic mean
Ozone (O_3)	0.12 ppm (235μg/m³) 0.08 ppm (157μg/m³)	1-hour average 8-hour average
Lead (Pb)	1.5μg/m³	Quarterly Average
Particulate (PM 10) (particles with diameters of 10 micrometres or less)	50μg/m³ 150μg/m3	Annual arithmetic mean 24-hour average
Particulate (PM 2.5) (Particles with diameters of 2.5 micrometres or less	15μg/m³ 65μg/m³	Annual arithmetic mean 24-hour average
Sulphur dioxide (SO_2)	0.03 ppm (80μg/m³) 0.14 ppm (365μg/m³) 0.50 ppm (1300μg/m³)	Annual arithmetic mean 24-hour average 3-hour average

Note: ppm = parts per million;
μg = micrometre;
m³ = cubic metres;
mg = milligrams.

Source: US Environment Protection Agency, 1999; Sterman and Matson, 2003

The assumption to solve Equation 3 is that the same volume of air enters and exits the room per unit time, as described in the following relation:

$$Q = Q_R \cdot (1 - R) \tag{4}$$

In the case where the filtration device is installed in the supply airflow channel (see Figure 11.3), the mass balance equation is the following:

$$(Q \cdot C_o + R \cdot Q_R \cdot C_i) \cdot (1 - E_f) + S = Q_R \cdot C_i \Rightarrow \tag{5}$$

$$C_i = \frac{Q \cdot C_o \cdot (1 - E_f) + S}{Q + E_f \cdot R \cdot Q_R}$$

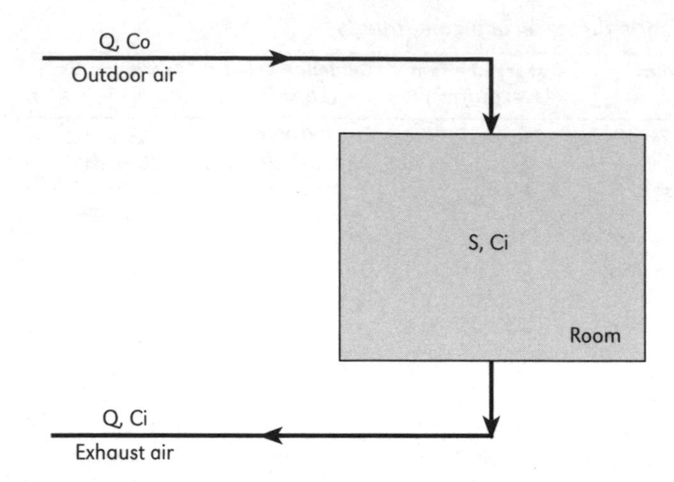

Figure 11.1 *Steady-state modelling without return air*

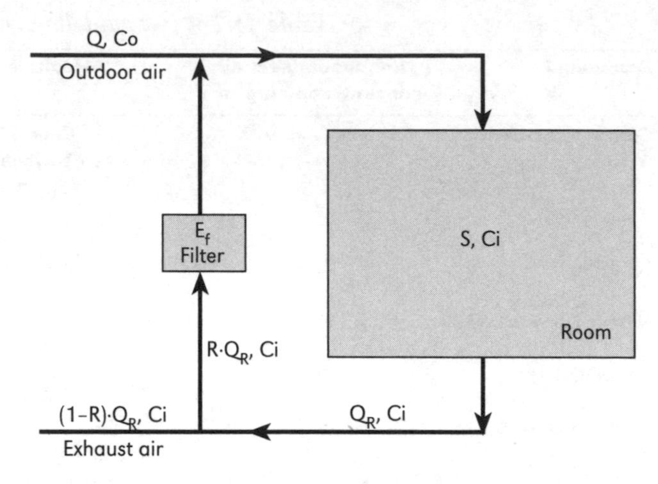

Figure 11.2 *Steady-state modelling with return air and filtration device installed in the return air channel*

Unsteady-state models

Unsteady-state models take into account the time dependence of parameters such as indoor and outdoor pollutant emission rates and removal rates of pollutants, as well as supply, return and exhaust air rates.

For the indoor pollutants emission rates, those that are related to the occupants' activities vary during the day according to the occupation of the building, while those that are related to the construction materials may have constant, slow or rapid decay according to the type of material.

Concerning the variation of the outdoor pollutants' emission rates, they depend upon the sources, the time periods of emission and the emission strength. Variation in emissions from street automobile traffic, industrial processes and other sources is also strongly related to the weather conditions that determine the distribution of the pollutants in the nearby area to the source area.

The installation of filtration devices or various air cleaners directly affects the concentration of pollutants. The reduction of pollutant concentration certainly concerns the channels where the previously mentioned devices are located and depends upon the pollutant load (thus, there is a time dependence as the pollutant load is variable). Furthermore, the chemical decomposition of certain pollutants may vary with time.

Finally, the airflow rates of the ventilation system in a building may vary. As infiltration is variable and depends upon the pressure and temperature difference between the indoor and the outdoor environment, this affects the airflow rates of the ventilation system (supply, return and exhaust). Furthermore, when variable air volume (VAV) systems are installed, the airflow rates of the ventilation system are changeable and are adjusted to the indoor and outdoor air temperature and humidity in zone or building level.

There are various unsteady-state mathematical models that take into account the time variation of the previously mentioned parameters. Such a model is developed by Rodriguez and Allard and concerns a transient multi-zone model, which assumes that pollutants are well mixed in each zone and that the pollutant transfer occurs through the air exchange between zones.

On the other hand, it is possible to describe the variation of pollutants' concentration under transient conditions by using the simple box model, which is the simplest approach. In this case, the following equation can be used:

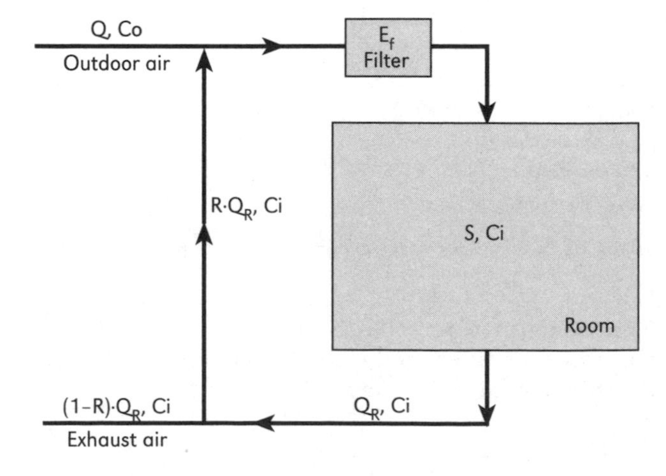

Figure 11.3 *Steady-state modelling with return air and filtration device installed in the supply air channel*

$$V\frac{dC_i}{dt} = Q(C_o - C_i) + S \qquad (6)$$

where Q is the airflow rate of the outdoor and exhaust air; C_o is the pollutant concentration in entering air; S is the pollutant emission rate (mass emission rate); C_i is the indoor air pollutant concentration; V is the volume.

Synopsis

The effect of indoor air quality (IAQ) on the health and comfort of a building's occupants is very important as people spend a great deal of their time in the indoor environment. Various syndromes are connected to poor IAQ levels and many researchers have shown that building materials play a significant role in causing these problems. Additionally, the ventilation strategy of the building, the amount of fresh air introduced from the outdoor environment and the operating conditions of the building (e.g. if smoking is allowed) are also important parameters for IAQ conditions.

The purpose of this chapter is to give an overview of issues related to IAQ. The first part presents sick-building syndrome (SBS) and building-related illness (BRI), which are related to IAQ. The chapter also presents the various sources of indoor and outdoor pollutants, together with the international standards and requirements for IAQ. Furthermore, the reduction of indoor pollutants' concentration by controlling indoor and outdoor air pollutants is discussed. Another important issue for a good IAQ design concerns the theoretical estimation of the indoor pollutant concentration, a topic that is covered in the last part of this chapter where some simple calculation models that can be used in simulating pollutant concentration levels in the indoor environment are presented.

References

Alevantis, L. and Xenaki-Petreas, M. (1996) *Indoor Air Quality in Practice*, Energy Conservation in Buildings Series, University of Athens, Greece

ASHRAE (American Society of Heating, Refrigerating and Air-conditioning Engineers) (1996) *HVAC Systems and Equipment*, ASHRAE, Atlanta, Georgia, US

ASHRAE (1997) *Fundamentals*, ASHRAE, Atlanta, Georgia, US

ASHRAE (2001) *Standard 62: Ventilation for Acceptable Indoor Air Quality*, ASHRAE, Atlanta, Georgia, US

du Pont, P. and Morrill, J. (1989) *Residential Indoor Air Quality and Energy Efficiency*, ACEEE, Washington, DC, US

EPA (1991) *Indoor Air Facts No 4 (revised): Sick Building Syndrome (SBS)*, www.epa.gov/iaq/pubs/sbs.html

Lester, J., Penny, R. and Reynolds, G. L. (eds) (1992) *Quality of the Indoor Environment*, IAI, London

Liddament, M. (1996) *A Guide to Energy Efficient Ventilation*, AIVC, Coventry

Maroni, B., Seifert, B. and Lindvall, T. (eds) (1995) Indoor Air Quality, Elsevier Science, The Netherlands

Martin, A. (1995) *Control of Natural Ventilation*, BSRIA, Bracknell

Morawska, L., Bofinger, N. D. and Maroni. M. (eds) (1995) Indoor Air: An Integrated Approach, Pergamon Press, Oxford

Rodriguez, E. A. and Allard, F. (1992) 'Coupling Comis Airflow Model with other transfer phenomena', *Energy and Buildings*, vol 18, pp147–157

Santamouris, M. (ed) (2001) *Energy and Climate in the Urban Build Environment*, James & James, London

Sherman, M. H. and Matson, N. E. (2003) *Reducing Indoor Residential Exposures to Outdoor Pollutants*, Lawrence Berkeley National Laboratory, Report Nb LBNL-51758, Berkeley, California, US

Spengler, J., Samet, J. and McCarthy, J. (eds) (2000) *Indoor Air Quality Handbook*, McGraw-Hill, New York, US

USEPA (US Environmental Protection Agency) *National Air Toxics Assessment*, www.epa.gov/ttn/atw/nata/index.html

USEPA (1991) *Building Air Quality: A Guide for Building Owners and Facility Managers*, Craftsman Book, Co, Carlsbad, US

WHO (World Health Organization) (1999) *Air Quality Guidelines*, WHO, www.who.int/environmental_information/Air/Guidelines

Recommended reading

1 USEPA (1991) *Building Air Quality: A Guide for Building Owners and Facility Managers*, Craftsman Book, Co, Carlsbad, US

This book deals with the quality and safety of the indoor air environment and what is commonly referred to as sick building syndrome. The text has been developed in order to resolve existing indoor air quality problems as well as to prevent future problems. Additional topics include air quality measurements, HVAC systems, mould, asbestos and radon. A complete evaluation section allows the reader to evaluate existing conditions in a building in order to determine the source of indoor air problems. The book is separated into four main sections. The first part presents the basics of IAQ issues and the second part explains how to prevent IAQ problems. Possible solutions are presented in the third part of the book, while in the final section IAQ survey forms are included.

2 Morawska, L., Bofinger, N. D. and Maroni. M. (eds) (1995) *Indoor Air: An Integrated Approach*, Pergamon Press, Oxford

Previously, the most common approach to monitoring, health risk assessment and management of indoor air quality was to consider each atmospheric pollutant individually. Today, there is a realization that both the comfort and the combined health risk may depend not only upon the concentrations of individual pollutants, but also upon the complexity of the interactions between all constituents in the air. In addition, any mitigation process aimed at one particular pollutant can also affect other pollutants. Integrated management strategies should take these processes into consideration and aim at an overall upgrading of the indoor environment. This book has a unique character and focus and includes a selection of papers presented at the International Workshop on Indoor Air: An Integrated Approach, covering a comprehensive description of indoor air, strategies for integrated health risk assessments and strategies for controlling and managing the indoor environment.

3 Maroni, B., Seifert, B. and Lindvall, T. (eds) (1995) *Indoor Air Quality*, Elsevier Science, The Netherlands

This book brings together much more than toxicology as it describes physical and chemical hazards, and many medical and psychological consequences, and welds them into approaches to better design and manage structures. Contributions from specialists (medical doctors, architects, engineers, chemists, biologists, physicists and toxicologists) in various disciplines concerned with indoor air quality are presented in order to provide comprehensive coverage of all aspects, including building design, health effects and medical diagnosis, toxicology of indoor air pollutants, and air sampling and analysis. Some chapters contain formative and educational messages for architects, engineers and public health professionals; but the high cost of this reference may deter its use in educational settings. Extensive references for each section recommend it as a reference book for those in any of the associated fields.

Activities

Activity 1

Sick building syndrome (SBS)

Sick building syndrome is one of the general categories of diseases that are related to IAQ. The problems that concern SBS may be localized or can extend to the whole building. Furthermore, the symptoms are often related, with complaints from building occupants of lethargy, headaches, difficulty in concentrating, runny noses and other symptoms with unknown causes. Frequently, these symptoms disappear when the occupants leave the building. Describe and explain same of the major factors that are related to SBS.

Activity 2

IAQ problem solution

During and after building investigation procedures, various problems related to IAQ can be identified. In order to solve these problems and the possible complaints by a building's occupants, identify some common actions and procedure that can improve IAQ levels.

Activity 3

IAQ design

During an IAQ design project, there is a sequence of major phases that designers should, more or less, follow. Give a short description of the phases that should be included in an IAQ design project.

Activity 4

IAQ design

The control of indoor and outdoor pollutant concentrations can be achieved by using various systems and techniques. Give a brief description of the available options that can help to control indoor and outdoor pollutant concentrations.

Answers

Activity 1

The operation of the ventilation system strongly affects the IAQ levels in a building. The first important parameter concerns the appropriate outdoor airflow rates that should be used to ventilate indoor spaces. The use of the proper rates – primarily according to national and international standards – can maintain acceptable levels of IAQ and reduce any SBS symptoms. Another important parameter is related to the distribution of the air in indoor spaces through the ventilation system. An ineffective distribution of ventilation air can lead to spaces that are 'isolated' and, eventually, not ventilated. The appropriate design and operation of the ventilation system can reduce air pollutants and increase indoor air quality. Finally, one of the most important factors concerns the maintenance of the ventilation system (primarily ducts and filtration systems): a procedure that can increase or reduce (when the process is not accurate) IAQ levels. The maintenance of the ventilation system is also connected with the existence of biological contaminants that can accumulate in ducts, humidifiers and drain pans (or where water has collected on ceiling tiles, carpeting or insulation). Appropriate cleaning of the ventilation system can rapidly reduce the symptoms caused by this category of pollutants.

Of course, the quantity and the type of indoor contaminants strongly affect the existence or the absence of SBS symptoms. The majority of indoor air pollutants come from sources located inside the building. For example, the high concentration of volatile organic compounds (VOCs) can cause chronic and acute health effects, and some are known carcinogens. In general, the construction and the use of the building, and the activities that take place in the indoor environment, affect IAQ levels. And the appropriate sizing, design and operation of a ventilation system can efficiently control the indoor contaminants and increase the IAQ of a building.

Since the role of the ventilation system is to introduce outdoor air into the building, the exterior air that enters indoor spaces can be a source of indoor air pollution as it transfers outdoor contaminants to the inside. Therefore, the study of outdoor sources is an important part of an IAQ study and sometimes requires the installation of specialized filters in order to 'trap' outdoor contaminants.

Activity 2

Generally, the reasons that poor IAQ occurs may not be clear; the interaction and the interrelations of the involved parameters are often complicated. Nevertheless, there are some 'standard' actions that can improve IAQ. These kinds of actions include the routine maintenance of HVAC systems and the periodic cleaning or replacement of filters. Equally important is the replacement of water-stained ceiling tiles and carpeting, while the institution of smoking restrictions is highly recommended in order to improve IAQ levels. Of course, the 'isolation' of pollutant-emitting materials (such as paints, adhesives, solvents and pesticides) and the proper ventilation of storage spaces are also important actions from the IAQ point of view. Finally, the use of materials that are pollutant sources during periods of non-occupancy and the delay of the occupation time in new or renovated areas in order to allow building materials to off-gas pollutants can be very effective under certain conditions.

When it is necessary, the increase of outdoor air ventilation rates can directly improve IAQ; but in this case one should respect the economic parameters as the total energy consumption of the building is also increased. Generally, an HVAC system should be operated to at least meet the design standards, and this very often is sufficient for the quality of indoor air. When there are important pollutant sources, the local ventilation system, if it exists, can be used to transfer the exhaust-contaminated air to the outdoor environment.

Furthermore, the use of various filters and/or devices for air cleaning can be a suitable solution, but may present certain limitations. Certainly, the education of the occupants, management and maintenance personnel is important in order to avoid and solve IAQ problems.

Activity 3

The main phases that the designer should follow during an IAQ design project can be summarized by the following points:

- Determine the IAQ design targets, the overall purpose of the project and the IAQ required levels. During this phase, the spaces that require special protection or contain possible sources of contaminants should be identified.
- Identify site characteristics related to IAQ, a procedure that includes the right location of the various parts of the ventilation system, the survey of possible outdoor pollutant sources and the investigation of the local climatic characteristics of the building site.
- Determine the overall approach related to environmental control, which concerns the definition and the design of the ventilation and distribution system in relation to the whole HVAC installation and operation.

- Provide a general description of the building from the structural–material point of view and a more detailed description of the HVAC system and its sub-systems, as well as the control levels of the ventilation system.
- Describe the building construction and cleaning materials and their properties in relation to IAQ.
- Determine ventilation system options and provide an evaluation by using calculation methods, such as box model analysis.
- Identify and analyse target materials for IAQ evaluation.
- Identify the chemical contents and emissions of the target materials.
- Identify the cleaning and maintenance requirements.
- Review chemical data for the presence of strong odorants, irritants, acute toxins and genetic toxins.
- Calculate the concentration of dominant emissions.
- Compare the calculated concentrations with the IAQ requirements.
- Select products and determine installation requirements.
- Determine ventilation system requirements during various operational conditions. Identify the location of the ventilation system parts and integrate them within a building management system.

Activity 4

The control of outdoor air pollutants can be achieved by using the following systems and techniques:

- Use filters in the mechanical ventilation system to prevent pollutants coming from the outside to the inside.
- Carefully move air intakes away from pollutant sources, when possible.
- Use air quality-controlled fresh air dampers, which close the intakes during the peaks of outdoor air pollution.
- Decrease the infiltration by increasing the building air tightness in order to more accurately control the indoor environment.
- Provide direct source control that can be achieved by eliminating avoidable pollutants through restricting pollutant emissions.
- Use a local ventilation system to directly extract the pollutants from their source (ventilation at source).
- Provide an appropriate design of the main ventilation system in order to meet the ventilation requirements during its minimum operation.

12

Applied Energy and Resource Management in the Urban Environment

Sašo Medved

Scope of the chapter

The great concentration of people in cities requires a very large supply of energy, water, food and other matters essential to people's everyday lives. In general, we can talk about 'the inputs' – renewable and non-renewable natural sources – and the 'outputs' that result from their energetic and material conversions, and which are then discharged as waste into the air, water or soil in gaseous, liquid or solid state. Reducing energy and mass flows is one of the main goals in the process of sustainable urban design. On the other hand, implementing technologies for renewable energy source conversion, systems for achieving a non-pollutant water supply and reducing the amount of waste are important goals in continuously enlarged urban regions in developing countries.

The purpose of this chapter is to describe the importance and quantity of energy and material flows, and the impact of these flows on the environment, together with the technologies that can be used to create sustainable cities. The chapter is divided into three main sections. The first section deals with energy sources, including renewables and the rational use of energy. The second section deals with water supply and treatment; and the third section describes material flows in cities and waste treatment.

Learning objectives

At the end of this chapter, readers will be able to:

- evaluate the quantities and environmental impacts of energy and mass flows in cities;
- compare different non-renewable and renewable energy sources;
- recognize the advantages of central supply systems;
- be aware of the importance of a reliable water supply and measures for protecting water sources;
- estimate the amount of waste in cities and compare technologies for their treatment.

Key words

Key words include:

- energy sources;
- fossil fuels;
- renewable energy sources;
- central supply systems;
- district heating and cooling;
- transportation in cities;
- water supply and treatment;
- waste and waste treatment.

Introduction

The population in urban areas has been growing continuously, resulting in a decreasing rural population for the last few decades. Cities have become bigger and what once used to comprise suburbs has become an integral part of cities. It is estimated that today more than 80 per cent of Europeans live in cities.

The large numbers of inhabitants in cities require a very intensive supply of energy, water, food, etc. A city can be presented as a 'black box' with different input and output flows of natural resources (land, energy and materials; see Figure 12.2). The development of cities and the standard of living, health and culture of their inhabitants depend upon a sufficient and continuous flow of resources.

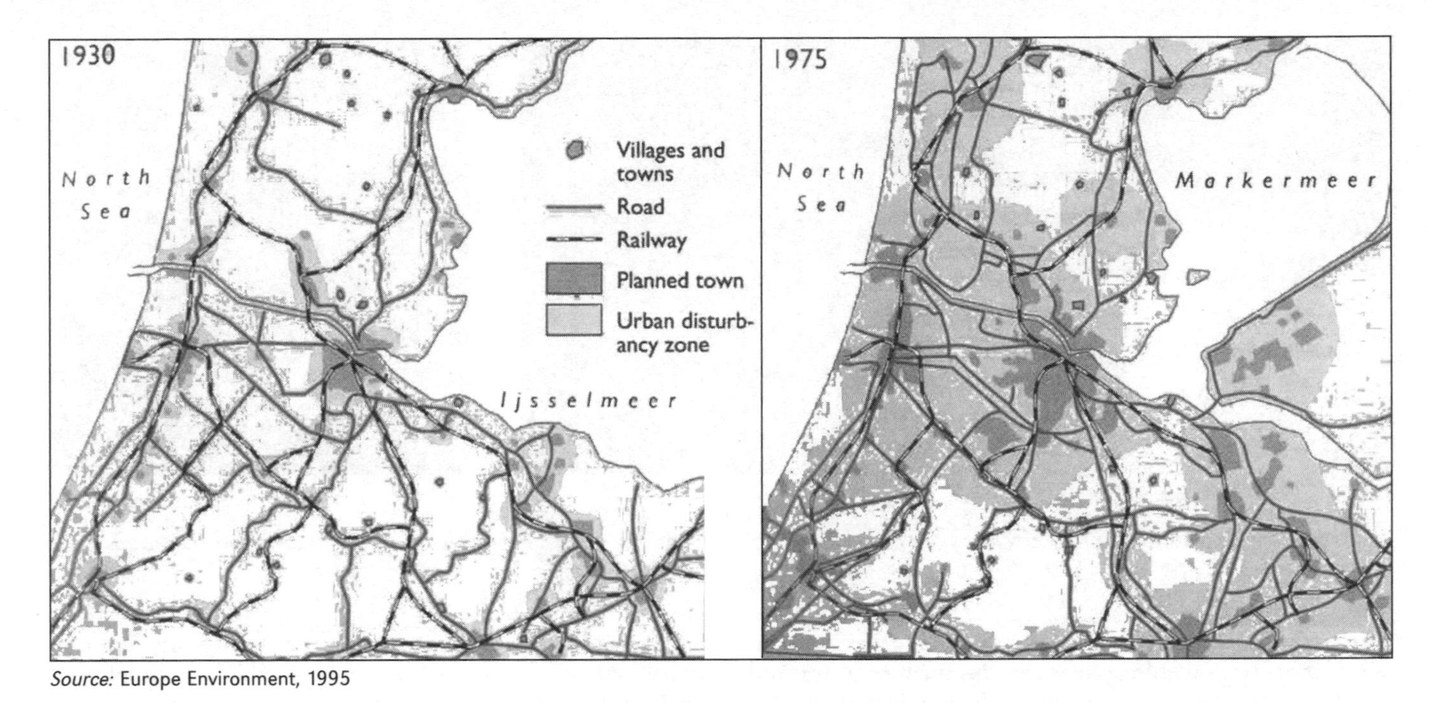

Source: Europe Environment, 1995

Figure 12.1 *Modern cities occupy substantially more surface area than they used to in the past, and 80 per cent of the population in the EU lives in cities*

The influence of cities on their surrounding environment can be expressed in different ways. Ecological footprints can be very useful, representing the necessary area of biological active land and water to provide all requirements for inhabitants and absorption of human-produced emissions (such as carbon dioxide and waste). For example, the ecological footprint of London is 120 times greater than the area of the city itself. And this is a general figure for all cities (EnerBuild RTF, 2000).

Decreasing cities' footprints is an important task on the way to achieving sustainable cities. Besides improving the economic and sociological quality of life in cities, the rational use of land, energy and materials is also an important step in this effort. Measures for rational use of land have been presented in previous chapters. Problems and some solutions for the rational use of energy and materials such as water and waste are presented at the end of the chapter.

Energy sources

Renewable and non-renewable energy sources

Our standard of living and culture depends upon a sufficient and continuous energy supply. In order to ensure its energy needs, mankind uses various kinds of energy conversion technologies to convert energy into required forms – such as heat, electricity and mechanical work. Fuels can be divided into two groups: renewable and non-renewable fuels. The former have the characteristic of renewing themselves in nature. According to their origins, they are divided into:

- solar radiation, emitted by the sun, which can be converted into heat and electricity, and which results in the energy of waves, wind, water energy and biomass;
- the planetary energy of the moon and the sun, which together with the Earth's kinetic energy cause tidal action in oceans and seas;
- heat emerging from the interior of the Earth towards its surface, and which is known as geothermal energy.

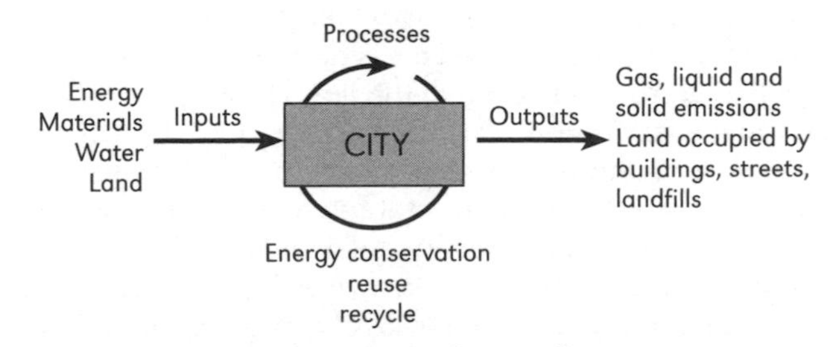

Figure 12.2 *Energy and mass flows in cities*

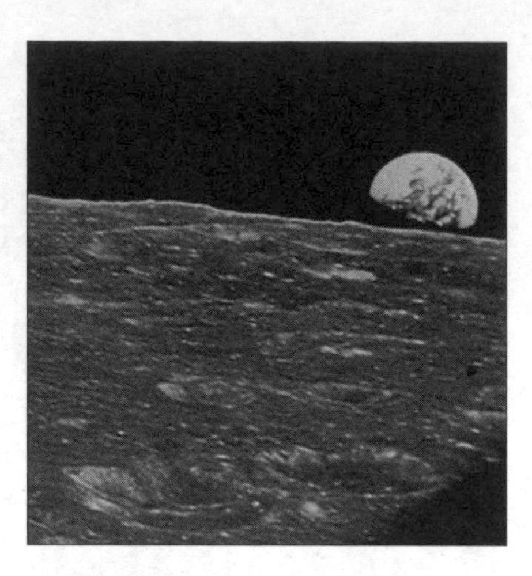

Source: Medved and Novak, 2000

Figure 12.3 *Renewable sources of energy are solar radiation, the gravitational force of the moon and the sun, and the heat within the Earth*

The sun's radiation represents the main energy source on the planet. Renewable sources in the form of water and wind energy represented the only source of energy besides human and animal mechanical work until the Industrial Revolution began. However, the rapid development of civilization that we have been facing during the last two centuries has only been possible due to the intensive use of natural (non-renewable) energy sources: fossil fuels and nuclear power.

Fossil fuels – coal, petroleum and gas – represent the various forms of energy accumulated during the past evolutionary era of the Earth. Coal is the most widespread fossil fuel on Earth. It took several millions years of high temperatures, pressure and chemical processes to turn the trees and other plants buried among the sediments and alluvia into coal. Beside carbon and hydrogen, coal consists of sulphur, nitrogen, ash and vapour. At larger depths, different kinds of heat reactions took place because of greater temperatures and pressure. Vapour partially extracted sulphur, oxygen and nitrogen; at the same time, organic substances started to decompose into liquid molecules. This is how petroleum came into existence. It consists of carbon, hydrogen, sulphur and nitrogen. Oil is a mixture of hydrocarbons and is distilled in oil refineries before use. This is how petrol and fuel oil, as well as gases, among which the best known are propane and butane, are produced. Both gases liquefy at relatively low pressure and occupy only 1/260 of the volume that they occupy in the gaseous state, which enables simple transportation to consumers by road tankers. The mixture of propane and butane is also known as liquefied gas (LG). At even larger depths, gaseous fossil fuel has been formed: natural gas that contains carbon, hydrogen and nitrogen. This liquefies at very high pressure and with high energy consumption; therefore, it is transported by gas pipelines. In Europe, a vast network of pipes for gas transportation has been constructed. Aggregation fuels also differ in the amount of chemically bound energy per mass unit, which is called calorific value, and in their environmental impacts through their transformation.

An important energy source is also represented by the nuclear fuels. The best known are uranium, plutonium and thorium. In peacetime energetic conversions, only the uranium isotope U^{235} is used. Nuclear energy is released by splitting the atom nuclei. The atom nucleus splits into two lighter nuclei after it has captured a neutron; new neutrons are also released, which launches a chain reaction. A nuclear reaction can be controlled in a nuclear power plant if it takes its course in water, heavy water or even in graphite providing that some of the formatted neutrons are absorbed and that very dangerous γ radiation is absorbed in an appropriate shield.

The rapid growth of civilization, over the past two centuries, was largely due to intensive consumption of fossil fuels. Even today, three-quarters of the world's total energy supply is produced from fossil fuels; the share of nuclear energy is also significant. In less developed countries, the share of renewables – principly, biomass and water energy – are of primary importance. With regard to energy, we distinguish between:

- primary energy in fossil and nuclear fuels, and renewables;
- secondary or end-use energy, which is produced from the primary energy sources by converting their internal energy into other forms that are needed by

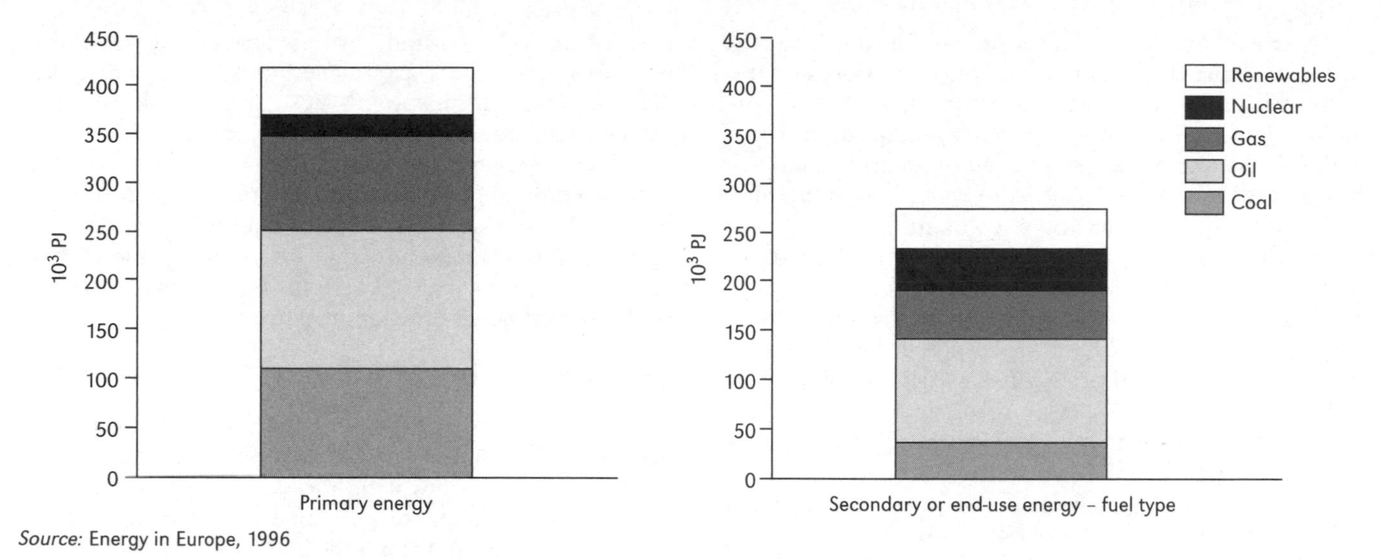

Source: Energy in Europe, 1996

Figure 12.4 *Primary and secondary or end-use energy use in the world in 1996*

consumers (heat, electricity and solid, liquid and gaseous fuels);

- useful energy emitted from numerous installations, such as heating and cooling systems, lamps and other devices.

Energy conversion and environmental impacts

The conversion of chemical energy into other forms has a strong impact on the environment, both in urban and rural areas. The most important transformation of fossil fuels is combustion. This is a process in which chemical energy is transformed into heat by the process of oxidation. Through this process products (air pollutants) are emitted into the atmosphere in the form of exhaust gases. Among the emitted pollutants from the burning of fossil fuels, those that pose the greatest threat to the environment are carbon dioxide (CO_2), carbon monoxide (CO), nitrogen oxides (NO_x), sulphur oxides (SO_x) and suspended particles. The amount of emitted matter depends upon the composition of fuel and the quality of the appliance in which combustion takes place.

CO_2 and CO result from the combustion through oxidation of the carbon from the fuel. CO_2 does not have a harmful effect on people by itself as long as the concentration is less than 1000 to 1500 ppm. However, it increases the greenhouse effect in the atmosphere (the thermal insulation encircling the Earth). Concentration of this gas had changed little (between 180 and 280 parts per million, or ppm, which is the unit used to measure very small concentrations; 1 ppm represents a mixture of 1 volume unit of gas in 1 million volume units of air) over the last 1000 years, until the Industrial Revolution began at the end of the 18th century. Since then, however, this concentration has been continuously increasing, already reaching 350 ppm in 1990. Alongside CO_2 concentration, air temperature has also increased. During the last 100 years, it has risen by 0.3 to 0.6 Kelvin (K). CO_2 does not allow long-wave Earth radiation to spread into the universe, thus increasing the greenhouse effect. Gases that have this effect are called greenhouse gases. Beside CO_2, these are methane, NO_x and a compound of fluor, chlorine and carbon. The consequences of the greenhouse effect are thought to be higher sea levels, which would cause massive floods over large coastal areas, and climate change in, for the time being, fertile regions. CO, which affects the capability of oxygen transportation in the blood, represents a much greater threat to people. When inhaled, it links immediately with haemoglobin to form carboxihaemoglobin (COHb). A concentration of 10 per cent of COHb in the blood will cause a headache; a 50 per cent concentration, however, can be lethal.

Nitrogen oxides released by the burning of fossil fuels originate from the nitrogen in fuel and from air, needed for the combustion process. Altogether, they come in various forms and are labelled NO_x. In the atmosphere, they react with hydrocarbons under the presence of the sun's radiation into hazardous photochemical smog. In London, from 12 to 15 November 1952, exceptionally high concentrations of NO_x SO_2 and solid particles were registered. During this period the rate of deaths caused by respiratory organ difficulties and heart problems increased by 160 per cent above the average value.

Nitrogen oxides, together with sulphur dioxide (SO_2), cause acid rain; they are dissolved in water drops and thus make diluted acid. Besides forest fires, acid rain is the main cause of forest decline. Acid precipitation reacts with the calcium carbonate present in limestone and thus forms gypsum, which dissolves in water and is therefore rinsed by rainfall. It has been estimated that, in urban areas, the annual damage due to acid raid corrosion amount to several tens of billions of Euros.

The burning of fossil fuels also creates solid and liquid particles of different shapes, sizes and chemical composition. These can comprise dust particles emitted by different materials (the size of dust particles is 1 to 100μm), fumes caused by the evaporation of metals (the size varies between 0.03 and 1μm), aerosols (mist and sprays with particle sizes of between 0.5 and 3μm) or smoke emitted by the process of fossil fuel burning containing 0.05 to 1μm sized solid particles. Suspended particles of between 0.5 to 10μm in size are particularly dangerous to human respiratory organs. These are labelled as PM10 (particulate matter, 10μm). They are inhaled into the lungs where they can cause lung diseases.

Volatile organic compounds (VOCs) are any organic compound that evaporates readily into the atmosphere. In car exhaust gases, there are as many as 100 different VOC substances, such as benzene and 1.3 butadiene. Benzene is an organic chemical that makes up about 2 per cent of petrol. The main source of any benzene in the air is from vehicle exhaust emissions. It is a genotoxic carcinogen (i.e. a substance that causes cancer). It is not present in vehicle fuel, but is formed by chemical reactions when fuels are burned and is emitted in the exhaust. In the atmosphere, VOCs react with nitrogen oxides on hot summer days to form ozone (smog).

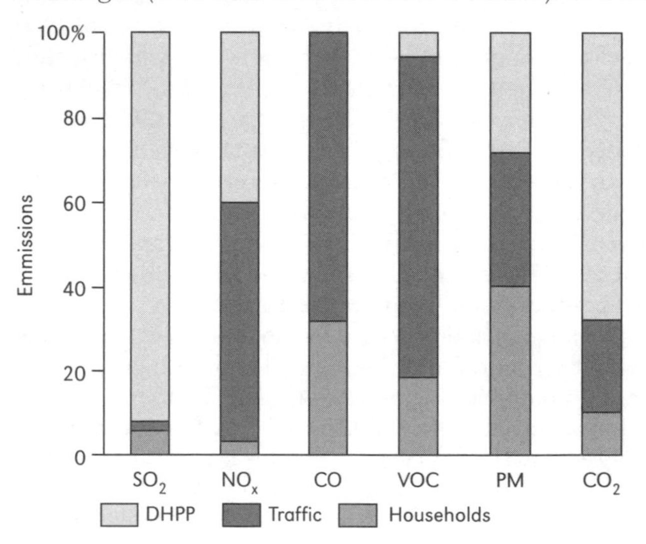

Source: Kraševec, 1998

Figure 12.5 *Emissions from households, traffic and a co-generation district heating power plant (DHPP) in a mid-sized city (300,000 inhabitants)*

As can be seen from Figure 12.5, emissions of volatile organic compounds from traffic are more than ten times higher than those from district heating power plants (DHPP), while emissions of CO are a hundredfold higher. Traffic also causes almost twice the emissions of nitrogen oxides regarding all other sources.

Emissions control

Because fossil fuels represent the main source of our energy supply, they cannot be replaced with environmentally friendlier sources within a short period of time, for both technological and economic reasons. However, fossil fuels also differ in their environmental impact. They should contain more hydrogen and less carbon, as well as less sulphur and nitrogen; hydrogen burns out into harmless water vapour. Solid fossil fuels are built up of long molecules of hydrocarbons, containing a large number of carbon atoms; gaseous fossil fuels, on the other hand, are composed of short molecules which have three to four times more hydrogen atoms than carbon ones. Therefore, through heat production, we substitute solid fossil fuels with liquid ones; in transport, more gaseous fossil fuels are used instead of liquid fuels. Furthermore, liquid fuels produced from biomass are being introduced (e.g. bio-ethanol, bio-methanol and bio-diesel) with a smaller impact on the environment. In the next 30 to 50 years, hydrogen, which only emits water while being burned, should be in massive use.

Besides heat, the burning of fossil fuels results in formatting the products, which are released into the atmosphere through exhaust gases. The emission of harmful substances in exhaust gases can be effectively reduced by elimination of sulphur dioxide and dust particles in a process known as gas cleaning. For 'de-sulphuration', 'wet' methods are generally employed through which water-diluted limestone ($CaCO_3$) is injected into the flow of the exhaust gases. Sulphur dioxide in chemical reaction is transformed into environmental friendly gypsum, which is then deposited in landfill. Solid particles (aerosols), however, are eliminated through electrostatic filters and fabric filter bags. In vehicles powered by internal combustion engines, catalytic converters are used to enable simultaneous oxidation of unburned hydrocarbons and CO into CO_2 and reduction of NO_x into N_2. In the case of diesel engines a special filter can be used to eliminate the particles from exhaust gases.

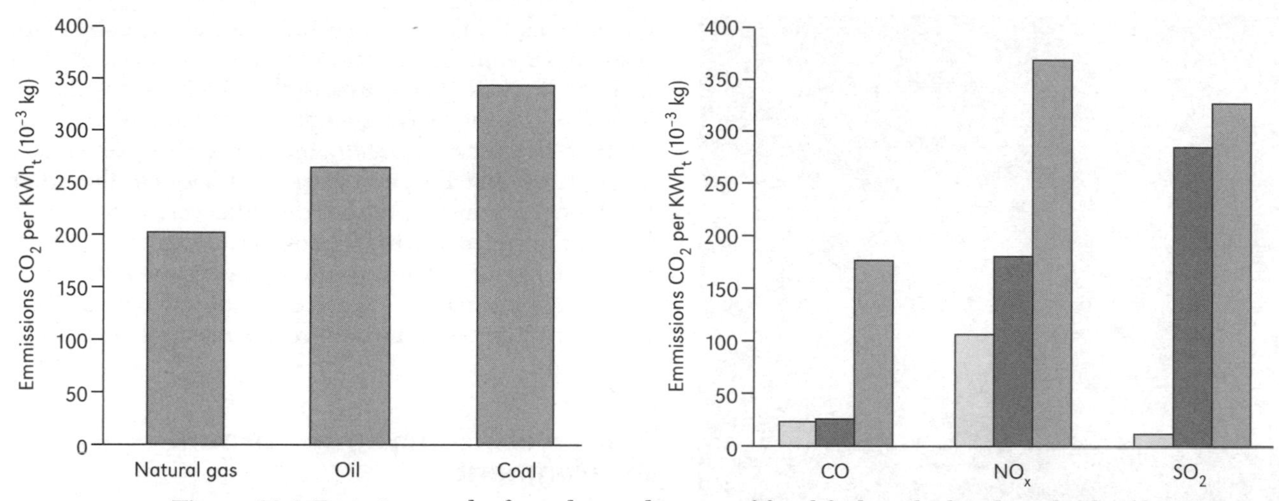

Figure 12.6 *Emission results from the combustion of fossil fuels, calculated per kWh of heat*

The most effective way of reducing environmental impacts is through energy saving, the use of efficient conversions of fossil fuels and their substitution with renewable sources of energy. Some possible solutions are discussed below.

Energy use in cities

The most important energy sources in cities are also fossil fuels, which are converted into heat, mechanical work and electricity. In some countries, nuclear power also plays an important role in energy supply (in France and Switzerland), and water energy is important in others (Austria and Slovenia).

Use of energy in cities depends upon the climate, building and transport, industrialization, life habits and standards of living. If we look at the energy consumption across different sectors, we can see that the biggest consumer in cities is the domestic sector. This is a result of energy consumption for heating and cooling, dense construction and a larger number of home appliances than in rural areas. A lot of energy in cities is spent on transport, too. Since 1970, the consumption of energy in

Fabric filters
Clean gas
Polluted gas
Large particles
Landfill

Source: Energy in Europe, 1996

Figure 12.7 *With fabric filters, 98 per cent of particles smaller than 1 to 5μm are removed*

Note: The rate of SO_2 elimination in this power plant is 94 to 95 per cent; the amount of used limestone is up to 16 tonnes per hour; and the amount of eliminated plaster and solid particles reaches 70 tonnes per hour.

Figure 12.8 *In a thermal power plant with 275 megawatts (MW) power, the wet method for 'de-sulphuration in the exhaust gases' washer and electrostatic filters are used for particle removal*

domestic and commercial sectors has increased by more than 50 per cent and by 10 to 15 per cent in transport in almost all cities. The latter is the result of population growth and the increasing number of cars. Statistics show us that in cities the average number of passengers amounts to only 1.4 persons per vehicle. On the other hand, energy consumption in the urban industrial sector has been reduced due to technological restructure. In the domestic sector, heating buildings consumes the most energy (approximately 75 per cent); hot water production requires 10 per cent and the rest is used for household appliances.

Energy efficiency in the urban environment

There are several possibilities for reducing energy consumption in cities, and for reducing environmental pollution and improving the quality of life. Among them are:

- demand-side management;
- introduction of central supply systems;
- integration of renewable energy sources within the energy supply chain.

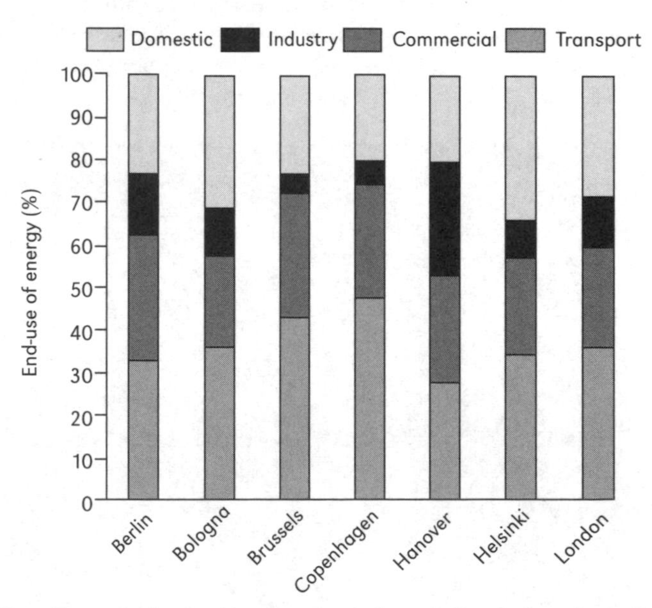

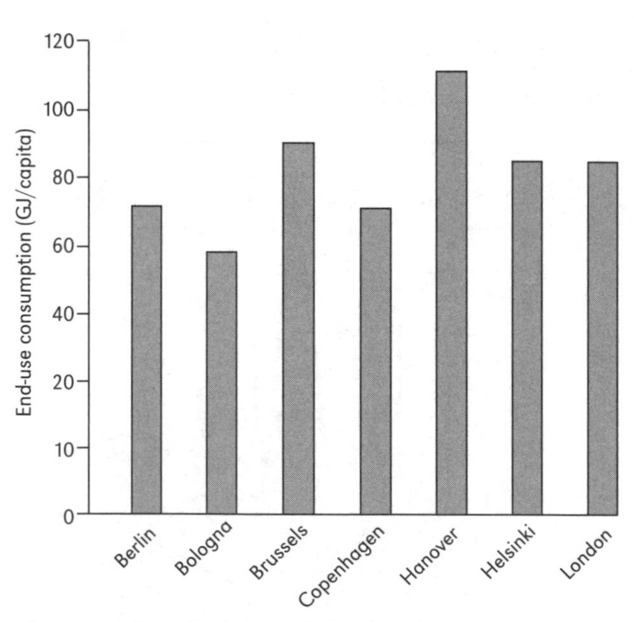

Note: The results for the cities are quite similar; only Hanover stands out due to the greater share of industry and, therefore, the greater per capita consumption.

Source: Europe Environment, 1995

Figure 12.9 *Relative end-use of energy consumption in cities for different sectors* (left) *and absolute end-use of energy consumption per capita* (right)

Demand-side management

The continuously increasing use of energy in cities can be fulfilled not only by 'supply-side' actions, which increase energy supplies, but even more efficiently by 'demand-side' actions or management (DSM). This action can be implemented for all resources or energy flows in cities. This section concentrates on demand-side energy management. Demand-side management of other resources – such as water and material flows – is presented in the following chapters.

Demand-side management encourages consumers to change their energy use and to use energy-efficient appliances, equipment and buildings. There are four basic DMS strategies:

1 *Peak-load clipping* seeks to reduce energy consumption at the time of daily peak. Examples include utility or user control of appliances such as air conditioning or water heating, or a timer for water heating.
2 *Valley filling* smoothes the load and improves the economic efficiency of systems. For example, electrical vehicles can be charged at night when electricity use is lower then during the day.
3 *Load shifting* can be accomplished through measures such as thermal storage; heat or cold can be provided and stored in a form of sensible or latent energy when consumption is low and can be used during high consumption periods.
4 *Practice energy conservation.* Lower consumption of energy sources can be achieved by individuals and cities, as a whole. As tenants, we can affect energy use, for example, with improved thermal insulation of our building; another possibility is more efficient ventilation with heat recovery, and buying the most energy-efficient appliances. For the assessment of a building's energy efficiency, energy labelling is also effective.

Transport use is another area where individuals can save energy by changing their life-style habits. It is a known fact that in the European Union (EU) 75 per cent of all private car drives are shorter than 8km in length, and one-third of them do not reach 1.6km. In order to overcome these short distances, it is more convenient to walk, cycle or to use public transportation. In Table 12.1 the energy required and the transportation emissions are given for various forms of transport.

On the city level, energy efficiency is highlighted by a wide range of promotion campaigns and design and planning strategies, which support and promote environmentally 'friendly' technologies and public awareness in both the domestic and transportation sector – for example, efficient energy transformations, planning and using district energy, cycle-path networks integrated with urban planning, and efficient public transportation.

Demand-side energy management systems have several advantages, such as reduced cost to consumers, pushing new capacity further into the future, reducing environmental impacts, and providing flexibility in free-market conditions. Implementation of DSM programmes requires new skills and costs money; but costs are usually lower than the cost of new production units.

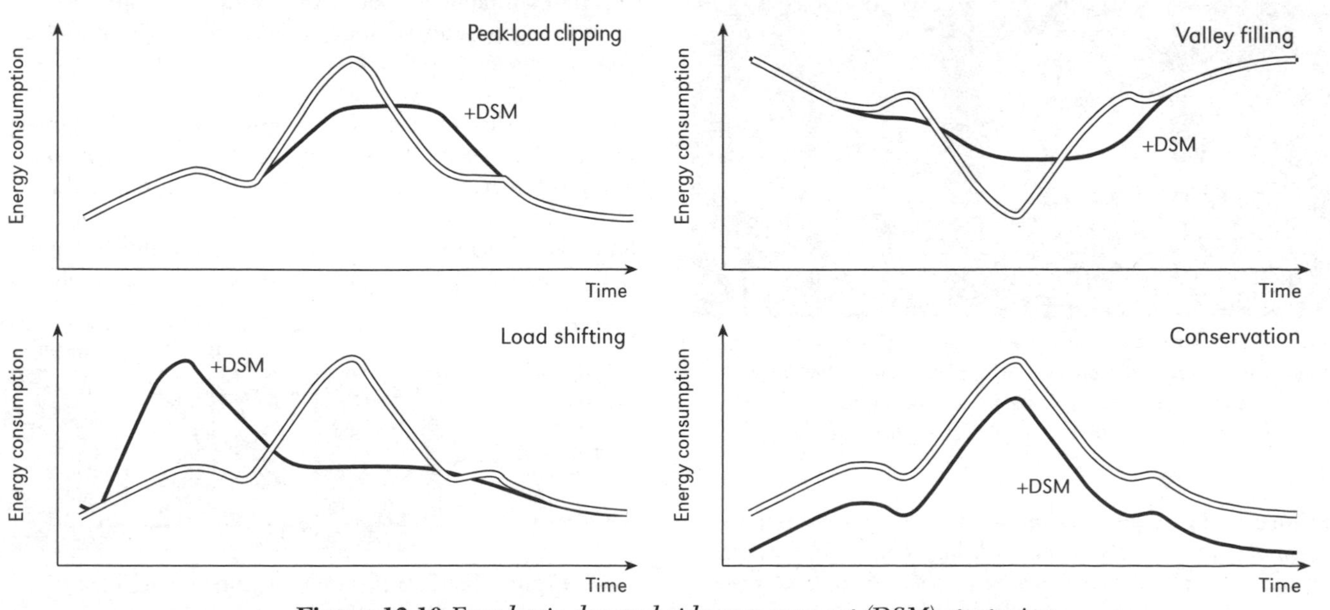

Figure 12.10 *Four basic demand side management (DSM) strategies*

Table 12.1 *Energy needed and transportation emissions for passenger and freight transport*

Passenger transport	MJ/km/person	g CO_2/km/person	Freight transport	MJ/tonne/km	g CO_2/tonne/km
Bicycle	–		Pipeline	0.25	–
Bus	1.0	35–60	Inland waterway	0.32	40–70
Train	1.5	40–80	Train	0.40	40–50
Car	3.2	130–200	Lorry	1.37	200–300
Plane	4.2	160–450	Plane	21.7	1160–2150

Source: Masters, 2001

Central-supply systems

Central-supply systems are one of the most important services in cities. They transport energy in the form of gas; electricity; heat; or water from a central source to single buildings, and collect the dray or black water and even the waste from individual buildings and transport them to the central treating plants. Taking into account the huge amounts of energy used in residential and commercial buildings, and the impacts of energy conversion processes on the environment, district heating and cooling systems have great advantages compared with individual supply (see Chapter 13).

District heating and cooling systems

District heating and cooling systems transport heat and/or cold via steam or hot or chilled water pipelines. The systems consist of three primary components: the central plant, the distribution network and the user heating system.

Note: Buildings like this consume at least 200 to 300kWh of heat per square metre of residential area in a year, while buildings that are well insulated, with low-emission windows, and with heat recovery by ventilation, consume only 40 to 50kWh heat per square metre per year.

Source: Zupan, 1995

Figure 12.11 *Older building with no insulation* (left) *and thermal picture of the same buildings* (right)*; the brighter spots show the thermal bridges, spots where heat loses are particularly large*

The heat source in a central plant can be a boiler or a refuse incinerator, or renewable energy sources such as geothermal and solar energy. The boiler may be fired by any fossil fuel – coal, gas, oil or biomass. The piping network conveys the energy between the source and the buildings. Usually, the pipeline consists of a combination of pre-insulated and field-insulated pipes in concrete tunnels or in direct buried application. The third component of district heating and cooling systems is a consumer system, which includes heating and cooling systems in buildings. In most cases, the district pipelines and building service system are separated by a heat exchanger.

The district heating and cooling systems are expensive and they can be used in case of the following:

- High thermal load density: this is characteristic of areas with high buildings and high population density in cold climates. In such cases, the capital investment for a distribution system, which usually costs 50 to 75 per cent of the total investment, will be covered.
- High annual load factor: this occurs when the district system operates with constant power throughout the whole year.

As a result, district systems are cost-effective in densely populated (at least 50 buildings per hectare) urban areas, for high-density building clusters with high thermal loads or in industrial complexes. These district systems have significant thermodynamic, economic and environmental advantages compared with individual systems, as follows:

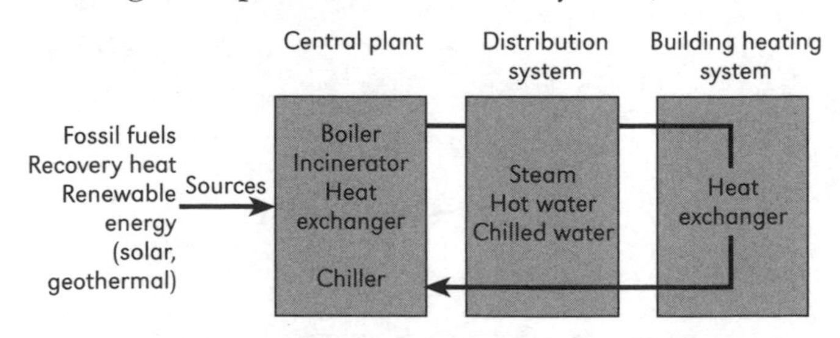

Figure 12.12 *Major components of a district heating system*

Figure 12.13 *Pipelines are, in most cases, pre-insulated and buried in the ground*

Figure 12.14 *Heat sub-station of the building, with 40 flats; on the left is a heat exchanger, which connects the district heating system with the one in the building*

- A larger central plant can achieve higher thermal efficiency than smaller ones; the differences are even larger in the case of partial load operation because larger plants operate with more heat sources, which can switch off separately.
- Larger units can be adapted to the use of various fossil fuels, thus making it possible to utilize the least expensive fuel. Furthermore, sources of waste heat and renewable energy can be exploited.
- District heating and cooling systems have the advantage of measuring energy consumed by users with simple remote heat meters, as well as the possibility of efficient remote surveillance of the system's operation.
- Heating and cooling systems that are operated by personnel can be reduced because less mechanical equipment in buildings requires less maintenance.
- Usable space in buildings increases when boiler- or chiller-related equipment and fuel storage are no longer needed.
- Emissions from central plants are easier to control; air pollution caused by the transportation of fuels to consumers and ash removal is reduced, as is the danger of spillage of liquid fossil fuels in case of road accidents.
- In the case of district cooling systems, the leaking of environmentally unfriendly refrigerants can be monitored readily in a central plant.

District heating systems

One of the most common classifications of the district heating system is based on different temperatures and mediums of transport fluid. Hot water or steam can be used as transport fluid. The latter is common in industrial areas because steam can be used not only for heating or cooling, but for industrial processes too. Hot water systems are divided into three temperature ranges: high-temperature supply comprises temperatures over 175° Celsius; medium-temperature supply comprises temperatures in the range of 120° to 175° Celsius, and low-temperature systems comprise temperatures of 120° Celsius or lower. The medium-temperature systems are common for district heating of residential areas because of lower pressure, lower thermal loses and lower leakage losses. They are designed for a high temperature drop (50 to 60K) in the building's heating system heat exchanger in order to reduce the flow rate of the transport fluid and pumping power.

The thermodynamic and economic efficiency of district heating systems can be increased by introducing co-generation: simultaneously producing electricity and heat. Large amounts of heat at lower temperatures, which are emitted into the surroundings by thermal power plants, are the reason that the primary energy efficiency rate of fossil, biomass and nuclear fuels only reaches 30 to 40 per cent. Waste heat, however, can be applied to district heating. Energy efficiency by co-generation can rise up to 80 or 85 per cent. Buildings, however, need

Note: The distant heating systems can operate using solar collectors instead of fossil flue boilers.

Figure 12.15 *District heating and power plant and part of 156km long supply pipelines of district heating system in mid-sized city; hot water system (130° Celsius/70° Celsius) provide heating for 45,000 residental units* (left)*; recently, the 24,000 cubic metre large hot water storage was built with a thermal capacity of 850MWh for peak-load clipping* (right) *instead of enlarging the thermal capacity of the boilers*

heating only in winter; therefore, the all-year energy efficiency of the primary fuels' energy amounts to around 55 per cent. In smaller co-generation systems, the diesel- or gas engine-driven generator can be used. Heat can be recovered from the exhaust, cooling and lubrication engine system. Large units have steam turbines and heat can be recovered from expanded steam leaving the turbine.

As cities grow continuously, one of the major difficulties in planning district heating systems is predicting future heat consumption. When the system is already built, there are still several options for connecting new settlements to an existing district heating system:

- by promoting and implementing demand-side management policy;
- by planning low-energy buildings with low-temperature heating systems (buildings can be connected to return pipeline because the temperatures of transport fluid are high enough for heating such buildings);
- by integrating heat storage (heat storage decreases daily peaks of heat consumption and provides higher and more economical electricity production in co-generation plants during the day time).

District cooling systems

Modern buildings are cooled in two basic ways: with local and central cooling (or air-conditioning) systems. These appliances need electricity for their operation and they primarily use environmentally unfriendly refrigerants. In cities, where there is greater building density and because of the heat island effect, and due to the increased need for cooling these buildings, district cooling systems can be cost-effective.

There are two types of district cooling systems: open and closed. With open systems, cold water from deep pits and larger sea depths (500 to 1000m), which is pumped onto the surface, flows through heat exchangers. Water is cooled in a central water pipeline. In this case, the cold source water temperature should not exceed 5° Celsius.

For example, seawater has been used to cool the central system in Stockholm since 1995.

In closed systems, cooling energy can be produced in a central plant. In the central unit, peak electricity consumption is lower, as cold can be effectively stored. Cooling energy can be transported by cold water, ice slurry or brine to buildings. Through the use of slurry and saltwater, the temperatures of the cooling medium are lower; therefore, smaller quantities are required and pumping costs are lower. Supply temperatures are normally around 5° to 7° Celsius, with return temperatures of between 10° and 15° Celsius. These low temperatures are necessary for cooling and the dehumidification of indoor air.

When the high temperature source or waste heat (>110° Celsius) is available during the summer, heat can be turned to cold with an absorption cooling system. Instead of hazardous refrigerants, which contain chlorine and fluor, binary solutions of the substances are used whereby one of them absorbs the vapour of the other. This liquid is called absorbent; the substance it absorbs, on the other hand, is known as cooler. The most commonly used substances are water and ammonia, or water and lithium bromide. The co-generation systems can be upgraded with an absorption cooling system. In this case, the system is referred to as triple generation (simultaneous production of heat, cold and electricity). Therefore the waste heat can be utilized throughout the year and the efficiency of the co-generation system is much higher.

Although, district cooling systems are most widespread in the US, applications also exist in Europe. In Sweden, the first system was built in 1992. Today, however, over 20 district cooling systems, covering a length of 85km, are in operation. In France, five systems are in operation; the largest is in Paris, with a cooling power of 220MW. In Germany, local systems of district cooling are the most common. They use absorbtion cooling systems, which are connected to the hot water network of district heating.

Integrating renewable energy sources

Unlimited use and great potential are the main characteristics of renewable energy sources. Most of them present very little or no danger to the environment. They can be used by individuals or they can be part of a larger central supply system, such as biomass and thermal solar systems for district heating systems or solar cells and wind turbines for electricity generation.

Biomass

With a 14 per cent share of the world's energy supply, biomass qualifies as the most important renewable energy source. In European alpine and some Scandinavian countries, the share of biomass in the primary energy supply reaches almost 20 per cent, while the European average amounts to only 2 to 5 per cent. For direct usage and biomass processing, various processes are used: burning; biological conversion with natural processes such as anaerobic fermentation; fermentation and composting, which occurs through microbes and enzymes; as well as thermal-chemical processes – for instance, pyrolysis, liquefaction or gasification. Among these processes, the most widespread is burning of the solid biomass and heat production for heating buildings.

Modern biomass solid fuels are made through processing the remains of forest biomass, agricultural plants and energetic plants (e.g. willow, poplar and Chinese reeds). Before being burned, they are processed into various shapes: chopping pieces, pellets, briquettes or bales. The advantages are that these fuels are easier to transport, provide better efficiency of boiler installations and feature lower emissions during burning. In some cities, a distribution system similar to the oil supply system has already been established.

When burning wood biomass, besides common pollutants, different organic compounds (C_xH_y) and small amounts of heavy metals (mercury, lead and chrome) contained in wood occur. Modern technologies for burning wood biomass have automatically regulated fuel and air intake, as well as automatically removing the ash, and there are filters for eliminating unburned particles from exhaust gases. These kinds of installations are expensive; it is better to build bigger units for district heating. Denmark provides a good example. After introducing environmental coal taxes in 1990, 37 district heating operators decided to substitute coal with wood chips.

Various types of liquid fuels can also be made from biomass, among which bio-ethanol and bio-diesel are the most important for motor vehicles. Bio-ethanol is produced through the fermentation of plants containing sugar, starch or cellulose. Fermentation is a natural chemical process, where microbes in water and plants in solution use sugar and emit residues in the form of alcohol and CO_2. The amounts of alcohol produced from different agricultural plants are given in Table 12.2.

The best known case of using bio-ethanol is a programme in Brazil, where 70 per cent of the world's production occurs. Measurements of environmental pollution have shown that air pollution in Rio de Janeiro and São Paolo sank by one-quarter between 1978 and 1983.

Source: Singh, 1998; *Renewable Energy World*, 1999

Figure 12.16 *Wood pellets* (left) *and transportation of the baled straw from storage into a boiler in a central plant of a small-sized district heating system* (right)

Bio-diesel, however, is produced by processing vegetable oil. It is produced by compression of seeds of oil rape, soya, sunflower, peanuts and other plants that have oil in their seeds. Natural oil can be added to fossil diesel fuel or is processed in ester – for example, rapeseed oil methyl ester (RME). RME contains about 7 per cent oxygen; therefore, the emissions are lower. It is also biodegradable. Up to 900 litres of fuel can be produced per hectare of agricultural land sown by the oil plants. Over 50 RME refineries operate in Europe and in numerous countries it is already possible to buy bio-diesel in petrol stations.

Solar-assisted district heating

Solar energy can by converted to heat using a device called the solar collector. Together with a thermal storage, piping network, pump and control unit, this comprises an 'active thermal solar system'. The most popular hot water heating solar systems are solar collectors mounted on the roof or façade of the building. Solar collectors can also be used for heating settlements using district hot water heating systems. Solar systems can operate with a central or divided field of solar collectors, and with smaller, short-term (daily) or bigger, seasonal heat storage.

Central field systems are composed of a large number of large-panelled flat-plate collectors, which are placed on the ground on one spot. This kind of system is used in the Danish town of Marstal, where the largest solar heating system for district heating with 8000m^2 of solar collectors is used. As a rule, solar heating with divided collector fields has roof-integrated solar collectors.

Systems with daily heat storage are designed in such a way that on a sunny summer's day, the entire required amount of hot water is prepared; in winter, on the other hand, about 10 to 20 per cent of the heat needed for domestic heating is produced. The size of the field

Table 12.2 *Amounts of alcohol produced from different agricultural plants*

Plant	Annual production of bio-ethanol		
	tonne/hectare/year	litre/tonne	litre/hectare
Sugarcane	50–90	70–90	3500–8000
Sweet corn	45–80	60–80	1750–5300
White beet	15–50	90	1350–5500
Wheat	4–6	340	1350–2050
Rice	2.5–5	430	1075–2150

Note: 1 hectare = 10,000 square metres.

Source: Medved and Novak, 2000

Source: Renewable Energy World, 1998

Figure 12.17 *Vehicles can be supplied with bio-diesel fuel in numerous filling stations*

reaches between 0.03 to 0.05 square metres of solar collectors per square metre of the building heated area.

With the seasonal heat storage systems, heat is stored between summer and winter in water reservoirs. Because of seasonal heat storage, the SC field can be bigger (0.2 to 0.3m^2 per 1m^2 of the heated building area); the container volume, on the other hand, amounts to 2m^3 per each 1m^2 of the SC. The portion of the building's heating and hot water supply is between 50 and 80 per cent.

Note: A total of 1260 buildings are connected to the district heating system.
Source: Large Scale Solar Heating, 1999

Figure 12.18 *The solar collectors' field of the biggest solar system for district heating in the Danish town of Marstal; the white cylindrical vessel, in the lower left corner next to the solar collectors' field, is a 2000m^3 large heat storage, which is sufficient for one day's heat accumulation*

There are more than 50 solar-assisted district heating systems that operate in Europe with a surface SC of between 500 and 8000m^2 and with a total heat power of over 40 megawatts. Experience shows that large solar systems can be successfully complemented with otherwise cheaper renewable energy, such as biomass and waste heat.

Heating with geothermal energy

Heat that is stored within the Earth is called geothermal energy. It originates from the natural radioactive decay of uranium, thorium and potassium isotopes. Therefore, geothermal energy is constantly being renewed. Heat emerges from the Earth's core in various ways. Geothermal water originates from atmospheric precipitation, which penetrates deep into porous layers where they heat up. These layers are known as aquifers. The average temperature gradient is 30° Celsius per kilometre of depth, but in areas where tectonic plates are joined it can reach up to 100° Celsius per kilometre.

Geothermal energy is exploited in various ways. The oldest form represents the capture of thermal springs. Rainfall, which trickles through porous rock into great depths, warms up and comes back to the surface through natural cracks in the ground or artificially made boreholes. It comes to the surface in the form of hot water, seldom as steam. After energetic exploitation, it must be pumped back into aquifers in order to prevent thermal pollution of the surface water and to maintain the pressure potential of the aquifer. In areas where no aquifers are to be found, non-porous rock at larger depths is crushed by explosion. After that, water is diverted into a formed artificial aquifer, and afterwards pumped out again as hot water or steam. This method of geothermal energy exploitation is called 'hot dry rock'. Geothermal energy can also be pumped using vertical heat exchangers in pipe form, which are called vertical borehole heat exchangers. Because of lower temperatures, borehole heat exchangers are used as a heat source for heat pumps.

In cities, geothermal energy is used for various purposes, such as road heating, greenhouse heating for food production and as an energy source for district heating. In 1930, Reykjavik started to use district heating. Later on the system was expanded and today the entire city is heated this way. Geothermal energy is also used in Paris, where seven systems of district heating were built between 1980 and 1985. In Switzerland, where the use of heat pumps is quite common, several thousand borehole heat exchangers are used. However, except in Italy, there are no thermal springs in Europe with temperatures high enough that would be sufficient for producing electricity.

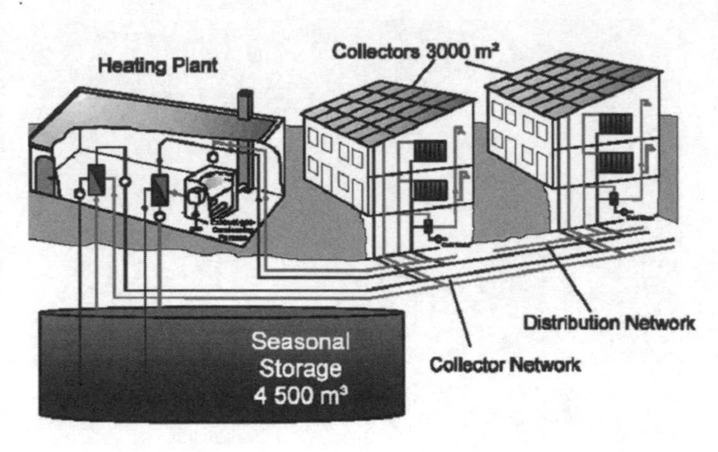

Note: One hundred and twenty-three single-family houses have solar collectors installed on their roofs with a total amount of 3000m^2 of surface; together with the buried 4500m^3 water heat storage, 50 per cent of heat required for domestic heating and hot water supply is produced by the solar system.

Source: Large Scale Solar Heating, 1999

Figure 12.19 *One of the first solar systems for settlement heating built in Germany operates in Hamburg*

The potential of geothermal energy varies according to the geological structure underground. Thus, it represents a significant source of renewable energy, first of all, in Iceland, Austria, Germany, France, and in the countries along the Panonian lowland in central Europe. Its usage, however, is associated with some environmental impacts due to sulphur dioxide (SO_2), hydrogen sulphide (H_2S) and nitrogen dioxide (NO_2) emissions, which are removed from geothermal water before its energetic exploitation, and because of saline fluids in it, which are corrosive.

Photovoltaic systems

By converting heat into mechanical work, electricity is generated in thermoelectric and nuclear power plants. Today, two-thirds of electricity is produced using this method. Solar thermal power plants function in a similar way by using mirrors to focus solar radiation. The energy of the sun, however, can be converted into electricity without thermal processes by using devices called photovoltaic cells. They are made of semiconductors, mostly from silicon. Photovoltaic cells consist of two layers of silicon, each with different electromechanical characteristics, which are connected to an outside electric circuit through which the generated low-voltage electric current is transported. Because of their fragility and smallness, photovoltaic cells are linked to modules and these are further linked to systems. Single appliances, a building or a settlement can be powered by using photovoltaic systems. The former are the low-power consuming devices and machines, such as lamps, parking meters and radio transmitters, and the latter are buildings, away from the public grid, and where we do not want to use fossil fuels,

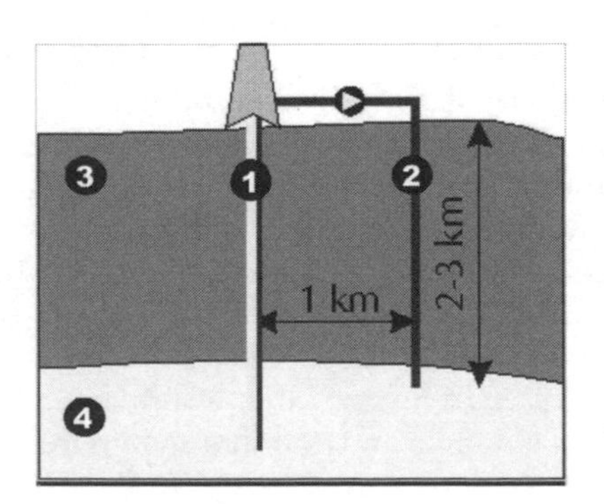

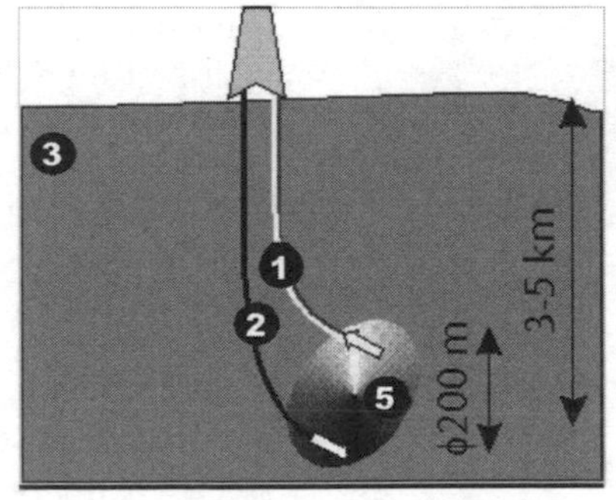

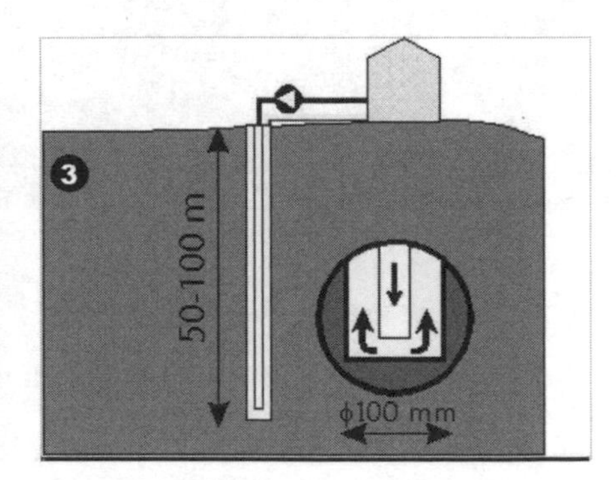

Figure 12.20 *Different ways of exploiting geothermal energy: thermal aquifers* (left), *hot dry rock* (middle) *and borehole heat exchanger* (right)

Source: Singh, 1998

Figure 12.21 *Reykjavik is the only capital that provides space heating entirely with geothermal energy*

which pollute the environment. For example, all buildings within national parks within Slovenia should be equipped with photovoltaic systems.

Larger systems of solar cells are called photovoltaic power plants. Buildings with integrated photovoltaic energy systems are of special interest for the electricity generated in cities. These systems use specially designed modules of solar cells in the shape of tiles, façade panels, shading devices and windows. They are used as a substitute for the common elements of the building's envelope, and in this way reduce the system installation costs. Furthermore, modules have a large life span and they do not need a lot of maintenance, as is the case with most construction elements of the building's envelope. Through electricity generation, solar cells do not cause any environmental pollution in terms of emissions and noise, and this is of extreme importance for cities.

Figure 12.22 *Luminaries with solar cell in the city park*

According to a forecast, the electric power of photovoltaic systems in Europe should increase from today's 52 megawatts to 2000 megawatts and as much as 65 per cent of all modules should be integrated within buildings' envelopes by 2010.

Wind energy

The use of wind energy for transportation, water pumping and milling has been known for thousands of years. In Europe, 50,000 small wind turbines were in operation since 1850. After the invention of the steam engine, their numbers rapidly declined, until recently. Large high-efficiency wind turbines have become cost-effective as electricity generation units. They have also become larger and cheaper. The modern wind turbine is 150m tall, has a rotor diameter of 100m and provides 3 megawatts of electrical power. Several wind turbines are connected in wind farms. They can be located on a city's surrounding hills and on shores or (off shore) on the sea, and are connected to the main electricity supply grid.

Note: Shading can be managed and glare controlled by choosing the right density of installed solar cells within window glass.

Source: European Directory of Renewable Energy Suppliers and Services, 1994

Figure 12.23 *The modules of solar cells can be made in the form of tiles, window glazing and façade panels*

Source: Singh, 1998

Figure 12.24 *Off-shore wind farm*

In the future, wind turbines will be constructed in two ways – as small units on industrial and office buildings or as large units as an integral part of specially designed buildings. Wind turbines can be combined with photovoltaics (PV) to ensure more constant electricity production because solar and wind energy complement each other: PV in summer; wind in summer and winter; PV during the day; wind at night.

Source: EnerBuild RTD, 2002

Figure 12.25 *Building-integrated wind turbines*

Water resources and management

Water represents the basis of life. Human health and the development of mankind depend a great deal upon water quality and a sufficient water supply. The science that studies the hydrological cycle on Earth is called hydrology. The processes of water movement in the lithosphere (land) are treated by hydrogeology; those of water and water vapour circulation in the atmosphere by hydrometeorology. Due to the radiation of the sun that reaches the Earth's surface, water circulates continuously. This circulation is called the hydrological cycle. It is estimated that around 23 per cent of the sun's radiation is consumed by this cycle. The amount of water involved is uncertain. Land is divided into two areas: arid or dry regions, and humid or wet regions. Water does not flow away from dry areas because precipitation evaporates before inland surface waters can be formed.

In cities, water and water surfaces have numerous functions. They improve microclimate and level daily temperature oscillations; they enabled transportation, as well as tourist and sporting activities; they render possible the production of mechanical work and electricity; and they provide storage of water for fire protection, to mention just a few functions. One of the most important functions is, of course, as a city's drinking water supply.

The total amount of water on Earth is estimated to be 1.36 10^9km^3; but only 3 per cent of this quantity occurs in the form of freshwater liquid. Water availability, as well as water quality, are also important factors. The main water source for water abstraction varies a great deal from country to country. In Spain, Belgium, The Netherlands and Finland there is surface water. In Switzerland, Slovenia and Denmark, on the other hand, there is groundwater. In Malta, an important portion of drinking water is provided through desalination of sea water. In Europe, it is estimated that 53 per cent of water is used in industry, 26 per cent in agriculture, and 19 per cent in households. Water use in Europe has increased from 100km^3 in 1950 to 560km^3 in 1990, which represents about 20 per cent of the available sources. Local conditions, however, can differ a great deal in terms of critical circumstances.

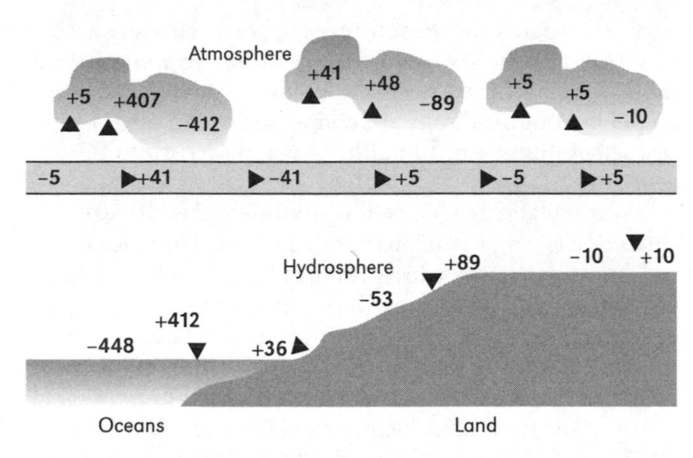

Figure 12.26 *Estimated amount of water within the hydrological cycle in millions of cubic metres (atmosphere, hydrosphere, arid regions, humid regions, lithosphere, oceans, land)*

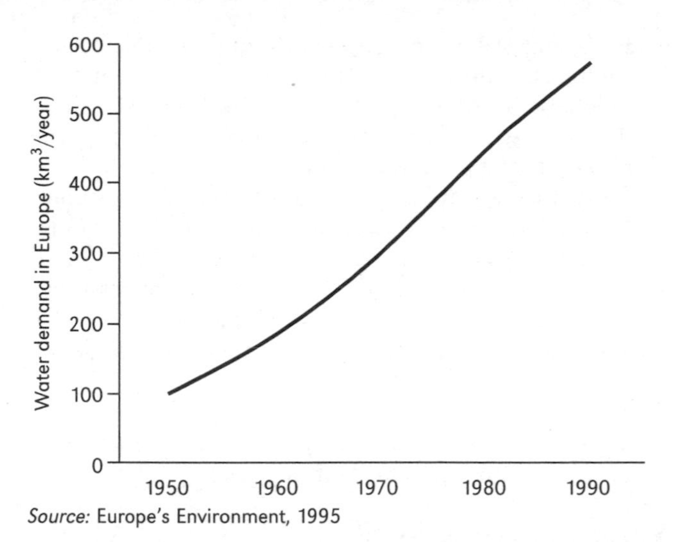

Figure 12.27 *Increasing water use in Europe during the last 50 years*

Water supply in cities

Water streams in cities are either natural (surface water or groundwater) or artificial (waterworks and sewage system). A city's water supply depends upon its waterworks. The increasing use of water is the result of a growing population and industrialization, as well as of broader use (e.g. for use in agriculture). Statistical evaluations show us that the use of drinking water in cities depends upon their size. Higher use in larger cities is an effect of a smaller number of people in lodgings, a higher number of appliances, as well as of leaking water works – between 25 and 30 per cent in France, and even 50 per cent in Spain and the UK, in particular cases.

More than 65 per cent of the population in Europe depends upon groundwater as a source of water supply. In the rural surroundings of many cities, however, water abstraction through pumping is greater than groundwater inflow, resulting in decreasing levels of groundwater. This affects vegetation, as well as the bearing capacity of the ground in the city, where building damage may occur.

Water management

Due to increasing pollution of the water circulating in the hydrological cycle in the atmosphere, the sources of drinking water are less and less usable. Water pollution is the result of numerous human activities, such as uncontrolled municipal and industrial waste disposal; the use of nitrates and pesticides in agriculture; the discharge of noxious substances; as well as the effect of natural geological characteristics, which are linked to wash-out of metals, minerals and salts, and the eruption of sea water into fresh groundwater. The problem is even more urgent because of very slow natural replenishment of groundwater. The speed of the groundwater replenishment in the majority of the cases does not exceed a few metres per month, or even per year.

Therefore, the protection and prudent use of existing water sources is necessary. In this way, enlarging of existing water sources can be avoided. Various measures are possible:

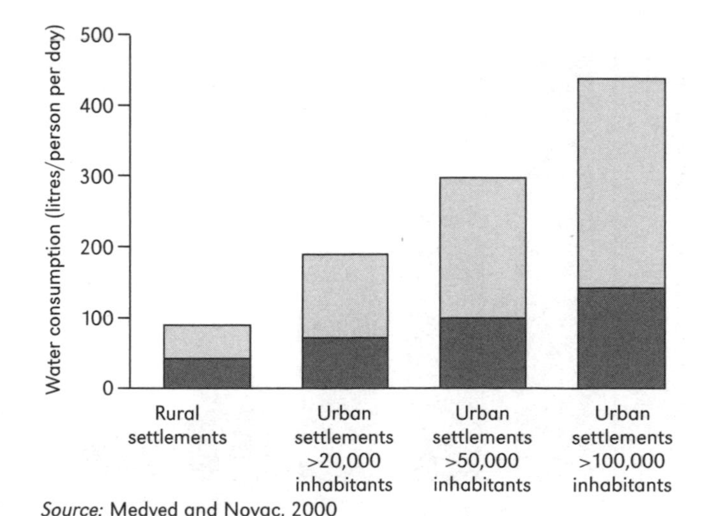

Figure 12.28 *Daily use of freshwater per capita: bottom represents minimum values; top represents maximum values*

- Reducing water consumption through technical fixes, such as the installation of percolators on water taps; toilets with variable quantities of flushing water (low-flow toilets); public campaigns and promotions; the implementation of water metering.
- Substituting drinking water with grey water – more than 60 per cent of water use in households can be replaced with clean waste water (grey water) and rainwater where there is no need for sanitary concerns, such as water for flushing toilets, laundry and car washing, garden watering and residence cleaning (Figure 12.29).
- Tightening up water networks, which is not just important for water savings but also for energy savings that are required for pumping.

Waste water treatment

After use, water is polluted; therefore, waste waters from households and industry are collected and drained off by sewage systems. Moreover, many city and building areas are covered with material that is impermeable to water (e.g. roofs and roads). As a result, groundwater must be collected and drained away. Roughly speaking, there are two types of sewage systems – mixed and separate systems. In the first case, we collect and drain away both waste water and rainfall; in the second, we do this separately. The latter is especially effective when non-polluted rainfall is collected (for instance, from buildings' roofs), which can be led away into underground streams, surface streams or reused (grey water). However, this way of managing waste water is rather expensive and therefore is not used very often. On the other hand, over 95 per cent of buildings are connected to the mixed-type sewage systems – in big cities, even more (up to 99 per cent).

Waste water has caused many infectious diseases. It contains a lot of organic matter as a result of food leftovers and human excreta, which represent the food for micro-organisms and rodents, as well as due to the presence of phosphate (the consequence of using purifiers and washing powders in households). For this reason, in modern settled cities waste water is purified in municipal waste-water treatment facilities (which were first built in 1960 and 1970) before being discharged into rivers, lakes or seas. Waste water is cleaned mechanically (removing sand and degreasing), chemically (neutralizing the swill) and biologically (decomposing organic ingredients). The capacity of purifying plants is measured by the population equivalent (PE) – that is, a unit of water burdening per capita per day. A schematic diagram of a modern, small waste treatment facility is given in Figure 12.30.

Material flows in cities

There is a constant inflow into cities of huge quantities of very diverse matter that we require for our survival (food) or for producing goods. Transformation of this matter and the production of goods emit part of these substances into air or water; a portion of unused matter remains as

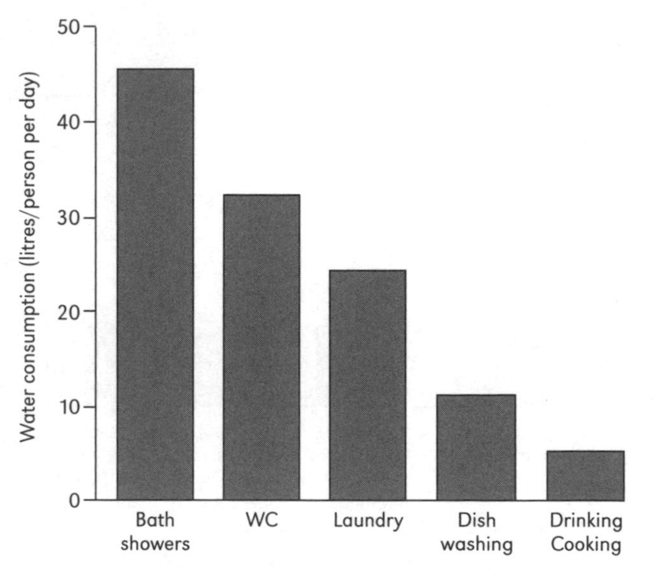

Source: Medvec and Novac, 2000

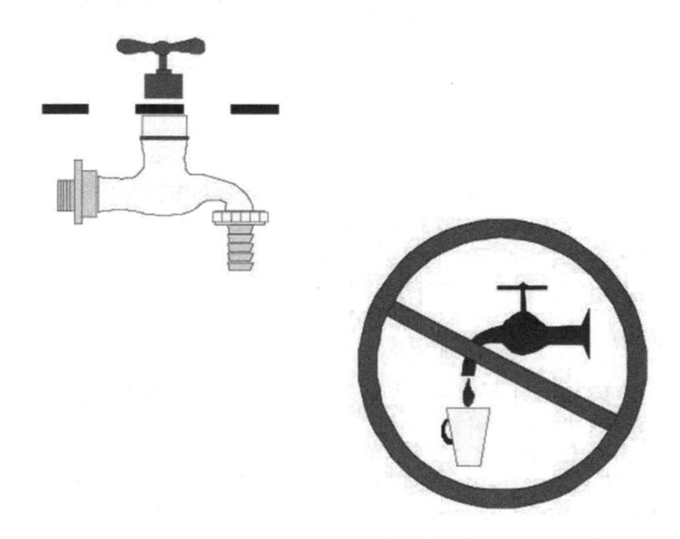

Figure 12.29 *Water consumption in households for different needs; all sites provided with rainwater should be marked with a sign that makes clear that water is not potable*

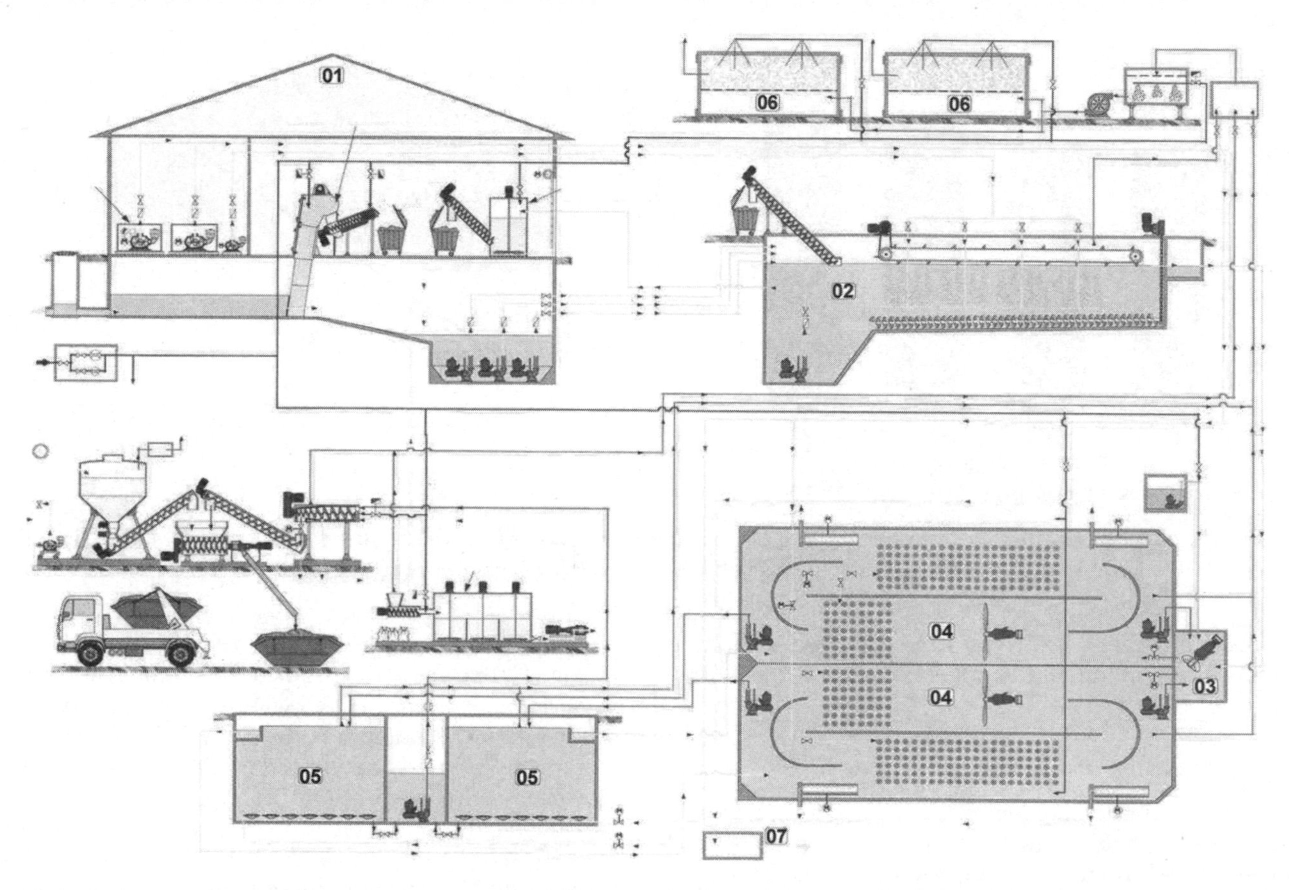

Note: Mechanical parts are removed from waste water on screen (01); in a grit chamber and grease catcher (02) sand is collected at the bottom and the grease skimmed. Biological decomposition of organics by micro-organisms takes place in two aeration tanks (04) into which air is blown; micro-organisms consume organics as their food. Sludge is being removed periodically, dewatered in a sludge digester (05) and disposed of. Purified water is being pushed into an outflow shaft and from there into the nearby river. The treatment plant shown is of a closed type due to the vicinity of a settlement, so all appliances are situated in closed spaces under the ground. Air, which contains smelling matter, is pumped into a bio-filter (06), thus preventing the stench from spreading into the surroundings.

Source: Water, 1999

Figure 12.30 *Municipal purifying plant with the capacity of 8000 population equivalents (PE)*

leftovers. The former are known as emissions, the latter as waste. Waste represents liquid, gaseous or solid state materials, which the owner does not want to preserve or use. As a result of all human activities, waste matter also originates in urban areas. According to its origins in the urban environment it can be physiological (human and animal excreta), a product of human activity (households, restaurants, offices, industry, remains of demolished buildings, etc.) or natural (leaves, grass, etc.). We call it municipal waste. Increasing amounts of municipal waste are evident in all countries, and its quantity correlates to the standard of living.

In developed countries, 250 to 500kg per capita of municipal waste is produced in a year. The correlation between standard of living and waste production in cities shows that, because of a higher living standard, the amount of waste is increasing, above all, as a result of food leftovers and packaging. The structure of municipal waste in which organic waste and (despite recycling campaigns) paper products abound confirms this fact.

Besides municipal waste in urban areas, hazardous waste is also produced, such as batteries, paint, oil and medicaments, as well as industrial waste and refuse from hospitals and research institutes. Although the amount is smaller, it represents a much greater threat to the

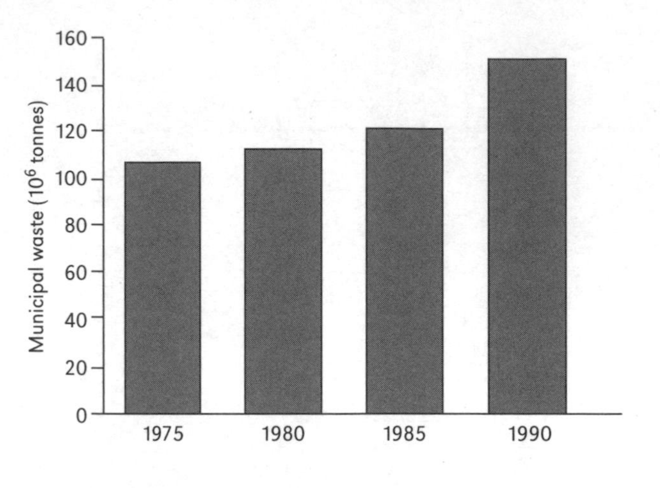

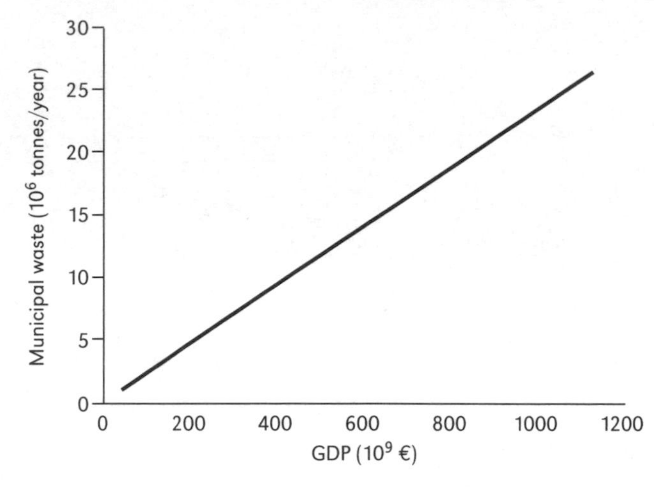

Source: Europe Environment, 1995

Figure 12.31 *Municipal waste generation in Organisation for Economic Co-operation and Development (OECD) countries* (left)*; municipal waste generation in relation to gross domestic product (GDP)* (right)

environment and therefore must be collected separately. Furthermore, it demands special transport and final treatment.

Waste treatment

There are various technologies in use for municipal waste treatment. A huge amount of waste contains matter that can be reused or recycled in raw materials. Thus, the volume of the waste is reduced, as well as the energy consumption for the production of raw materials. In Table 12.3, possible reduction of energy consumption and pollution through the recycling of varied materials is shown.

The amount of waste that is recycled varies a great deal from country to country. It is estimated that in Europe between 20 per cent (Norway, UK) and 60 per cent (The Netherlands) of waste paper is recycled; up to 60 per cent of aluminium is recycled; and between 20 (Norway, Greece, UK) to 63 per cent (The Netherlands) of glass is recycled. Re-use and recycling represent the waste treatment methods through which the least emissions are caused.

Municipal waste also contains organic substances, which can disintegrate into compost in composting facilities and then be reused as a fertilizer in gardening and agriculture. This method of waste treatment occurs in Spain, Portugal, Denmark and France, where between 6 and 21 per cent of collected municipal waste is composted. Composting is especially effective when used in households in their own environment.

In large cities, another effective method for waste treatment is waste incineration. The main purpose of waste incineration is the effective reduction of its volume, to about 20 per cent of the former content. The technology originates from the UK, where, more than 100 years ago, waste incineration was developed. In modern incinerators the released heat is used for generating electricity. However, incinerators are rather expensive and therefore only big facilities (in which the average annual amount of burned waste reaches 100,000 to 200,000 tonnes) are being built. Yet, through waste incineration heavy metal emissions occur (cadmium, lead, mercury, chromium,

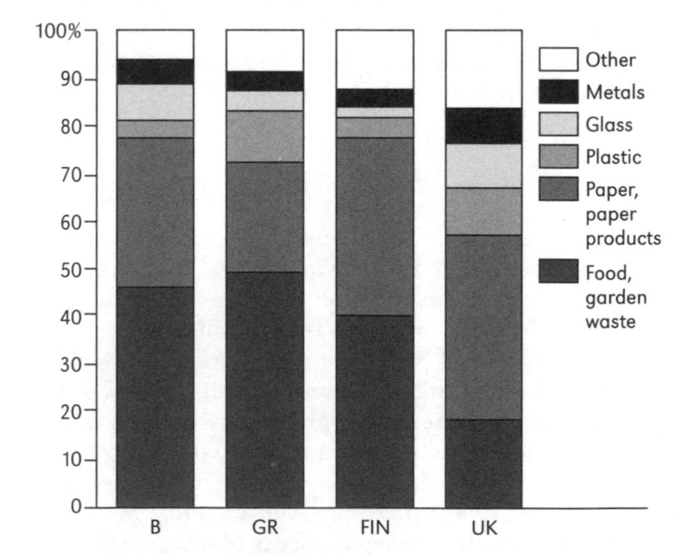

Source: Europe Environment, 1995

Figure 12.32 *Structure of municipal waste in selected European countries*

Table 12.3 *Possible reduction of energy consumption and pollution by recycling of various materials*

	Aluminium (%)	Steel (%)	Paper (%)	Glass (%)
Impact on energy				
consumption	90–97	47–74	23–74	4–32
air pollution	95	85	74	20
water pollution	97	76	35	-
mining industrial waste	-	97	-	80
water usage	-	40	58	50

Source: Europe Environment, 1995

zinc, etc.). Furthermore, there are also emissions of unburned residues and gases, which cause the acidification of precipitation (such as HCl, SO_2 and NO_x). These emissions can be substantially reduced in contemporary purification plants. However, this matter is also found in ash and scoria, which occur in large quantities (300 to 500kg per tonne of dry burned waste) after burning and gas purification. The proportion of municipal waste treated in incinerators is greater in countries with developed systems of waste treatment and smaller in countries where this does not occur.

In larger incineration plants, the released heat can be used for steam production and electric power generation. First of all, the waste is ground and mixed up; in this way the combustible compounds are distributed evenly. The waste is then fed into the boiler, where it burns through the addition of air and fossil fuels. The hot exhaust gases are cooled using water that turns into vapour and which expands in a turbine, thus producing mechanical work needed for electricity generation. The exhaust gases are cleaned in semi-dry scrubbers, in which sulphur oxides are eliminated, and in fabric or electrostatic filters, which retain solid particles. The cost of the energy, produced by incineration, is today quite high. Firstly, due to the low cost of waste removal and the high cost of the incinerators and, secondly, because of the very strictly conducted combustion process as a consequence of the severe regulation of the permissible exhaust gas emissions. However, waste removal is going to be much more expensive in the future. Moreover, new incineration technologies with better efficiency will be available. In spite of rigorous emission regulations and comprehensive emission controls, local public opinion is important and is very vocal about expressing its concern about emissions of dioxin and heavy metals. Nevertheless, until 2010, in Europe the rate of municipal waste incineration is projected to grow from today's 22 per cent up to 30 per cent. Analyses show that greater composting of organic waste and increased recycling should also increase the caloric value of waste.

The most widespread way of treating municipal waste is disposal on a landfill. Sanitary landfills are public structures with special requirements in relation to site selection, preparative arrangements, operation, closing, and final completion and restoration. During the preparation phase, rainwater and leakage are drained into the disposal purification plant. The bottom of the landfill and its sides are boarded with an impermeable layer (e.g. clay), which is covered with plastic liners. Waste is spread out in thin layers and compacted simultaneously with various machines. Within the pressed waste, decay of the organic matter (through anaerobic fermentation and without the presence of oxygen) takes place. A landfill gas is released during the process, consisting mainly of methane (45 to 55 per cent) and CO_2.

In uncontrolled waste disposals (open dumps), landfill gas soars in the atmosphere and thus increases the greenhouse effect. Methane is the most dangerous greenhouse gas besides CO_2. Landfills are, worldwide, the main cause, next to agriculture, of greenhouse gas emissions. Therefore, the landfill gas is collected and burned on atmospheric burners, called 'the torch'. In this way, the spreading of odours into the surrounding landscape and the danger of explosion is reduced. The share of landfilled municipal waste varies between countries: 50 per cent in Denmark and France, 93 per cent in the UK and 100 per cent in Slovenia. In larger disposal sites, we can use

Table 12.4 *Share of the incinerated municipal waste in various countries and the European average share*

Country	Percentage of incinerated municipal waste
Austria	11%
Germany	36%
UK	7%
The Netherlands	35%
Denmark	48%
France	42%
EU (average)	22%

Source: Europe Environment, 1995

Source: http://europa.eu.int/comm/energy_transport/atlas

Figure 12.33 *A contemporary waste incinerator in Birmingham, UK*

landfill gas for the production of heat and electricity. In 1970, it was first used in the US, and soon afterwards in Europe. At the design stage, it is important to estimate the potential of landfill gas. Although chemical processes start right away after waste is deposited, it takes many years to reach the maximum of the natural production of the methane. Later on, the reaction process slows down as the organic matter content diminshes. Although methane is produced for 50 to 100 years, the quality of the landfill gas is only suitable for energy recovery during the first 10 to 20 years.

Sanitary landfills occupy a large surface area and they have a great impact on the appearance of the landscape. In the long term, the extent of landfilled, biodegradable municipal waste should strongly decline (possibly by 75 per cent until 2010): for these wastes, composting seems to be more than adequate. This will affect the potential of the landfill gas in the future. Nevertheless, the power generated from these installations for the production of electricity is expected to rise from today's 700 megawatts to 1400 megawatts.

Influence of waste treatment technologies on the environment

Regardless of the selected waste treatment technology, we cannot avoid its impact on the environment.

Fortunately, everyone can influence the amount and type of waste in all phases of the life cycle of matter or products. The elementary way of solving waste problems is by reducing their quantity. With product design, designers can use methods of product assessment throughout the whole life cycle of a product. This method

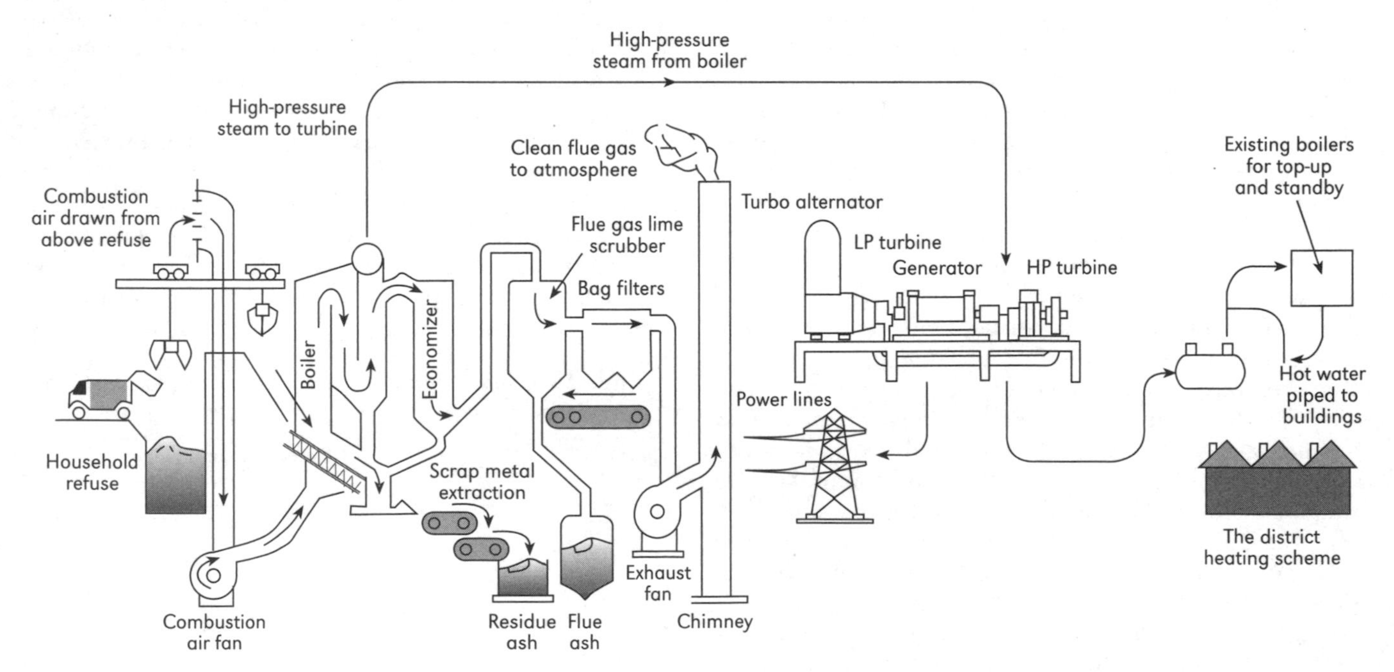

Source: http://europa.eu.int/comm/energy_transport/atlas

Figure 12.34 *Schematic diagram of a typical municipal waste incinerator with energy recovery*

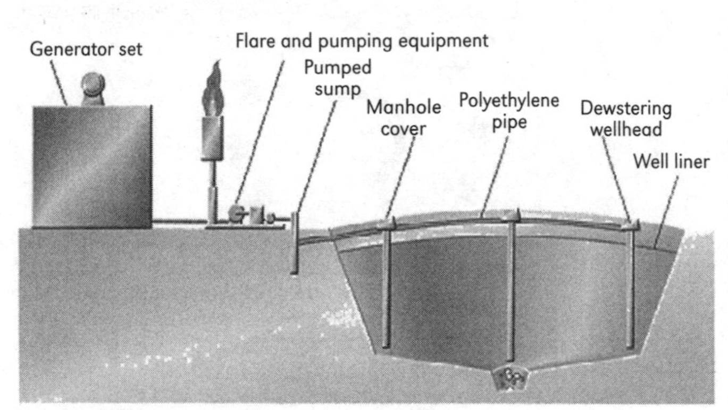

Source: http://europa.eu.int/comm/energy_transport/atlas

Figure 12.35 *Schematic diagram of a sanitary landfill*

Note: One thousand tonnes of municipal waste from the nearby city (approximately 340,000 inhabitants) is landfilled daily. Two piston engines coupled with electric generators have 1.2 megawatts of power. For the gas capture, there are 64 vertical perforated pipes, with a diameter of 200mm. If the quality of the gas is not suitable for the piston engines, it burns automatically on the torch. This disposal site, in relation to its scale, represents a typical urban municipal sanitary landfill with energy recovery from the landfill gas.

Figure 12.36 *Engines and generators where landfill gas is being used for heat and electricity production*

is regulated by the group standard ISO 14000 and is known as life-cycle assessment (LCA). With environmental labelling of products, we can offer customers the possibility of product choice in relation to its environmental friendliness.

Individuals can make a significant contribution to waste reduction and recycling. This process can be very effective, but has to be supported by the city community through regaining public consciousness, promotion and

Table 12.5 *Impact of waste treatment technologies on the environment*

Technology of the waste treatment	Air	Water	Soil	Ecosystems	Urban environment
Landfills	Emissions of methane (CH_4) and carbon dioxide (CO_2), and odours	Increasing rates of salts, heavy metals, and organics in the groundwater	Accumulation of the hazardous material in the soil	Risk of hazardous matter intake into the food chain	Exposure to hazardous matter
Composting	Emissions of CH_4 and CO_2, and odours			Risk of hazardous matter intake into the food chain	
Incineration	Emissions of sulphur dioxide (SO_2), nitrogen oxide (NO_x), hydrochloric acid (HCl), carbon monoxide (CO), CO_2, dioxin, heavy metals (zinc, lead, copper, arsenic) Emissions	Excretion of hazardous matter on water surfaces	Flying ash and disposal of residues remaining after the exhaust gas purification	Risk of hazardous matter intake into the food chain	
Recycling	Emissions of dust	Release of waste water	Appearance disturbances		Noise

effective organization of waste treatment. Removal of waste should be organized in relation to population density, and transport routes should be carefully planned. Experience from a small town with a population of 50,000 shows us that a reduction of 2.8 to 2.5 litres of fuel per tonne of transported waste is possible (Grmek and Medved, 2000).

Synopsis

Population numbers in urban areas have been continuously growing, resulting in a decreasing rural population over the last few decades. The great concentration of people in cities, however, requires a very intensive supply of energy, water, food and other resources that are required in people's everyday life. Decreasing the city's ecological footprint is an important task on the way to achieving sustainable cities. In addition to improving the economic and sociological quality of life in cities, the rational use of land, energy and materials is an important step in this grand effort.

Source: www.eco-label.com

Figure 12.37 *European daisy, which is used to label environmentally friendly products; among the criteria there is also the possibility of recycling and natural decomposition*

References

ASHRAE (American Society of Heating, Refrigerating and Air-conditioning Engineers) (1996) *HVAC Systems and Equipments*, ASHRAE, Atlanta, Georgia, US

Atkinson, G., Dubourg, R., Hamilton, K., Munasinghe, M., Pearce, D. and Young, C. (1997) *Measuring Sustainable Developments*, Edward Elgar Publishing Limited, Cheltenham

Cross, B. (ed) (1994) *European Directory of Renewable Energy Suppliers and Services: Photovoltaics in Buildings*, James & James, London

Directorate-General for Energy (DG XVII) (1996) *Energy in Europe: A Scenario Approach*; ECSC-EC-EAEB, European Commission, Brussels

'Ecological footprints' (nd), www.earth.day.net/footprints.stm

EnerBuild RTD (nd) *Wind Enhancement and Integration Techniques to Enable the Productive Use of Wind Energy in the Built Environment*, ERK6-CT-1999-2001 'ENERBUILD' Energy Research Group Dublin, Research Directorate General European Commission

Energetika Ljubljana (1995) *District Heating in Ljubljana*, JP Energetika, Ljubljana, Slovenia

Grmek, M. and Medved, S. (2001) 'Implementation of Directive IPPC', Faculty for Mechanical Engineering, Ljubljana, Slovenia

Grubler, A. (1988) *Technology and Global Change*, Cambridge University Press, Cambridge www.hawaii.gov/dbedt/ert/dsm_hi.html

Hewitt, M. and Hagan, S. (eds) (2001) *City Fights: Debate on Urban Sustainability*, James and James, London

International District Energy Association (nd), www.districtenergy.org

International Energy Agency (nd) www.iea.org

International Energy Agency (nd) *District Heating and Cooling, Including the Integration of CHP, Annex VI*, www.iea.org/textbase/publications/index.asp

Kitanovski A. and Poredo? A. (2000) *District Cooling*, SITHKO, Ljubljana

Kraševec, R. E. (1998) *Ljublana's Health Profile*, Institute of Public Health of the Republic of Slovenia, Slovenia

'Landfill gas' (nd) www.europa.eu.int/comm/energy_transport/atlas

Lechner, N. (1991) *Heating, Cooling, Lighting: Design Methods for Architects*, John Willey and Sons, Toronto

Lewis, J. O. (1999) *A Green Vitruvius: Principles and Practice of Sustainable Architectural Design*, James and James Science Publishers, London

Masters, C. M. (1991) *Introduction to Environmental Engineering and Science*, Prentice Hall International Editions, Englewood Cliffs, New Jersey, US

Medved, S. and Novak, P. (2000) *Environmental Engineering and Renewable Energy Sources*, University of Ljubljana, Faculty of Mechanical Engineering, Ljubljana, Slovenia

'Municipal waste' (nd) www.europa.eu.int/comm/energy_transport/atlas

Municipality of Ljubljana (2000) *Water*, Periodic Publication, Municipality of Ljubljana, Ljubljana, Slovenia

Pogacnik, A. (1999) *Urbanistical Planning*, University of Ljubljana, Faculty of Civil Engineering, Ljubljana, Slovenia

Renewable Energy World (1998) *From the Field to the Fast Line: Biodiesel*, James & James, London

Renewable Energy World (1999) *Biomass: Does Renewable Mean Sustainable?* James and James Science Publishers, London

Singh, M. (1998) *The Timeless Energy of the Sun*, Sierra Club Books and UNESCO, San Francisco, California, US

Stanners, D. and Bourdeau, P. (ed) (1995) *Europe's Environment: The Dobris Assessment,* European Environment Agency, Copenhagen

Sustainable Urban Design (nd) *Energy: General Information*, EC, Energy Research Group, University College Dublin, Ireland

Dalenbäck, J.-O. (1999) 'European large-scale solar heating network thermie project', DIS/1164/97, Chalmers University of Technology, Goteborg, http://wire.ises.org/wire/doclibs/EuroSun98.nsf/id/1A880E3F0CED45FCC1256771003045F0/$File/DalenbaeckIII28.pdf

Twidell, J. and Weir, T. (1986), Renewable Energy Resources, E. & F. Spon Ltd, New York, US

Zupan, M. (1995) *Thermal Insulation of Residential Buildings: IR Approach*, ZRMK, Ljubljana, Slovenia

Recommended reading

1 Stanners, D. and Bourdeau, P. (ed) (1995) *Europe's Environment: The Dobris Assessment,* European Environment Agency, Copenhagen
Europe's Environment is a detailed and comprehensive review of the state of the environment in Europe. The book is clearly structured and its information is accessible. The topics are divided into chapters dealing with the state of the environment in eight different fields, including air, water, soil, nature, wildlife and urban areas. Chapters describe the pressures on the environment, such as emissions and waste, and examine the source of environmental pressures that arise from human activities in different sectors, including energy and resource management.

2 Atkinson, G., Dubourg, R., Hamilton, K., Munasinghe, M., Pearce, D. and Young, C. (1997) *Measuring Sustainable Developments*, Edward Elgar Publishing Limited, Cheltenham
The focus of this book is on the theory and practice of economic development as viewed from the perspective of sustainability. The authors are not overly concerned with detail, but focus on the implications of sustainability for development policy. The book shows that traditional discussions and analyses of savings and investments at the micro-economic level can be greatly enriched by integrating the environment with the macro-economic picture.

3 Hewitt, M. and Hagan, S. (eds) (2001) *City Fights: Debate on Urban Sustainability*, James & James, London
This book presents aspects from a wide variety of disciplines, with the aim of developing sharp ideas about making better and more sustainable cities according to environmental, social and economical terms. The different views described in the book bring into focus the complexity and diversity of the issues involved. It is a useful source of knowledge to all with a common interest in the future of cities.

4 Twidell, J. and Weir, T. (1986) *Renewable Energy Resources*, E. & F. Spon Ltd, New York
Renewable Energy Resources covers a subject of increasing technical and economic importance. It is divided into numerous chapters and describes different renewable energy sources. Each chapter begins with a fundamental theory from a physical science perspective, then considers applied examples and developments. Chapters then conclude with a set of problems and solutions. The book is intended both for basic study and for application.

Activities

Activity 1

Describe the most important energy sources for a city's energy supply.

Activity 2

Which emitted pollutants released by the burning of fossil fuels are most dangerous for human health and the environment?

Activity 3

Describe four basic demand-side energy management strategies.

Activity 4

Describe central supply systems for heating and cooling in the urban environment.

Activity 5

Describe what you know about the use of renewable energy sources in the urban environment.

Activity 6

Describe the importance of the water supply in the urban environment.

Activity 7

Which waste treatment technologies are primarily used in cities?

Answers

Activity 1

Energy sources can be divided into two groups: renewable and non-renewable fuels. The former have the characteristic of renewing themselves in nature. According to their origins, they are divided into:

- solar radiation, emitted by the sun, which can be converted into heat and electricity, and which originates wave, wind and hydro energy and biomass;
- the planetary energy of the moon and the sun, which together with the Earth's kinetic energy cause tidal action;
- heat emerging from the interior of the Earth towards its surface, and which is known as geothermal energy.

However, the rapid development of civilization that we have been facing during the last two centuries was only possible due to the intensive use of natural (non-renewable) energy sources: fossil fuels and nuclear power. Fossil fuels – coal, petroleum and gas – represent the various forms of energy accumulated during the past evolutionary era of the Earth. Coal is the most widespread fossil fuel on Earth. It took several millions years to turn the trees and other plants buried among the sediments and alluvia under high temperatures, pressure and chemical processes into coal. Beside carbon and hydrogen, coal consists of sulphur, nitrogen, ash and vapour. At larger depths, different kinds of heat reactions took place because of greater temperatures and pressure. Vapour partially extracted sulphur, oxygen and nitrogen; on the other hand, organic substance started to decompose into liquid molecules. This is how petroleum came into existence. It consists of carbon, hydrogen, sulphur and nitrogen. Oil is a mixture of hydrocarbons and is distilled in oil refineries before use. At even greater depths, gaseous fossil fuel has been formed: natural gas that contains carbon, hydrogen and nitrogen. It liquefies through very high pressure; therefore, it is transported by gas pipelines. Nuclear fuels are an important source of energy. The best known are uranium, plutonium and thorium. In peacetime, only uranium isotope U^{235} is used. Nuclear energy is released by splitting the atom nuclei. The atom nucleus splits into two lighter nuclei after it has captured a neutron; new neutrons are also released, which launches the chain reaction.

Activity 2

Among the emitted pollutants from the burning of fossil fuels, the most threatening to the environment are carbon dioxide (CO_2), carbon monoxide (CO), nitrogen oxides (NO_x), sulphur oxides (SO_x), dust particles and volatile organic compounds (VOCs). For a complete answer, describe the influence of each pollutant on humans and the environment.

Activity 3

Four basic demand-side energy management strategies are:

1. Peak-load clipping seeks to reduce energy consumption at the time of daily peak. Examples include utility or user control of appliances such as air conditioning or water heating, or a timer for water heating.
2. Valley filling smoothes the load and improves the economic efficiency of systems. For example, electrical vehicles can be charged at night when electricity use is lower then during the day.
3. Load shifting can be accomplished through measures such as thermal storage; heat or cold can be provided and stored in a form of sensible or latent energy when consumption is low and can be used during high consumption periods.
4. Practice energy conservation. Lower consumption of energy sources can be achieved by individuals and cities, as a whole. As tenants, we can affect energy use, for example, with improved thermal insulation of our building; another possibility is more efficient ventilation with heat recovery, and buying the most energy-efficient appliances.

Activity 4

District heating and cooling systems transport heat and/or cold as steam, hot or chilled water using pipelines. The system consists of three primary components: the central plant, the distribution network and the user's heating systems. The heat source in a central plant can be a boiler or a waste incinerator, or renewable energy sources such as geothermal energy, biomass or solar energy. A piping network conveys the energy between the source and the buildings. In most cases, the district pipelines and the building service system are separated by a heat exchanger.

District heating and cooling systems are expensive and they can be used in case of high thermal load density.

In this case, district systems are cost-effective in densely populated (at least 50 buildings per hectare) urban areas, for high-density building clusters with high thermal loads or in industrial complexes. These district systems have significant thermodynamic, economic and environmental advantages compared with individual systems, as follows:

- A larger central plant can achieve higher thermal efficiency than smaller ones. The differences are even larger in the case of partial load operation because larger plants operate with more heat sources, which can switch off separately.
- Larger units can be adapted to the use of various fossil fuels, thus making it possible to utilize the least expensive fuel. Furthermore, sources of waste heat and renewable energy can be exploited.
- District heating and cooling systems have the advantage of measuring energy consumed by users with simple remote heat meters, as well as the possibility of efficient remote surveillance of the system's operation.
- Heating and cooling systems that are operated by personnel can be reduced because less mechanical equipment in buildings requires less maintenance.
- Usable space in buildings increases when boiler- or chiller-related equipment and fuel storage are no longer needed.
- Emissions from central plants are easier to control; air pollution caused by the transportation of fuels to consumers and ash removal are reduced, as is the danger of spillage of liquid fossil fuels in case of road accidents.
- In the case of district cooling systems, the leaking of environmentally unfriendly refrigerants can be monitored readily in a central plant.

There are two types of district cooling systems: open and closed. With open systems, cold water from deep pits and larger sea depths (500 to 1000m), which is pumped onto the surface, flows through heat exchangers. Water is cooled in a central water pipeline. In this case, the cold source water temperature should not exceed 5° Celsius. For example, seawater has been used to cool the central system in Stockholm since 1995.

In closed systems, cooling energy can be produced in a central plant. Chilled water can be produced by an electricity-driven compressor. In the central unit, peak electricity consumption is lower, as cold can be effectively stored. Cooling energy can be transported by cold water, ice slurry or brine to buildings. Through the use of slurry and saltwater, the temperatures of the cooling medium are lower; therefore, smaller quantities are required and pumping costs are lower. Supply temperatures are normally around 5° to 7° Celsius, with return temperatures of between 10° to 15° Celsius. These low temperatures are necessary for cooling and the dehumidification of indoor air.

Activity 5

Modern biomass solid fuels are made through processing the remains of forest biomass, agricultural plants and energetic plants (e.g. willow, poplar and Chinese reeds). Before being burned, they are processed into various shapes: chopping pieces, pellets, briquettes or bales. The advantages are that these fuels are easier to transport, provide better efficiency of boiler installations and feature lower emissions during burning. In some cities, a distribution system similar to the oil supply system has already been established.

Various types of liquid fuels can also be made from biomass, among which bio-ethanol and bio-diesel are the most important for motor vehicles. Besides being the most popular hot water heating solar systems, with solar collectors mounted on the roof or façade of the building, solar collectors can also be used for heating settlements using district hot water heating systems. Solar systems can be projected with a central or separate field of solar collectors, and with smaller short-term (daily) or bigger seasonal heat storage.

The energy of the sun, however, can be converted into electricity without thermal processes by using devices called photovoltaic (PV) cells. Single appliances, a building or a settlement can be powered through PV systems. Larger systems of solar cells are called photovoltaic power plants. Buildings integrated within photovoltaic energy systems are of special interest for electricity generated in cities. These systems use specially designed modules of solar cells in the shape of tiles, frontage panels, shading devices and windows. They are used as a substitute for the common elements of the building's envelope, and in this way reduce the system installation costs. Furthermore, modules have a long life span and they do not need a lot of maintenance, as is the case with construction elements of the building's envelope.

Geothermal energy is exploited in various ways. The oldest form represents the capture of thermal springs. Rainfall, which trickles through porous rock into great depths, warms up and comes back to the surface through natural cracks in the ground or artificially made boreholes. It comes to the surface in the form of hot water, seldom as steam. After energetic exploitation, it goes back into aquifers in order to prevent thermal pollution of the surface water and to maintain the pressure potential of the aquifer. In areas where no aquifers are to be found, non-porous rock at larger depths is crushed through the use of hydraulics. After that, water is diverted into a formed artificial aquifer, and afterwards pumped out again as hot water or steam. This method of geothermal energy exploitation is called 'hot dry rock'. Geothermal energy can also be pumped using vertical heat exchangers in pipe form, which are called vertical borehole heat exchangers. Because of lower temperatures, borehole heat exchangers are used as a heat source for heat pumps.

Large high-efficiency wind turbines have become cost-effective as electricity generation units. They have also become larger and cheaper. The modern wind turbine is 150m tall, has a rotor diameter of 100m and provides 3 megawatts of electrical power. Several wind turbines are connected in wind farms. They can be located on a city's surrounding hills and on shores or (off shore) on the sea, and are connected to the main electricity supply grid.

When burning wood biomass, besides common pollutants, different organic compounds (CxHy) and small amounts of heavy metals (mercury, lead, chromium) contained in wood occur. Modern technologies for burning wood biomass have automatically regulated fuel and air intake, as well as automatically removing the ash, and there are filters for eliminating unburned particles from exhaust gases. These kinds of installations are expensive. It is better to build bigger units of district heating. Denmark provides a good example. After introducing environmental coal taxes in 1990, 37 district heating operators decided to substitute coal with wood chips.

The exploitation of geothermal energy is associated with some environmental impacts due to SO_2, H_2S and NO_2 emissions, which are removed from geothermal water before its energetic exploitation, and because of corrosive saline fluids in it.

Activity 6

Water pollution is the result of numerous human activities, such as uncontrolled municipal and industrial waste disposal; the use of nitrates and pesticides in agriculture; the discharge of noxious substances; as well as the effect of natural geological characteristics, which are linked to wash-out of metals, minerals and salts, and the eruption of sea water into fresh groundwater. Therefore, the protection and prudent use of existing water sources is necessary. In this way, the expensive expansion of existing water sources can be avoided. Various measures are possible:

- Reducing water consumption through technical fixes, such as the installation of percolators on water taps; toilets with variable quantities of flushing water (low-flow toilets); public campaigns and promotions; the implementation of water metering.
- Using relatively small portions of drinking water for drinking and washing; more than 60 per cent of water use in households can be replaced with cleaned waste water (grey water) and rainwater where there is no need for sanitary concerns, such as water for flushing toilets, laundry and car laundry, garden watering and residence cleaning.
- Tightening up water networks, which leads to water savings and will also reduce the amount of energy required for pumping.

Activity 7

There are various technologies in use for municipal waste treatment. Waste contains matter that can be reused or recycled in raw materials. Thus, the volume of waste is reduced, as is the energy consumption for producing raw materials. Municipal waste also contains organic substances, which disintegrate into compost in composting facilities, and are then reused as a fertilizer in gardening and agriculture. This method of waste treatment occurs in Spain, Portugal, Denmark and France, where between 6 and 21 per cent of collected municipal waste is composted. Composting is especially effective when used in households in their own environment. In large cities, another effective method is also waste incineration. The main purpose of waste incineration is the effective reduction of its volume to about 20 per cent of the former content. Yet, through waste incineration heavy metal emissions occur (e.g. cadmium, lead, mercury, chromium and zinc). Furthermore there are also emissions of unburned residues and gases, which acidify precipitation (e.g. HCl, SO_2 and NO_x). These emissions can be substantially reduced in contemporary purification plants; however, ash and scoria still remain in large quantities after burning and gas purification.

The most widespread way of managing solid waste, which prevails in cities, is still disposal in a landfill site. Sanitary landfills are very exacting public structures with special requirements in relation to site selection, preparatory arrangements, operation, closing, and completion and restoration. Within the pressed waste, decay of the organic waste occurs through anaerobic fermentation. A landfill gas is released during the process. It consists mainly of methane (45 to 55 per cent) and CO_2.

13

Economic Methodologies

Vassilios Geros

Scope of the chapter

The aim of this chapter is to cover the principal methods of economic evaluation that can be used to calculate the economic impact of an investment project. By applying these methodologies, one can estimate the economic effectiveness of various building-retrofitting scenarios, as well as the efficiency of different types of building components. For example, some of the methodologies concern cost–benefit evaluation methods (i.e. the comparison between the receipt and the disbursement of an investment during a certain time period in order to evaluate the economic efficiency of the investment).

Learning objectives

Upon completing a study of this chapter, readers will be able to:

- evaluate the effectiveness of a building project from an economic point of view;
- compare various alternatives scenarios and decide which one is more cost-effective;
- perform various economic-related calculations, from simplified to more detailed approaches.

Key words

Key words include:

- economic methodologies;
- discount techniques;
- non-discount techniques;
- life-cycle cost method;
- net savings method;
- internal rate of return method;
- discounted payback method;
- net cash-flow method;
- simple payback method;
- unadjusted rate of return method.

Introduction

The evaluation of a building project by using economic methodologies can be an important tool in designing buildings. These methods provide the means to investigate various economic aspects of an investment project (e.g. insulation of the roof or replacement of the shading devices). Therefore, they permit evaluations of various scenarios and help the designer to decide which of the alternative solutions is economically appropriate.

In this chapter, two categories of methods are presented. The first category covers discount techniques, while the second one concerns non-discount techniques. The main difference between these two is the fact that the first category takes into account the time value of money (i.e. how the value of money changes through time: an amount received or paid at some point in the future is not worth as much as the same amount today). The second category does not consider this parameter. Generally, the application of this type of technique is simpler and gives a quick estimation of the economic performance of an investment.

Economic methodologies

It is important to clarify the term 'economic methodologies'. This term refers to well-defined approaches in order to understand various economic phenomena. These methods are adapted from various fields of economics and other related areas, and are used to investigate the economic phenomena associated with buildings.

Table 13.1 *A simple cash flow example that compares two alternative solutions*

Year	Cash receipts (+) Cash disbursements (–)	Alternative scenario 1 (Euros)	Alternative scenario 2 (Euros)
0	Equipment cost	-9000	-11000
	Installation cost	-2000	-2500
1	Maintenance cost	-700	-1200
	Energy savings	+1500	+2200
2	Maintenance cost	-700	-1200
	Energy savings	+1500	+2200
3	Maintenance cost	-700	-1200
	Energy savings	+1500	+2200
4	Maintenance cost	-700	-1200
	Energy savings	+1500	+2200

Therefore, it is useful to summarize the major building energy issues in order to review these methods. These issues primarily include economic feasibility (financing decisions, purchase decisions and design/sizing decisions), as well as economic impact (employment, environment and energy-related impact).

Questions such as the following should also be answered. Do the examined systems, components or techniques have a cost-effectiveness potential (i.e. are savings higher than costs over the long run)? What types of designs and/or sizes are potentially more cost-effective? Is it possible for a system to be financed? What are the impacts on the environment and the energy aspects?

Generally, the costs and benefits are the main parameters that describe an investment project when the investment is realized. In this simplified case, the cash flow is the sum of the cash receipts (this parameter can be considered positive) and the disbursements (contrary to receipts, this parameter can be considered negative) during one or more time periods – frequently, the lifetime of the project. If the user wants to compare different alternatives by using this basic approach, the cash flow for each alternative must be estimated. But the important and sometimes quite difficult and time-consuming part of the procedure is to have an accurate economic evaluation of the necessary cash flows. This determination also requires the evaluation of both direct and indirect cash flows.

Another aspect that is important when comparing alternatives is to determine if there is enough money for the initial cost. In this case, an alternative scenario can be rejected due to a high investment cost, even though it can provide great savings.

On the other hand, it is essential to have acceptable cash flows during each year of the evaluation. In this case, it is important to estimate precisely the various cash flows for each time step (e.g. yearly) of the examination period, such as operating and maintenance costs, as well as energy savings.

Table 13.1 presents a simple example where two alternative scenarios are compared. This table summarizes the cash flows for the two examined situations. According to this example, after the purchase and the installation of the equipment the cash flows are limited only to maintenance costs and the cash receipts due to the energy savings.

Discount techniques

Using an index called a discount rate, it is possible to compare the various investment alternatives. The advantage of this index is that it considers the time value of money (i.e. the fact that the value of money changes with time).

In this category of methods, money that is due to be received now is worth more than money due to be received far in the future. Because of the time value of money, the difference in the timing of cash flows can make one investment more attractive than another.

The discount techniques use the discount rate, which is chosen according to the type of the investment project. The use of these methods allows the comparison of cash flows at different times. For example, if the discount rate is d, it is equivalent to receive F today or $F(1 + d)$ in one year:

$$F_j(0) = \frac{F_j}{(1+d)^j} \tag{1}$$

where j is the year; d is the discount rate; $F_j(0)$ is also called present value of a future account of income.

Table 13.2 presents an example of the use of the discount rate. In this example, the discount rate is considered equal to 3 per cent and the initial cash flow equates to €100. According to this example, the initial amount of money after four years is equal to €112.6.

Table 13.2 *An example of the use of the discount rate*

Year	Cash flow (Euros) (discount rate = 0.03)
0	100
1	100 • (1 + 0.03) = 103.00
2	100 • (1 + 0.03)2 = 106.10
3	100 • (1 + 0.03)3 = 109.30
4	100 • (1 + 0.03)4 = 112.60

The discount rate should be chosen carefully because this value strongly influences the result and, therefore, the comparison between the alternative solutions with very different net cash flows. Furthermore, it is possible to use a different discount rate for each time step of the calculation in order to increase the accuracy of the method.

Life-cycle cost method

A method for evaluating the economic performance of a system or a component is the life-cycle cost (LCC) method. This method includes the sum of the relevant percentage and future costs that concern an energy system or component, minus any positive costs (e.g. salvage values). This summation is performed for the current or annual value of money over the studied period. The various costs include the facility's energy costs, equipment and/or system costs, maintenance costs, and repair and replacement costs. The LCC can be calculated by using the following formula:

$$LCC_{A1} = \sum_{j=0}^{N} (C_{A1} - B_{A1})_j \,/\, (1 + d)^j \qquad (2)$$

where C_{A1} is the cost in year j for the system (A_1) being evaluated ($j = 0$ indicates the cost at the beginning of the period); B_{A1} is the benefits (positive cash flows) in year j for the system (A_1) being evaluated; d is the discount rate.

The use of the LCC method permits one to determine if a project is cost-effective. Furthermore, it gives the possibility of finding which combination of the project's components that meet the performance requirements can minimize the long-term costs. In order to evaluate various alternatives, it is possible to compare their LCC, or to compare the LCC of an energy system or component with the base case. In this case, if the performance of the compared systems is equal, the one that has the lower LCC is the most cost-effective. Consequently, by using the procedure that compares the alternative solutions, it is possible to size and/or design an energy system or component. The objective of this method is to minimize the overall LCC for a facility by finding the appropriate combination of sub-systems/components.

The use of the LCC method is more suitable when the economic analysis focuses on the cost rather than on the benefits. Additionally, when the project's budget is limited, the use of this method gives the possibility of rejecting projects or increments to projects in order to have higher savings due to investment for the completion of the projects (although they have lower LCC than the other alternatives).

An example that compares a hypothetical base-case scenario with an alternative one is presented in Table 13.3 and Figure 13.1. According to this example, the alternative scenario 1 has the greatest LCC and therefore is more cost-effective than the base-case scenario.

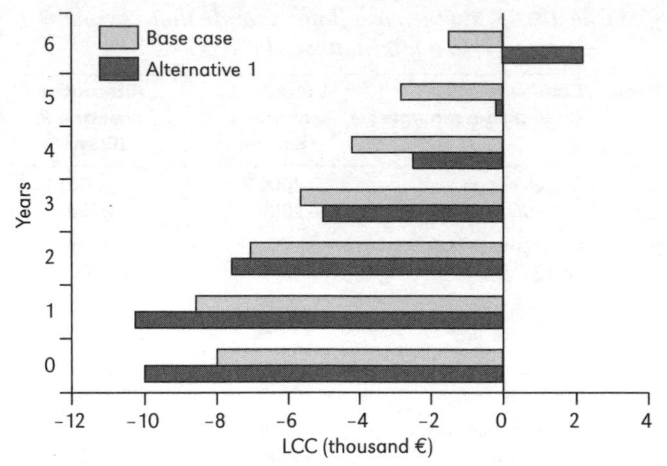

Figure 13.1 *A graph showing the comparison of two alternative solutions by using the life-cycle cost (LCC) method*

Table 13.3 *An example of the life-cycle cost (LCC) method that compares two alternative solutions*

Year	Base-case scenario		Alternative scenario 1	
	Net cash flow (Euros) ($C_{A(bc)} - B_{A(bc)}$)	LCC($_{bc}$) (Euros) (d = 3%)	Net cash flow (Euros) ($C_{A1} - B_{A1}$)	LCC$_1$ (Euros) (d = 3%)
0	-8000	-8000.00	-10,000	-10,000.00
1	-600	-8582.52	-200	-10,194.17
2	1600	-7074.37	2800	-7554.91
3	1600	-5610.14	2800	-4992.51
4	1600	-4188.56	2800	-2504.75
5	1600	-2808.39	2800	-89.44
6	1600	-1468.42	2800	2255.51

Net savings or net benefits method

Another method that can be used to evaluate the economic performance of a building's system or component is the net savings (NS) or net benefits (NB) method. The NS or NB method can be used to decide if a project is cost-effective. The use of this method determines the net difference between two alternative systems, in present or annual, as follows:

$$NS_{A1:A2} = LCC_{A2} - LCC_{A1} \quad (3)$$

where LCC_{A1} is the life-cycle cost for the system A_1; LCC_{A2} is the life-cycle cost for the system A_2.

or

$$NB_{A1:A2} = \sum_{j=0}^{N} \frac{(B_{A1} - B_{A2})_j - (C_{A1} - C_{A2})_j}{(1 + d)^j} \quad (4)$$

where: B_{A1} are the benefits (positive cash flows) for the system A_1; B_{A2} are the benefits (positive cash flows) for the system A_2; C_{A1} is the cost for the system A_1; C_{A2} is the cost for the system A_2; d is the discount rate.

A project is cost-effective if *NS* or *NB* is positive. As in the previous method, this one can be used to find the best project design and size. Furthermore, the net savings or net benefits are calculated for each design and size. The highest result of *NS* or *NB* method indicates the optimal choice.

Table 13.4 demonstrates the use of the net savings method. This example is the same as in the life-cycle cost method, but includes the NS calculation procedure and compares the base case scenario (bc) with an alternative solution (A1). The $NS_{\text{BaseCase: Alter1}}$ represents the net savings of these two scenarios.

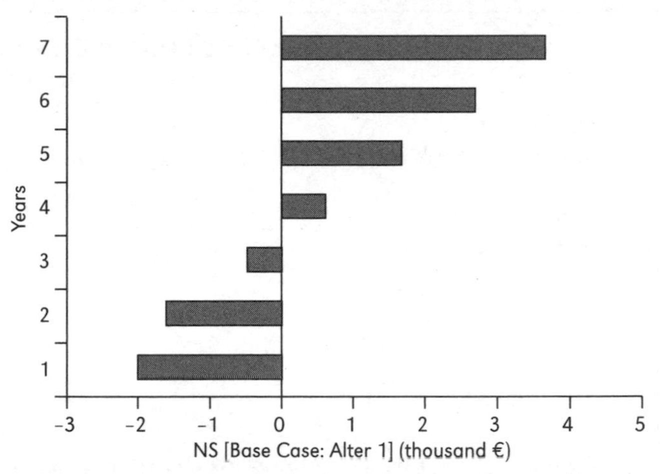

Figure 13.2 *A graph showing the comparison of two alternative solutions by using the net savings (NS) method*

Internal rate of return method

The current method can also be utilized to evaluate the economic performance of a building's system or component. The internal rate of return (IRR) method can be used to find the discount rate that gives a net present value equal to zero when applied to the cash flows of a project. The main concept of the method is focused on calculating the minimum internal rate of return at which the investment is still cost-effective. There are two versions of this method. The first one, the unadjusted version, requires the solution of the following equation for the interest rate i:

$$\sum_{j=1}^{N} \frac{(B_{A1} - B_{A2})_j - (C_{A1} - C_{A2})_j}{(1 + i)^j} - (C_{A1o} - C_{A2o}) = 0 \quad (5)$$

where i is the interest rate; C_{A1o} is the cost at the beginning of the period for the system A_1; C_{A2o} is the cost at the beginning of the period for the system A_2.

Table 13.4 *An example of the LCC method that compares two alternative solutions*

Year	Base-case scenario		Alternative scenario 1		
	Net cash flow (Euros) $(C_{A(bc)} - B_{A(bc)})$	$LCC_{(bc)}$ (Euros) (d = 3%)	Net cash flow (Euros) $(C_{A1} - B_{A1})$	LCC_1 (Euros) (d = 3%)	$NS_{\text{BaseCase: Alter1}}$ (Euros)
0	-8000	-8000.00	-10,000	-10,000.00	-2000.00
1	-600	-8582.52	-200	-10,194.17	-1611.65
2	1600	-7074.37	2800	-7554.91	-480.54
3	1600	-5610.14	2800	-4992.51	617.63
4	1600	-4188.56	2800	-2504.75	1683.82
5	1600	-2808.39	2800	-89.44	2718.95
6	1600	-1468.42	2800	2255.51	3723.93

BOX 13.1 SOURCE CODE IN C LANGUAGE FOR PERFORMING THE NECESSARY CALCULATIONS OF THE IRR METHOD

```
// file cflow_irr.cc
// author: Bernt A Oedegaard

#include <cmath>
#include <algorithm>
#include <vector>

#include 'fin_algorithms.h'

const double ERROR=-1e30;

double cash_flow_irr(vector<double>& cflow_times, vector<double>& cflow_amounts) {
// simple minded irr function. Will find one root (if it exists.)
// adapted from routine in Numerical Recipes in C.
   if (cflow_times.size()!=cflow_amounts.size()) return ERROR;
   const double ACCURACY = 1.0e-5;
   const int MAX_ITERATIONS = 50;
   double x1=0.0;
   double x2 = 0.2;

// create an initial bracket, with a root somewhere between bot,top
   double f1 = cash_flow_pv(cflow_times, cflow_amounts, x1);
   double f2 = cash_flow_pv(cflow_times, cflow_amounts, x2);
   int i;
   for (i=0;i<MAX_ITERATIONS;i++) {
      if ( (f1*f2) < 0.0) { break; }; //
      if (fabs(f1)<fabs(f2)) { f1 = cash_flow_pv(cflow_times,cflow_amounts,  x1+=1.6*(x1-x2)); }
      else {f2 = cash_flow_pv(cflow_times,cflow_amounts,  x2+=1.6*(x2-x1)); };
   };
   if (f2*f1>0.0) { return ERROR; };
   double f = cash_flow_pv(cflow_times,cflow_amounts, x1);
   double rtb;
   double dx=0;
   if (f<0.0) {                          rtb = x1;  dx=x2-x1;    }
   else {                                rtb = x2;  dx = x1-x2;    };
   for (i=0;i<MAX_ITERATIONS;i++){
      dx *= 0.5;
      double x_mid = rtb+dx;
      double f_mid = cash_flow_pv(cflow_times,cflow_amounts, x_mid);
      if (f_mid<=0.0) { rtb = x_mid; }
      if ( (fabs(f_mid)<ACCURACY) | | (fabs(dx)<ACCURACY) ) return x_mid;
   };
   return ERROR;  // error.
   };
```

This equation may have multiple solutions or no solutions. The adjusted version of the equation, to be solved for i, is:

$$\sum_{j=1}^{N} \frac{[(B_{A1} - B_{A2})_j - (C_{A1} - C_{A2})_j](1 + r_j)^{N-j}}{(1 + i)^N} - (C_{A1o} - C_{A2o}) = 0 \quad (6)$$

where r_j is the reinvestment rate.

The decision rule is to accept a project if its IRR exceeds the investor's minimum acceptable rate of return (or its IRR is higher than the alternatives).

As previously mentioned, the IRR is the discount rate that makes the current value of an income cash-flow total equal to zero. Generally, there is no closed-form solution for the IRR. In order to find a solution, it is necessary to adopt a repetitive method. The basic procedure for these methods is to start with an initial value for the IRR and to perform the calculation. According to the result of this calculation (how close to zero is the result of the equation?) a different IRR value is used and repeated until the equation is as close to zero as desired.

In order to apply the IRR method, it is possible to use the following source code in C, as presented in Box 13.1.

This approach will find only an interest rate. If there is more than one IRR solution, one can graph the current value as a function of interest rates, and use the graphical representation to evaluate when an investment meets the profit requirements.

In the case of a single initial investment without any further investments, Equation 6 is simplified to the following, when considering only one system:

$$IRR = \left(\frac{C_{A1}}{C_{A1o}}\right)^{1/j} - 1 \quad (7)$$

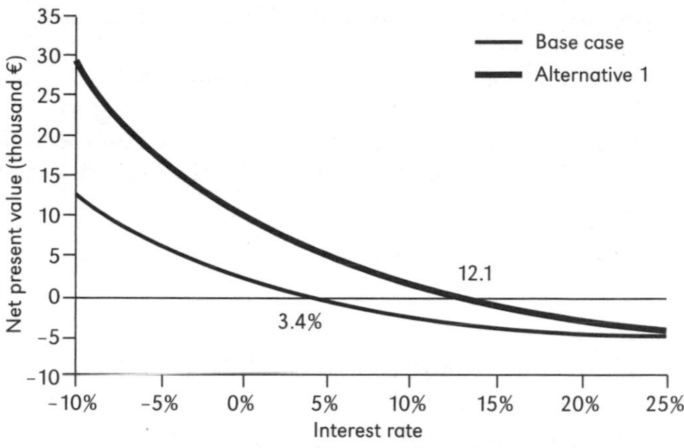

Figure 13.3 *A graph showing the comparison of two alternative solutions by using the internal rate of return (IRR) method*

In Table 13.5 and Figure 13.3, an example of the IRR method is presented for a ten-year time period. According to this example, the use of the alternative scenario is more profitable because the IRR is about three times the IRR of the base case.

Discounted payback method

The payback period is the necessary length of time in which to recover the initial cash flow. The payback period can be calculated by using both discount and non-discount techniques. This section presents the discounted payback method (the non-discounted technique is presented in the following section). For the comparison of two different building systems or components, the method determines the time, according to the time step of the calculations, required for the cumulative difference in future cash flows of one energy system relative to an

Table 13.5 *An example of the internal rate of return (IRR) method that compares two alternative solutions*

Year	Base-case scenario		Alternative scenario 1	
	Net cash flow (Euros) ($C_{A(bc)} - B_{A(bc)}$)	$IRR_{(bc)}$	Net cash flow (Euros) ($C_{A1} - B_{A1}$)	IRR_1
0	-8000		-10,000	
1	-900		-1000	
2	1200		2300	
3	1200		2300	
4	1200		2300	
5	1200	3.4%	2300	12.10%
6	1200		2300	
7	1200		2300	
8	1200		2300	
9	1200		2300	
10	1200		2300	

alternative system to just equal the difference in their initial investment costs.

The method requires finding the minimum solution value of *PB*: the number of the time periods (e.g. years), for which the first part of the following equation is equal to the difference of the costs at the beginning of the period:

$$\sum_{j=1}^{PB} \frac{(B_{A1} - B_{A2})_j - (C_{A1} - C_{A2})_j}{(1 + d)^j} - (C_{A1o} - C_{A2o}) \qquad (8)$$

In order to calculate the payback period for a single building energy system or component, the following equation can be applied:

$$\sum_{j=1}^{PB} \frac{(B_{A1} - C_{A1})_j}{(1 + d)^j} - C_{A1o} \qquad (9)$$

For the simple payback method, $d = 0$.

This methodology can be used as an index in order to estimate the project's cost-effectiveness. When the payback period of a project is less than the expected project's life, then this project can be considered as cost-effective.

It is important to clarify that the payback period method does not provide a complete economic analysis. The project with the shortest payback period may not be the best investment and may not provide the expected return. Additionally, the method is quite defective as it ignores the cash flow after the payback period, which is an additional economic benefit.

Table 13.6 and Figure 13.4 represent an example of the discounted payback period. In this example, the base-case scenario of a project is compared with an alternative solution. According to this example, the initial cost of the base case is recovered during the eighth year after the initial investment. On the contrary, the alternative scenario is more effective and gives a payback period of about six years.

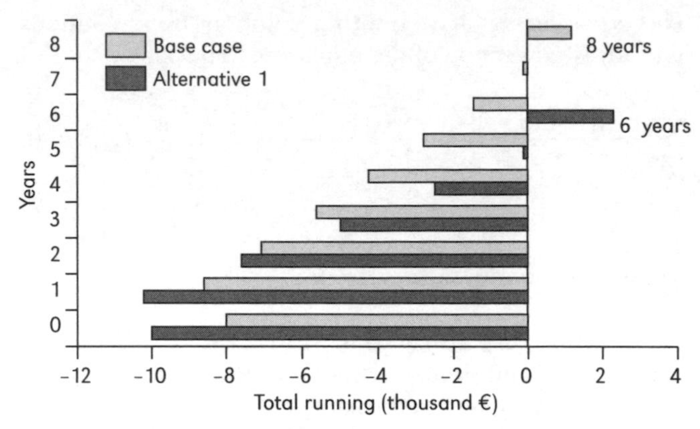

Figure 13.4 *A graph showing the comparison of two alternative solutions by using the discounted payback method*

Non-discount techniques

Non-discount techniques are simpler to use than discount techniques. An investment project that concerns a building's energy system or component has an initial cost (purchase and installation of equipment), plus other expected cash outflows associated with the investment. Additionally, the economic analysis of the project should take into account the various cash inflows, which are any current or expected revenues or savings directly associated with the investment. By using these two parameters and without considering the time value of money, it is possible to investigate the economic performance of a system or component.

Table 13.6 *An example of the discounted payback method that compares two alternative solutions*

Year	Base-case scenario		Alternative scenario 1		
	Net cash flow (Euros) ($C_{A(bc)} - B_{A(bc)}$)	Running total (d = 3%)	Net cash flow (Euros) ($C_{A1} - B_{A1}$)	Running total (d = 3%)	Payback period
0	-8000.00	-8000.00	-10,000.00	-10,000.00	
1	-600.00	-8582.52	-200.00	-10,194.17	
2	1600.00	-7074.37	2800.00	-7554.91	
3	1600.00	-5610.14	2800.00	-4992.51	
4	1600.00	-4188.56	2800.00	-2504.75	
5	1600.00	-2808.39	2800.00	-89.44	
6	1600.00	-1468.42	2800.00	2255.51	←Alternative 1 (sixth year)
7	1600.00	-128.44			
8	1600.00	1211.53			←Base case (eighth year)

Three main methods are presented: the net cash flow, the payback method and the unadjusted rate of return method.

Net cash-flow method

The net cash flow is a simple methodology that summarizes the various cash flows that occurred during the realization of an investment project and permits the economic evaluation of the project. This method also compares various alternative projects and decides which one is the most cost-effective.

Generally, a project has both receipts (cash inflows) and disbursements (cash outflows), which provide a calculation of net cash flow. The net cash flow is the sum of cash inflows and outflows that take place during the same studied period and during the same time intervals (e.g. yearly). Benefits may be gained through energy savings, cost savings during operation, increased productivity, etc.

When energy related flows are part of the calculation (e.g. energy savings), it is necessary to convert energy to cash units in order to take into account these parameters to the calculation of the net cash flow. In this case, the evaluation of the energy cash flows may require considering a specific inflation rate g for the energy prices. The inflation rate may be used in order to correct other periodic costs, such as operating costs. The energy cash flow, which occurs at year j, can be calculated by using the following formula:

$$B_j = E \cdot c \cdot (1 + g)^j \quad (10)$$

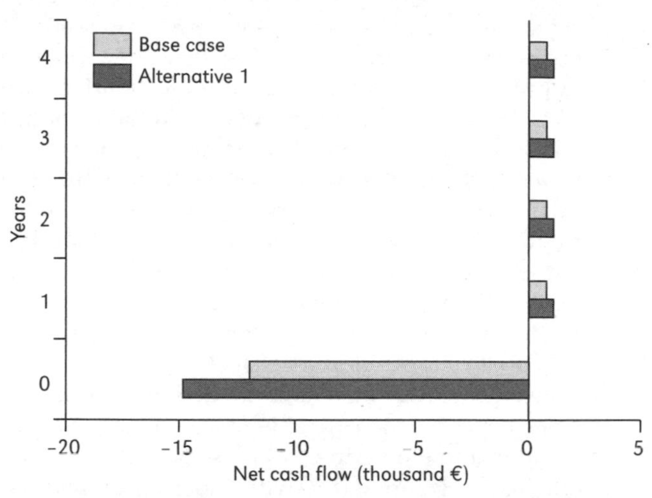

Figure 13.5 *A comparison between two alternative solutions by using the net cash flow method*

where j is the year; E is the energy saving; c is the unitary energy price; g is the specific inflation rate

Table 13.7 and Figure 13.5 present a simple example of the method.

Simple payback method

The simple payback method is like the discounted method without taking into account the time value of money; therefore, it is less accurate. It is a technique that estimates the amount of time it takes the net cash flows of an investment to recover from the initial investment costs.

Table 13.7 *Net cash flows of the two alternatives*

Year	Cash receipts (+) Cash disbursements (–)	Alternative scenario 1 (Euros)	Alternative scenario 2 (Euros)
0	Equipment cost	-9000	-11,000
	Installation cost	-2000	-2500
	Initial cost	-1000	-1250
	Net cash flow	-12,000	-14,750
1	Maintenance cost	-700	-1200
	Energy savings	+1500	+2200
	Net cash flow	+800	+1000
2	Maintenance cost	-700	-1200
	Energy savings	+1500	+2200
	Net cash flow	+800	+1000
3	Maintenance cost	-700	-1200
	Energy savings	+1500	+2200
	Net cash flow	+800	+1000
4	Maintenance cost	-700	-1200
	Energy savings	+1500	+2200
	Net cash flow	+800	+1000

By using this method, one can accept a project investment if the estimated payback period is less than the economic life of a project. When various alternative solutions are compared, the most efficient solution is the one that has the shortest payback period. Unfortunately, the simple payback method does not account for the savings after the end of the payback period. But this method is helpful for a 'first-cut' analysis of a project as a screening tool. If a project does not have a payback period less than a specified period, then parameters such as the risk due to changing technology can possibly reject the project.

There are two types of payback methods. The first one is the payback with equal annual savings, while the second one concerns the case of unequal annual savings.

The following equation can be used for the calculation of the payback period (PB) with equal annual savings:

$$PB = \frac{C_{A1o}}{B_{A1} - C_{A1}} \quad (11)$$

where C_{A1o} is the initial investment cost; $B_{A1} - C_{A1}$ is the annual operating savings (e.g. the annual net cash inflows), which are constant during each time step of the project (benefits minus costs).

If the annual cash flow differs from time step to time step, the payback period can be estimated by using the following equation in order to find the value of *PB* so that:

$$\sum_{j=1}^{PB} (B_{A1} - C_{A1})_j = C_{A1o} \quad (12)$$

where *j* is the time step of the studied period.

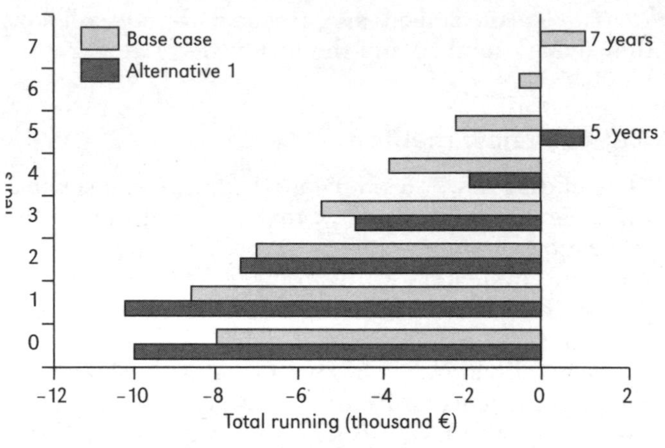

Figure 13.6 *A comparison of two alternative solutions by using the simple payback method*

Table 13.8 and Figure 13.6 give an example of the application of the simple payback method. This example uses the same net cash flow as in the case of the discounted payback method. As shown, the payback periods of the base case and the alternative scenarios are reduced compared with the discounted method (for the base case from eight to seven years, and for the alternative scenario from six to five years). The absence of the discount rate from the present method reduces its accuracy; but the simplicity of the methodology gives rapid results.

Unadjusted rate of return method

The unadjusted rate of return (URR) method is a capital budgeting technique in which one can quickly estimate a simple rate of return in order to evaluate an investment opportunity. This method uses the average expected income from the investment and expresses it as a percentage of the initial investment cost.

Table 13.8 *An example of the simple payback method that compares two alternative solutions*

Year	Base-case scenario		Alternative scenario 1		
	Net cash flow (Euros) ($C_{A(bc)} - B_{A(bc)}$)	Running total (d = 3%)	Net cash flow (Euros) ($C_{A1} - B_{A1}$)	Running total (d = 3%)	Payback period
0	-8000.00	-8000.00	-10,000.00	-10,000.00	
1	-600.00	-8600.00	-200.00	-10,200.00	
2	1600.00	-7000.00	2800.00	-7400.00	
3	1600.00	-5400.00	2800.00	-4600.00	
4	1600.00	-3800.00	2800.00	-1800.00	
5	1600.00	-2200.00	2800.00	1000.00	←Alternative 1 (fifth year)
6	1600.00	-600.00	2800.00		
7	1600.00	1000.00	2800.00		←Base case (seventh year)

The decision rule requires that a project must have a minimum URR to be acceptable. When comparing alternative scenarios, the one with the higher URR is the most effective.

The following simple formula can be used in order to estimate the unadjusted rate of return:

$$\sum_{j=1}^{N} (B_{A1} - C_{A1})_j / N \qquad (13)$$

The primary advantage of the unadjusted rate of return is its simplicity. The primary disadvantage is that the technique fails to take into account the time value of money.

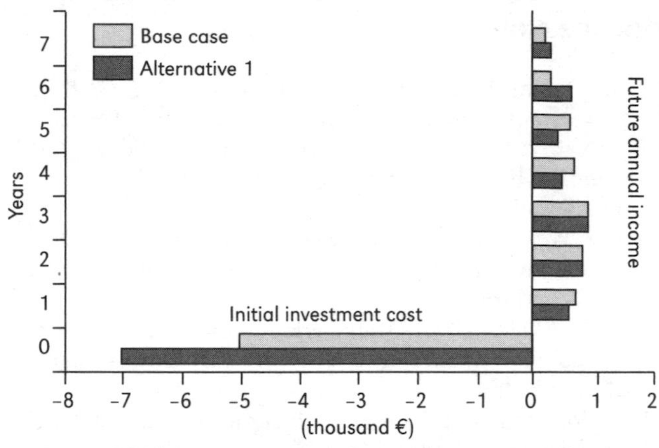

Figure 13.7 *A comparison of two alternative solutions by using the unadjusted rate of return (URR) method*

Table 13.9 *An example of the unadjusted rate of return (URR) method that compares two alternative solutions*

Year	Base-case scenario		Alternative scenario 1	
	Initial cost (–) and future annual net income)+) (Euros)	$URR_{(bc)}$	Initial cost (–) and future annual net income)+) (Euros)	URR_1
0	-5000		-7000	
1	700		600	
2	800		800	
3	900	12.00%	900	8.37%
4	700		500	
5	600		400	
6	300		600	
7	200		300	

Synopsis

The economic methodologies for evaluating a building project are important tools when investigating the efficiency of a project. The application of various technologies, design approaches and systems in order to improve the indoor conditions in a building and to reduce its energy consumption is usually evaluated only from the energy point of view; but the economic approach is sometimes the main criterion for selecting a project from the various alternatives.

In this chapter some basic economic methodologies are presented. These methodologies are separated into two main categories. The first category refers to methods that consider the time value of money, while the second one does not include this parameter. In general, the first category uses an index called discount rate, which considers the fact that the value of money changes with time. By using this index, it is possible to compare different cash flows that occur at different times. This category of methods is more complicated than the second one, but the results are more accurate.

The second category, called non-discount techniques, is simplest and the relevant methods can be used for a quick estimation of the economic efficiency of an investment project. These kinds of techniques do not consider the time value of money and simply take into account the various cash inflows and outflows in order to investigate the economic performance of a system or component.

References

Atrill, P. and McLaney, E. (1994) *Accounting II, Unit 3: Investment Decisions*, Open Learning Foundation Enterprises Ltd, www.eds.napier.ac.uk/flexible/OLF/materials/bs/MANAGEMENT%20ACCOUNTING/06unit3.pdf

Brown, W. B. (1999) *Problems in Microeconomics*, Michigan State University Department of economics, www.bus.msu.edu/econ/brown/pim/

Cottrell, M. D. (2000) 'Managerial accounting, Chapter 22: Capital investment decisions', www.panoptic.csustan.edu/2130/22/

Hazardous Waste and Toxics Reduction Program (2000) *Cost Analysis for Pollution Prevention*, www.ecy.wa.gov/pubs/95400.pdf

Johnson, R. E. (1990) *The Economics of Building: A Practical Guide for the Design Professional*, Wiley-Interscience, New York

Lott, C. (2002) 'The investment frequently asked questions (FAQ)', www.invest-faq.com/articles/

McCracken, M. (1998–2000) 'Teach Me Finance: Basic Finance Concepts', www.teachmefinance.com/

Odegaard, B. A. (1999) *Financial Numerical Recipes*, www.finance.bi.no/~bernt/gcc_prog/algoritms/algoritms/

Peterson, P. P. (2001) 'Capital budgeting techniques', Florida State University, www. garnet.acns.fsu.edu/~ppeters/fin3403/pp01/day20.ppt

Polistudies (1998) *A Multimedia Tool for Buildings in the Urban Environment*, SAVE Programme, Commission of the European Community, XVII/4.1031/Z/96-121, Brussels

West, R. E. and Kreith, F. (1988) *Economic Analysis of Solar Thermal Energy Systems*, MIT Press, Cambridge, Massachusetts and London

Recommended reading

1 West, R. E. and Kreith, F. (1988) *Economic Analysis of Solar Thermal Energy Systems*, MIT Press, Cambridge, Massachusetts and London
This book is recommended because of the economic methodologies that are presented. Although these methodologies are oriented towards thermal energy systems, the main concept for a building project remains the same. This book reviews the spectrum of economic methods that have been developed from the mid 1970s to the mid 1980s in order to analyse the feasibility of solar systems. The text also demonstrates how the use of these techniques has influenced federally sponsored research, development and demonstration techniques. Additionally, the book reviews applications analysis, net energy analysis and cost requirements for active and passive heating and cooling, for electric power generation and for industrial process heat. The change in the costs of solar systems over time is one indication of programme success, and the book includes a useful summary of the cost histories of various solar programmes.

2 Johnson, R. E. (1990) *The Economics of Building: A Practical Guide for the Design Professional*, Wiley-Interscience, New York
In this book, both an introduction to economic principles as they relate to building design and a practical guide to putting these principles to effective use are presented. The book brings together a variety of specialized topics that are relevant to building economics, including cost estimation, life-cycle costing, cost indexes, capital budgeting, decision analysis and real-estate feasibility analysis. It develops these concepts within the framework of an integrated approach to design and management decision-making, simplifying where appropriate, but never at the expense of intellectual content. Incorporating a number of sample spreadsheet models, *The Economics of Building* is a practical resource and guide to the financial assessment of planning, design and management decisions about buildings.

3 Atrill, P. and McLaney, E. (1994) *Accounting II, Unit 3: Investment Decisions*, Open Learning Foundation Enterprises Ltd, www.eds.napier.ac.uk/flexible/OLF/materials/bs/MANAGEMENT%20ACCOUNTING/06unit3.pdf>
This report comprises material from the Business Studies Programme, a distance learning course of the Napier University at Edinburgh. In this report various economic aspects are presented; basic concepts and terms are also explained. The text also clarifies how to use various economic techniques in order to make investment decisions. Furthermore, examples are given for each economic methodology in order to make clear its application.

Activities

Activity 1

Simple payback period

An energy study of an office building requires the installation of new shading devices on the south side of the building. The total purchase cost plus the installation is €6500, while the annual cooling load reduction is 9kWh/m^2 and the maintenance cost is considered null. Calculate the simple payback period of the shading system using the following additional information:

- The total floor area of the building is 1000m^2.
- The total coefficient of performance (COP) of the cooling system is 1.6.
- The cost of the electrical energy is €0.15 per kWh.

Activity 2

Life-cycle cost (LCC) method

An educational institution considers reducing the energy consumption for cooling. One of the measures that the energy study of the building proposes is the installation of ceiling fans in order to increase the thermal comfort zone of the building. This measure requires the installation of one ceiling fan per 35m^2. If the floor area of the building, where the fans are going to be installed, is 3850m^2 and the electrical energy consumption is 12.2kWh per m^2 per year and 11kWh per m^2 per year before and after the installation of the fans, respectively (the second value includes the electrical energy consumption for operation of the ceiling fans), then use the life-cycle cost method in order to evaluate the economic efficiency of the scenario. The total installation cost per ceiling fan is €40, while the discount rate is 3 per cent (cost of the electrical energy is €0.15 per kWh).

Activity 3

Unadjusted rate of return (URR) method

In a fast-food restaurant, the requirements for cooling vary quite rapidly during the daily operation of the building, following the peak of customer occupancy. A central heat pump unit equipped with an air distribution ductwork are the main elements of the cooling system installation. The designer examines the use of two solutions in order to reduce the energy consumed for cooling. The first solution proposes the replacement of the one-stage compressor by a variable speed one in order to adjust the cooling capacity of the system to the variant cooling requirements of the building. The second solution suggests the installation of an evaporating cooling system that is used to reduce the temperature of the introduced fresh air. This measure can be quite effective as the amount of fresh air that is required is quite important due to the high occupancy of the indoor spaces. For this specific building, when the previous mentioned measures are not taken into account, the annual electrical energy consumption for cooling is close to 90kWh/m^2, while the total surface area is 150m^2. The solution that requires the replacement of the compressor (first solution) reduces the annual consumption for cooling by 15 per cent and the installation cost is €10,000 (no additional maintenance cost is taken into account). The solution that proposes the evaporative cooling of the fresh air (second solution) reduces the electrical energy for cooling by 21 per cent, has an installation cost of €4000 and a maintenance cost of €150 per year. If the cost of the electrical energy is €0.15 per kWh, use the unadjusted rate of return method to evaluate which is the most cost-effective solution.

Activity 4

Net savings (NS) method

A supermarket located in an individual building is being retrofitted and various measures are examined in order to reduce the energy consumption of the building. The following two proposed measures should be evaluated according to the net savings method.

The first scenario requires the replacement of the existing fluorescent lamps with more efficient ones. This measure reduces the number of lamps that are installed in order to achieve the same lighting levels as before retrofitting occurred. In the initial installation, 1253 lamps of 36 watts per lamp comprised the lighting system, while after the retrofitting the number of lamps is 1085 at the same wattage. Therefore, the installed lighting power is decreased and this reduces the energy consumption for lighting. This also influences the cooling load of the building – which is decreased as the internal gains due to lighting are reduced. On the other hand, the heating load is increased for the same reason due to the reduction of the internal gains. The energy study of the building shows that the annual total electrical consumption for heating is 64,669kWh per year, for cooling 77,215kWh per year and for lighting 24,6552kWh per year (the HVAC system fully consumes electrical energy), when the actual lighting system is used. When the more efficient lamps are considered, the corresponding annual reduction is 10 per cent for lighting and 4 per cent for cooling, while the consumption for heating is increased by 5 per cent. The installation cost is €6500, while no maintenance cost is considered.

The second scenario examines the possibility of installing ceiling fans on the roof of the supermarket. This kind of system can increase the comfort zone of the occupants and reduce the cooling requirements of the building. For this measure the installation cost is close to €1500 and the reduction of the cooling energy consumption is 7 per cent of the actual situation. Furthermore, the cost of the electrical energy is €0.15 per kWh and the discount rate is 3 per cent.

Answers

Activity 1

The annual cash flow due to the cooling energy reduction (energy savings) is calculated by multiplying the cooling load with the floor area of the building and the cost of the electrical energy consumption, divided by the coefficient of performance of the cooling system:

9[kWh/m²/year] 1000[m²]·0.15[€/kWh]/1.6

= €844 per year

Table 13.10 *Net cash flow and running total per year*

Year	Net cash flow (€)	Running total (€)	Payback period
0	-6500.00	-6500.00	
1	844.00	-5656.00	
2	844.00	-4812.00	
3	844.00	-3968.00	
4	844.00	-3124.00	
5	844.00	-2280.00	
6	844.00	-1436.00	
7	844.00	-592.00	
8	844.00	252.00	8 years

The payback period of the shading system is eight years.

Activity 2

The annual reduction of the electrical energy consumption is calculated by the following formula:

(12.2[kWh/m²/year] – 11[kWh/m²/year]) 3850 [m²] =

4620kWh/year

Therefore, the annual financial gain due to the operation of the ceiling fans is:

4620[kWh/year] 0.15[€/kWh] = €693 per year

The installation cost of the ceiling fans is:

3850 [m²] / 35 [m²] 40[€] = €4400

The life-cycle cost (LCC) method can be calculated by the following formula:

$$LCC = \sum_{j=0}^{N} (C - B)_j / (1 + d)^j$$

where C is the cost in year j for the system being evaluated; B is the benefits in year j for the system being evaluated; d is the discount rate.

Table 13.11 *Net cash flow and life-cycle cost per year*

Year	Net cash flow (€) ($C - B$)	LCC (€) (d = 3%)
0	-4400	-4400.00
1	693	-3727.18
2	693	-3073.97
3	693	-2439.77
4	693	-1824.05
5	693	-1226.26
6	693	-645.89
7	693	-65.51
8	693	514.87

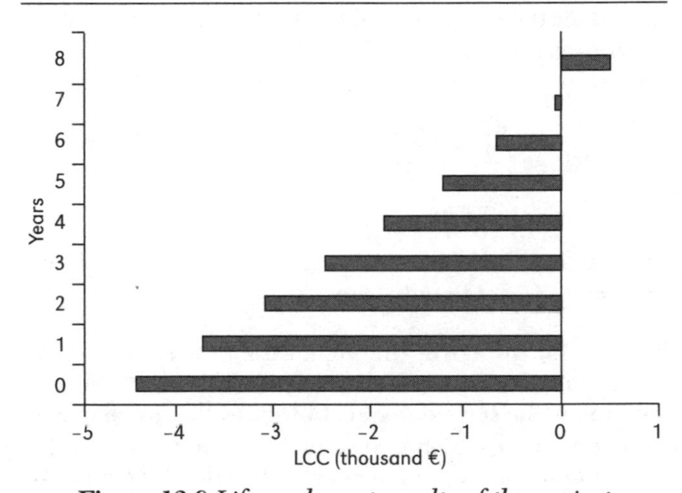

Figure 13.8 *Life-cycle cost results of the project*

Activity 3

For the first solution, the annual gain due to the reduction of the electrical energy for cooling is:

90[kWh/m²/year] 150m² 0.15 0.15[€/kWh]=

€303.75 per year

For the second solution, the annual gain is the following:

90[kWh/m²/year] 150m² 0.21 0.15[€/kWh]=

€425.25 per year

Table 13.12 *Results of the alternative scenarios for the unadjusted rate of return method*

Year	First solution Initial cost (–) and future annual net income (+) (€)	URR_1	Second solution Initial cost (–) and future annual net income (+) (€)	URR_2
0	-10,000		-4000	
1	303.75		275.25 (425.25-150.00)	
2	303.75		275.25 (425.25-150.00)	
3	303.75	3.04%	275.25 (425.25-150.00)	6.88%
4	303.75		275.25 (425.25-150.00)	
5	303.75		275.25 (425.25-150.00)	
6	303.75		275.25 (425.25-150.00)	
7	303.75		275.25 (425.25-150.00)	

As the annual cost and benefits are constant after the year of the installation, one can choose any period of year in order to calculate the URR for each scenario. For the current situation, seven years have been chosen as the period for the calculation of the URR by using the following formula:

$$\text{URR} = \frac{\sum_{j=1}^{N} (B - C)_j / N}{C_0}$$

where C is the cost in year j for the system being evaluated; B is the benefits in year j for the system being evaluated; C_o is the initial cost.

The higher the URR, the higher the cost-effectiveness. Therefore, according to the results, the most effective solution is the second one that requires the installation of an evaporative cooling system in order to reduce the temperature of the fresh air that is handled by the HVAC system.

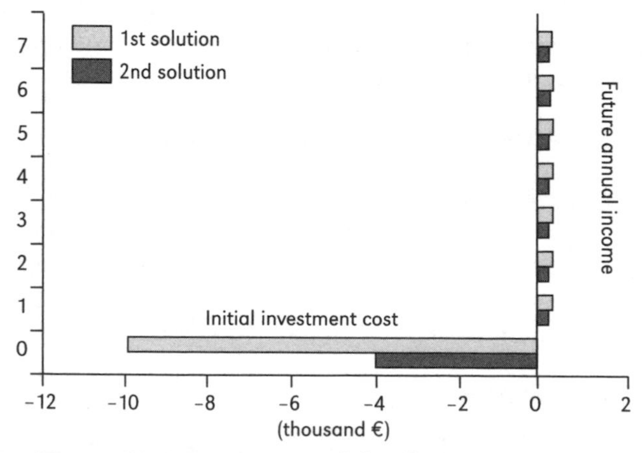

Figure 13.9 *Comparison of the alternative scenarios for the unadjusted rate of return method*

Activity 4

The heating, cooling and lighting electrical energy consumption for before the retrofitting situation is as follows:

$$64{,}669[\text{kWh/year}] + 77{,}215[\text{kWh/year}] + 246{,}552[\text{kWh/year}] = 388{,}436\text{kWh/year}$$

For the first possible solution (replacement of lamps), the annual electrical consumption is:

$$(64{,}669[\text{kWh/year}]\ 1.05) + (77{,}215[\text{kWh/year}] \cdot 0.96) + (246{,}552[\text{kWh/year}]\ 0.9) = 363{,}926\text{kWh/year}$$

Therefore, the annual financial gain due to the energy conservation is:

$$(388{,}436[\text{kWh/year}] - 363{,}926[\text{kWh/year}])\ 0.15[€/\text{kWh}] = €3676.50$$

For the second examined solution (ceiling fans), the annual electrical consumption is:

$$64{,}669[\text{kWh/year}] + (77{,}215[\text{kWh/year}]\ 0.93) + 246{,}552[\text{kWh/year}] = 383{,}031\text{kWh/year}$$

And the annual financial gain is:

$$(388{,}436[\text{kWh/year}] - 383{,}031[\text{kWh/year}])\ 0.15[€/\text{kWh}] = €810.75$$

Table 13.13 *Results of the alternative scenarios for the net savings method*

Year	Solution 2		Solution 1		
	Net cash flow (€) $(C_{A2} - B_{A2})$	LCC_{A2} (d = 3%)	Net cash flow (€) $(C_{A1} - B_{A1})$	LCC_{A1} (d = 3%)	$NS_{A2:A1}$ (€)
0	-1500.00	-1500.00	-6500.00	-6500.00	-5000.00
1	810.75	-712.86	3676.50	-2930.58	-2217.72
2	810.75	51.35	3676.50	534.87	483.53
3	810.75	793.30	3676.50	3899.39	3106.09
4	810.75	1513.64	3676.50	7165.91	5652.27
5	810.75	2213.00	3676.50	10,337.29	8124.30
6	810.75	2891.99	3676.50	13,416.30	10,524.32

According to the net savings (NS) method, for each alternative scenario the life-cycle cost is calculated as follows:

$$LCC = \sum_{j=0}^{N} (C - B)_j / (1 + d)^j$$

where C is the cost in year j for the system being evaluated; B is the benefits in year j for the system being evaluated; d is the discount rate.

The NS is then calculated by using the following equation for the scenarios A1 and A2:

$$NS_{A1:A2} = LCC_{A2} - LCC_{A1}$$

Both solutions have a very quick amortization period; but the second one is the most cost-effective, with higher positive cash flows.

14

Integrated Building Design

Koen Steemers

Scope of the chapter

This chapter provides an overview and demonstration of the interrelationships between architectural and technical parameters. The technical performance characteristics that are referred to include heating, ventilation, cooling and lighting, and they are linked to considerations of urban planning, building form, façade design and fabric design.

Learning objectives

Upon completing this chapter, readers will be able to:

- understand the interrelationships between design and technical parameters;
- describe the complexities and interactions.

Keywords

Keywords include:

- integrated design;
- design methodology.

Introduction

This chapter argues that for the successful performance of buildings, it is essential to consider all aspects that affect energy use – from planning to detailed materials specifications. These aspects have been discussed in detail in previous chapters. Here the emphasis is on integrated design. This implies an understanding of the relative impacts of each parameter – both those determined by design and those that can be described as technical – in order to achieve a balanced and holistic strategy.

At a most fundamental level, an example of integrated design is one in which the use of passive strategies is exploited to reduce the reliance on conventional mechanical services. Thus, for example, shading devices reduce the reliance on mechanical cooling, and natural lighting strategies can limit the need for artificial lighting energy demand.

One strategic aim of the integrated approach is to avoid conflicts between architecture and technology. This requires close collaboration between architect and engineer at the beginning of the design process. This is contrary to the common approach where an architect designs a building first and then an engineer is expected to make it work through the application of services (and the use of energy to 'correct' poor design decisions). If energy considerations – discussed broadly in Chapter 3 and in more detail in other chapters – are not integral to the design solution, it becomes difficult to improve the energy saving potential through the application of technology alone. Thus, if a design does not integrate natural ventilation strategies, then more energy-intensive mechanical systems may be the only recourse without fundamentally changing the building design.

It has been argued that design integration is critical, and that the means to achieve this is through the early and effective collaboration of the design team. The next step is to describe more precisely what the procedure is, or might actually be.

One could broadly outline three stages in integrated environmental design:

1. Define the problem and boundary conditions.
2. Develop strategies and options in response to the criteria.
3. Apply tools and knowledge to evaluate the performance of the strategies.

Three examples are provided below to demonstrate this approach:

1 In determining how to achieve appropriate indoor air quality (IAQ) one of the key boundary conditions will be the outdoor environment. Urban air pollution levels that are known, or can be predicted, and exceed internationally or regionally set standards, will have a significant impact upon the appropriate design response. The use of natural ventilation on a street side may be unacceptable, and thus an appropriate ventilation system (probably mechanically driven) with filtration will comprise one strategy. Another may be to draw the air for ventilating the building from somewhere away from the pollution source (assuming that this has been determined), such as a vegetated courtyard. The performance of such options then needs to be tested and assessed using tools and expertise before the appropriate design strategy can be identified.
2 The need to achieve certain levels of ventilation for cooling a building will be determined by the temperature conditions in the urban microclimate. Proposals to integrate, for example, thermal mass, ground pipes, evaporative cooling and ventilation stacks will influence whether natural ventilation is sufficient, or whether mechanical systems will be required. The performance of ventilation strategies can be evaluated using computer modelling, which will inform the ultimate building solution.
3 The aim of reducing electrical lighting as a means of decreasing energy demand can be very effective. However, it requires the design team to clearly define the problem – notably, the presence of obstructions to light, the potential conflict between daylighting and solar shading, and problems with glare. The response will involve a careful balance between key design parameters, such as glazing ratios and plan depth, to ensure the availability of natural light. A critical condition that is required is the effective control of artificial lighting in response to daylight.

These brief examples highlight the need for integrated thinking and design team collaboration. The following section will develop a broader overview of such interrelationships between architecture and technique.

An integrated building design system

This chapter aims to outline a structure and methodology for an integrated building design system (IBDS) in an urban context. It sets out to provide a framework of working that demonstrates and reminds the design team of the range of issues and interactions through the design process. It should not be considered a rigid process, but, rather, as a means of raising awareness of the integration implications of a range of environmental and design parameters. IBDS is broadly based on the design stages that are explained in more detail in Chapter 2.

The IBDS proposed here can be broken down into four main sections:

1 principles of low-energy design;
2 pre-design context;
3 building design; and
4 building services.

Principles of low-energy design

This part of IBDS considers the roles of the key environmental design principles and the associated building physics that will impact upon the design. The focus here is on those factors that determine the energy performance of the building's form and fabric, and the related comfort issues, and thus includes:

- passive solar design;
- daylighting;
- natural ventilation;
- comfort.

This brief list is by no means exclusive and additional or alternative aspects could be included that are of particular relevance to the project in hand. However, it is proposed that the above factors are central to the context of energy-efficient urban design.

Each aspect – which can be further broken down into sub-categories – will have an impact upon strategies adopted for the building's design and services, and provides the necessary principles upon which to base decisions. The purpose of including these principles is that they are central to explaining the physical mechanisms that link design decisions with performance consequences.

Pre-design context

Any project will have a number of pre-determined design constraints. These are determined by the site, the client and the planning authorities, and thus include the following:

- site climate and context;
- the building brief;
- local building and planning regulations.

Again, additional pre-design aspects could be included if desired. Each of the above key factors will have a significant impact upon the design from the outset and are largely fixed, although some manipulation and negotiation is occasionally possible under each category. Thus, for example, the urban context is largely a given; but changes to the site boundary may be negotiated. Similarly, the client may change the building brief as a result of site analysis, and some negotiation may be possible with planning authorities in order to obtain exemption from certain regulations.

Building design

At the core of IBDS lie the building design considerations. The primary parameters can broadly be defined as:

- urban planning;
- building form;
- façade design;
- building fabric.

Not only will these variables be influenced by the principles and pre-design issues already outlined, but there will be strong interdependencies within this group of design concerns. For example, the building form – whether terraced or courtyard or deep plan, etc. – will impact upon the overall layout, but will also influence the decisions related to the façade design and building fabric. These considerations will, furthermore, have a bearing on the appropriate choice of building services, outlined below.

Building services

The above sections on 'Principles of low-energy design' and 'Building design' focus primarily on the passive design strategies. However, in any given context it is more than likely that buildings will need to rely, to a certain extent, on mechanical systems in order to ensure that comfort conditions are maintained. Here we consider such systems as auxiliary (i.e. the aim is to minimize reliance on them and thus reduce the energy demand). The following four categories are considered:

1 heating;
2 cooling;
3 mechanical ventilation;
4 artificial lighting.

It is clear that building design decisions should determine the appropriate building services strategies. At a simple level, if a deep plan is adopted, then increased mechanical ventilation – possibly even cooling – as well as artificial lighting is necessary. This may be offset against reduced solar gains or heat loss, and requires the principles of low-energy design to be rigorously applied.

The integrated building design system

The aim of IBDS methodology is to demonstrate how the various factors described above interact and – more importantly – how they can be integrated successfully and holistically in order to achieve low-energy urban building design. The IBDS is depicted below in Figure 14.1.

Clearly, design is an iterative process and the strategy outlined here should not be considered as a simplistic linear process. The main purpose is to increase an awareness and understanding of interrelationships that exist in the design process. It can be used as a framework for design team discussions at the various key design stages, as well as a design tool at any given stage (whether outline design or construction detailing). The system inevitably needs to be sufficiently general to enable local conditions, expertise and individual procedures to be incorporated, and should not be used in a deterministic manner or in isolation.

Figure 14.2 provides a simple overview of the structure. The highlighted (grey) area is the building-related procedure, which will be the focus of IBDS. The following schematics will first address building design issues – broken down into a number of sub-categories – and the relationships to other design parameters and to issues of low-energy principles (Figure 14.2). This is followed by a schematic of building services issues in a similar manner. Finally, an overall matrix of all the key parameters will be shown to demonstrate the integrated interrelationships between each.

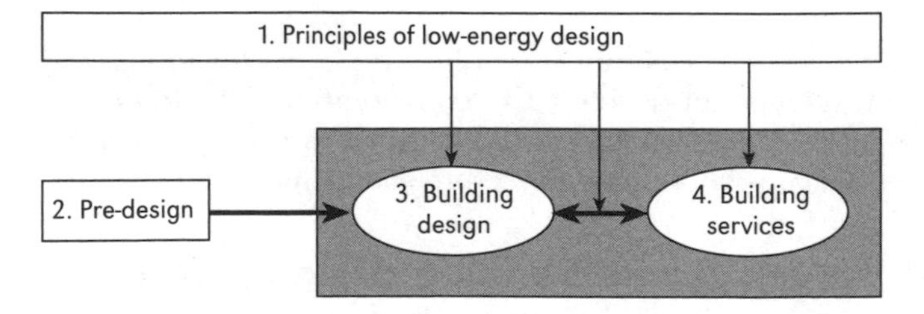

Figure 14.1 *Schematic layout of overall integrated building design system (IBDS) stages and relationships*

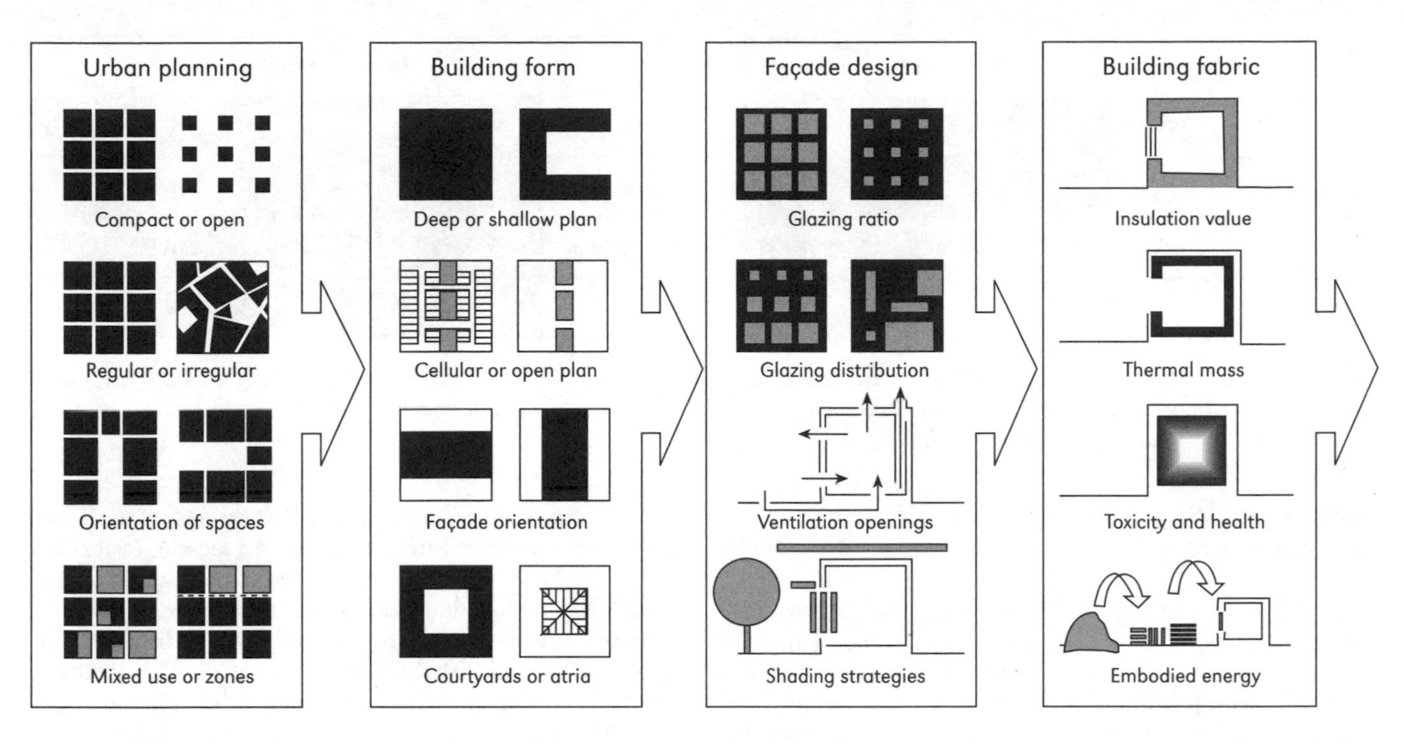

Figure 14.2 *Schematic layout of building design-related issues; the primary sub-categories of each main design consideration are depicted*

		Urban				Form				Façade				Fabric			
		Compact or open	Regular or irregular	Orientation of spaces	Mixed use or zoned	Deep plan or shallow	Cellular or open plan	Façade orientation	Courts or atria	Glazing ratio	Glazing distribution	Ventilation openings	Shading strategies	Insulation value	Thermal mass	Toxicity and health	Embodied energy
Urban	Compact or open					■		■	■	■	■	■	■	■			
	Regular or irregular			■				■		■							
	Orientation of spaces		■		■	■	■	■		■			■		■		
	Mixed use or zoned			■		■	■	■	■	■		■	■		■		
Form	Deep plan or shallow	■		■	■		■	■	■	■	■	■		■	■	■	
	Cellular or open plan			■	■	■		■		■	■	■	■		■		■
	Façade orientation	■	■	■	■	■	■			■	■	■	■		■		
	Courts or atria	■			■	■				■	■	■	■	■			
Façade	Glazing ratio	■	■	■	■	■	■	■	■				■	■	■	■	
	Glazing distribution	■				■	■	■	■			■	■	■			
	Ventilation openings	■			■	■	■	■	■		■				■	■	
	Shading strategies	■		■	■		■	■	■	■	■				■		
Fabric	Insulation value	■				■			■	■	■				■		■
	Thermal mass			■	■	■	■	■		■		■	■	■			■
	Toxicity and health					■				■		■					
	Embodied energy						■							■	■		

Figure 14.3 *Matrix of building design issues showing environmental interrelationships between parameters*

	Urban	Form	Façade	Fabric
Urban	25%	69%	63%	19%
Form		50%	94%	44%
Façade			38%	44%
Fabric				38%

Figure 14.4 *Diagram of strength of links, expressed as a percentage of 'interconnectedness' between building design variables*

	Urban	Form	Façade	Fabric	
Passive solar	38%	63%	69%	58%	57%
Daylighting	56%	69%	63%	17%	51%
Ventilation	50%	69%	25%	25%	42%
	48%	67%	52%	33%	

Figure 14.6 *Diagram of interconnectedness between design and low-energy strategies*

Interrelationships between design parameters

Each filled box in the matrix in Figure 14.3 indicates a possible interaction between the two variables. Thus, for example, the urban design factor of 'compact or open' urban planning will have an impact on building form aspects, such as whether it is deep plan or not, what the implications are for façade orientation (and level of obstruction), as well as whether courts or atria are likely to be possible. Having gone through such an exercise, it is now easy to see which group of building design parameters have a strong link to other groups. A simplified version of the chart is in Figure 14.4.

The strongest relationships between design variables are those related to building form and façade design (94 per cent). Thus, design decisions about one should significantly affect the decisions made about the other. For example, the basic orientation of form will affect the glazing ratio and distribution in the façade. Other strong interrelationships are found between urban planning and form (69 per cent), as well as urban planning and façade design (63 per cent). Even within the categories, strong dependencies are found – particularly for building form (50 per cent). Thus, decisions about, for example, plan depth will be influenced by plan arrangement (i.e. whether open plan or cellular).

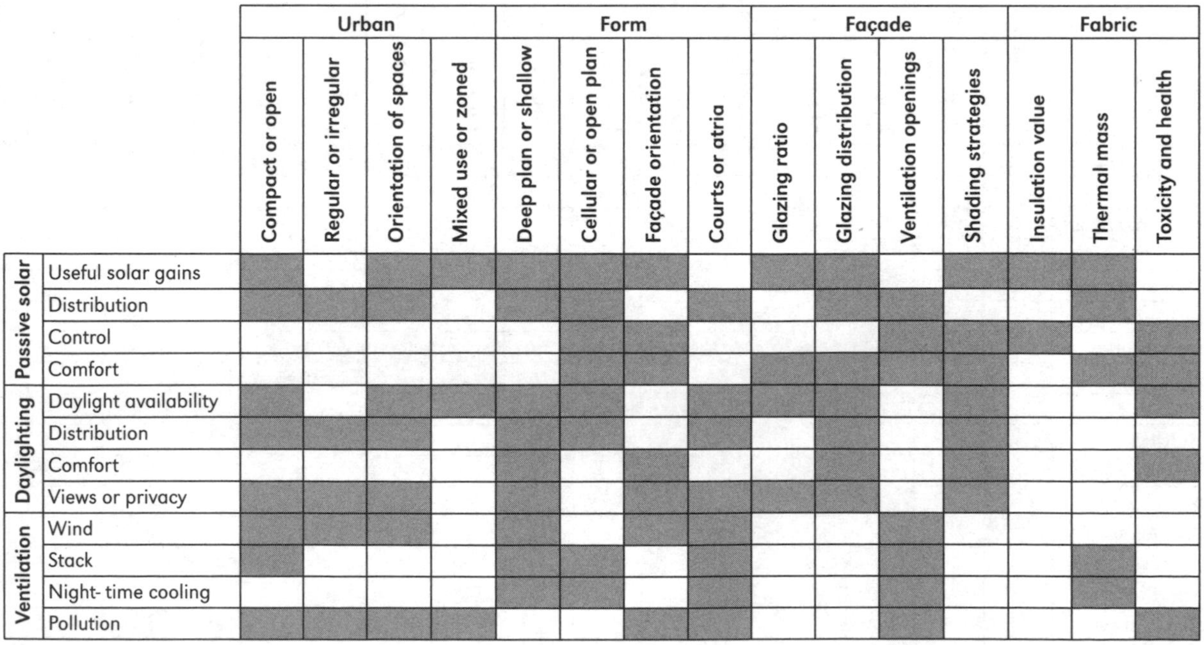

		Urban				Form				Façade				Fabric		
		Compact or open	Regular or irregular	Orientation of spaces	Mixed use or zoned	Deep plan or shallow	Cellular or open plan	Façade orientation	Courts or atria	Glazing ratio	Glazing distribution	Ventilation openings	Shading strategies	Insulation value	Thermal mass	Toxicity and health
Passive solar	Useful solar gains	■		■	■	■	■	■		■	■		■	■	■	
	Distribution	■	■	■		■	■		■		■	■			■	
	Control						■	■				■	■	■		■
	Comfort						■	■		■	■	■	■		■	■
Daylighting	Daylight availability	■		■	■	■	■		■	■	■		■			■
	Distribution	■	■	■		■	■		■		■		■			
	Comfort					■		■			■		■			■
	Views or privacy	■	■	■		■		■	■	■	■		■			
Ventilation	Wind	■	■	■		■		■	■			■				
	Stack	■				■	■		■			■			■	
	Night- time cooling					■	■		■			■			■	
	Pollution	■	■	■	■			■	■			■				■

Figure 14.5 *Matrix of building design issues and related environmental performance parameters*

Design parameters versus low-energy strategies

Figure 14.5 indicates the potential links between design parameters and some key passive-energy strategies, where each filled box suggests a connection. As before, it is useful to draw a simplified matrix to communicate the key interrelationships.

It can be seen from Figure 14.6 that building form has the strongest role to play in terms of low-energy design strategies overall (67 per cent). Façade design is the next most significant issue (52 per cent), particularly for passive solar (69 per cent) and daylighting design (63 per cent). Urban planning, too, has an important role, though the building fabric is of primary importance in terms of the thermal (passive solar) potential (58 per cent). This reflects the importance of thermal mass and insulation. The building fabric is only weakly connected to daylight and ventilation strategies.

Of the energy strategies, all have significant implications for design, particularly passive solar strategies (57 per cent), followed by daylighting (51 per cent) and ventilation design (42 per cent).

	Urban	Form	Façade	Fabric	
Heating	44%	44%	50%	17%	39%
Cooling	69%	75%	50%	67%	65%
Lighting	13%	75%	25%	0%	28%
	42%	65%	42%	28%	

Figure 14.8 *Strategic relationships between design and services*

Design parameters versus environmental systems

Figure 14.8 highlights detailed connections between design and services aspects. These are presented more strategically in Figure 14.7.

Figure 14.7 demonstrates the strong potential links between cooling (air conditioning or natural, mechanical, mixed mode, etc.) and urban building form, as well as fabric aspects. As a result, cooling has a significant connection to design parameters overall.

Of the design parameters, it is building form that tends to influence the environmental services strategies most significantly. The weakest links are between fabric

		Urban				Form				Façade				Fabric		
		Compact or open	Regular or irregular	Orientation of spaces	Mixed use or zoned	Deep plan or shallow	Cellular or open plan	Façade orientation	Courts or atria	Glazing ratio	Glazing distribution	Ventilation openings	Shading strategies	Insulation value	Thermal mass	Toxicity and health
Heating	Fuel/plant type				■							■		■	■	
	Emitters								■	■	■	■				
	Distribution	■	■	■	■	■		■	■			■				
	Location			■	■	■	■		■	■	■	■				
Cooling	Air conditioning versus natural ventilation	■		■	■	■	■	■		■		■	■	■	■	■
	Mechanical versus natural ventilation			■	■	■	■	■	■	■		■	■	■	■	■
	Mixed mode		■	■	■	■	■					■			■	
	Integration	■	■		■	■	■		■			■			■	
Lighting	Manual/automated				■	■	■	■			■		■			
	Lamps/luminaries				■	■	■		■							

Figure 14.7 *Schematic of the interrelationships between building design and services*

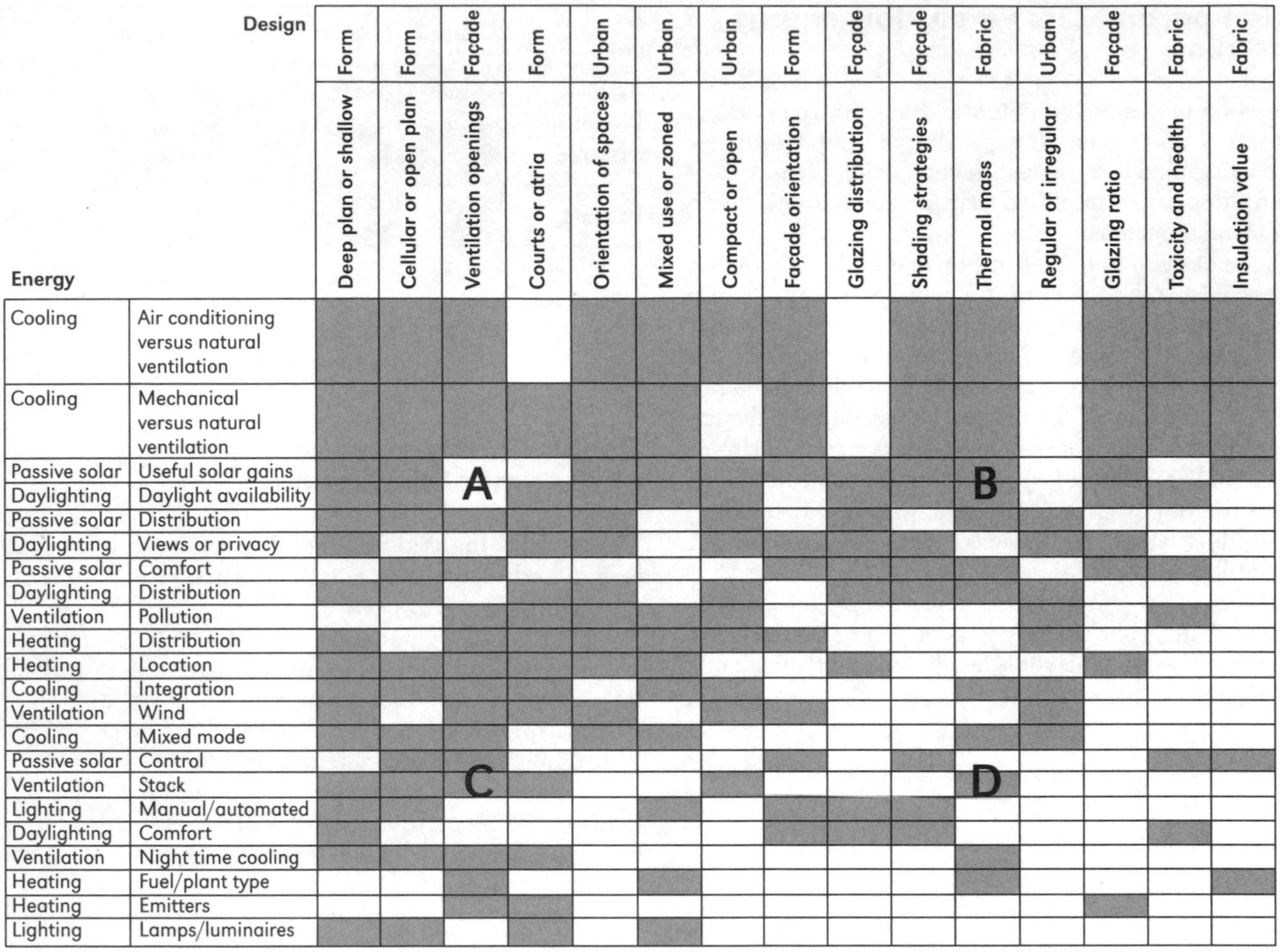

Energy	Design	Deep plan or shallow	Cellular or open plan	Ventilation openings	Courts or atria	Orientation of spaces	Mixed use or zoned	Compact or open	Façade orientation	Glazing distribution	Shading strategies	Thermal mass	Regular or irregular	Glazing ratio	Toxicity and health	Insulation value
		Form	Form	Façade	Form	Urban	Urban	Urban	Form	Façade	Façade	Fabric	Urban	Façade	Fabric	Fabric
Cooling	Air conditioning versus natural ventilation															
Cooling	Mechanical versus natural ventilation															
Passive solar	Useful solar gains															
Daylighting	Daylight availability			A								B				
Passive solar	Distribution															
Daylighting	Views or privacy															
Passive solar	Comfort															
Daylighting	Distribution															
Ventilation	Pollution															
Heating	Distribution															
Heating	Location															
Cooling	Integration															
Ventilation	Wind															
Cooling	Mixed mode															
Passive solar	Control															
Ventilation	Stack			C								D				
Lighting	Manual/automated															
Daylighting	Comfort															
Ventilation	Night time cooling															
Heating	Fuel/plant type															
Heating	Emitters															
Lighting	Lamps/luminaires															

Note:
A Many links between design and energy strategies
B Some energy implications of design strategies
C Some design implications of energy strategies
D Few links between design and energy strategies.

Figure 14.9 *IBDS matrix of key parameters, indicating zones of levels of interaction between building design and energy strategies*

parameters and lighting, in particular (it should be noted here that fabric design does not include interior finishes, which could entail adding another parameter to the IBDS methodology).

Design parameters versus energy strategies

If one combines the design variables with both the passive and active energy strategies, then it becomes possible to rank the strength of interrelationships. The following matrix in Figure 14.9 shows this in detail.

Figure 14.9 lists the various parameters, whether design related or energy related, according to the frequency of interrelationships between each category. This methodology can be applied to any key set of parameters as set by the design team. For this matrix, at the top of the design list, in terms of the variables that have the greatest links and implications for energy and services strategies, are the following:

- deep or shallow plan;
- cellular or open plan;
- ventilation design;

- courts or atria;
- orientation.

The primary environmental issues are as follows:

- the need for air conditioning versus natural ventilation;
- mechanical versus natural ventilation;
- solar gains;
- daylight;
- distribution of solar gains.

Area A

One can thus conclude that, for example, the plan organization (deep versus shallow, cellular versus open plan) is strongly associated with primary decisions about the ventilation strategies (mechanical versus passive), as well as other key low-energy strategies (solar and daylight). This implies that such strategies require careful and fully integrated design solutions, particularly in urban and building form terms.

Area B

A number of key environmental strategies will be influenced by the design considerations in this section – consisting typically of façade and fabric design aspects. Thus, the opportunity to naturally ventilate (as opposed to air condition or mechanically ventilate) is primarily dependent upon building form and urban planning, but will also rely upon appropriate façade and fabric design (i.e. shading, mass, glazing ratio, etc.).

Figure 14.10 is a simplified version of the more complex matrix presented in Figure 14.9.

Area D

Conversely, the IBDS matrix also indicates the environmental strategies that are only linked to a few design parameters and which have little effect on other environ-

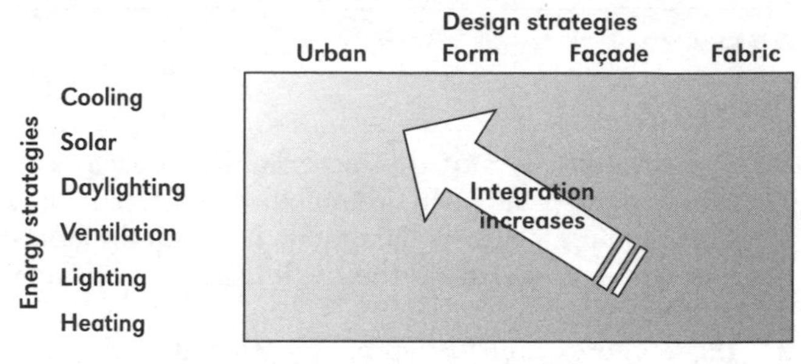

Figure 14.10 *A simplified matrix indicating the hierarchy of interrelationships of the key energy and design parameters*

mental strategies. For example, the choice of lighting and heating appliances has relatively little bearing on design strategies.

Area C

This zone of the matrix indicates those design strategies that have some environmental potential. For example, plan form will have implications for stack ventilation options, but will not be affected significantly by some façade and fabric design considerations.

Synopsis

The IBDS methodology provides a flexible system for assessing the interrelationships and levels of integration of design parameters for low-energy design in an urban context. The method is flexible in that additional and alternative parameters can be included in the analysis. Thus, if the emphasis of a project shifts to include, for example, interior planning issues (such as interior finishes, visual and thermal comfort, etc.) or wider urban issues (such as the microclimate, transport, green space, etc.), these can be incorporated by the design team in the IBDS method. However, the variables presented here are considered to be the primary ones.

Activities

Activity 1

What is the role of design team collaboration in determining successful integrated building design?

Activity 2

Develop your own simplified matrix to demonstrate the links between façade design, plan and section, and daylighting design.

Activity 3

Describe what and how design factors may influence a natural ventilation strategy?

Answers

Activity 1

1 One strategic aim of the integrated approach is to avoid conflicts between architecture and technology. This requires a close collaboration between the architect and engineer from the beginning of the design process.

2 If the energy considerations are not integral to the design solution, it becomes difficult to improve the energy saving potential through the application of technology alone. Thus, if a design does not integrate natural ventilation strategies, for example, then more energy-intensive mechanical systems may be the only recourse without fundamentally changing the building design.

3 It has been argued that design integration is critical, and that the means to achieve it is through the early and effective collaboration of the design team.

Activity 2

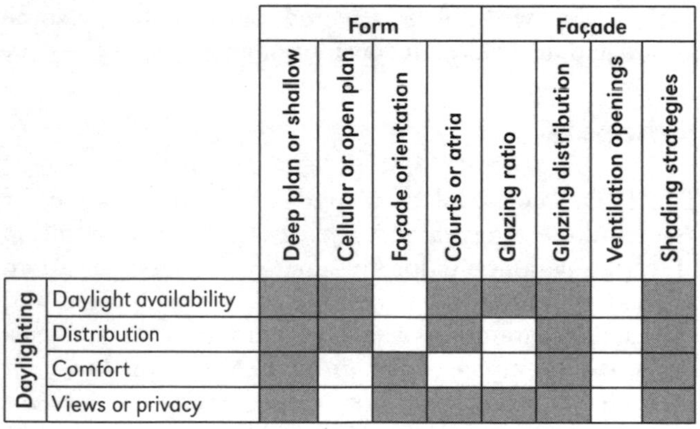

		Form				Façade			
		Deep plan or shallow	Cellular or open plan	Façade orientation	Courts or atria	Glazing ratio	Glazing distribution	Ventilation openings	Shading strategies
Daylighting	Daylight availability	■	■		■	■	■		■
	Distribution	■	■		■		■		■
	Comfort	■		■			■		■
	Views or privacy	■		■	■	■	■		■

Figure 14.11 *Example of a matrix showing links between form and façade design, and daylighting criteria*

Activity 3

- Plan form: a deep, compact plan makes natural ventilation more difficult to achieve easily as the ventilation paths need to be integrated.
- Orientation: spaces with solar gains (particularly if highly glazed) or facing the prevailing wind direction will require specific ventilation strategies that respond to these factors.
- Zoning of uses: certain space uses will require higher or lower ventilation rates.
- Cellular plans make passive cross-ventilation difficult; special air paths will need to be integrated.
- Atria can be used to provide stack effect or ventilation preheat passively, and courtyards can be a source of relatively quiet, clean air.
- The location of ventilation openings will determine the potential stack forces that can be exploited.
- The building fabric can reduce the cooling loads (thermal insulation and mass) for passive strategies.

Index

Page numbers in *italic* refer to Tables, Figures and Boxes.